Taphonomy studies the transition of organic matter from the biosphere into the geological record. It is particularly relevant to zooarchaeologists and paleobiologists, who analyze organic remains in the archaeological record in an attempt to reconstruct hominid subsistence patterns and paleoecological conditions. In this user-friendly, encyclopedic reference volume for students and professionals, R. Lee Lyman, a leading researcher in taphonomy, reviews the wide range of analytical techniques used to solve particular zooarchaeological problems, illustrating these in most cases with appropriate examples. He also covers the history of taphonomic research and its philosophical underpinnings. Logically organized and clearly written, the book is an important update on all previous publications on archaeological faunal remains.

VERTEBRATE TAPHONOMY

CAMBRIDGE MANUALS IN ARCHAEOLOGY

Series editors

Don Brothwell, *University of London*
Graeme Barker, *University of Leicester*
Dena Dincauze, *University of Massachusetts, Amherst*
Ann Stahl, *State University of New York, Binghamton*

Already published

J.D. Richards and N.S. Ryan, *Data processing in archaeology*
Simon Hillson, *Teeth*
Alwyne Wheeler and Andrew K.G. Jones, *Fishes*
Peter G. Dorrell, *Photography in archaeology and conservation*
Lesley Adkins and Roy Adkins, *Archaeological illustration*
Marie-Agnès Courty, Paul Goldberg and Richard MacPhail, *Soils and micromorphology in archaeology*
Clive Orton, Paul Tyers and Alan Vince, *Pottery in Archaeology*

Cambridge Manuals in Archaeology are reference handbooks designed for an international audience of professional archaeologists and archaeological scientists in universities, museums, research laboratories, field units, and the public service. Each book includes a survey of current archaeological practice alongside essential reference material on contemporary techniques and methodology.

VERTEBRATE TAPHONOMY

R. Lee Lyman

Department of Anthropology
University of Missouri-Columbia

Published by the Press Syndicate of the University of Cambridge
The Pitt Building, Trumpington Street, Cambridge CB2 1RP
40 West 20th Street, New York, NY 10011-4211, USA
10 Stamford Road, Oakleigh, Victoria 3166, Australia

First published 1994

Printed in Great Britain at the University Press, Cambridge

A catalogue record for this book is available from the British Library

Library of Congress cataloguing in publication data

Lyman, R. Lee.
Vertebrate taphonomy / R. Lee Lyman.
    p.   cm. – (Cambridge manuals in archaeology)
Includes bibliographical references and index.
ISBN 0 521 45215 5 (hard). – ISBN 0 521 45840 4 (pbk.)
1. Animal remains (Archaeology).   2. Taphonomy.   3. Vertebrates.
I. Title.   II. Series.
CC79.5.A5L96 1994   93–28675
930.1′0285–dc20   CIP

ISBN 0 521 452155 hardback
ISBN 0 521 458404 paperback

SE

To Barbara, John, and Michael

CONTENTS

xii   *Contents*

FIGURES

xiii

TABLES

xx

PREFACE

When I started my studies of vertebrate faunal remains recovered from archaeological sites over twenty years ago, I had no idea what taphonomy was nor was I particularly concerned about what are today typically asked questions concerning the preservation and formation of the archaeofaunal record. But as I read the zooarchaeological literature while completing my doctoral dissertation in the mid-1970s, I found an increasing number of papers dealing with taphonomic issues. The fact that since then it has become increasingly difficult to keep up with the ever growing literature on taphonomy is something of a mixed blessing. It is a mixed blessing because (a) we are constantly realigning the relation between what we *want* to learn and what we *think we can* learn from the vertebrate faunal remains we recover from archaeological sites, and thus our conclusions tend to be much more strongly founded than even a decade ago (this is good), and (b) it is nearly impossible for any one analyst to conceive of all of the logically possible taphonomic problems that a single reasonably sized assemblage of vertebrate remains might present. The latter is not bad; it just means a taphonomist's and zooarchaeologist's (and thus my) job is much more difficult now than it was a mere decade ago. Simply put, the analysis of zooarchaeological remains is no longer the simple, straightforward task that it was in the 1960s or 1970s. Taphonomic research has found a home in zooarchaeology, and it is here to stay.

Today, the number of zooarchaeologists who simply identify the bones, tally them up, and write a report about what prehistoric hominids were eating, is diminishing. Most reports on zooarchaeological remains written in the past ten years contain a more or less detailed consideration of at least a few taphonomic issues. This book is about how taphonomic questions might be analytically addressed and, sometimes, answered. It is a book that I wanted to write ten years from now. However, when Ann Stahl talked to me in the Spring of 1991 about the possibility of writing it, I realized, upon reflection, that now (from May 1991 until January 1993) was just as good a time as later. In fact, the more I thought about it, the better the idea of writing it now became. Many of my friends and professional colleagues were working hard on important taphonomic problems, and virtually all of them were eager to tell me what they were working on and what they were learning. Writing the book would, I decided, be easy because of all of these wonderfully knowledgeable people, and there weren't more of them than I could keep track of with a little effort. Any value this book has is a tribute to all of those people who knowingly and unknowingly helped me with putting it together. For being a friend and taphos colleague as I wrote this book, I thank Diane Gifford-Gonzalez, Donald K. Grayson, Stephanie D. Livingston, Fiona Marshall, Dave N. Schmitt, and Mary C. Stiner. I especially thank Lee Ann Kreutzer for finding and sending me a couple of reprints at the last minute, and keeping me informed about her studies of bone density. Many other people have helped me over the years by reviewing some of my manuscripts and by always being ready to share ideas and reprints. For help in many ways taphonomic and zooarchaeologic, I thank Anna K. Behrensmeyer, Robert L. Blumenschine, Robson Bonnichsen, Luis A. Borrero, Virginia L. Butler, Gary Haynes, Jean Hudson, Eileen Johnson, Richard G. Klein, Curtis W. Marean, Duncan Metcalfe, Richard Morlan, James F. O'Connell, Paul W. Parmalee, James Savelle, Pat Shipman, Gentry Steele, and Lawrence C. Todd. There are, to be sure, many others whose talks I have heard and whose papers I have read; they have, no doubt, influenced my thoughts more than I realize.

Permission to reprint some of the illustrations that are critical to the volume was provided by several individuals doing important taphonomic research. To these individuals I can only say "I owe you one:" Peter Andrews, Anna K. Behrensmeyer, Lewis R. Binford, J. D. Currey, Diane Gifford-Gonzalez, Brian Hesse, Eileen Johnson, Lee Ann Kreutzer, Larry G. Marshall, Richard H. Meadow, Stanley Olsen, T. B. Parsons, Richard Potts, and Mary C. Stiner.

I have been given many opportunities to analyze and study archaeofaunal remains over the years. Without that breadth and depth of experience, this book would be much less than it is, and, I probably would not have written it. Frank C. Leonhardy and Carl E. Gustafson initiated my interest in bones, and Frank gave me the assemblage on which I cut my teeth. I am deeply saddened that his untimely death prevented his being here to see what he helped create. My early interests were fine tuned by Donald K. Grayson, who provided me access to several unique collections (including the Mount St. Helens crispy elk) and who knew when to let me figure out I was headed in the wrong direction and when to not waste time and tell me I was wrong. Other friends who provided boxes of bones for me to study include Kenneth M. Ames, David R. Brauner, Richard L. Bryant, Terry Del Bene, David T. Kirkpatrick, Dennis E. Lewarch, Michael J. O'Brien, Kenneth C. Reid, and Richard E. Ross. In particular, Jerry R. Galm has, over the past decade, seen to it that I didn't go more than six months without receiving a box of bones in the mail; thanks, Jerry, for ensuring that I didn't have to suffer withdrawal.

Many people helped in small but important ways. Gail Lawrence, Amy Koch, and Rob Dunn helped with some of the early word processing. Eugene Marino and Paul Picha helped with correspondence via the fax machine. Virginia L. Butler, James Cogswell, Dolores C. Elkin, Donald K. Grayson, and Paul Picha variously helped me obtain several hard-to-find articles and books, and kept me up-to-date with what was coming off the presses. Michael B. Schiffer was always ready to visit and offer encouragement; thanks, Mike, for a copy of the tomato book. Gregory L. Fox made sure I went fishing occasionally during the early writing phases, and the students in my zooarchaeology class made sure I identified the big errors in some of my early reasoning. Ann Stahl and Anna K. Behrensmeyer provided helpful comments on an early draft of several chapters and thereby made sure I was on the right track. Ann subsequently read the entire manuscript, doing all of us a major service by ensuring that the more cumbersome sentences were revised. Linden Steele printed some of the photographs. The University of Missouri-Columbia, Department of Anthropology helped get some figures reproduced.

Finally, without the faith in my ability shown by my wife, Barbara, and the distractions provided by my sons John and Mike, it never would have been finished.

January 26, 1993

ACKNOWLEDGEMENTS

For permission to reproduce figures, I thank the following:

Figure 2.5, Richard Meadow; Figures 2.6, 4.4, and 4.14, Brian Hesse and Taraxacum Press; Figure 2.7, Anna K. Behrensmeyer and The Paleontological Society; Figure 2.8, Peter Andrews and The Royal Anthropological Institute; Figure 3.2, Diane Gifford-Gonzalez and Academic Press; Figure 3.3, Diane Gifford-Gonzalez and The Center for the Study of the First Americans; Figures 4.1 and 4.3, T. S. Parsons and W. B. Saunders Company; Figure 4.2, J. D. Currey and Edward Arnold Ltd.; Figures 4.9, 4.10, 4.11, 4.12, and 4.13, Stanley J. Olsen and the President and Fellows of Harvard College, Peabody Museum; Figure 6.1, Anna K. Behrensmeyer and Plenum Press; Figure 6.5, Anna K. Behrensmeyer and Harvard University, Museum of Comparative Zoology; Figures 6.10, 6.21a, and 6.22, Center for the Study of the First Americans; Figures 7.4 and 8.2, Academic Press; Figure 7.5, Academic Press, Ltd.; Figures 8.1a, 8.13a and 8.13b, Society for American Archaeology; Figure 8.4, Larry G. Marshall and The Center for the Study of the First Americans; Figure 8.6, Eileen Johnson and Academic Press; Figure 9.4, Richard Potts and Aldine de Gruyter; Figure 9.7, Lewis R. Binford and Academic Press.

1

WHAT IS TAPHONOMY?

Only a small part of what once existed was buried in the ground; only a part of what was buried has escaped the destroying hand of time; of this part all has not yet come to light again; and we all know only too well how little of what has come to light has been of service for our science.
(O. Montelius 1888:5)

Introduction

Taphonomy is the science of the laws of embedding or burial. More completely, it is the study of the transition, in all details, of organics from the biosphere into the lithosphere or geological record. These definitions were given by the Russian paleontologist I. A. Efremov (1940) who coined the term from the Greek words *taphos* (burial) and *nomos* (laws). Taphonomy is, however, important not only to paleontologists, but to archaeologists, especially zoo-archaeologists and paleoethnobotanists, who study the organic remains making up part of the archaeological record. That importance has come to be widely recognized in the past 20 or 30 years. Taphonomy is now seen as important because it is often taken to connote that the zooarchaeological and ethnobotanical records are biased if some non-human-related processes have affected the condition or frequencies of biological remains. While that perception is often correct, I will show that this perception is frequently incorrect.

The reason archaeologists should be concerned with taphonomy is that it involves the formation of what is often a major part of the archaeological record. If the archaeological record is those modern traces of past human or hominid behaviors, then the discarded remains of meals such as mammal bones and plant parts constitute a portion of the archaeological record. Thus, taphonomic research involving the zoological and botanical portions of the archaeological record involves "the study of processes of preservation and how they affect information" contained within these parts of the record (Behrensmeyer and Kidwell 1985:105).

Granting the preceding, the reason for this book's existence should be self-evident. What is perhaps not so evident, however, is the reason such a book is appearing now given that archaeology has been practiced within a scientific paradigm for over 100 years (e.g., Trigger 1989). In order to assess why taphonomy is now seen as important, and to help explore why taphonomic research of the late twentieth century appears the way it does, the first part of

1

Chapter 2 reviews the history of taphonomic research and assesses its current status. The second part of Chapter 2 presents a personal view of the structure of taphonomic inquiry. The history and current status of taphonomic research allow me to take up the topic of Chapter 3, what is variously called actualistic research, ethnoarchaeology, middle-range research, or neotaphonomy, and how this relates to identifying the formational dynamics of a zooarchaeological record from its modern static traces. Chapter 4 is devoted to a review of vertebrate skeletal tissues and skeletons, and a discussion of how to quantify faunal remains. Chapters 5 through 12 are devoted to describing many of the commonly employed taphonomic analytic techniques. In Chapter 13 materials from earlier chapters are integrated and synthesized to provide a framework for performing intensive and extensive taphonomic analyses.

It is important here to introduce some basic terms and concepts that are used throughout this volume. The next few sections of this first chapter, then, are devoted to reviewing some of the basics of zooarchaeological analyses and how taphonomic research contributes to those analyses. That background leads to a consideration of the kinds of contributions taphonomic research can make to archaeology in general. The final topic of Chapter 1 is a discussion of what this book is meant to be, what it is not, and why.

On the analysis of archaeological faunal remains

> We must first eliminate causes of error, and discover what Nature can do to bones submitted to her action.
> (H. A. Breuil 1938:58)

Analyses of archaeological faunal remains have been undertaken at least since the late nineteenth century in North America (Robison 1978), and probably for at least 50 years prior to that time in Europe (Morlot 1861). While once scarcely more than a subsidiary endeavor, archaeological site reports now regularly contain a section on recovered faunal remains, often written by a specialist, and many more independently published and in-depth studies of faunal remains are being prepared by specialists in zoology and archaeologists with zoological training (Lyman 1979a). This reflects the holistic approach of archaeologists trying to understand and explain the totality of human history.

There are two basic goals to analyzing prehistoric faunal remains: reconstruction of hominid subsistence patterns, and reconstruction of paleoecological conditions (Hesse and Wapnish 1985; Klein and Cruz-Uribe 1984). The former has been characterized as an attempt "to explain, in the form of predictive models, the interface that existed between prehistoric human populations and the faunal section of the biotic community" (Smith 1976:284). This goal is anthropological in orientation as it addresses topics such as human diet, animal resource procurement strategies, or predator–prey relationships. Analytic goals are attained using anthropological and ecological principles in analysis and interpretation (Rackham 1983; e.g., Lyman 1992b). Analyses of paleoecological conditions use zoological and ecological data, methods, and theory (Dodd and Stanton 1981; King and Graham 1981) to reconstruct faunal turnover and succession, paleoenvironmental history, and zoogeographic history (e.g., Grayson 1987).

The two distinguished goals are not mutually exclusive. Both require taxonomic identification of faunal remains, which necessitates adherence to zoological method and theory. Data interpretation requires use of ecological principles whether those concern habitat preferences of taxa or determining available biomass or meat. Interpretation of a single assemblage of faunal remains recovered from an archaeological site may accomplish either or both goals (King and Graham 1981) because, in part, analytic techniques overlap. Distinction of the two goals is useful for discussion purposes, but is not mandatory to actual analysis.

Basic concepts

In this volume I focus on vertebrate remains. Research on invertebrate taphonomy is largely, but not entirely, found in the paleontological literature. My remarks are applicable to the remains of virtually all animal taxa, and many are also applicable to plant remains. I restrict discussion and examples in this volume largely to mammal remains for the simple reason that more taphonomic research has concerned mammals than any other vertebrate taxonomic group; non-mammalian vertebrates are covered in some detail in Chapter 12.

Taphonomy is generally construed as focusing on the postmortem, pre-, and post-burial histories of faunal remains. Burial is considered to be a stage intermediate to pre- and post-burial histories due to the potentially destructive and disruptive nature of burial processes (e.g., Dixon 1984; Kranz 1974a, 1974b). Various arrangements of taphonomic agents and processes have been posited in the form of models depicting a general taphonomic history (see Chapter 2). Generally, a bone may be gnawed, buried, exposed, reburied, re-exposed, broken, transported, and reburied prior to recovery (see the Glossary). Realistic sequences of taphonomic factors may therefore require the inclusion of loops. A general chronology of taphonomic agents and processes affecting animal remains is called a *taphonomic history* or *taphonomic pathway*. A *taphonomic agent* is the source of force applied to bones, the "immediate physical cause" of modification to animal carcasses and skeletal tissues (Gifford-Gonzalez 1991:228), such as gravity, a hyena, or a hominid. A *taphonomic process* is the dynamic action of an agent on animal carcasses and skeletal tissues, such as downslope movement, gnawing, or fracturing (relative to the agents listed in the preceding sentence). A *taphonomic effect* or *trace* is the static result of a taphonomic process acting on carcasses and skeletal tissues, the physical and/or chemical modification of a bone. As we will see, taphonomic analysis involves identifying and/or measuring taphonomic effects, and on that basis identifying and/or measuring the magnitude of effects of taphonomic processes and agents.

A taphonomic history begins when one or more members of a biotic community die. Postmortem processes that affect carcasses and skeletal tissues constitute the major part of a taphonomic history. Recovery of faunal remains is a potentially biasing factor because it affects the collected assemblage through differentially moving and sampling it. What the collector perceives as pertinent observations may significantly influence which fossils are collected and which data are recorded, and consequently influence the final analytic results. A large literature already exists on this crucial topic (Gamble 1978). Zooarchaeologists have become much more aware of the stratigraphic and sedimentary contexts of animal remains and the potential taphonomic significance of such geological and contextual data (e.g., Rapson 1990). As a result, more care is taken in the recovery of animal remains today than in the past.

A *fauna* is some specified set of animal taxa found in a geographic area of some specified size, kind, and location at some specified time (Odum 1971:366–367). For example, one can specify a modern intertidal fauna of the Pacific Rim, a prehistoric terrestrial fauna of Europe, and a Pleistocene mammalian fauna of Colorado. Zoologists study faunas by observing living animals. Paleontologists and zooarchaeologists study faunas by analyzing fossils. I have had several zooarchaeologists tell me "fossils are mineralized animal remains," or "fossils are older than 10,000 years." I find neither of these criteria in definitions published by paleontologists (see the Glossary). I use the term *fossil* in this volume to denote any trace or remain of an animal that died at some time in the past (ascertaining the age of animal death is a separate problem).

A *fossil record* is some set of remains of organisms, either or both plants and animals, having a geological mode of occurrence in some defined geographic space and geological context (i.e., with a delineated spatial distribution) (modified from Lyman 1982a, 1987e). A fossil record consists of those observable phenomena such as the particular bones in a particular stratum. A *fossil fauna* consists of those taxa represented by the fossil record at a specific locality. The term fossil fauna serves to emphasize the taphonomic distinction between a living fauna and a fauna represented by fossils. While the term fossil fauna as defined here is virtually synonymous with the terms *local fauna* and perhaps *faunule* (Tedford 1970), the first term emanates from the taphonomic perspective of this volume while the latter two emanate from a paleoecological perspective.

I distinguish two kinds of fossil faunas: those without, and those with, spatially associated cultural materials, or *paleontological faunas* and *archaeofaunas*, respectively. The distinction is not meant to imply whether or not humans had a role in the taphonomic history of a particular fossil assemblage. Analytically categorizing a particular fossil record as constituting a naturally or culturally deposited set of faunal remains is a major part of taphonomic analysis in zooarchaeology (e.g., Avery 1984; Binford 1981b; Potts 1984; Turner 1984). What is meant by these two terms is simply whether artifacts are or are not spatially associated with the faunal remains.

Goals of taphonomic analysis in zooarchaeology

Subsistence studies, by the nature of their research questions, require knowledge of the formation of the archaeofaunal record (Lyman 1982a; Maltby 1985b; Medlock 1975; Rackham 1983). Similar knowledge is important to paleoecological research but for different reasons (Behrensmeyer and Hill 1980; Gifford 1981; Grayson 1981). Subsistence studies require that the fossils constituting the archaeofauna be sorted into at least two categories: those deposited as a result of human (subsistence and other) behaviors, and those naturally deposited (Binford 1981b; Thomas 1971). Culturally deposited fossils must be qualitatively and quantitatively representative of the fauna exploited, and quantification techniques must produce accurate relative abundances of economically important taxa (Grayson 1984; Lyman 1979b). Paleoecological studies need not have representative samples of humanly exploited taxa, but may require representative samples of prehistorically extant faunas. Sample requirements are flexible in the sense that they have certain tolerance limits. For example, a bison kill site probably does not include all taxa exploited by a group of people yet it can be studied and analyzed. Similarly, a zooarchaeologist may focus only on the microfauna and ignore larger taxa in an archaeofauna. Both depend on the research questions being asked. Sample representativeness is relative to some population which in turn is dictated by the research goal. The representativeness of a sample of faunal remains is controlled by the sample's taphonomic history, the sampling techniques used to collect the sample, *and* the research questions being asked of the sample.

Gifford (1981) distinguishes two basic goals of taphonomic research: (1) "stripping away" the taphonomic overprint from the fossil record to obtain accurate resolution of the prehistoric biotic community, and (2) determining the nature of the taphonomic overprint in order to be able to list the precise taphonomic mechanisms responsible for a given fossil assemblage, enabling the writing of taphonomic histories. The latter goal is similar to studying the formation of the archaeological record (Schiffer 1987), and may lead to conclusions regarding the human behaviors that created the fossil record. The former goal is seen as a necessary step towards paleoecological analysis because the target of analysis requires knowledge of the prehistoric biotic community. For example, as Graham and Kay (1988:227) note, "the definition of taphonomic pathways is not the ultimate goal of paleobiological or anthropological studies, but [it is] a means to factor out biases and further our fundamental understanding of past ecosystems and human cultures."

Determination of the exact taphonomic history of a particular fossil assemblage is frequently attempted by zooarchaeologists who wish to know which taxa were exploited by hominids and the relative proportions in which they were exploited. Many interpretations therefore involve outlines of the suspected human behaviors that resulted in the fossil assemblage. For example, Wheat's (1972) description of the hunting and butchery process evidenced at the Olsen-Chubbuck bison kill site is simply a narrative model of the suspected taphonomic history of that site's fossil record.

The two goals of taphonomic research are not mutually exclusive. Stripping away the taphonomic overprint requires that the overprint be known. Once the taphonomic overprint is known, the prehistoric biotic community can, theoretically, be determined by analytically reversing the effects of the taphonomic processes. Of course, this procedure requires the assumption that the sample of fossils is representative of the biotic community, or that it can be analytically made representative of that community. This assumption has been analytically controlled in cases where an archaeofauna is directly compared with a paleontological fauna in close geographic and temporal proximity to one another (e.g., Briuer 1977), and in cases where two or more geographically and temporally adjacent archaeofaunas are compared (e.g., Grayson 1983; Guilday *et al.* 1978). The covert assumption of such comparative analyses is that because each fossil assemblage has undergone a more or less unique taphonomic history, similar but independant interpretive results derived from the assemblages are thought to represent prehistoric reality. That is, the analyst can have greater confidence that taphonomic processes have not totally obscured all indications of a prehistoric biotic community when all examined fossil assemblages indicate the same community.

As we shall see, the goals of taphonomic analysis can be much more finely distinguished. Here I have shown that the two general goals of taphonomic analysis are easily aligned with the two basic goals of zooarchaeological analysis, and that the two kinds of analysis are, and in fact must be, synergistic. That is, determining which animal taxa were eaten by prehistoric peoples, and how much of each animal taxon was eaten, is surely a taphonomic problem. Taphonomy thus presents a challenge to zooarchaeologists that can be simply phrased as "What are these bones doing in this site?"

The challenge of taphonomy

Understanding how taphonomic processes affect quantitative measures of faunal remains is a major challenge facing zooarchaeological research (Gilbert and Singer 1982; Holtzman 1979; A. Turner 1983). Quantitative measures such as taxonomic abundances, meat weights, and frequencies of particular skeletal elements are all affected by taphonomic processes (Badgley 1986a; Grayson 1984). Not only are quantitative data important in many analyses, but so are the distributions of bones and taxa within a site (Grayson 1983; Lyman 1980; Wheat 1972). Taphonomic processes may obscure distributional contexts, unrelated elements may become spatially associated, or related elements may lose their spatial association (Hill 1979b). The second major challenge in zooarchaeological research is, then, ascertaining the meaning of the distributions of faunal remains. The third major challenge is determining how and why the recovered faunal remains differ from the biotic community in which they originated. This question is particularly important to paleoecologists as well as zooarchaeologists. Other, more finely distinguished challenges are easily conceived, but these major ones tend to underpin virtually all of those narrower challenges.

Taphonomy's contribution to zooarchaeology

Examples make clear the nature of the contribution of taphonomy to zooarchaeology. For instance, archaeologists have, beginning in the late 1970s, adopted optimal foraging theory from ecology as an explanatory device (see Bettinger 1991 for a review). Part of that theory demands measuring the breadth of the niche exploited by the subject forager, in this case humans. Typically, niche breadth is measured by tallying the number of plant and animal taxa exploited. Thus, the remains of plants and animals found in archaeological sites must minimally be sorted into two categories: those representing taxa that were exploited, and those representing taxa that were not exploited by people. This entails examining the remains for indications that humans accumulated and deposited certain of the remains and indications that the other remains were naturally deposited. If some of the bones have butchering marks on them, then it is reasonable to suppose that the taxon or taxa represented were accumulated and butchered by people. In this case, the butchery marks are the indication of a human agent in the taphonomic history of those remains. If the nearly complete and partially articulated skeleton of a burrowing animal such as a gopher is found in a krotovina in a site, it is likely that this individual was naturally deposited and did not form a part of the human occupants' diet. Here, the degrees of skeletal completeness and articulation, the behavior of the taxon, and the context of the remains are the relevant taphonomic traces leading to the inference that these animal remains were naturally deposited. These are, as we shall see, simplistic examples. Modern taphonomic analysis is seldom so straightforward.

An archaeologist interested in measuring the dietary breadth of some prehistoric human group must distinguish between culturally and naturally deposited animal remains. If this is not done, then the analyst has simply measured the taxonomic richness (number of species) represented by the sample of animal remains, and not dietary breadth. Inferences that humans broadened their dietary niche, or narrowed it, or did not alter it through time, will surely be inaccurate without such taphonomic analyses.

In paleoecological analysis, animal remains can grant insights to the climatological and floral environment to which a human group adapted. If the remains of an animal taxon that today prefers cool-moist environments is found in a site located in a warm-dry habitat, those remains potentially indicate that the site may have been occupied during a time of cooler and moister conditions (assuming the remains were deposited at the time of human occupation). That indication can, however, be false if the animal remains were not locally derived. Did, perhaps, a far-ranging predator (human or not) collect those remains some significant distance away from the site and then transport those remains and deposit them there? Or, did the taxon in question actually live on or very near, say, < 1 km, from the site? Are the remains corroded from the digestive acids of some carnivore? If so, then perhaps, but not necessarily, the remains came from a significant distance away. Do the remains represent complete skeletons, or selected portions of skeletons? Is the taxon represented so large that it could not have been transported whole to the site? Producing answers to these and related questions involves taphonomic analysis and provides the data necessary to make inferences about the local or distant origin of the remains in question.

One example of how taphonomy has contributed to our understanding of human prehistory involves the debate over whether our Plio-Pleistocene ancestors were hunters or scavengers. The literature on this topic has grown to immense proportions in the last decade (e.g., Binford 1981b, 1984b, 1988b; Blumenschine 1986a, 1986b, 1987; Bunn and Kroll 1986, 1988; Potts 1984, 1988; Shipman 1986a, 1986b). All of that literature involves detailed taphonomic analysis. As a result of that research, the 1950s consensus that early hominids were hunters has been changed to one of perceiving them as individuals who probably hunted small game and scavenged large game, although opinions differ on the precise magnitude of the dietary contribution of both hunting and scavenging. This topic remains one of the most discussed in the literature of the 1990s.

Terminology used in this book

I follow several conventions in describing data used in examples in this volume. These are not formalized but rather are ad hoc conventions. "P" stands for proximal, "D" for distal. Abbreviations for particular skeletal elements vary between investigators, and I have not attempted to be systematic in my use of these; rather I have in many cases applied the original investigator's abbreviations with appropriate definitions. An *assemblage* of fossils is some analytically defined set of faunal remains usually, but not always, from a particular spatiotemporal context. I use the term "bones" frequently as a generic term for bones, teeth, horns, antlers, etc. Finally, I use the terms "hominids" and "humans" interchangeably.

In many cases I employ some very basic statistical tests. Throughout, the significance levels are denoted by P, Pearson's parametric correlation coefficient is denoted by r, Spearman's rank order correlation coefficient is denoted by r_s. Discussion and descriptions of these can be found in any introductory text on statistics. In this age of personal computers, I suspect we will see an increasing number of statistical analyses of taphonomic data. The quantitative aspects of zooarchaeological materials are a subject that could readily fill another volume; many of them are discussed in detail by Grayson (1984). An introduction to the basic quantification units of zooarchaeological research is provided in Chapter 4 of this volume, and additional comments are provided in Chapter 8.

What this book is and what it is not

> The objectives of taphonomic analysis are very varied and its methods are eclectic, being governed in part by disparate characteristics of different types of fossil assemblages, and in part by the nature of [research] problems it is called upon to address.
> (R. D. K. Thomas 1986:206)

This book is meant to review many of the potentially useful and informative analytic techniques taphonomists have developed to help solve particular zooarchaeological problems. It is not meant to provide a set of algorithms for solving conclusively all or any potential interpretive problems; it is not a cookbook in the sense that following a particular recipe will produce a tasteful or even edible product. The heart of the volume lies in the following chapters, which are meant to introduce the novice's mind to, and refresh the expert's memory concerning, the diversity of variables that must be considered and the plethora of analytic techniques that might make up a detailed taphonomic analysis. As the volume's title indicates, only vertebrate taphonomy is considered. For recent synopses of invertebrate taphonomy, see the papers introduced by Thomas (1986) and the volumes edited by Allison and Briggs (1991b) and Donovan (1990). Due to my own limitations, my review of the available taphonomic literature is largely restricted to that portion of it published in English. I took a zooarchaeological perspective in writing this volume, and thus zooarchaeologists will, I hope, find much of value in its pages. I hope as well that vertebrate paleontologists may find some of the material useful.

Nowhere in this volume will a detailed taphonomic analysis of a particular zooarchaeological collection be found, although various collections are described to exemplify the results of particular taphonomic processes or are used to illustrate how a particular analytic tool works. Simply, this book is a review of many of the analytic techniques used in the 1980s and early 1990s to help determine the taphonomic history of bone collections. Because archaeological taphonomy is a rapidly developing field, there is no doubt that some of the techniques reviewed here will not be in use ten years from now, and techniques not yet developed and thus not described here will be developed in the future. As I neared finishing what I thought would be a reasonably complete first draft, I continued to encounter newly published articles and to find references to articles published years ago that I had not previously been aware of. Because it was necessary to the completion of this volume, I stopped reviewing and incorporating new data and ideas in December of 1992. Thus, with few exceptions, the references cited herein were published prior to that date. The volume is in some ways, then, incomplete and in other ways it is out of date. I take these facts alone to indicate that taphonomic research is reaching, and perhaps will continue to enjoy for some time, a period of florescence. This volume is simply one mark of this period.

There is no clearly stated, explicit paradigm for taphonomic research and no rules for how to do it (Thomas 1986), except perhaps those under the umbrella of uniformitarianism. Taphonomic research in prehistoric contexts has few criteria for assessing the validity of a solution to a taphonomic problem. It is not always clear how to determine if a particular analytic technique was the appropriate one for a particular problem. Instead, the results of taphonomic research are often evaluated from the perspective that those results should be replicable if another analyst uses the same data and analytic procedures.

This is not a volume on techniques of zooarchaeological analysis, although it should be clear that much of what is described here often does (and should always) make up major portions of modern zooarchaeological research. Several good volumes on zooarchaeological analysis exist (e.g., Davis 1987; Hesse and Wapnish 1985; Klein and Cruz-Uribe 1984), and these can be consulted in conjunction with use of this volume.

This volume is *not* a pictoral or descriptive essay. There are illustrations here, but I have kept the number of photographs to a minimum because there are now available so many excellent descriptive volumes containing numerous photographs of variously modified bones that I could not hope to replicate the extent or quality of their coverage. The reader is encouraged to study closely the volumes and photographs found in Andrews (1990), Binford (1981b), Brain (1981), Haynes (1991), and White (1992). I have chosen in this volume to focus on what an analyst might do once the modifications have been recognized in a fossil assemblage. Therefore the majority of the illustrations I have chosen to include here are examples of graphs and charts intended to show how various taphonomically modified fossil assemblages may appear when graphed in a particular way. I find it much more mentally stimulating to manipulate taphonomically modified bones analytically than simply to describe some as carnivore chewed and some as burned, and it is such analytical manipulation that in fact allows us to meet the challenges of taphonomy head-on.

Finally, I have certainly not covered all possible topics that might be covered in a volume of this nature. With minimal exceptions I do not delve deeply into the chemistry of fossilization nor do I discuss the effects of sampling and differential recovery. The book is longer than I intended as it is; to have included additional topics would of course have resulted in a longer book.

There are two excellent ways to learn a subject. One is to teach it, the other is to write a book about it. Having said that, how can I expect anyone to read this book? I can have that expectation (and hope) because the next best way to learn a subject is to read about it; not just one article or one book, but as many as you can lay your hands on. I learned a lot because I read a lot in order to write this book. Other writers have said many of the same things said here, often times better. I offer this book as a starting point, then, to those with interests in vertebrate taphonomy.

2

THE HISTORY AND STRUCTURE OF TAPHONOMY

Taphonomy is now in its salad days.
(P. Dodson 1980:8)

A brief history of taphonomic research

One hallmark of a scientific discipline's maturity is the publication of a history of it. Several brief histories have been written about taphonomy (Cadée 1990; Dodson 1980; Olson 1980). Although informative, these have been largely limited to the relation of taphonomy to paleontology. Perhaps because taphonomy developed first in paleontology, which has as its focus the study of biological evolution and paleoecology, when archaeologists borrowed the concept they also borrowed the connotation that the fossil record is probably *biased*. That is, information on the ecology and morphology of animals is lost or altered between the time of an organism's death and the time its remains are recovered and studied (Dodson 1980; Lawrence 1968, 1971). North American paleobiologists initially adopted Efremov's version of taphonomy – that the fossil record is incomplete and therefore biased – rather than the German version which focused on reconstruction of past environments via detailed paleobiological analysis (Cadée 1990:9–13). It is not surprising, then, that North American zooarchaeologists see the zooarchaeological record as biased. Paleobiologists today often worry about the "fidelity" (Kidwell and Bosence 1991) and completeness (e.g., McKinney 1990) of the fossil record; lack of either denotes a biased fossil record with regard to how accurately paleobiotas are reflected. But because bias is relative, not all assemblages of animal remains are biased in all ways.

Another hallmark of the maturity of a discipline or research topic is the publication of books devoted to it. Several books have appeared (Allison and Briggs 1991b; Behrensmeyer and Hill 1980; Donovan 1990; LeMoine and MacEachern 1983; Shipman 1981b), the first ones at the same time that the histories of taphonomic research appeared. Importantly, two major review articles (Behrensmeyer 1984; Gifford 1981) and two major books (Behrensmeyer and Hill 1980; Binford 1981b) appeared about a decade ago and, with the other books published at this time, presented the taphonomic perspective explicitly to an archaeological audience. Several years ago a newsletter devoted to taphonomic research was initiated by two paleobiologists (Plotnick and Speyer 1989, 1990; Plotnick and Walker 1991). Special symposia (another historical landmark of a discipline's development) with zooarchaeological taphonomy as their foci have been convened and the proceedings from them published in an effort to synthesize the current state of our knowledge (Bonnichsen and Sorg 1989; Dixon and Thorson 1984; Hudson 1993; Solomon *et al.* 1990; see also Nitecki and Nitecki 1987). This brief synopsis gives one cause to wonder "Why is taphonomy so important in archaeology today?" The question becomes particularly cogent when the nineteenth-century archaeological literature is examined.

Nineteenth-century zooarchaeological taphonomy

> Knowing the history of one's discipline can save one a good deal of unnecessary originality. It can also give one a great many good ideas, for the past never says things quite the way the present needs them said.
> (P. Bohannan and M. Glazer 1988:xv)

Grayson (1986) documented that research not unlike taphonomy has a deep history in archaeology. The nineteenth-century debate over the meaning of broken stones thought to represent a Tertiary-aged "eolithic" culture resulted in much research directed towards identifying physical attributes of the stones that unambiguously signified human modification. Grayson (1986) notes that William Buckland early in the nineteenth century, and Charles Lyell, Edouard Lartet, and others during the second half of that century were concerned with the agents and forces which had created marks on bones recovered from various prehistoric contexts. In remarkable anticipation of much modern research, Buckland (1823:37–38), for example, wrote:

> I have had an opportunity of seeing a Cape hyaena at Oxford . . . I was enabled also to observe the animal's mode of proceeding in the destruction of bones: the shin bone of an ox being presented to this hyaena, he began to bite off with his molar teeth large fragments from its upper extremity, and swallowed them whole as fast as they were broken off. On his reaching the medullary cavity, the bone split into angular fragments . . . he went on cracking it till he had extracted all the marrow . . . this done, he left untouched the lower condyle, which contains no marrow, and is very hard. The state and form of this residuary fragment are precisely like those of similar bones at Kirkdale; the marks of teeth on it are very few . . . these few, however, entirely resemble the impressions we find on the bones of Kirkdale; the small splinters also in form and size, and manner of fracture, are not distinguishable from the fossil ones . . . there is absolutely no difference between them, except in point of age.

The significance of this observation was not lost on many prehistorians working in the nineteenth century. For example, 50 years after its original publication the above citation was quoted at greater length by Dawkins (1874:281–283).

Tournal (1833), citing Buckland's research, took the idea of hyenas as agents of bone accumulation in caves one step further. He extended the analogy from traces of gnawing on bones signifying hyenas in particular to signifying carnivores in general.

> The manner in which the broken and gnawed bones of different species are accumulated is easy to conceive. One has, so to speak, surprised nature in the act, when one has observed in our time the charnel houses of hyenas and other carnivorous animals which carry their prey into grottoes in order to eat them, and which at length accumulate immense quantities of gnawed bones, belonging to all sorts of animals. Now, in the case which concerns us, the identity is perfect, since with the hyenas are found the bones which they have gnawed, and even their coprolites. At the same time it has been observed that in the caverns the bones accumulated in the most remote passages. What has just been said for the hyena bone caverns applies equally to caverns which contain less ferocious or smaller-sized carnivores.
> (Tournal 1833 [1969:87])

Note that Tournal extended the analogy by looking not just at the gnawing damage on the bones of prey, but also at the locations of the bones and their associations with the remains and traces of their predators.

The fact that some of the early taphonomic literature was published in European journals in non-English languages might have made some of that information inaccessible to Americans, but two facts make it clear that this did not hinder taphonomic research in the New World. First, Morlot's (1861) "General views on archaeology" was published in English by the Smithsonian Institution. He summarized some European research on the destruction of bird bones by canids. As well, Morlot (1861:300) noted "nearly all the cartilaginous and more or less soft parts of the [deer] bones have been irregularly subtracted," an observation suggested as well by the avian taphonomy research he cited. That the structure or density of skeletal elements would influence whether or not they survived attritional processes was also noted by Dawkins (1869:207) who wrote that "the stone-like molars of the Mammoth would survive the destruction of all traces of the bones of the smaller animals, and remain in many instances as the sole evidence that Postglacial mammals ever dwelt in the area where they were found." Such observations on the inherent properties of faunal remains mediating taphonomic processes anticipated research that was not to be taken up with rigor until 100 years later.

The second reason I suspect there was transoceanic exchange of ideas on taphonomy involves explicit reference to the avian taphonomy cited by Morlot by American archaeologists such as Wyman (1868), who was concerned with understanding the formation of what are today termed shell middens. Further, Wyman (1868:578) expanded the range of taphonomic attributes one might examine by noting "on looking over the specimens of our collections, marks of teeth of animals were frequently noticed, some of them of such size as might be made by dogs, but others by a much smaller animal, as a cat or mink." Uniquely, Wyman (1868) suggests that the size of bone fragments was a result of humans breaking the bones so that the fragments could be cooked in ceramic vessels with orifices of small size or limited capacities.

Buckland's approach, involving examination of the attributes of modern bones that had been modified by known taphonomic processes in order to have an analog for interpreting prehistoric specimens, was extended to include the study of human agents as well. Lartet (1860 [1969:122]), for example, believed that the "very deep incisions" he observed on various bones were the result of them having been cut with "an instrument having an edge with slight transverse inflections, so as to produce, by cutting obliquely through the bone, a plane of section somewhat undulated." Lartet (1860 [1969:123]) "satisfied [himself], by comparative trials on homologous portions of existing animals, that incisions presenting such appearances could only be made in fresh bones," and he obtained "analogous results by employing as a saw flint knives, or [stone] splinters with a sharp chisel-edge." Such experimental work to derive analogs for interpreting the remains of past human activities was to become a hallmark of modern archaeology (Chapter 3).

Peale (1871:390) reported that bones of prey of North American Indians are "roasted on the coals or before the fire, then split with a stone hatchet, and in some cases with a wedge driven between the condyles when the bone has these terminations. [Sometimes bones] are broken into small fragments and boiled in water until all the marrow which they contain and the grease which adheres to them are separated, and rise to the surface, when they are skimmed off" and stored. Lord Avebury (1913:321; better known as Sir John Lubbock) used ethnographic information from Denmark, Africa, and the Arctic, as an analogy to explain the condition of some prehistoric mammal remains. He wrote, "some of the ancient stone hammers and mortars were no doubt used for this purpose, and the proportions of the different bones afford us, I think, indirect evidence that a similar custom [of breaking bones for marrow extraction] prevailed among the ancient inhabitants of southern France." Lord Avebury apparently believed fragmentation destroyed some bones, a theme picked up in later work. He also discussed (a) fragmentation patterns distinctive of a human fracture agent, (b) the influence of fragmentation on the taxonomic identifiability of vertebrate remains, (c) the differential destruction of long bone epiphyses and diaphyses, and destruction of vertebrae and ribs by canids, (d) the correlation of ontogenic age of an ungulate skeleton and the structural density of bones, and (e) the correlation of structural density of bone parts with the nutritional value of those parts for canids (i.e., less dense bones contained more blood, marrow, and grease and thus were both more nutritious and more prone to destruction by bone-gnawing carnivores). This list of taphonomic topics is easily found in the literature of the 1980s.

The range of taphonomic topics covered by nineteenth-century prehistorians, while not completely documented in this section, is sufficiently broad that I have to wonder what happened to this line of inquiry in the first half of the twentieth century; it seems largely to have disappeared from the published record during that period. It probably disappeared due to the shift from the nineteenth-century concern for detecting a human presence in the archaeological record to the twentieth-century focus on establishing the temporal relations of archaeological collections (Daniel 1981; Trigger 1989; Willey and Sabloff 1980). Whatever the reason, it is difficult to find comments on taphonomic issues in the archaeological literature published during the first six decades of the twentieth century that echo the concerns of the nineteenth century. In the early twentieth century the tendency was to conclude that people were the sole taphonomic agent one had to account for during analysis and interpretation of archaeofaunal collections. Taphonomic research in the early and middle twentieth century is largely found in the paleontological literature.

Taphonomy in the early and middle twentieth century

> There is great need for better understanding of the factors that act between the living fauna and the preservation of part of it in the fossil state, as well as factors involved in the formation of fossil deposits in general.
> (G. G. Simpson 1961:1683)

Weigelt (1927; English translation 1989) suggested the term *biostratinomy* (originally biostratonomy) which now is taken to denote "the study of the environmental factors that affect organic remains between an organism's death and the final burial of the remains" (Lawrence 1979a:99). Weigelt's (1927/ 1989) volume has been called "the first major work on vertebrate taphonomy" written by "the first naturalist to mount a full-scale research effort to document processes of vertebrate death, decay, disarticulation, transport, and burial, and to determine their relevance to fossil preservation" (Behrensmeyer and Badgley 1989:vii, viii). Later, Müller (1963) used the term *fossildiagenese* (now simply *diagenesis*) to denote "fossilization events that take place after the final burial of organic remains" (Lawrence 1979b:245). The relationship of taphonomy, biostratinomy, and diagenesis was often shown as in Figure 2.1 during the middle of the twentieth century (e.g., Lawrence 1968; Noe-Nygaard 1977).

Richter (1928) defined *aktuo-paläntologie* as the science of the origin and present-day mode of formation of future fossils (Warme and Häntzschel 1979). Actualistic paleontology is "the application of the uniformitarian principle to paleontological problems" (Warme and Häntzschel 1979:4; see Chapter 3). Richter (1928) also distinguished the importance of the causes of death and their direct consequences, burial mode, and alterations of the animal carcasses prior to diagenesis (Warme and Häntzschel 1979:5–6). It was Efremov (1940), then, who subsumed all of these stages between the death of an organism and the recovery of its remains by a paleobiologist into the field we today label *taphonomy*. Perhaps because Efremov published his new term and its definition in a geological journal, and most previous work on the subject had been also published in such outlets, there were, during the first half of the twentieth century, what Koch (1989) has called two parallel traditions for the development of taphonomy: a paleontological tradition, and an archaeological one.

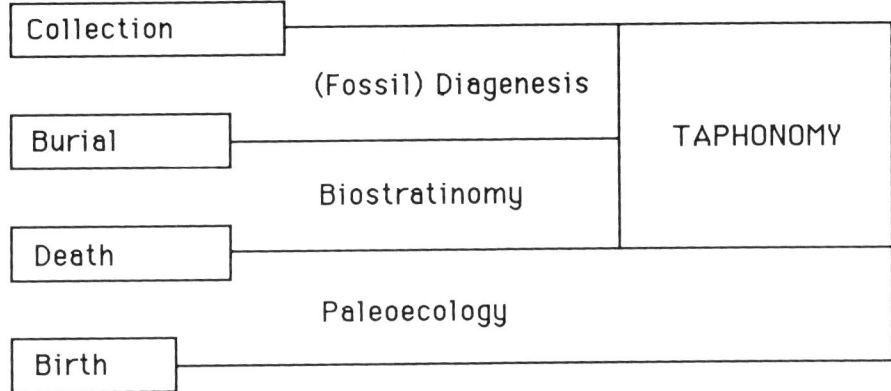

Figure 2.1. General relations of the subdisciplines of taphonomy relative to an animal's life, death, and scientific recovery (after Lawrence 1968:1316, Figure 1).

Their coalescence in the 1970s was a major historical event in the development of taphonomy as a research field in its own right, but first I consider each tradition in turn.

Paleontological tradition
This tradition of taphonomic research is "as old as paleontology" (Cadée 1990:17). Cadée (1990:4–5) argues that "taphonomic reasoning in the truest sense" was perhaps used first by Leonardo da Vinci in the late fifteenth–early sixteenth centuries, although his notes were not widely available until after his death. Cadée also notes that Niels Stensen (often referred to as N. Steno) made major contributions to taphonomic reasoning during the seventeenth century. The establishment of paleontology as a scientific discipline in the early nineteenth century resulted in the publication of numerous taphonomic observations, many of which were founded on actualistic research (see below). By the end of the nineteenth century, Germany was the center of paleontological research (Rudwick 1976), and it was here that much taphonomic research was performed at this time, although the German school found little support outside of Germany in the early twentieth century (Cadée 1990:9–13). This is important because the work of authors such as Weigelt (1927/1989) and Schäfer (1962/1972) is included in the German school.

Müller (1951) presented a major synthesis of much of the paleontologically related taphonomic research to that point in time, especially the biostratinomic aspects. Schäfer (1962/1972) summarized an extensive body of research concerning the disintegration of marine organisms, especially invertebrates. The influence of these and other works on English-speaking paleontologists might not have been significant due to language barriers (Behrensmeyer and Badgley 1989; Gifford 1981:370), but Olson (1952, 1958), Clark *et al.* (1967), Lawrence (1968), and Voorhies (1969) were influenced by the European research and their English language publications helped breach the language barrier. As well, increasing interest in the Americas in taphonomic research and issues during the 1970s is attributed by Behrensmeyer and Kidwell (1985) to the 1972 publication of the English translation of Schäfer's (1962/1972) volume on actuopaleontology. Because most of these early writings were written by paleontologists interested in paleoecological issues, the tendency in this early literature was to focus on the fossil record as potentially biased in terms of how well it reflected the actual paleoecology of the biotic community (Figure 2.2).

Recognition that the fossil record is potentially biased resulted in many analysts attempting to detect and compensate for those biases. Shotwell's (1955) quantitative technique of measuring how complete the skeletons of individual taxa were in order to distinguish taxa that lived near the site of fossil recovery (taxa making up this *proximal community* should have the most complete skeletons) from taxa that lived farther away is a classic example of developing an analytic technique to compensate for the taphonomic processes of bone accumulation and concentration. Shotwell's technique was adapted to the important zooarchaeological question of distinguishing taxa that owed their presence in sites to human activities (cultural bone) from taxa whose remains had been deposited by natural processes (Thomas 1971; Ziegler 1973). While there are serious problems with both the approach advocated by Shotwell for paleoecological analysis and the way the approach is applied to archaeological assemblages (Grayson 1978b; Holtzman 1979), recognition of that fact has only come with increases in our knowledge of the influence of taphonomic factors on such quantitatively based analytic techniques.

Toots (1965c) examined the modern disarticulation of mammal skeletons in a rare actualistic study of the middle twentieth century. He also discussed an analytic technique for describing and interpreting the orientation of fossils (Toots 1965a, 1965b), pointing out that orientation processes were not necessarily biasing and that orientation data were rather valuable to paleoecological studies. Both the study of disarticulation of skeletons (e.g., Hill 1979a, 1979b; Hill and Behrensmeyer 1984, 1985) and of fossil orientation (e.g., Kreutzer 1988) have been followed up recently by archaeologists interested in zooarchaeological and taphonomic problems, and they often stress the value of such data to research rather than as signs of potential bias.

Explicit recognition of the effects of sampling on paleontological research came in the middle of the twentieth century (Krumbein 1965; McKenna 1962; Voorhies 1970), just as it did in archaeology (Binford 1964; Ragir 1967). Given the potential effects of sampling and sample size on measures of relative taxonomic abundances, a critical variable in paleoecology, analytic techniques to contend with such effects were developed (e.g., Sanders 1968). Some of these were adopted by zooarchaeologists (e.g., Styles 1981). But like Shotwell's technique, these techniques for controlling sampling effects have subsequently been shown to be faulty in some situations (e.g., Grayson 1984; Tipper 1979).

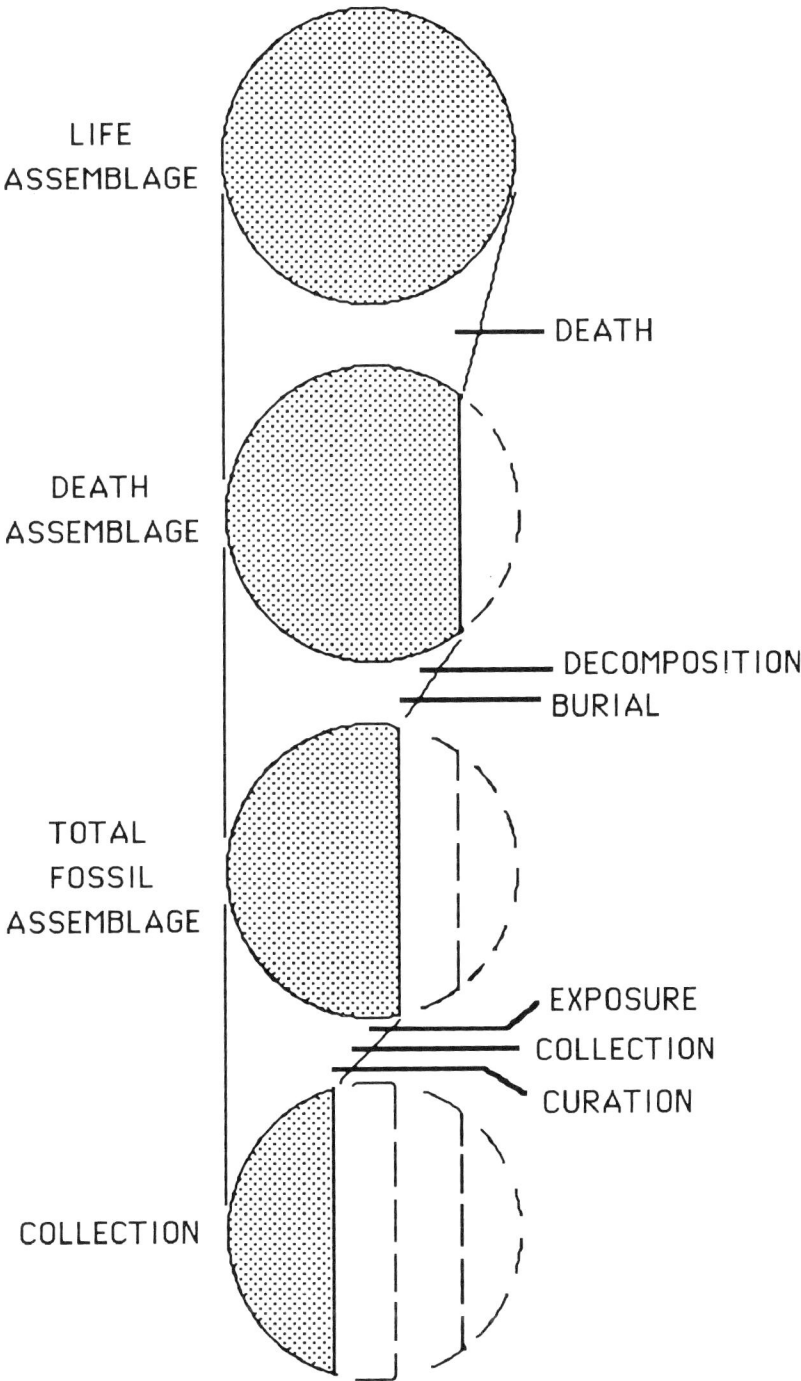

Figure 2.2. Modeled taphonomic history of a biotic community or life assemblage (after Clark and Kietzke 1967:117, Figure 53).

Again, what is underscored here is not the inaccuracy of techniques designed to deal with complex taphonomic issues, but rather the fact that we continue to learn in the 1990s about that complexity and the limits of our analytical approaches.

Today taphonomy is very much a part of paleontology. In a recent synthesis of paleobiology (Briggs and Crowther 1990) approximately 100 pages are devoted to taphonomy. In fact, paleobiologists borrowed a term from mining geology, *fossil-Lagerstätten*, to label strata that are sufficiently rich in fossils or in which fossils are preserved in conditions sufficient "to warrant exploitation, if only for scientific purposes" (Seilacher 1990:266). Similarly, "*taphofacies* consist of suites of rock characterized by particular combinations of preservational features of the contained fossils" (Brett and Speyer 1990:258). *Obrution deposits* are "conservation Lagerstätten" that consist of rapidly buried, and thus exceptionally well preserved, articulated fossils (Brett 1990b:239), and "*concentration Lagerstätten*" are dense concentrations of fossils that may consist of multiple depositional events (Seilacher 1990:267). The florescence of taphonomic terminology indicated by these and other labels, and the recent publication of two edited volumes devoted to taphonomy for paleontologists (Allison and Briggs 1991b; Donovan 1990) is an effective indication of the continually growing importance of taphonomy in paleontology.

Archaeological tradition
What would today be considered taphonomic research for archaeological purposes was rare in the early part of the twentieth century, but some important studies were undertaken, such as Duckworth's (1904) observations that small streams could disperse the bones of a horse skeleton. Martin (1910) experimentally broke mammal bones in an attempt to discern criteria denoting a human agent of fracture. Breuil (1932, 1938, 1939) and Pei (1932, 1938) were also concerned with how bones might be modified (broken, incised) by hominid and other taphonomic agents. Perhaps because some of this research was published in non-English languages in European journals, it seems to have had little impact on American archaeology. Also, the research focus in the Americas at this time on culture history issues (Willey and Sabloff 1980) made these early taphonomic studies irrelevant to the concerns of most practicing archaeologists.

The importance of taphonomic research for zooarchaeology was brought out clearly with Dart's (e.g., 1949, 1956b, 1957, 1960) publications on his proposed osteodontokeratic culture of the australopithecines based on collections of mostly bovid remains recovered from limestone caves in South Africa. Dart was convinced that the way some bones were broken and the relative abundances of skeletal parts indicated hominids were the taphonomic agent largely responsible for the faunal remains. Dart (1956a; Hughes 1954) utilized actualistic data in his arguments and actively contributed to discussions about the artificial status of modifications to bones (e.g., Dart 1958). Regardless of the eventual outcome of the debate over whether or not the South African specimens were variously modified, accumulated, and deposited by hominids (the controversy is not dead; e.g., Brain 1989; Wolberg 1970), Dart's ideas were provocative and served as a major catalyst for the development of explicit taphonomic research in the service of archaeology (e.g., Bonnichsen 1975; Brain 1967a, 1967b, 1969, 1974, 1981; Hill 1976; Isaac 1967; Read 1971; Read-Martin and Read 1975; Shipman and Phillips 1976; Shipman and Phillips-Conroy 1977).

This awakening interest in archaeological taphonomy was reinforced in the Americas with the report of an excessively old bone tool from near Old Crow in the Yukon of northwestern Canada (Irving and Harington 1973). That the specimen in question was a tool was never in doubt; its age was, however, the subject of much discussion and it has since been shown to be of middle to late Holocene age (Nelson *et al.* 1986). But the original report, like Dart's, resulted in a flurry of paleontological, archaeological, and actualistic research (e.g., Bonnichsen 1973, 1979; Johnson 1985; Lyman 1984b; Morlan 1980), some of which was remarkably similar to that of Peale (1871) 100 years earlier (e.g., Zierhut 1967). Interestingly, as with the unsettled status of an early Pleistocene African osteodontokeratic culture, the reality of a middle to late Pleistocene American culture founded on similar tool materials is not yet settled (e.g., Irving *et al.* 1989; Morlan 1988).

A new level of awareness of the importance of taphonomic processes was attained by the late 1960s and early 1970s. Michael Voorhies' (1969) seminal monograph on vertebrate taphonomy had just been published, along with the important work of Clark *et al.* (1967). The term taphonomy now appeared in the title of some paleontological studies (e.g., Boyd and Newell 1972; Dodson 1971) whereas it was not even listed in the index of Kummel and Raup's 1965 *Handbook of Paleontological Techniques*. Given the new level of awareness born in the late 1960s, it is not surprising to find synopses of procedures for recording taphonomic data published during the early 1970s. One was authored by two paleoanthropologists interested in hominid evolution (Hill and Walker 1972); a second was authored by two paleontologists (Munthe and McLeod 1975). Both articles focus on vertebrate remains and are quite similar in their recommendations of what constitutes important taphonomic data (Table 2.1). In many respects the list of data is an accurate reflection of what is typically recorded today, although many of the variables are now more explicitly defined.

Some zooarchaeologists produced explicit statements about relevant taphonomic data even though they did not use the term "taphonomy" in their writings. Guilday *et al.* (1962:65), for example, report that many white-tailed deer (*Odocoileus virginianus*) bones from an archaeological site they studied had been gnawed by carnivores, and suggest the less "durable" bones of

Table 2.1 *Kinds of taphonomic data that should be recorded for vertebrate fossil remains recommended by Hill and Walker (1972) and Munthe and McLeod (1975). Distinction of field and laboratory data by Lyman*

FIELD DATA
1. Geographic locality of fossil collection site (mapped, photographed)
2. Conditions and time of collection (visibility, weather, date)
3. Collection methods and possible biases
4. Stratigraphic position of fossils, including general geology and topography of collection site and lithology and sedimentology of stratum (strata) containing fossils
5. Horizontal and vertical distribution of fossils – three dimensional point provenience of each individual specimen
6. Associated plant and invertebrate fossils
7. Degree of disarticulation and tooth loss
8. Orientation of individual bones and/or articulated skeletons – azimuth of long axis – declination of long axis
9. Flexion or extension of articulated specimens

LABORATORY DATA
10. Taxonomic identification
11. List of elements present
12. Attributes of individual specimens
    a. cracks and flaking (weathering)
    b. fractures
    c. crushing
    d. abrasion
    e. color
    f. root etching
    g. distortion and deformation
    h. gnawing marks
    i. weight
    j. shape (measured as a maximum length:maximum width ratio)
    k. area (measured as the product of maximum length × maximum width)
13. Taxonomic abundances
14. Size range of taxa
15. Ontogenic age of individual animals

ontogenically young deer are probably relatively rare because of that gnawing. Bonnichsen and Sanger (1977) emphasize that bones are not randomly distributed in sites (contra Ziegler 1973) and urge greater attention to the context and association of faunal remains. The published proceedings of the 1974 Archaeozoological Conference includes 47 articles contributed by 49 authors from around the world, yet the terms taphonomy, biostratinomy, and diagenesis are not listed in the index (Clason 1975). Important taphonomic topics are nonetheless discussed in those articles, including patterns of bone fracture, traces of injury to animals recorded in their bones, bone fragment sizes, and gnawing and accumulation of bones by carnivores.

Taphonomic factors had seldom been discussed by zooarchaeologists prior to 1970. Cornwall (1956) focused on the effect fragmentation would have on the identifiability of animal remains. His only taphonomic observations were that naturally deposited bones "will have suffered the ravages of scavengers and of decay, so that only the more massive and durable parts are likely to have been preserved," culturally deposited animal remains are likely to have been "smashed for their marrow or to obtain industrial material," and "in favorable circumstances the material is found much as it was discarded by man, but very often natural agencies have further affected it, so that, again, only the more resistant fragments are available" (Cornwall 1956:184). Chaplin (1971) devoted two and a half pages to "the limiting factors of the archaeological evidence." He emphasized the need to determine how bones had been accumulated and deposited, stating "the time has surely come to apply some more conclusive tests to the problem" (Chaplin 1971:121). He was also worried about how representative of the complete site deposit the collected faunal remains were, but offered only the suggestion that the analyst assume the available sample was a "fair cross section" of what was actually in the deposit. The term taphonomy was not used in either Cornwall's or Chaplin's book, although both clearly were concerned about taphonomic histories and the potential biases created by such histories. The same can be said for most of the journal articles published between 1900 and 1970. Because these articles seem to be directed more toward educating archaeologists about the value of faunal remains for addressing anthropological and zoological questions, it is perhaps not surprising that the authors did not mention that taphonomic factors might decrease the value of those remains by somehow biasing them (e.g., Chaplin 1965; Daly 1969; Gilmore 1949; White 1953c, 1956; Wintemberg 1919; Ziegler 1965).

Medlock's (1975) review of "faunal analysis" in North America contained a thorough review of the significance of processes influencing what a zooarchaeologist might recover from a site. He pointed out that, as of the early 1970s, zooarchaeologists had variously interpreted variations in faunal assemblages as resulting from "temporal or cultural differences, differential preservation, butchering differences, or functional differences among sites" (Medlock 1975:224). Medlock emphasized the need for developing analytic techniques for distinguishing the effects of these variables, techniques that themselves should be verified with experimental work. He presented a model for the formation of a zooarchaeological record that included virtually all variables in models later formally proposed as depicting a generalized taphonomic history (Figure 2.3).

More accessible to archaeologists was the series of papers on the taphonomy and paleoecology of Plio-Pleistocene hominid sites in Africa by Behrensmeyer (1975a, 1975b, 1979; Boaz and Behrensmeyer 1976). She focused not only on the biasing factors but also on the paleoecological data that could be derived from detailed taphonomic analyses, and she worked towards building strong

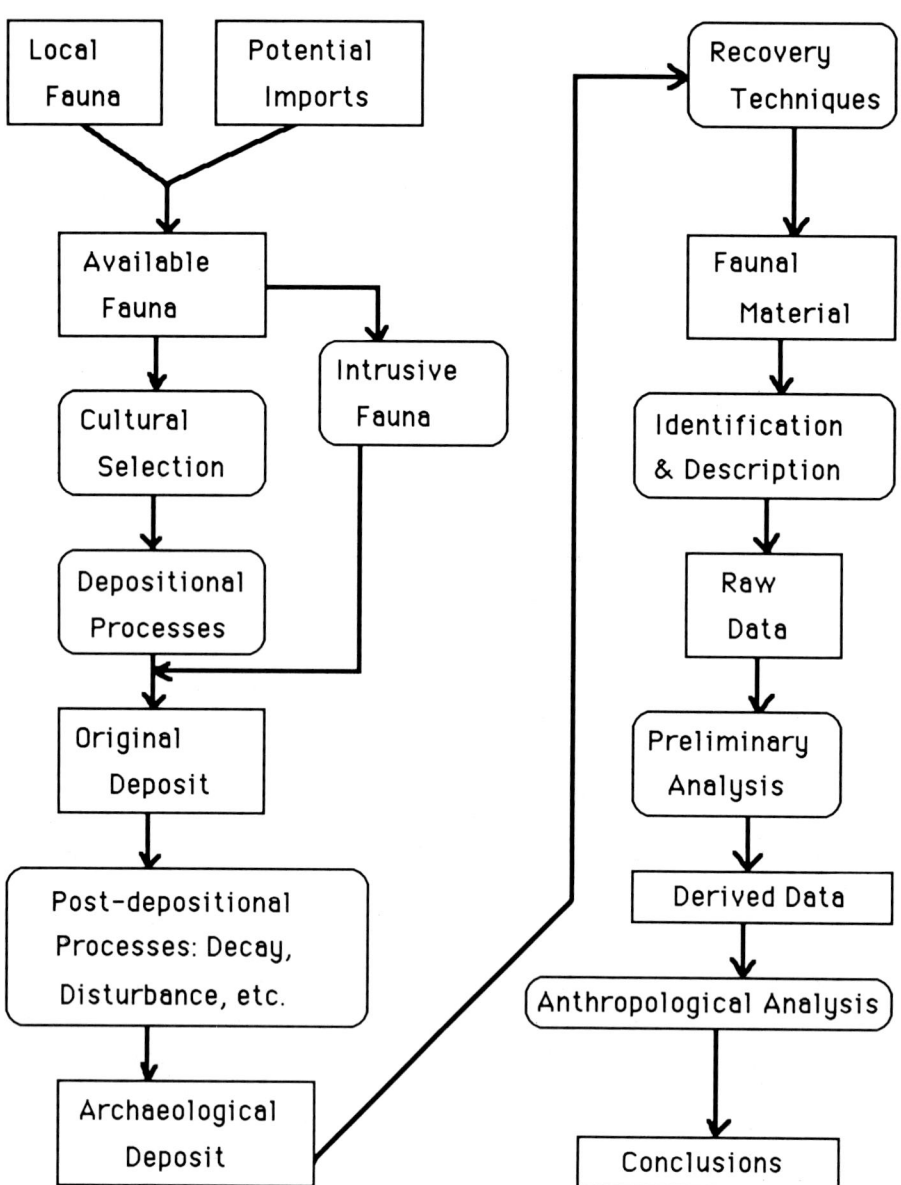

Figure 2.3. Medlock's (1975) model of the taphonomic history of a faunal assemblage (after Medlock 1975:225, Figure 2).

analogical arguments by performing various actualistic and experimental work (Behrensmeyer 1978, 1981; Behrensmeyer and Boaz 1980; Behrensmeyer *et al.* 1979; Gifford and Behrensmeyer 1977). No less important contributions were made by several individuals working on the taphonomy of Plio-Pleistocene and other sites in Africa (e.g., Brain 1969, 1974, 1976; Gifford 1977; Hill 1978, 1979a, 1979b; Isaac 1967; Shipman and Phillips 1976; Shipman and Phillips-

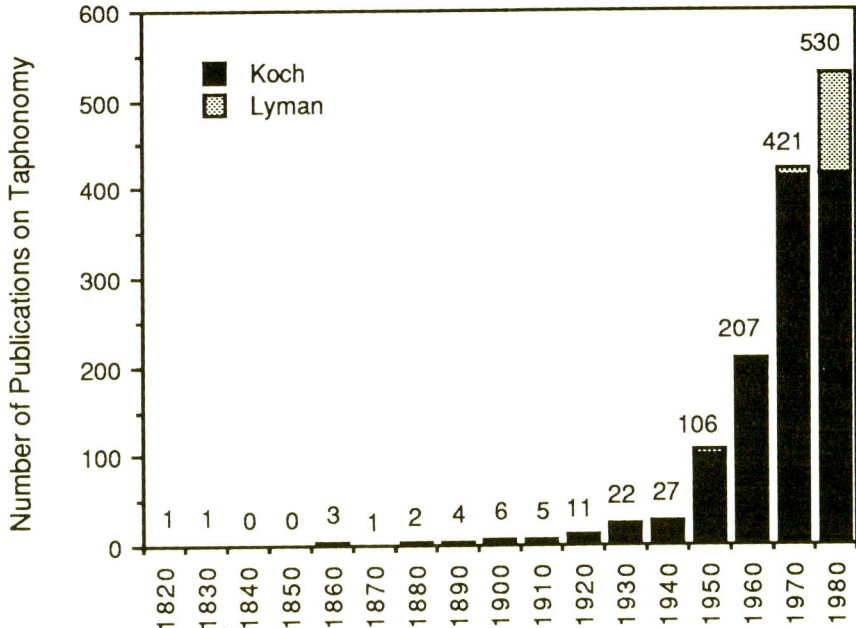

Figure 2.4. Frequencies of titles of taphonomic literature per decade (data from Koch 1989, supplemented by Lyman).

Conroy 1977; Shipman and Walker 1980). Much of this research was presented at the Burg Wartenstein Conference in 1976 (Boaz 1980), the proceedings of which were published in 1980, under Behrensmeyer's co-editorship (Behrensmeyer and Hill 1980).

Researchers working in the Near East (e.g., Gilbert 1979), Europe (Noe-Nygaard 1977; Payne 1972a, 1972b), and the Americas (e.g., Binford and Bertram 1977; Briuer 1977; Read 1971) also published taphonomic analyses. Paleontologists published much taphonomic research in the 1970s and 1980s; Behrensmeyer and Kidwell (1985:109) report an "average publication rate in taphonomy of nearly 50 articles per year" between 1975 and 1984, inclusively. Koch's (1989) bibliography of taphonomy provides a good sample of references through early 1987, which when supplemented with the nineteenth-century titles cited above and titles for 1987–1989, illustrates well the increase in taphonomic research (both that related to paleontology, and that related to archaeology) that came with the 1970s (Figure 2.4). From the 1940s to the 1950s the number of titles nearly quadrupled; perhaps that increase is due to the post-World War II increase in all scientific endeavors. The number of titles then doubles from the 1950s to the 1960s, and doubles again from the 1960s to the 1970s. There is only a 25.8% increase from the 1970s to the 1980s; if the publication rate for the 1970s (42.1 titles per year) held in the 1980s, about 675 to 700 titles should be recorded for the 1980s rather than the 530 plotted in Figure 2.4. I suspect many titles are not included as there are 27 titles on taphonomy published in the first four issues of the journal *Palaios* which are not tallied in Figure 2.4, and others (e.g., Noe-Nygaard 1989; Plotnick and Speyer 1989; Wilson 1988) were discovered after the figure was completed. Further, I have tended to exclude publications in non-English languages. No titles more recent than 1989 are included in Figure 2.4, but a perusal of the bibliography for this volume will show the rate of publication of taphonomy articles has not significantly abated in the early 1990s. Data presented by Russell (1992) indicate the rate of publication of titles on taphonomy, as compiled in the GEOREF data base (i.e., titles published in geological and paleontological journals), seems to have increased between 1982 and 1988, but then leveled off from 1989 through 1991. Whether or not the publication of articles and books on taphonomy continues at its present rate into the late 1990s remains to be seen.

In North America, the 1960s witnessed a shift from studying culture history to a focus on human behavior as it was reflected in the archaeological record (Binford 1968; Willey and Sabloff 1980). Interest in taphonomic issues in the service of archaeology came with the awakening realization in the late 1960s and early 1970s that the archaeological record was not a perfect reflection of human behavior (Ascher 1961a, 1961b, 1968; Binford 1964, 1977; Isaac 1967; Schiffer 1972, 1976). Extracting the dynamics of human behavior from a static archaeological record would not be a straightfoward process. In conjunction with a growing awareness of what zooarchaeological research could contribute to our knowledge of human prehistory in general (e.g., Chaplin 1965; Daly 1969), the discipline of archaeology was ready to adopt taphonomic research as an important and crucial part of its analytical toolkit (see also Binford 1981a; Schiffer 1983, 1985).

Recent taphonomic research: the 1980s and 1990s

> Research reports in archaeozoology, paleoanthropology, paleontology, and paleo-botany are considered incomplete without some discussion of the taphonomic history of the collection.
> (C. P. Koch 1989:1)

In her overview of taphonomy and its relevance for archaeological research Gifford (1981) points out that taphonomic data are important for detecting potential biases in the zooarchaeological record, and notes that taphonomic data can provide significant paleoecological information. Not all authors writing at that time were well attuned to the latter fact. For example, Hill (1978:88) writes, "taphonomy ultimately deals with the differences that exist between an assemblage [of fossils] and the community of animals from which it came. These differences constitute the bias in many fossil assemblages and are due to various taphonomic causes." Hill's description of taphonomy is written from a paleontological perspective and may not have been read or completely appreciated by many zooarchaeologists. In a paper perhaps read by more zooarchaeologists Meadow (1981) focuses on potential biases that can influence archaeological interpretations based on quantitative data and presents a model of the formation of the zooarchaeological record reminiscent of models published by paleobiologists (Figure 2.5). The decreasing size of the symbols used by Meadow is symbolic of the loss of information that occurs during a taphonomic history, as it is for Clark *et al.* (1967; see Figure 2.2).

The symbolic loss of information indicated in Figure 2.5 eventually came to denote that archaeological faunal remains are potentially (and probably) biased. Hesse and Wapnish (1985:19) are explicit about such symbolism in their depiction of a taphonomic history (Figure 2.6) and say, "the sizes of the boxes arranged along the vertical [temporal] axis reflect the quality of information available at various points in time." They elaborate that symbolism by dividing each box into two distinct sets of information: C or cultural information, and N or natural information (Figure 2.6). Importantly, their diagram implies natural information will, over time, increasingly mask or remove cultural information that might otherwise be derived from a bone assemblage. While probably true at least some of the time, this must be assessed for each assemblage. Finally, it is unusual to find cultural processes listed as "biases," yet Hesse and Wapnish (1985) are correct in doing so for the simple reason that bias, as noted earlier, is relative.

In the 1980s paleontologists were very aware of what could be learned from detailed study of the traces of taphonomic processes. In fact, a symposium titled "The Positive Aspects of Taphonomy" was held in 1984 as part of the annual meeting of the Geological Society of America (Thomas 1986). Behrensmeyer and Kidwell (1985:105), two paleobiologists, proposed a "new working definition for the field as *the study of processes of preservation and how they affect information in the fossil record*" because such a definition encompasses not only the more traditional foci of taphonomic research – loss and bias of information – but what they call the "positive contributions" made by taphonomic processes. Their model of a taphonomic history did not, for instance, utilize symbols depicting information loss; rather, it simply listed categories of taphonomic processes in an implied temporal order (Figure 2.7). A structurally different depiction of a taphonomic history was published that same year by a paleontologist and a zooarchaeologist (Andrews and Cook 1985). That illustration (Figure 2.8) also did not symbolically indicate information loss or bias. Interestingly, neither of these last two illustrations include symbolism for the positive contributions of a taphonomic history. Positive contributions include the addition of information to animal remains, such as traces of the predator that exploited the animal (e.g., gnawing marks) and the processes that moved (e.g., abrasion and rounding due to fluvial action), sorted (e.g., selective removal of particular skeletal elements due to fluvial action), and oriented the bones (e.g., fluvial action causing the long axis of some skeletal

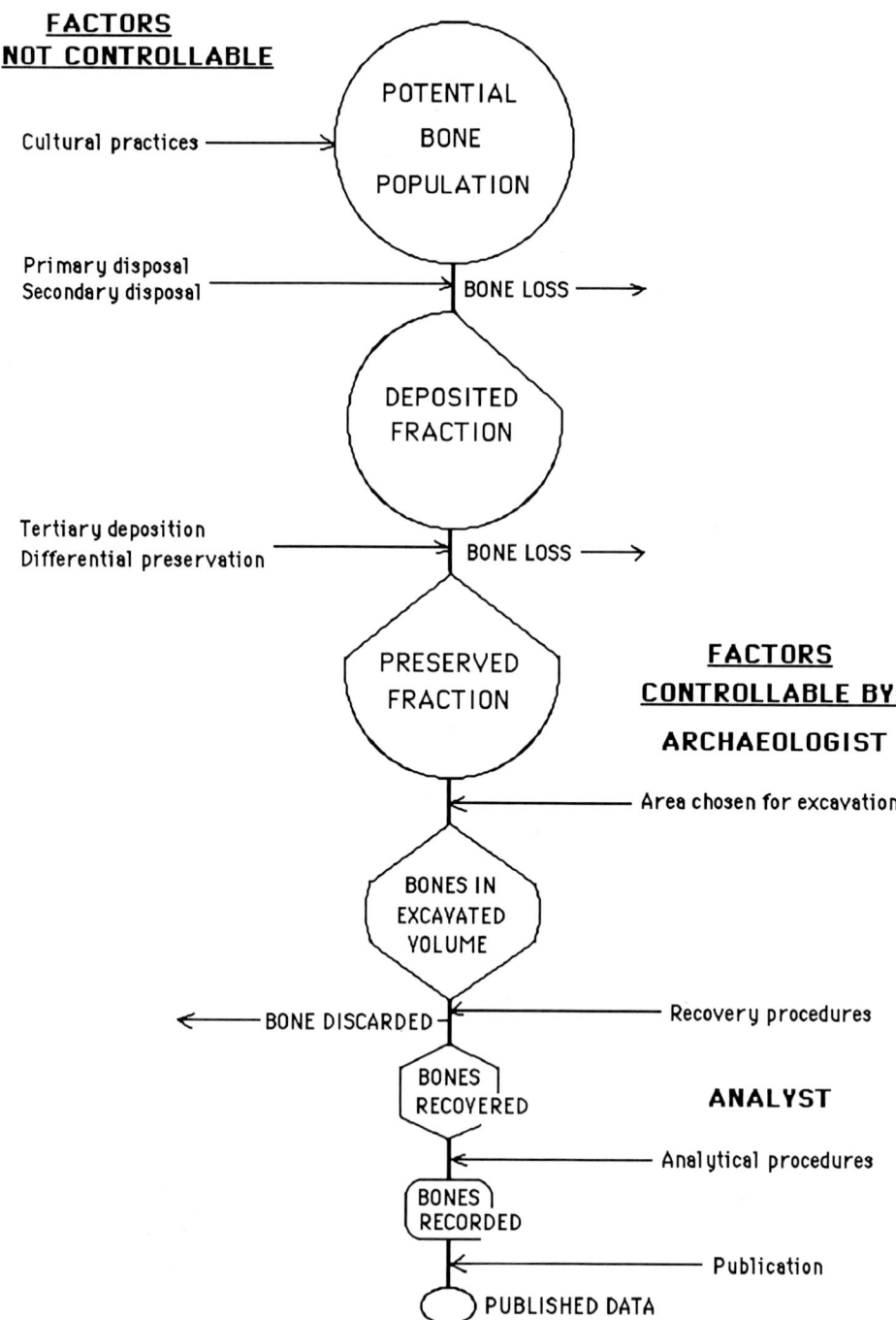

Figure 2.5. Meadow's (1981) model of the taphonomic history of a faunal assemblage. The decreasing size of the symbol (from top to bottom) denotes the loss of information through taphonomic time (after Meadow 1981:Figure 1, courtesy of the author).

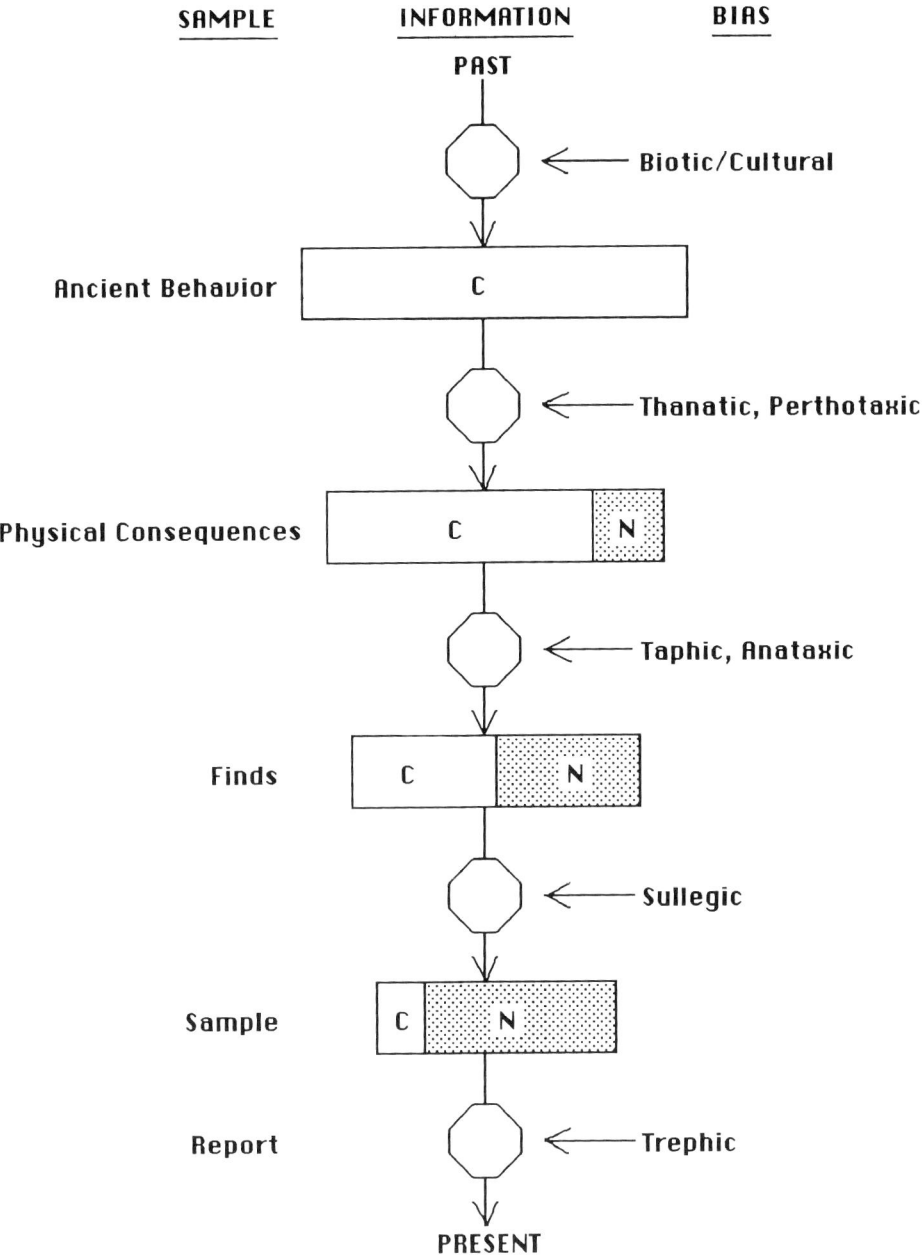

SAMPLE INFORMATION BIAS

PAST

←—— Biotic/Cultural

Ancient Behavior C

←—— Thanatic, Perthotaxic

Physical Consequences C N

←—— Taphic, Anataxic

Finds C N

←—— Sullegic

Sample C N

Report ←—— Trephic

PRESENT

Figure 2.6. Hesse and Wapnish's (1985) model of a taphonomic history of a zooarchaeological assemblage of faunal remains. Note how the natural factors (N) effectively remove the cultural information (C) displayed by the assemblage. Compare to Figure 2.5. (after Hesse and Wapnish 1985:19, Figure 9, courtesy of the authors and Taraxacum Press).

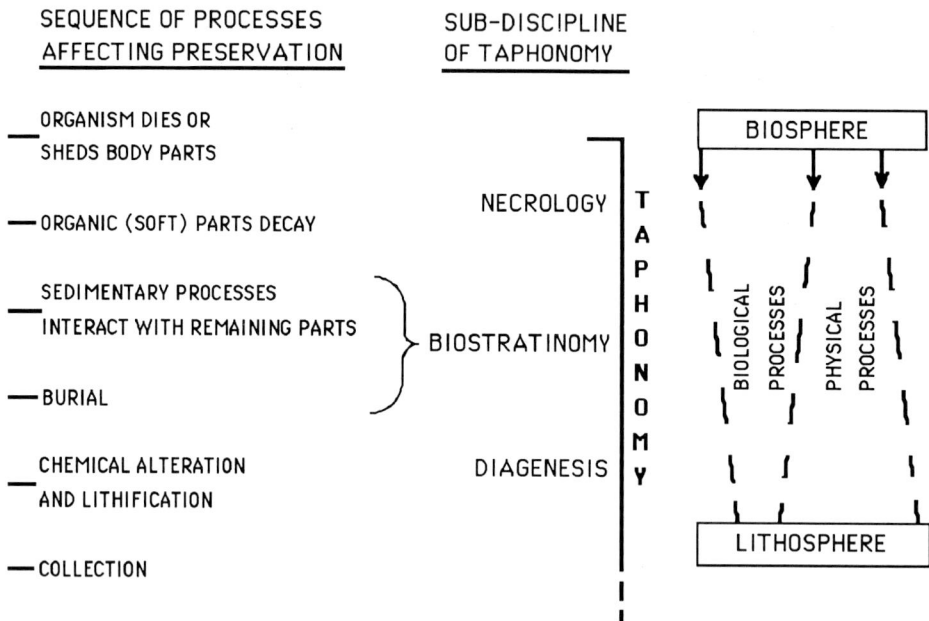

Figure 2.7. Behrensmeyer and Kidwell's (1985) model of a taphonomic history with relations of subdisciplines of taphonomy indicated (after Behrensmeyer and Kidwell 1985:107, Figure 2; courtesy of the authors and The Paleontological Society). Note the lack of symbolic loss of information.

elements to display patterned orientations). The positive aspects of taphonomic processes are appreciated by paleontologists (e.g., Allison 1991; Wilson 1988), but what about zooarchaeologists?

Among archaeologists, Bonnichsen (1989:2) thinks Behrensmeyer and Kidwell's (1985) definition is a good modification to Efremov's (1940) original one, but proposes that their new definition "be explicitly extended to include the goal of learning about human adaptive systems" because he believes the new definition is limited to "the enhanced understanding of past environments that one gains through taphonomy." This is similar to Koch's (1989:2) suggestion that "archaeologists and paleoanthropologists may find it more useful to restate Behrensmeyer and Kidwell's definition as the study of the processes of preservation and modification, and how they affect geological, biological, and cultural information in the geological record."

The significance of an accurate definition of taphonomy resides in its specification of the field of inquiry. *What* do taphonomists study, and *why* do they study that material? Efremov (1940:85) originally defined taphonomy as "the study of the transition (in all its details) of animal remains from the biosphere into the lithosphere." While expansion of this definition has correctly come to encompass the transition of any organism from the biosphere to the lithosphere, it is not at all clear how the original definition differs

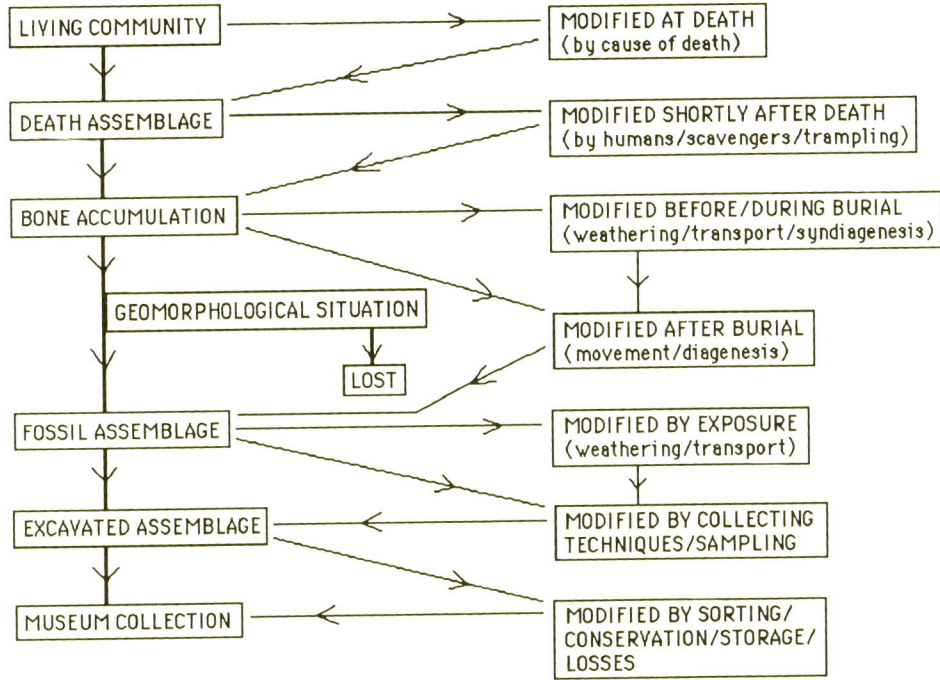

Figure 2.8. Andrews and Cook's (1985) model of a taphonomic history showing stages of modification (after Andrews and Cook 1985:689, Figure 7, courtesy of the authors and The Royal Anthropological Institute).

significantly from Behrensmeyer and Kidwell's (1985) new definition or Bonnichsen's modification of that new definition *when the definitions alone are considered* without the appended explanations of those definitions. That is particularly so when one reads a little further in Efremov's seminal paper and finds the statement that "the passage from the biosphere into the lithosphere occurs as a result of many interlaced geological and biological phenomena. That is why, when this process is analyzed, the geological phenomena must be studied in the same measure as the biological ones" (Efremov 1940:85). Thus, taphonomy is "the science of the laws of embedding [and unites paleontology] both with geology and biology into one general geo-biological historical method of study" (Efremov 1940:93).

The reason for the suggestions that Efremov's original definition be modified is that he, being a paleontologist, was concerned with the paleobiologically related biasing aspects of taphonomic processes. He noted, for example, that fossil assemblages may bear little resemblance to the communities of living organisms from which they were derived because fossil accumulations were "accidental selections" conditioned in some cases by the selective action of fluvial processes for bones of larger taxa and against the bones of smaller taxa. And he wondered why fossils were sometimes found in dense concentrations and at other times as isolated and scattered single individuals or bones. Only by controlling for such bone accumulation and dispersal factors or understanding their causes, Efremov reasoned, could we produce accurate reconstructions of paleocommunities and evolutionary histories. That is, Efremov clearly conceived of his "new branch of paleontology" as the study of the transition of organisms from being alive to being the fossils (*what* a taphonomist studies) recovered by a paleontologist. Such study could inform the researcher when a fossil assemblage provides a clear reflection of prehistoric biotic communities and when a fossil assemblage provides a distorted or biased reflection of a prehistoric biotic community (*why* a researcher does taphonomic research, according to Efremov).

But *bias* is a phenomenon that cannot be isolated; that is, it must be given a context in order that its nature can be defined and thus recognized. To say someone or something is "biased" tells us little about that person or object. To determine how an individual is biased demands that we know the referential context: "biased with regards to what?" A complete, unmodified bone is not a biased indicator of the taxon it represents, nor is a complete tooth. Similarly, a broken bone represented only by its distal end is not a biased indicator of the skeletal element represented nor of the taxon represented. However, it might be impossible to determine the sex of the individual represented by that distal half of a skeletal element, and in that sense the bone is "biased" in terms of the sexual information that can be extracted from it; in this case the specimen is biased against sexing due to the fact that it displays none of the sexually diagnostic morphometric features biologists have found to denote reliably the sex of individual animals. In the same way, but from a purely taphonomic perspective, the pubis of the innominate of some artiodactyls displays notable differences in the robustness and position of the ilio-pectineal eminence. A prehistoric artiodactyl pubis can be sexed by examination of the robustness and position of that eminence *if that part of the pubis is well preserved*; if the specimen has been weathered and bone material has exfoliated from the bone surface, or if rodent or carnivore gnawing (or any of a myriad of other processes) has removed the eminence or obscured its position, then the specimen is biased against sexing.

These examples make it clear that bias in a taphonomic sense is relative to the question being asked of the fossils (Wilson 1988). A particular mammalian fossil assemblage may accurately reflect the taxa living in the area at the time the fossils were deposited (be taxonomically unbiased; this was in part the kind of bias Efremov was most concerned about), but consist of a disproportionate number of teeth (be biased in terms of the relative abundances of skeletal parts). Thus the context or referent must be specified when it is said a fossil assemblage is or is not biased; not specifying the context results in the statement that a fossil assemblage is or is not biased having an ambiguous meaning at best, and no meaning at worst.

The notions that fossil assemblages are potentially biased and taphonomy is concerned solely with such biases has had an influence on how some archaeologists conceive taphonomy. Given Efremov's original definition, taphonomic processes are those energy-wielding agents, whether geological, climatological, or biological, that affect animal (and plant) remains. Humans and their hominid ancestors are animals that can be considered biological taphonomic agents acting at least in the realm of biostratinomy and perhaps in the realm of necrology as well. Thus Hiscock's (1990:44) suggestion that taphonomy be defined as "the study of the transformation of objects, and their spatial relationships, after they leave a living system and before they are recovered by the scientist," seems too narrow as it appears to ignore necrology. Hiscock (1990:35) seems to believe that while taphonomic processes remove information from the fossil record, the information those processes add concerns "the [taphonomic] processes themselves" and not generally the paleoecology, for instance, of the animal represented. Wilson (1988), for one, would argue the last is inaccurate (see also Behrensmeyer and Kidwell 1985). And Colley (1990a:59), for another, would counter Hiscock's suggestion by pointing out that "the study of what people do with bones is often of great interest in its own right" and such studies are the focus of zooarchaeologically oriented taphonomic research.

Based on the preceding, it should be clear that statements like, it is difficult to demonstrate that "taphonomic rather than hominid agencies are responsible for [a bone] assemblage" (Speth 1991:37), or that disarticulation of some animal carcasses "is probably attributable to taphonomic processes rather than butchering [by humans]" (Landals 1990:139), or similar statements implying that human and/or hominid interactions with animal carcasses are not taphonomic, are incorrect. Such statements are rare, and I suspect they reflect a discipline-centric view of zooarchaeologists primarily interested in human behavior; thus any non-human process is biasing and therefore taphonomic. (Reed [1963:210–211] turns this notion around and argues the "cultural filter" biases the faunal record because humanly accumulated faunal remains will not accurately reflect the paleofauna.) The implicit reasoning seems to be that because human activities are the subject of interest, they cannot be biasing and thus are not taphonomic because taphonomy concerns how the fossil record is biased. The notion that taphonomy concerns only the biases to a fossil record comes from taphonomy's conceptual roots in paleontology where it was recognized that fossil assemblages seldom represent completely undistorted pictures of past biological communities. In archaeology, the context of concern is generally human behavior, and while archaeological assemblages of animal remains may well be biased with regards to that behavior, this does not mean humans are not taphonomic agents. This, in the final analysis, is the reasoning underpinning Bonnichsen's and Koch's modifications of Behrensmeyer and Kidwell's definition of taphonomy.

On the structure of taphonomy: a personal view

> Each [taphonomic] event has unique characteristics that require a particular approach. [Due to historical variability] useful analyses must select carefully from among variables and apply only those that are pertinent to the unique situation. (E. C. Olson 1980:9–10)

Taphonomic histories

The paleontological fossil record has been formed totally by natural processes including geological and biological processes. These natural processes act upon the available organisms (which are in turn conditioned by such natural factors as topography, substrate, vegetation, and climate) and affect the addition to, maintenance in, and subtraction from the paleontological fossil record of organisms and their remains.

The archaeological fossil record is potentially formed not only by the same natural processes as the paleontological fossil record, but also by human processes. An archaeological site consists of cultural and natural objects that are added, spatially arranged, and preserved and/or destroyed by various human and natural processes. Human behaviors that result in the formation of a fossil record have been labeled the *cultural filter* (Reed 1963; Daly 1969). Human processes that affect potential additions to the fossil record include selective hunting (Smith 1979; Wilkinson 1976) and butchery practices (Binford 1978; Noe-Nygaard 1977, 1987). Variation in both the archaeological and the paleontological fossil records is created by varying the additions, the means of addition, the mechanisms that distribute faunal remains, and the means of maintenance and subtraction of animal remains from the fossil record.

The distinction between paleontological faunas and archaeofaunas is based on characterizations of their respective lack of associated artifacts or presence of associated artifacts (Chapter 1). It must be emphasized, however, that these characterizations are simplistic. Some fossil records may have no spatially associated cultural materials even though humans had an active role in their formation. In the absence of associated artifacts, attributes of bone modification attributable to human activities are cited as evidence of human intervention. One example of this involves mastodon (*Mammut americanum*) bones in North America (Fisher 1984a, 1984b; Gilbow 1981; Gustafson *et al.* 1979). In these cases, modifications to bones which are inexplicable given natural processes are cited as evidence of human taphonomic agents even though no (or a few possible) artifacts are associated with the faunal remains (see Haynes and Stanford [1984] for similar arguments regarding North American late Pleistocene *Camelops* sp.). However, it is not clear as yet which of these cases actually represent humanly modified carcasses (e.g., Graham *et al.* 1983; see Haynes 1991 for a review of proboscidean taphonomy). This in turn has resulted in detailed comparative analyses of mastodon remains clearly associated with artifacts and similarly aged mastodon remains with no associated artifacts in attempts to derive bone modification criteria that unambiguously signify a hominid taphonomic agent (e.g., Fisher 1987; Graham and Kay 1988).

The literature on attributes of bone modification is expanding rapidly (Bonnichsen and Sorg 1989; Hudson 1993; and references therein) yet debate abounds over the precise meaning of many attributes (e.g., Johnson 1982, 1985 versus Lyman 1984b; and Shipman 1981a; Shipman and Rose 1983a, 1983b; Olsen and Shipman 1988 versus Eickhoff and Herrmann 1985; Behrensmeyer *et al.* 1986, 1989; Fiorillo 1989; Oliver 1989). Many of the attributes of bone modification used to distinguish culturally deposited from naturally deposited faunal remains are, when used alone, equivocal. Multiple kinds of data are necessarily used regularly to help to distinguish analytically the two types of fossil records.

Taphonomic histories are reconstructed from the abundance, distribution, and modification attributes of fossils. The objects in a site, their frequencies, physical attributes, spatial loci and associations, and geological and cultural associations are all that are observable in the fossil record. A scientific approach to taphonomy must realize what the empirical phenomena of the fossil record are, and produce a model that permits expectations to be phrased concerning fossil assemblage content and distribution; i.e., the archaeologically visible fossil record. Such a model would ideally be universally applicable and yet specific enough to grant insights to particular taphonomic pathways. A first step to model building involves understanding the basic structure of taphonomic processes and effects. As defined in Chapter 1, a *taphonomic process* is the dynamic action of some source of force or the physical cause of modification to animal carcasses and skeletal tissues. A *taphonomic effect* is the static result or trace of a taphonomic process that acted on and modified carcasses or skeletal parts and tissues. I begin by considering some general characteristics of taphonomic processes, and then turn to some general categories of taphonomic effects.

Processes that form the fossil record vary along three dimensions. The OBJECT dimension consists of three categories: addition, subtraction, and maintenance of objects. Animal remains may be added to, removed from, or maintained within a particular spatially defined assemblage of remains. The SPATIAL dimension consists of movement, and nonmovement. Faunal remains may be moved within (and perhaps removed from) or not moved within a particular spatial unit (the boundaries of which define the assemblage). The MODIFICATION dimension consists of two possibilities: a bone may be modified from its natural, living state (e.g., be broken, burned, mineralized), or it may not be so modified (other than perhaps drying). A complete skeletal element that retains its anatomical integrity and features, if not fossilized, can be considered to be not modified. Taphonomic processes create variability between and within particular fossil records. Because the major expressions of taphonomic effects in archaeological contexts are assumed to be largely attributable to human processes, random variation in fossil assemblage content, distribution, and modification attributes is not anticipated, an anticipation borne out by studies of archaeological faunas (e.g., Lyman 1978; Meadow 1978; Noe-Nygaard 1977; Pozorski 1979; Rapson 1990; Stiner 1990a). However, natural processes also create non-random patterns in the archaeological record (Binford 1981a, 1981b; Brain 1969; Haynes 1980a, 1980b; Hill and Behrensmeyer 1984, 1985). The first step in analysis then, is to identify patterns associated with natural processes. This may be accomplished by comparing the fossil record under study to models of the variable(s) under scrutiny, such as the relative frequencies of skeletal elements in complete skeletons. Subsequent to this comparison, the analyst can begin to assess the meaning of any detected patterns by comparing the fossil record to patterns displayed by neotaphonomic records created by various known taphonomic processes. Brain's (1967b, 1969) comparison of bone part frequencies in ethnoarchaeological contexts with frequencies of bone parts from South African caves is a classic example of this comparative approach.

Taphonomic histories are initiated when an animal dies (the mode of mortality can be a significant taphonomic variable; see Chapter 5). Soft tissues may then be removed, bones may become disarticulated, scattered, buried, fossilized, may erode or chemically deteriorate away, and may eventually be recovered. Of course, various taphonomic processes may or may not act simultaneously on a bone or carcass, and may or may not affect particular carcasses or bones. The set of potential effects of taphonomic processes may be arranged into four categories: disarticulation, dispersal, fossilization, and mechanical modification. All of these are monitored from the starting point of a complete, unmodified skeleton. That is, we know what a living skeleton looks like, and what taphonomists are interested in is how that skeleton, or what is left of it, appears when it is recovered from a geological context.

Disarticulation refers to the anatomical disassociation of skeletal elements; it often is the first step in the loss of anatomical integrity of a skeleton. Disarticulation is related to soft tissue that functions to hold joints together (Dodson 1973; Hill 1980; Schäfer 1962/1972; Toots 1965c). Chemical or mechanical breakdown and removal of soft tissue ultimately results in disarticulation (e.g., Coe 1978; Micozzi 1986; Payne 1965). Because soft tissue anatomy varies from joint to joint, the process of disarticulation is extremely complex under natural conditions, but is not so complex as to preclude construction of models of natural disarticulation (Hill 1979b, 1980).

Dispersal of skeletal parts means the increase or the decrease of distance between bones. Dispersal of skeletal elements may precede, or be simultaneous with or subsequent to disarticulation, and is related to disarticulation because it concerns the spatial location of fossils. While disarticulation requires only a few centimeters of spatial disassociation of parts to destroy anatomical integrity, dispersal entails centimeters to kilometers (Hill 1979b). Models of dispersal have been constructed for fluvial transport (e.g., Behrensmeyer 1975b), human transport (e.g., Binford 1978), raptor transport (Plug 1978), transport by porcupines (Brain 1980), carnivore transport (Binford 1981b), and random processes (Hill 1979b). Most analyses of fossil assemblages that consider taphonomic issues assess whether or not the fossil assemblage has been transported to its recovery location from a near or a far source location. Shotwell (1955) developed an analytic technique to assess whether or not a fossil assemblage had been transported, and to assess which taxa in the assemblage were locally derived and which were probably intrusive or non-local. While Shotwell's (1955) technique was later adapted to distinguishing naturally from culturally deposited taxa in archaeofaunas (Thomas 1971), it has since been shown to contain serious flaws (Grayson 1978a, 1978b; Wolff 1973). Analysts still address this issue, using techniques such as assessing the degree of abrasion of bones (Behrensmeyer 1975b) to determine whether the assemblage or portions thereof have been fluvially transported to the collection locality, and experimental data to infer the agent of transport (Binford 1981b; Lyman 1985a).

Fossilization denotes the alteration of bone chemistry (Cook *et al.* 1961; Rolfe and Brett 1969). The type of sedimentary matrix in which the bone is deposited largely determines the particular types of fossilization processes. Secondary determinants include environmental conditions such as soil moisture regimes as determined by precipitation and temperature. Some fossilization processes, especially weathering (Behrensmeyer 1978), may result in fragmentation of bones.

Mechanical alteration denotes the structural and/or morphological alteration of the original living bone by mechanical or physical processes. Examples include fragmentation and abrasion. For instance, each bone in an animal is a complete, discrete object. The cause of the animal's death and/or postmortem factors may result in broken bone (Lyman 1984b, 1989b). Fragmentation, then, is the destruction of the original anatomical discreteness of a bone by generating multiple discrete objects from the original by mechanical or physical means, in this case by the loading of force on the bone. Abrasion is the modification of original bone morphology by the application of frictional forces to bone surfaces or edges (Olsen and Shipman 1988; Shipman and Rose 1988).

Discussion

Taphonomy is concerned with differences and similarities between fossils and organisms, and between a fossil record and the prehistoric fauna from which it derived. Concerning the latter, obvious differences include the presence of living organisms. Ecological and ethological studies of extinct taxa are, of course, impossible. Even those taxa with modern, living counterparts are not so easily dealt with when represented by fossils because studying living taxa presents certain difficulties (Coe 1980). Differences and similarities between living and fossil faunas present taphonomic challenges to paleoecological research because the ecological principles used to study living faunas are commonly used in paleoecological research (Van Couvering 1980; Western 1980). Consequently, the fossil record must be analytically reconstituted into a fossil fauna or the original biotic community to answer many research questions. As noted in Chapter 1, subsistence studies using archaeofaunas face similar analytic challenges (King and Graham 1981).

Clues for developing analytical techniques for meeting these analytic challenges can be found by considering the four categories of taphonomic effects (disarticulation, dispersal, fossilization, mechanical alteration) in light of the variability in taphonomic processes. For example, new or additional fossils cannot be added to an assemblage that is *in situ* without movement of the "new" fossils. The only conceivable way this may happen is if the assemblage moves to a new location and is deposited around the "new" fossils without the latter's movement, as in some fluvial settings (Boaz 1982). A common and readily conceivable way for new fossils to be added to an assemblage without movement of the former is for the sampling universe to be enlarged such that additional fossils are collected from new areas of a site.

The second important point deriving from comparison of processes and effects is that different processes can have similar effects; *equifinality* is a very real problem. Disarticulation, dispersal, and mechanical alteration all involve movement of the fossils while fossilization does not require movement. Finally, all effects and process categories concern frequencies (add, maintain, subtract) and distributions (move, non-move) of fossil categories. It is therefore pertinent to discuss possible techniques for measuring each taphonomic effect in the fossil record. I do this briefly here, and in more detail in later chapters.

Hill (1979b:744) concludes that "the determining controls of the [disarticulation] pattern are inherent in the anatomy of the dead animal itself and thus independent of the agents whereby it is realized." Disarticulation might be modeled by a rank ordering of the cross-sectional area of soft tissue surrounding joints. The basic analytic assumption might be phrased as: the greater the cross-sectional area of soft tissue associated with a joint the longer the joint will remain intact subsequent to the animal's death. This assumption of course presumes that soft tissues associated with each joint are qualitatively identical, which is unlikely (Hildebrand 1974; Romer and Parsons 1977). Thus the amount and type of connective tissue seem to be variables with significant influence on the disarticulation process. Study of disarticulation in the fossil record requires detailed data on the locations and spatial associations of faunal remains (Chapter 5).

Dispersal is a complex process minimally controlled by disarticulation, type and strength of dispersal mechanism, substrate, topography, and bone density, size, and morphology. Hill (1979b:269–270) hypothesizes that scattering is caused by processes that act randomly. Departures from the random pattern suggest non-randomly acting processes whose identity must be determined. Hill's (1979b) hypothesis could be used as the first null hypothesis to be tested with fossil data. Then, intrinsic properties of bones can be used to generate expectations regarding distributional patterns of fossils (e.g., Frostick and Reid 1983; Korth 1979). Clearly, data on bone location, orientation, and angle of dip should be recorded during field recovery, as well as sedimentological data indicating mode of deposition and turbation processes (Chapters 5–9).

Fossilization mechanisms are minimally dependant on climate, depositional matrix, and bone porosity. There apparently is no detailed model of fossilization comparable to Hill's (1979b) models of disarticulation and dispersal. Documented processes of fossilization (e.g., Rolfe and Brett 1969; Schopf 1975; Whitmer *et al.* 1989) indicate, however, that in order to study fossilization, data required include matrix chemistry and mineralogy, chemistry of the fossils and original chemistry of the bones, climatic (past and present) information such as temperature, precipitation, and ground water regimes, and a knowledge of the geologic and pedogenic processes that contributed to the formation of particular strata (Chapter 11).

Mechanical alteration seems to be largely controlled by bone structure and morphology, at both microscopic and macroscopic levels, and bone porosity and density. In order to measure mechanical alteration in the fossil record, the minimal requisite data are frequencies of fragment types (e.g., Bunn 1989; Todd and Rapson 1988) and whether or not fragments of a bone are associated, *in situ*, polished, abraded, or display other features (Chapters 8–10). For example, Klein and Cruz-Uribe (1984) suggest sediment overburden may crush more deeply buried bones; all else being equal, the analyst could measure fragment sizes to determine if fragments decreased in size with increasing depth (e.g., Lyman and O'Brien 1987).

Despite the pleas of several authors over two decades ago (Hill 1978; Hill and Walker 1972; Munthe and McLeod 1975) the kinds of data mentioned above (Table 2.1) were seldom published by analysts prior to the mid-1980s. Because abundances of fossil categories are important to many traditional analyses and interpretations, bone frequency data are nearly always published, and more is known about the taphonomy of frequencies of bone parts than virtually any other variable of the fossil record.

Summary and conclusion

> Taphonomy has come of age!
> (P. A. Allison 1991:345)

Taphonomic histories are usually complex. Taphonomic processes vary along three dimensions (objects [added, subtracted, maintained], spatial [moved, not moved], modification) while taphonomic effects may be arranged in four general categories (disarticulation, dispersal, fossilization, mechanical alteration). Consideration of the process and effect categories reveals the types of data required for taphonomic analyses. I have distinguished two basic goals of zooarchaeological faunal analysis (determination of human subsistence patterns and prehistoric ecological conditions) and two types of fossil records (archaeofaunas and paleontological faunas). All of these issues are found in one form or another in most taphonomic studies.

It is important to keep in mind throughout the remainder of this volume that taphonomic processes are *historical* and *cumulative*. That is, they have a direction through time in the explicit sense that effects of taphonomic processes which occur late in the history may depend on the effects of taphonomic processes which occurred early in the taphonomic history. A taphonomic history results in a fossil assemblage which may poorly reflect the quantitative properties of the biotic community from which the fossils derived. Taphonomic processes sometimes mimic and other times obfuscate their respective effects, thereby rendering the writings of taphonomic histories difficult. The history of taphonomic research, especially in archaeology, illustrates why taphonomic research at the end of the twentieth century appears the way it does. The complexities of taphonomic processes can be described by a small set of general processes and their respective general effects. This simple framework guides us towards recognition of data requisite to taphonomic analyses. My purpose here has been to provide a general framework that allows us now to turn to a general consideration of the practice and theory of taphonomic research.

3

TAPHONOMY IN PRACTICE AND THEORY

History suggests that the road to a firm research consensus is extraordinarily arduous . . . In the absence of a paradigm or some candidate for paradigm, all of the facts that could possibly pertain to the development of a given science are likely to seem equally relevant. As a result, early fact-gathering is a far more nearly random activity than the one that subsequent scientific development makes familiar.
(T. S. Kuhn 1970:15)

Introduction

The foundations for taphonomic research were laid in the nineteenth and early twentieth centuries with a focus on observations of modern processes that resulted in deposits containing bones with certain modifications (Behrensmeyer and Kidwell 1985). Early taphonomists followed the uniformitarianist approach used by geologists of the nineteenth and twentieth centuries. That approach and its present structure in the service of zooarchaeological taphonomy is reviewed in the second part of this chapter. Prior to that I review several examples of what I consider to be good taphonomic analyses. These illustrate what makes for strong conclusions and lead to a consideration of uniformitarianism and actualism as methodologies for studying the past. This in turn leads to a consideration of ethnoarchaeology and middle-range research. Finally, because actualism and middle-range research ultimately lead to analogical arguments, the structure of such arguments is described.

Examples of taphonomic analysis

The criteria I used to select the examples reviewed were simple. The analysis must be published in a generally available form so that the original can be consulted by interested readers. The analysis must have explicit hypotheses that were being tested, and explicit assumptions and methods. As well, the data must be available (generally in the published articles) for evaluation and additional analysis. In keeping with the distinction of paleontological and archaeological traditions of taphonomic research, I begin with two examples of taphonomic analysis in paleontological contexts, and then turn to examples from archaeological contexts.

41

The extinction of Irish elk

Extinction of the large-antlered Irish elk (*Megaloceros giganteus*) near the end of the Pleistocene has often been explained as resulting from their having become mired in bog mud and/or being drowned in marshes or bogs in part because their antlers were large, heavy, and cumbersome. Barnosky's (1985, 1986) analysis of remains of this large cervine is instructive. He excavated a part of Ballybetagh Bog near Dublin, Ireland, and studied specimens previously collected from it. His excavations covered 21 m² and ranged from 1.5 to 2.5 m deep. He examined 35 skulls, all males, from this bog, and compared them to various samples of Irish elk skulls from other localities. Some of the comparative specimens were on display in museums (large and complete specimens) and others were not on display (small and/or broken).

Barnosky (1985) lists six test implications of the miring-drowning due to large antlers hypothesis. Only one is met with available data: all individuals in the bog he sampled were males (females do not carry antlers). But the other five implications are not met. Antlers are smaller than average in Barnosky's sample; skeletons are not articulated nor complete; the bones are embedded in clay deposits too thin for the animals to have become mired; the deposits are not disturbed by trampling or struggling of these animals as they should have been by mired animals; and the bog waters were apparently shallow enough (as inferred from geologic data) to preclude drowning of upright animals. Note that the six test implications referred to require detailed morphometric, contextual, associational, and stratigraphic data.

Failing to confirm the miring-drowning hypothesis, Barnosky (1985, 1986) proposes and tests two alternative hypotheses. Pleistocene overkill by human hunters (Martin and Klein 1984) is quickly discarded because no artifacts have been found associated with remains of Irish elk, and "the few examples of modification to Irish elk bones reputed to have been inflicted by humans cannot be distinguished from other naturally created kinds of breaks, abrasions, gnaw-marks, or scratches" (Barnosky 1986:132). Further, the oldest archaeological evidence of humans in Ireland dates between 9000 and 8500 BP, while the Irish elk was extinct there ca. 10,600 BP. Tight chronological and stratigraphic control of the Irish elk remains described by Barnosky makes the argument for absence of a human taphonomic agent convincing, and thus the overkill hypothesis cannot be sustained.

The second alternative hypothesis is that "male Irish elk visited bogs more often than females did during winters, when unfit animals died and decomposed near the water's edge, in some cases on the ice, and were scavenged and trampled" (Barnosky 1985:340). Evidence bearing on the six test implications for the miring-drowning hypothesis are consistent with the winterkill hypothesis, as are four other test implications specific to the latter hypothesis. All elk died with antlers attached, suggesting an autumn-winter death season based on analogy with modern cervines. Mortality was demographically attritional, as it should have been if the Irish elk most susceptible to malnutrition-related death (the young and the very old) were dying. Barnosky's (1985:341) Irish elk tend to be small in body and antler size, suggesting "some combination of limited resources, malnutrition, or disease during fetal or postnatal growth." Finally, in modern cervines, "male mortality is greater than female mortality during winter, apparently because males, unlike females, eat little during the fall rut and enter the winter in poor condition," and males more often seek winter shelter in valley bottoms (near bogs) and thus many die near lake (bog) shores "because they need water and because they are easy prey on ice" (Barnosky 1985:343). This explains the overabundance of males relative to females in collections of Irish elk fossils. The test implications for the winterkill hypothesis underscore the necessity of age–sex demographic data, morphometric data, and the use of modern analogs as comparative bases in taphonomic analyses.

Barnosky not only provides many (but not all; e.g., the "scratches" on Irish elk bones are not described) relevant data, he considers three separate hypotheses. The winterkill hypothesis succeeds because its test implications are met whereas test implications of other hypotheses are not. Further, Barnosky (1985, 1986) considered a dozen test implications, some with negative evidence (paleontologically/archaeologically invisible test implications) and some with positive evidence (paleontologically/archaeologically visible test implications) to confirm his hypothesis.

Crocodilian scatology

Fisher (1981:263) attempted to establish "unambiguous signature" criteria of crocodilian consumption and digestion of microvertebrates via the controlled, long-term feeding of eight individual crocodiles and the collection and study of all scats and regurgitated materials. He suggests extrapolation of his actualistic data "to fossil crocodilians seems justifiable as a working hypothesis" (Fisher 1981:270). He justifies the extrapolation by reviewing the digestion of bones by other vertebrate predators and noting the unique attributes of bones subjected to crocodilian consumption and digestion; "only crocodilians are known to decalcify calcified tissues, while leaving their organic matrices intact" (Fisher 1981:270). He also describes suspected causal factors for the observed effects on bones consumed by crocodiles, such as variation in retention time in the stomach and bone robusticity influencing the extent of decalcification. He then examines a Paleocene fauna and, by carefully eliminating other hypothetical causes of decalcification such as abrasion and subaerial weathering, he presents a convincing argument that prehistoric crocodiles were the taphonomic agent that created the fossil assemblage.

Fisher's (1981) clearly stated approach to taphonomic analysis via the actualistic method (see below) is laudable and approximates that of others. He concisely documents the distinctiveness of a set of attributes displayed by a set of animal remains. He believes these attributes can be used as diagnostic criteria given suspected causal relations between the attributes of bone modification and particular taphonomic processes, and he eliminates other possible causes. Fisher (1981) clearly describes the origin of his explanatory premises and their attendant weaknesses by indicating several times that they are at present the best approximation, and further research may require their revision. Fisher's study is one of the best taphonomic studies to appear in recent years.

Archaeological examples

It is reasonable to wonder if zooarchaeologists with interests in taphonomic problems perform taphonomic analyses in a manner similar to the two paleontological examples just reviewed. That is so because it has been proclaimed that humans and their hominid ancestors are sufficiently distinct from other bone-processing organisms that special treatment is required when they are perhaps part of the taphonomic history of a bone assemblage (e.g., Binford 1981b). In my view, such arguments are taxon-centric and unwarranted. For example, in an analysis of an assemblage of cow (*Bos taurus*) bones recovered from the ground surface inside a cave in Nevada, I tested the hypothesis proposed by the excavator that the cattle bones represented the remains of a winter meal left by Native Americans early in the twentieth century (Lyman 1988a). I derived test implications from the hypothesis and searched the cow bones for evidence of them. I failed to find convincing evidence that people had anything to do with the cattle remains. Of the test implications derived, two clearly are not met (skulls were not fractured for brain extraction but rather were disarticulated along sutures; no butchering marks were found), three do not support the hypothesis of human intervention but neither do they refute it (some bones may have been cooked but many had been burned by natural fires; there was no evidence of differential transport or utilization based on economic utility of skeletal parts; tooth eruption and wear suggest a coarse diet and possible winter death due to malnutrition), and two were not met but are ambiguous in terms of indicating a human taphonomic agent.

   While it might be suggested that the hypothesis-testing procedure I used for the Nevada cattle remains was reasonable because the bone assemblage seems to represent a naturally deposited one, that does not imply the procedure is inappropriate for assemblages that were deposited by hominids. Further, while my conclusion that humans had little to do with the cow bones is in direct contrast to the excavator's, that does not mean I believe the excavator was stupid. I believe he was taphonomically naive, as we all were in the late 1960s when he did his analysis and offered his conclusion, and the difference in our conclusions simply "measures the increase in our understanding of taphono-mic processes" over the intervening years (Lyman 1988a:104).

Binford (1984b:10–14) argues that we must maintain a healthy scepticism about knowledge claims presented today because those claims are to some degree the product of our "knowledge of the moment;" thus, when we can demonstrate that a suspected cause–effect relation (for instance) is inaccurate "it is always because we have gained knowledge that was not available to [earlier workers]." This echoes Fisher's (1981) suggestion that the cause–effect relations he postulates are "working hypotheses" subject to revision on the basis of further research.

Binford (1984b) presents several suspected causal links between the location and orientation of butchering marks on various bones and joints (the effects of butchering), and the process of butchering. The relevant linkages or causal relations are founded on whether the animal carcass being butchered is fresh and supple or partially desiccated and rigid. Based on his actualistic research, Binford (1984b:71) writes "when a carcass is stiff, the joints are generally bound – the tissue has shrunk and locked the articulation into a fixed position, making manipulation of the joint impossible. This means that the orientation of cuts relative to the shape of bones will generally be in regular and determined places, rather than the [fresh carcass situation] in which the orientation of the cut shifts as the joint is flexed during dismemberment." Binford (1984b) uses this suspected causal relation to explain butchery marks he observed on bones from Klasies River Mouth Cave in South Africa site as indicating many of the bovid remains appear to have come from rigid, partially desiccated carcasses collected by the hominid occupants of the site.

In another exemplary analysis, Stiner (1990a, 1990b, 1991a) considers the demography of the population of large mammals represented by fossil collections from Middle Paleolithic sites in Italy. She reports that she "was both surprised and frustrated by the lack of clear logical connections between [a mortality] pattern and [its] cause" and that "a great disparity currently lies between the technically sophisticated means for constructing mortality profiles and the knowledge available to interpret these patterns" (Stiner 1991d:2). Stiner reviews much of the ecological literature on mortality, and describes how the behavioral variability of carnivorous predators influences the mortality patterns of their prey. She builds a model from available data which incorpor-ates causal links between predator and prey behaviors as causes, and their effects in terms of the kind of prey mortality that will result from certain interactions of predator and prey taxa. The model is actualistic because it is based on modern cases, and it becomes an analog when Stiner (1990a, 1990b, 1991a) uses it to help interpret the prehistoric record. But Stiner is appropria-tely cautious. She notes that the model is not an algorithm for deciphering demographic data derived from zooarchaeological remains. She points out its weaknesses, such as it being based on diachronic assemblages and thus it may not be applicable to relatively synchronic assemblages. The model cannot, by itself, "establish the cause of death or the agency responsible for creating a bone assemblage," and independent taphonomic research must help establish those causes and agencies (Stiner 1990b:341). The model is suggestive of predatory behaviors due to the suspected causal linkages of predator behaviors and prey mortality type. But Stiner (1990b:343) recognizes that additional neotaphonomic data "will undoubtedly improve and strengthen" the model.

Summary

The examples reviewed share several features of the foundations of taphonomic research. These include reference to modern cases wherein the relations of causes and effects are known or suspected, testing of hypothetical taphonomic histories, and the realization that all conclusions regarding taphonomic histories are the best accounts *presently available*. Conclusions may be modified as new data and/or concepts in either modern or prehistoric settings come to light. Simply, taphonomic research is a particular kind of scientific research. A taphonomist is neither omnipotent nor infallible. Theories and concepts are assessed in the empirical world, modified, and assessed again. To delve further into the enterprise of taphonomic research, we now turn to its epistemological underpinnings.

Uniformitarianism and actualism

> There are operations proper to the surface of this globe by which the form of the habitable earth may be affected; operations of which we understand both the causes and effects, and, therefore, of which we may form principles for judging of the past. (J. Hutton 1795; cited in Scharnberger *et al.* 1983:312)

Gifford (1981) outlines the common perception of the theoretical and methodological basis of taphonomic analysis. She advocates that we phrase our research goals and methods in terms of the "uniformitarian methodology and assumptions" which underlie our discipline (Gifford 1981:397). The basis of this advocacy can be found in the German aktüo-palaeontologie proposed by Richter (1928) and defined by him as the science of the origin and present-day mode of formation of future fossils, and the application of that knowledge to paleontological problems via the methodology of uniformitarianism (Warme and Häntzschel 1979). Uniformitarianism, in one form or another, permeates all aspects of geology (Hooykaas 1970; Schumm 1985; Watson 1969), paleontology (S. J. Gould 1965, 1967, 1979; Simpson 1970), paleoecology (Lawrence 1968, 1971; Nairn 1965; Scott 1963), archaeology (Ascher 1961a, 1961b; Binford 1981b; R.A. Gould 1980; Stiles 1977), archaeofaunal analysis (Lyman 1982a; Medlock 1975; Rackham 1983) and taphonomy (Brain 1981; Hill 1978, 1988; Shipman 1981b). It is therefore appropriate to consider first the concept of uniformitarianism and the related concept of actualism.

What is uniformitarianism?

> Modern [methodological] uniformitarianism has no substantive content–that is, it asserts nothing whatever about nature. Uniformitarianism must be viewed as telling us how to behave as scientists and not as telling nature how it must behave.
> (J. H. Shea 1982:458)

While the concept of uniformitarianism is usually attributed to James Hutton, it received greater success as a guiding scientific concept in the writings of Charles Lyell (Gould 1979, 1982, 1984; Haneberg 1983; Hooykaas 1970; Rudwick 1971; Wilson 1967). The detailed meaning and connotations of the concept have changed over the 160 or so years since its introduction (W. Whewell [1832] suggested the term in a review of Lyell's *Principles of Geology*), but its basic meaning has remained more or less the same. It is perhaps because of the evolution of the concept's meaning that much misunderstanding exists and various meanings are attributed to it. Here I review what uniformitarianism stands for today, and what some scientists (mostly geologists) think about the concept.

Uniformitarianism, as phrased by Lyell, consists of two major parts: a testable theory, and an analytic procedure or assumption (Gould 1965, 1979; Rudwick 1971). The theory entails two hypotheses. The first is labeled "gradualism" and suggests that rates of change have been uniform throughout time and that large results are not the product of sudden castastrophic causes, but the accumulated effect of innumerable minute changes (Gould 1979:126–127). The second hypothesis is labeled "nonprogression" and suggests that the configuration of the earth is in a dynamic steady-state; change is incessant but cyclic (Gould 1979:126–127). Labelled "substantive uniformitarianism" (Gould 1965), the theoretical part of uniformitarianism now seems false, and if "too rigidly held [becomes] an a priori assumption, stifling to the formulation of new hypotheses which may better explain certain data" (Gould 1965:226). The analytic procedure, labelled "methodological uniformitarianism" (Gould 1965), makes the past amenable to purely scientific explanation, and also consists of two parts. First, it assumes that natural laws are invariant in time and space. Second, it assumes that processes have been invariant in time and space, therefore past results may be properly ascribed to causes now in operation (Gould 1979:123, 126). The two parts of methodological uniformitarianism are requisite to inferring past dynamics; the analytic process involves the association of modern results with particular modern processes. When similar results, some formed in ancient times and others formed in modern times, are found, the inference is made that the processes were the same or at least similar in both the past and present cases.

Simpson (1963:24–25, 1970) suggests "immanent properties" are the "unchanging properties of matter and energy and principles arising therefrom" (including processes such as gravity) and "configurational properties" are the arrangements and organizations of the matter making up the world. The former is subsumable under methodological uniformitarianism, the latter is more or less synonymous with substantive uniformitarianism. It is, of course, the former that is our concern here.

Gould (1965:223) defines methodological uniformitarianism as "a procedural principle asserting spatial and temporal invariance of natural laws." But Simpson (1970:59) suggests "the invariability of natural laws is indecisive about such basic problems as their sufficiency or as to whether the actions of all historically relevant laws are currently ('actually') observable." As Watson (1966) makes clear, this is a problem of scale given, for instance, that humans live only about 70 years and thus one individual cannot observe the formation of an erosional feature the magnitude of the Grand Canyon. Watson (1966) points out that if the laws of erosion written by scientists are correct, then we can infer how the Grand Canyon and similar features were formed by reference to those laws.

Kitts (1977:63–64) discusses two principal aspects of methodological uniformitarianism. First, the grounds upon which the assertion of uniformity rests are not testable. If the grounds for the assertion are that the past was like the present (the natural laws we have written are temporally invariant), we cannot test them because we cannot observe the past. Second, methodological uniformitarianism imposes restrictions on statements made about the past, i.e., appeal to supernatural or unknown forces are precluded. Clearly, without this restriction one could make any statement at all about the past, including statements that cannot be empirically tested. Yet, Kitts (1977) continues, one must subscribe to methodological uniformitarianism in order to make historical inferences. "Because it is impossible for us to observe anything except the present, our interpretations of prior events must necessarily consist of inferences based upon present observations" (Hubbert 1967:29–30). Our assumptions must therefore be (1) natural laws are invariant with time, and (2) violation of natural laws by any non-natural mechanism is excluded from consideration.

Goodman (1967) and Shea (1982) argue that uniformitarianism is a term for how we do science. It involves the confrontation of empirical data with theory, and modification of the latter if necessary, in order to derive an explanation of the empirical items under study (see also Gould 1965, 1967; Watson 1966). This procedure is common to all sciences as it presumes natural laws are temporally invariant. All sciences involve induction and simplicity, two key aspects of methodological uniformitarianism (Goodman 1967; Gould 1967). *Simplicity* is the denial of processes or forces unique to the past (Gould 1965), which is to say that methodological uniformitarianism assumes natural laws are atemporal. Simplicity recognizes the fact that scientists write the laws, which are descriptions of how we think things work today (Goodman 1967). If we do not have an explanation for a phenomenon that was created in the past, then we have *not yet* observed the processes in operation today which create that kind of phenomenon (Scott 1963).

Salmon (1953) concludes that because we cannot demonstrate natural laws to be invariant in the past, our conclusions are *inductively* derived. Such conclusions cannot be shown to be true but rather "we can justify them by showing that they are useful in predicting and acting" (Salmon 1953:47). We can, Salmon (1953) suggests, justify inductive arguments without assuming any uniformity of nature. Thus, I believe, because our inductive conclusions are based on the assumption of uniformity of nature, our conclusions are acceptable in a scientific sense. I agree with Kitts (1977:13) that methodological uniformitarianism "may be viewed as a methodological device or convention that limits the generalizations used in primary historical geological explanations to statements that meet the empirical requirements set for any valid generalization in science, that is, that they may have been verified and have not been falsified here and now."

What is actualism?

> ... the animals died during a few weeks of drought, and their remains were accumulated around the few remaining pools of water- a theory that from an actuo-geological point of view is very well founded.
> (B. Kurtén 1953:69)

Actualism asserts spatial and temporal invariance of natural laws (Haneberg 1983), particularly those concerned with mechanical, chemical, and physical processes (but not behavioral ones). It thus is equivalent to methodological uniformitarianism. "It refers to uniformity [of immanent properties] in all four dimensions of space and time" (Simpson 1970:63). Actualism "denotes the methodology of inferring the nature of past events by analogy with processes observable in action at the present" (Rudwick 1976:110). Actuopaleontology is the expression of actualism pertaining to paleontological issues. Neotaphonomy (Hill 1978) and ethnoarchaeology (Stiles 1977) are similar expressions of actualism relevant to their respective fields of study. All of these terms may be subsumed under the concept of actualism.

Non-actualistic methods assume that changes may have been wrought by agents of *different kind* than those observable today, and include catastrophism in its classic sense (agents of different kind and intensity). Actualistic methods assume agents were of the same kind and of the same or different intensities (Hooykaas 1970). While including classic methodological uniformitarianism (agents of same kind and intensity), actualistic methods also allow variance in energy or intensity across time while denying temporally unique kinds of agents (Hooykaas 1970; Figure 3.1).

Are actualistic methods necessary to the study of historical phenomena? Many would answer "Yes" (Binford 1981b; Gifford 1981; Gould 1965; Simpson 1970; Watson 1969, 1976), while a minority would answer "No" (Scott 1963). The disagreement results from perceived shortcomings of the actualistic method, the degree to which actualistic methods are required and/or

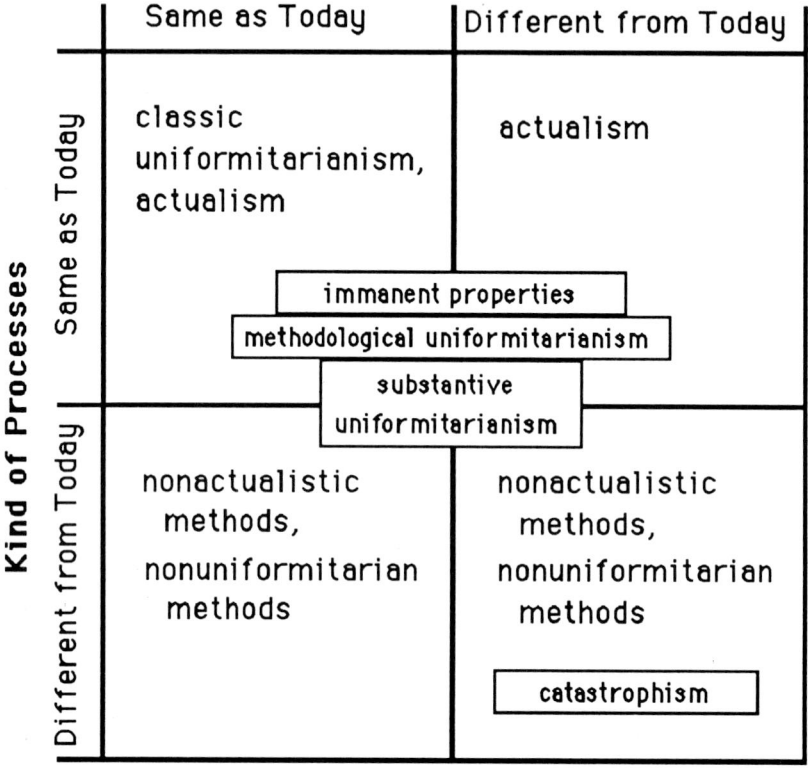

Figure 3.1. Intersection of different kinds and intensities of historic (taphonomic) processes defining uniformitarianism, actualism, and catastrophism as paradigms for explaining the past. Substantive uniformitarianism can encompass all four categories; methodological uniformitarianism and immanent properties assume processes of the same kind and of either the same or different intensity as observed today.

used to study past phenomena, and confusion of substantive and methodological uniformitarianism. Actualism is required and largely unquestioned when a particular fossil must be identified as a femur (Gifford 1981). The identification is unquestioned because of the necessary and sufficient causal relation between the genetic controls and ontogenic processes (immanent properties) resulting in the formation of the femur, and the femur itself. To argue that the natural history and behavior of living hyaenas are the same as those of extinct genera of hyaenas is, however, questionable because this transfer of modern ecological parameters to the past (Lawrence 1971) is based on sufficient causal relations or configurational properties and represents substantive uniformitarianism.

There are two major criticisms typically leveled against actualistic methods that are similar to those of methodological uniformitarianism. The first is that when the actualistic method is used, the analyst must assume that observable processes and their results are the same today as in the past (natural laws are temporally invariant), and this is an unprovable and therefore potentially false assumption (Binford 1981b:27; Simpson 1970:62–63). Advocates of the actualistic method indicate that the assumption must be warranted in order to use the method (Binford 1981b; Gould 1965). As we have seen, the assumption of temporal invariance of natural laws cannot be warranted because it cannot be tested; we cannot observe the past. Simpson (1970:62) argues that geological and paleontological observations on phenomena created over the entire history of the world are explainable using actualistic methods, and "that is the source and principle support of the canon of actualism" in those and other sciences. Of course, simply because it works does not make it so. We cannot present historical events as "instances of confirmation for [the assumption] we wish to test" if that assumption "has been presupposed in inferring the events" (Kitts 1977:79). Nonetheless, geologists, paleontologists, and archaeologists must, and do, grant the assumption, largely because it works, and there is no strong alternative as yet.

The second criticism suggests that exclusive use of the actualistic method as a basis of interpretation may conceal important insights to the past (Scott 1963; Gould 1980; Binford 1981b; Behrensmeyer 1988a). While this criticism might be confusing substantive with methodological uniformitarianism, and thus configurational with immanent properties, advocates of the actualistic method argue that when historical phenomena cannot be explained using actualistic methods, "a minimal inference is that knowledge of present processes is incomplete" (Simpson 1970:84). When actualistic and prehistoric data cannot be matched and the latter consequently not explained, "a maximal inference is that there have been past processes not now operative" (Simpson 1970:84). The unexplainable "residue" (Scott 1963:516) of phenomena seems to be small compared to the number and diversity of phenomena that are explainable using actualistic methods, but has served as the basis for a rejection of uniformitarianism (Gould 1980).

The second criticism can be phrased another way: When an unexplainable residue of phenomena remains the analyst must invoke ad hoc arguments to explain the residue and this "weakens the assumption" of temporal invariance of natural laws (Scott 1963:513). Advocates of the actualistic method argue that the occasional necessity of invoking ad hoc arguments is due largely to incomplete knowledge of present processes, not some internal weakness of the method. Even those recommending that actualism be abandoned realize that "there is no such thing as the final or ultimate interpretation – only better and better approximations of past reality" (Gould 1980:46).

It is one thing to criticize a method of analysis and entirely another to offer an equally (or more) productive alternative. While not a direct criticism of the actualistic method, a strong argument for abandonment of it would be the development of a method not subject to either or both of the two criticisms and yet capable of permitting explanations comparable to, better than, or more complete than those permitted by actualistic methods. Thus far, no alternatives have been outlined.

Actualism in archaeology and taphonomy

Solomon (1990:29) suggests that when taphonomic research is implemented the way Efremov intended, it involves "the seeking of *laws* that describe the processes by which the remains of animals become incorporated into the lithosphere, with equal emphasis given to the geological and biological formation of the site." She (Solomon 1990:29) argues that "what we do, in archaeology specifically, is actuopaleontology" as defined by Richter (1928). Solomon (1990:30–31) draws attention to perceived differences between Efremov's taphonomy, which she describes as law seeking and thus nomothetic, and actuopaleontology, which she describes as non-nomothetic and employing uniformitarian methods "to solve problems found at individual sites." Therefore, what taphonomists do is actuopaleontology and not taphonomy, and "to attempt to invoke a law which will explain what happened at a site is foolhardy. There is no such law" (Solomon 1990:31).

Solomon (1990:26) defines a *law* as a "correct statement of invariable sequence between specified conditions and specified phenomena; regularity of nature." Thus Solomon has identified a taphonomic (and semantic) red herring. To suggest that because taphonomy is a historical science, and that historical sciences do not or cannot use laws is false. It ignores the fact that biological evolution and the study of it via paleontology is an historical science involving the use of laws of natural selection or what Gould (1986:64) terms "nomothetic undertones." Further, one of the major proponents of studying the formational (including taphonomic) processes of the archaeological record has written that "formational processes exhibit regularities that can be expressed as laws" (Schiffer 1987:11). Cadée (1990:13) suggests that "the methods that Efremov proposed for taphonomical studies are not significantly different from those of Weigelt and Richter." Efremov (e.g., 1940:92) was not only aware of his predecessors (e.g., Efremov 1940:92), he noted that he was after "principles" rather than laws (Efremov 1958; after Cadée 1990), which I take to mean he was well aware of the difference between substantive and methodological uniformitarianism.

I suspect what Solomon (1990) has done is confused substantive and methodological uniformitarianism, and confused immanent and configurational properties. Immanent properties include those immutable physical and chemical reactions that occur with predictable results regardless of spatiotemporal context. Configurational properties, because they are context specific, are historical and mutable; because they concern human behaviors in archaeology, and behavior changes through time, we must test any substantive (configurational) laws of human behavior. What is important is the fact that while taphonomists speak of "taphonomic histories," taphonomic research is founded in methodological uniformitarianism and immanent properties, or actualism.

Use of the actualistic method is implicit throughout much or all of archaeological method and theory. One clear statement of its use is provided by Watson (1976:621):

> Archaeologists who propose laws of cultural evolution [and] of human behavior assume – as do other scientists – that the past was like the present, and although combinations and rates may have been different then from what they are now, the basic behavioral characteristics of men and material were not different then from what they are now. This may in fact be wrong. But if human nature and the environment were radically different in the past from what they are now, we assume that there has been lawlike change from the past to the present that can be derived and understood from its physical remains. This general uniformitarianism is the primary procedural or methodological assumption of archaeology, as it is of all other sciences.

Treatises on archaeological method and theory are quick to point out that if prehistoric dynamics are a goal of analysis, inferences regarding past dynamics must be founded on how the modern world works (Binford 1981b; Gould 1980). Taphonomists, because they of necessity deal with past dynamics, make similar arguments (Gifford 1977, 1981; Hill 1978, 1980). The arguments are straightforward: prehistoric dynamics are not observable, only their static results are, along with modern dynamics and their resultant static effects.

The problem thus becomes one of determining how the actualistic method is implemented in archaeology and taphonomy, and if it is implemented properly. The actualistic method of studying the past is superficially simplistic (e.g., Fisher 1981). Observe present processes in action, establish or postulate causal relations between particular processes or dynamics and particular static effects, and match the modern record to the past, inferring similar processes for similar effects on the basis of the causal relations. Is this how archaeologists and taphonomists perceive and use the actualistic method? I consider each in turn.

Archaeology and actualism

The most common form that the actualistic method has taken in archaeology is ethnographic analogy. This is reflected in statements like "in its most general sense interpreting by analogy is assaying any belief about nonobserved behavior by referral to observed behavior which is thought to be relevant" (Ascher 1961a:317). It is generally cautioned that analogs derived from ethnographic data be used merely as clues to general conditions, not particulars, because there is the potential of multiple modern analogs for particular archaeological phenomena (Ascher 1961a, 1961b). Pleas for ethnographic data with specific applicability to archaeological research have been met with a voluminous body of ethnoarchaeological data generated by archaeologists (e.g., Gould 1978; Kramer 1979; and Bonnichsen and Sorg 1989; Hudson 1993).

A classic example of ethnographic analogy is found in Binford (1967). Binford (1967) points out that analogic reasoning is inductive, and ethnographic analogy involves matching the observable archaeological record with the observed ethnographic record followed by the proposal that, when similar, both static records can be explained as resulting from the same processes. Such arguments can only be evaluated by monitoring concomitant variations of attributes in both static records, and if the variations are similar in both records, the inference is strengthened (Binford 1967). The implication here is that this concomitant variation indicates a causal relation between the monitored variables. Binford's (1967) inference regarding a particular archaeological phenomenon is questioned by Munson (1969), who provides an equally viable but alternative inference based on a different analog. This illustrates that the attributes both Binford and Munson have used are not, perhaps, causally related to dynamic processes which created the phenomena.

A different kind of analogic reasoning is provided by Schiffer (1976:162), who estimates the amount of lithic material in a site on the basis of what was recovered (multiplying the amount in the recovered sample by the reciprocal of the sampling fraction). Schiffer here assumes a uniformity between the density of lithic material in the sampled portion of the site, and the density of lithics in the unsampled portion, which in turn construes the density of lithic material as causally related to the amount of sampled space. Whether or not the sample population and the sampled space are each representative of their respective sample universe in a statistical sense is not known. And while it is true that as we dig more we tend to find more, it is not at all clear that lithics are randomly distributed throughout the site, therefore, it is not known if the variables involved are correlated. Of relevance to taphonomic studies is the similar practice of estimating the number of animals or bones in a kill site from the number recovered (e.g., Dibble and Lorrain 1968; Reher and Frison 1980). Seldom are these estimates justifiable because they entail configurational properties and substantive uniformitarianism.

Attempts to base analogs on causal relations are reflected in Ascher's (1961a) characterization of the "new analogy" as including boundary conditions (Stiles 1977; Watson 1979). Both the analog and the phenomenon under scrutiny must occur in similar environmental settings, or the technological levels of the two involved cultures must be similar. Analogies are even stronger, it is argued, if the modern analog is from a human group known to have descended from the archaeological group (e.g., Anderson 1969; Charlton 1981; Lange 1980); this implies that an analogous relation will be stronger and a causal relation more certain if the modern analog and the prehistoric subject are also homologous (i.e., share an ancestor). Boundary conditions are simultaneously construed as providing a means of controlling the choice of modern analogs in order to increase confidence in resultant inferences by improving the distinctiveness of the chosen attributes (Ascher 1961a), and as a device that results in limitations on possible inferences (Freeman 1968). Both of these factors are believed to impose a kind of proxy causal relation on the variables in the analog. If a prehistoric and an existing society occupy similar environments and utilize similar technologies, then behaviors must be similar due to the assumed direct relation of a particular technology for exploiting a particular environment. This reasoning rings with environmental determinism and is surely founded in substantive uniformitarianism and configurational properties.

To use actualism properly, Binford (1981b:26–27) argues that a "necessary causal relation" between a particular process and its result(s) must be established, a point Binford (1972) recognized in the debate with Munson (1969). Binford (1981b:26) states that "if we can isolate causal relations between things, and if we can understand such relations in terms of more general principles of necessity, such as the theories of mechanics, then we have a strong warrant for the inference of the cause from the observed effect." For Binford (1981b:26–29), a necessary causal relation is one that is "constant and unique," and establishment of such relations will allow us to specify archaeologically visible "signature patterns" that allow discrimination of one agent or process from all others. Establishing these kinds of relations faces two challenges: (1) is the relation in fact causal and not just correlational (a point I return to in the following section), and (2) was the process characteristic of the past? The latter clearly involves immanent properties, or what Binford (1981b:29) terms "eternal objects." The search for immanent properties or temporally invariant natural laws has been termed *middle-range research* by Binford (1977).

The concern for diagnostic criteria is an important one which has permeated many treatises on archaeological reasoning. Fritz (1972:137) implies that diagnostic criteria indicate the presence of particular past phenomena, but may be difficult to establish. He notes the actualistic method is requisite to explaining the archaeological record but there is an absence of explicit arguments attempting to demonstrate that a particular cluster of criteria are diagnostic of a particular past dynamic (Fritz 1972:138, 143). He labels these "arguments of relevance" the requirements of which involve undertaking actualistic research and establishing diagnostic criteria. Smith's (1977:607, 611) "plausibility considerations" are identical to arguments of relevance, and narrow the range of possible explanatory hypotheses. The explanatory hypotheses with the lowest prior probabilities or with the least distinctive criteria are not considered. If necessary, the final selection of one of several competing hypotheses is accomplished by examining concomitant variation of different variables in the archaeological and actualistic records. Of course, concomitant variation suggests that criteria were in fact distinctive but would not necessarily indicate that the relations between variables were causal or of Binford's "necessary" kind.

Gould (1980) recommends the disposal of uniformitarian arguments. He notes that while archaeologists generally assume that the processes which structure the ethnographic record have also structured the archaeological record, (1) this notion is self-limiting because we assume the very things we are trying to find out, (2) the notion is subject to the fallacy of affirming the consequent, (3) we have no good analogs for many adaptational forms of culture, and (4) uniformitarianism, which he does not define, is a "seductive notion" which presupposes that the modern world and the past are similar (Gould 1980:29–32). He argues that "while some processes may be subject to the principle of uniformitarianism, others may not," therefore analogic reasoning should only be used to look for contrasts or anomalies between the modern world and the past (Gould 1980:33–35). This "contrastive approach" is "analogy's last hurrah;" "analogues are better at informing archaeologists about what they do know or can expect to know" (Gould 1980:37). The characteristics Gould (1980) attributes to uniformitarianist arguments describe substantive uniformitarianism and configurational properties, and Gould's anomalies are explained using methodological uniformitarianism and immanent properties (see also Crawford 1982).

Gould (1980:37) outlines an alternative approach, beginning with the statement that "all scientific laws have irreducible properties of stating [causal] relations that are invariable in time and space. These laws are derived from observing regularities in time and space." Gould (1980:42) wants to develop general propositions "about human behavior that posit relations that are invariable in time and space and are susceptible to testing [because] the past can be perceived only in terms of our present-day ideas about it." Gould (1980:251) explicitly states two requisite steps and implies a third step to explanation of the archaeological record. First, observe and model adaptive behavior in contemporary societies. Second, establish convincing linkages between particular kinds of adaptive behavior and distinctive "archaeological signatures" in human residues that identify these kinds of behaviors; in other words, establish "necessary causal relations" in Binford's (1981b) terms, between processes and effects. And third, compare attributes the formation of which have been observed to modern static phenomena created in the past and, when similar, infer, on the basis of the causal relation, similar processes created both. While Gould (1980) has perhaps correctly characterized ethnographic analogy as an analytic procedure for the matching of forms and inferring similar causes, his suggested procedure for explaining the archaeological record is an accurate rendition of the actualistic method, methodological uniformitarianism, and the role of immanent properties (Wylie 1982a).

More recently, Gould (1990:48) softens his stance on actualism when he writes that (methodological) uniformitarianism operates at two levels in archaeological inference:

> the level of site-transformation [particularly natural] processes; and the level of real-world constraints on human behavior, as these are observed to operate in the present and are assumed to have operated in similar ways under similar circum-stances in the past ... At both of these levels, the use of general uniformitarian assumptions to build a bridge between the present and the past is comparable to that of the other historical sciences.

But, Gould (1990:50) is still hesitant to accept actualism because while "general uniformitarian assumptions play an essential role in the stepwise process of archaeological inference, they do not necessarily (or even usually) provide complete explanations." As we have seen, advocates of the actualistic approach would respond by suggesting that an incomplete explanation results from insufficient information on relevant modern processes (e.g., Charlton 1981; Lange 1980), or inadequately or inaccurately written natural laws.

Some archaeologists still call upon boundary conditions and historical relatedness to enhance the strength of analogical arguments (e.g., Charlton 1981). Others take a different approach. For instance, faced with the potential that the assumption that natural laws are temporally invariant might be false, Bailey (1983:3) suggests study of multiple sets of "independently verifiable data, [all of which are archaeologically visible] rather than to explain one set of [archaeological] data in terms of another set which is archaeologically invisible and can only be derived by extrapolation from a non-archaeological context." This sounds similar to Binford's (1967) suggested monitoring of concomitant variation of attributes in an attempt to ensure causal relations are being utilized in the analogy.

Murray and Walker (1988) advocate a refutation strategy when using analogies. They argue that others use a confirmationist, "verificationist," or "self-fulfilling prophecy" strategy (Murray and Walker 1988:266–267). A refutationist strategy attempts to show an analogically based conclusion is false "such that inability to refute them helps provide further justification for their acceptance." Such a strategy would, they argue, lead to significant break-throughs in research because it would open up "hitherto unimagined areas of potential knowledge" (Murray and Walker 1988:261, 283). This is an import-ant point warranting consideration in any analogically based argument, but it is also another characterization of good scientific research within the context of the actualistic method (Goodman 1967; Gould 1965; Shea 1982).

Many archaeologists now realize that analogical arguments do not specify all and only identities between phenomena. Following Wylie (1985), Murray and Walker (1988:262) correctly note that analogies are not "equivalences of identity; if they were so, the worked analogy would be superfluous." However, their misunderstanding of the distinction between substantive and methodolo-gical uniformitarianism (Figure 3.1) lead Murray and Walker (1988:279) to argue nonuniformitarian substantive theories may be necessary. In contrast, Hodder (1982:14) deals with his related substantive concern that "if we interpret the past by analogy to the present, we can never find out about forms of society and culture which do not exist today" by suggesting we must build stronger analogical/actualistic arguments.

Hodder (1982:14) states "because similarities in some aspects do not necessarily, certainly or logically imply similarities in others, we can never prove [analogically-based] interpretations." Hodder (1982:16) distinguishes *formal analogies* from *relational analogies*, noting that the former produce conclusions based on simple similarities between two objects, one of which is better or more fully known or understood than the other. Here, the conclusion takes the form that because the two objects share some properties visible or known for both, they also share other properties only known or visible for one. "Such analogies are weak," Hodder (1982:16) argues, because "the observed association of [shared] characteristics of the objects or situations may be fortuitous or accidental." For example, because one object is of bone and is also a fragment of a humerus does not necessarily mean the next bone object encountered will also be a humerus; to argue that second object is also a humerus would constitute a formal analogy. In contrast to formal analogies, then, Hodder (1982) suggests archaeologists should use relational analogies wherein associated attributes are interdependent or causally related. For example, to determine the function of a stone tool, archaeologists once simply examined the shape of artifacts. But because shape *may not* be directly related to tool function (a screwdriver shape denotes a screwdriver function until that screwdriver is used to pry open a can of paint, or a soup ladle would not work well for opening a can of paint), archaeologists have turned to formal attributes of artifacts interdependent with and causally related to tool function, such as use wear (e.g., Salmon 1981).

Hodder (1982:16, 18, 19) suggests strong formal analogies may be built by (a) noting that the more similarities two phenomena share, "the more likely are other similarities to be expected;" (b) using homologous phenomena (the direct historical approach) in building analogical arguments; (c) documenting multiple cases across many different instances where relevant attributes and processes are associated; and (d) the analyst could limit their conclusions to low levels, avoiding broad or general similarities. Relational analogies are stronger still because they explicitly involve "some necessary relation between the various aspects of the analogy;" that is, the associated attributes are thought to be "relevant" or causally related to the inferred properties (Hodder 1982:19–20). To improve such analogies, Hodder (1982:20–21) suggests we consider not only the "relevant causal links between the different parts of the analogy," but also the "contexts" of the parts of the analogy; that is, the cultural and archaeological context of the phenomena of interest must be considered as potentially causally related to the observed and inferred properties (Hodder 1982:24–25; see also Salmon 1981). Relevant linkages are theoretically informed judgements about which attributes should be causally related.

While other discussions of how to build analogies can be found in the archaeological literature, the ones I have reviewed are some of the major ones to have been published in the past two decades. What these discussions indicate is a set of differing ideas about how to use the actualistic method. Many of them also discuss to one degree or another the proper use of ethnoarchaeological data. Such data may serve as a detection device for recognizing significant patterns or relations of materials in the archaeological record not otherwise immediately apparent (Binford 1981b; Gould 1980, 1990; Kramer 1979). These data may also serve as a device by which significant, or causal, relations between materials in the archaeological record and the processes (human and natural) which resulted in that record can be inferred (Binford 1981b; Charlton 1981; Gould 1980, 1990; Kramer 1979; Stiles 1977; Tringham 1978). Finally, these data may serve as a source from which general explanatory models may be derived, the models subsequently being used to explain the archaeological record (Binford 1968; Charlton 1981; Gould 1980, 1990; Gould and Watson 1982; Lange 1980; Schiffer 1978; Stiles 1977; Tringham 1978). The first use can hardly be faulted. The second and third uses are simply statements of how the actualistic method is properly employed.

Archaeologists have long acknowledged the actualistic method as *the* means to explain humankind's past (e.g., Charlton 1981; Grayson 1986). Much of the polemic, while couched in the jargon peculiar to archaeology, contains all the basic elements of the method. The inescapable conclusion is, however, archaeological analogy often lacks the required causal relations between processes and effects (does not consist of relational analogies), and also lacks the diagnostic criteria some believe are requisite to use of the actualistic method. As Murray and Walker (1988) imply, diagnostic criteria in the sense of Binford (1981b) are not required; if such were available our arguments would not be analogical but rather identificational. Without the establishment of causal relations between processes and effects (relevant linkages), attributes of phenomena used as signature criteria are inductively derived empirical generalizations (Binford 1977, 1981b). Analogical explanation is only as sound as the empirical generalization, and data indicating that the empirical generalization is inaccurate are sure to be found (Willer and Willer 1973). If causal relations between processes and effects are known or suspected, we may consider inferences more highly probable (Binford 1981b:26–27), and exceptions should be much harder to find (Willer and Willer 1973). It is in the latter where Fisher's (1981) analysis of crocodilian taphonomy finds its strength, which brings us to a consideration of actualism in taphonomy.

Taphonomy and actualism

Hill (1978:88) suggests that actualistic taphonomy (what he labels *neotaphonomy*) "involves relevant experimentation or observations on the condition of modern vertebrate remains in various closely defined environments" and is "designed to test" hypothetical assertions about causes resulting in the effects observed in a fossil assemblage. Hill (1978:89) suggests that while actualistic observations must be made to account for differences between fossil assemblages and living communities, it is premature to attribute observed effects too rigidly to particular processes because the details are often poorly known; specifically, causal relations between processes and effects are unknown, and diagnostic criteria are unavailable. Differences between a fossil assemblage and the living community from which it derived can be subsumed under two basic categories: (1) differences in state (chemical, associational, locational, etc.) of fossil and living organics, and (2) differences in how ecological parameters are reflected by the fossil assemblage and the once living community. Knowledge of the former is requisite to discernment of the latter, and knowledge of the former is obtained by actualistic research leading to the establishment of causal relations and the definition of diagnostic criteria (Hill 1978:98).

Shipman (1981b:12) writes that if the "first law of taphonomy" is uniformitarianism of cause and effect, then the "second law of taphonomy" must be that only unique and "distinctive" aspects of an effect or result can be considered "diagnostic" of a particular cause or process. She correctly argues that "the occurrence of a past event can be deduced only by demonstrating that its effects differed from those of other similar events" (Shipman 1981b:12). Shipman has here set up the establishment of biconditional statements as the qualification for inferring past events. That is, her "diagnostic" signature criteria are of the "if X, and only if X, then Y" type, where X is the event of interest and Y is the distinctive signature of events of kind X. As Murray and Walker (1988) and Wylie (e.g., 1982a, 1982b, 1985) make clear, arguments of this sort are not analogies, they are identifications. But as we have seen, we seem to lack diagnostic signature criteria in Shipman's biconditional sense. I suspect that Shipman would agree that immanent properties are, ultimately, what we are after, and that our taphonomic conclusions are, in fact, analogy based.

Klein and Cruz-Uribe (1984) imply taphonomic analyses involve stripping away biases in the fossil record, and while valuable, can be contrasted with what they label the *comparative approach*. The latter involves controlled comparisons of different prehistoric assemblages, and when two assemblages differ in terms of one or more attributes of their respective faunal remains, a difference between the two assemblages in one or more non-faunal attributes such as associated artifacts or sedimentological contexts serves as a signal of the source of potential differences in the taphonomic histories. The comparative approach is a valuable one for pointing out ways to enhance our understanding, but in my view it rests soundly on the taphonomic approach with which it is contrasted by Klein and Cruz-Uribe (1984).

In one of the few treatments of the topic of which I am aware, Klein and Cruz-Uribe (1984:9) list several "limitations" to the actualistic method on which the taphonomic approach is founded (see also Cruz-Uribe 1991). The first is that observational conditions may affect results; for example, carnivores in zoos may process bones differently than carnivores in the wild. Second, some biological bone-modifying agents, including humans, no longer exist in contexts similar to prehistoric ones. Third, biological bone-modifying agents that are presently extinct cannot be observed. These three "limitations" concern substantive uniformitarianism and configurational properties, and not the methodological uniformitarianism and immanent properties of actualistic research. Thus, listing these limitations helps us focus on the latter rather than the former. The fact that many modern studies are temporally brief, that is, span only months or a few years, while most prehistoric faunal assemblages represent much longer time spans of accumulation is an important potential limitation, but seems to be a matter of the scale of observation like that alluded to earlier concerning the formation of the Grand Canyon. Finally, prehistoric assemblages have both post-depositional and post-burial histories whereas modern sets of animal remains typically do not have either, particularly the latter. Some might dispute the lack of a post-depositional history as many modern bone assemblages have been and are being studied after their deposition (e.g., Boaz 1982; Olsen and Shipman 1988). The most serious limitation involves studying post-burial processes; because they are buried the bones cannot be directly observed as they undergo diagenetic processes. Here, of course, is where experimental research could fill a major gap in actualistic research.

Gifford (1981:388–389) points out that actualistic observations will indicate the range of variation in fossil assemblages produced by similar agents under similar conditions, which in turn will suggest the most distinctive criteria to be used in the matching process. Note that this is not the same as the biconditional diagnostics of Shipman's "if, and only if" sort. Comparative study of many fossil assemblages will, Gifford (1981) suggests, aid in the search for distinctive criteria by indirectly controlling for stochastic variation; spurious correlations between causes and effects can be more readily detected and discarded. In order to establish that visible effects are diagnostic of particular causes, Gifford (1981:393–394) suggests the following procedure: (1) observe dynamic interactions between postmortem organic remains and processes that operate on them at the scale of the individual skeletal element; (2) establish the nature of cause–effect relations; (3) eliminate possible causes until each criterion has only one possible cause, or each criterion is diagnostic of a particular taphonomic process; (4) establish the expected range of variation in a diagnostic criterion; (5) test possible cause–effect relations and suspected diagnostic criteria with further observations by changing the scale of investigation to the level of fossil assemblage and "predicting the structure of assemblages produced by the action of specified processes." Gifford (1981:394) argues that "only if predictions pass this test of actualistic evaluation should they be employed in analysis of fossil materials." The testing procedure is implicitly perceived as part of establishing causal relations and echoes Binford's (1967) recommendations for testing ethnographic analogs by examining concomitant variation of attributes suspected to be causally related in actualistic settings. Both display apparent confusion between correlation and causation. It is perhaps because exhaustive testing is literally impossible that Gifford (1981) suggests elimination of other possible causes for particular effects as an alternative to testing. This is similar to Binford's (1981b) recommended procedure of argument by elimination. In both cases, the retreat to a position of eliminating some causes suggests "if, and only if" diagnostic criteria will be difficult to establish. But that is not a damning or fatal criticism of the actualistic method.

Gifford (1981:394) believes that "the gravest problem in actualistic research is assuming that a given process is a necessary and sufficient cause of an observable attribute when no such relation has actually been established." A sufficient cause is one that is capable of creating a particular result, but it is not the only one capable of producing that result. For example, a house may burn down for several reasons: a careless smoker who falls asleep in the house, an electrical short circuit, arson, or a lightning strike. All are sufficient to cause the house to burn down, but none of these is necessary to the house burning down as a spark from the fireplace in the house (another sufficient cause) might produce the same result. It is necessary, in our burning house example, that the smoker not awaken in time to put the blaze out before it is out of control, and that the ashes from whatever he is smoking contact some flammable material. Similarly, it is necessary for the arsonist to light flammable material, and for the blaze not to be discovered prior to its becoming uncontrollable for a burned-down house to result (see Salmon 1984 for further discussion of necessary and sufficient causes).

Klein and Cruz-Uribe (1984:9) suggest that "causation may be observed; it does not have to be inferred" when working in actualistic contexts. While superficially true, some philosophers would disagree (e.g., Salmon 1984). Requirements for establishing causal relations include (1) the cause and the effect must be contiguous in time and space, but sole use of this requirement can result in the "post hoc fallacy" (Salmon 1963:74) that temporal–spatial coincidence denotes a causal relation; and (2) there is some connection between a cause and an effect such that the two are regularly coincident, and this may involve the specification of necessary and sufficient conditions (Salmon 1984:211). Establishing necessary and sufficient causal relations is similarly a difficult matter at best. In short, actualistic research along the lines proposed by Gifford (1981) and exemplified by Fisher (1981) is used to *propose* such relations and linkages. "Causal statements are hypotheses" (Salmon 1963:88), and thus are capable of evaluation and refutation. Numerous taphonomic studies have, in fact, not established or even explicitly proposed causal linkages. For example, Mellet (1974) simply compared modern and fossil bones and inferred a similar taphonomic process on the basis of formal similarities of the bones. Dodson and Wexlar (1979), Shipman and Walker (1980), Hoffman (1988), and Kusmer (1990) go farther towards describing the effects of certain taphonomic processes including that discussed by Mellet (1974), but they offer only brief discussions of why the causes they monitor produce the effects they describe (see also Andrews 1990). Voorhies (1969) and Korth (1979) also summarize actualistic observations, but suggest causal reasons why some of the effects they observe result from various processes; their causal linkages include intrinsic properties of the bones such as structural density, size, and shape, and how those properties interrelate with the taphonomic processes they monitor. While I return to causal linkages below, it is important at this junction to note that as long as empirical generalizations (inductively documented instances of processes and effects without explicit proposed causal linkages) continue to serve as interpretive bases, debate over the precise taphonomic meaning of various attributes of bone assemblages will continue.

The fossil record is an effect or result of taphonomic causes or processes. To explain the effects and write taphonomic histories, taphonomists attempt to determine the causes. The method of determination involves the actualistic method. Antecedant statements (Binford 1981b; Gifford 1981; Hill 1978; Shipman 1981b) to this effect are therefore correct in intent. *Equifinality*, the property of allowing or having the same effect or result from different events, is poorly controlled in many cases, however, because of a lack of causal linkages between the attributes of interest. As recent reviews (e.g., Bonnichsen and Sorg 1989; Hudson 1993) show, while actualistic data are being collected at ever increasing rates, taphonomy still regularly lacks such linkages. The result is that many analyses employ a set of criteria which are suspected to be diagnostic of particular taphonomic processes but which are in fact empirical generalizations. Interpretations based on these empirical generalizations are published, only to have a "cautionary tale" published a few years later that questions the employed criteria's distinctiveness based on other actualistic data that also lack any explicit statement about causal relations. Actualistic research cannot simply record input (kinds and numbers of bones) to and output (modifications to kinds and numbers of bones input) of taphonomic processes. It must also be designed to investigate causation. It must seek answers to *why*, given particular inputs and processes, a particular output results.

Because establishing causal relations and linkages is difficult (Salmon 1984), they are often simply proposed as "working hypotheses" (Fisher 1981). Relational propositions of causes and effects abstracted from actualistic observations can then be phrased to make purely logical models (Binford 1981b), much as the example by Stiner (1990a, 1990b, 1991b) reviewed earlier illustrates.

> Abstraction is a matter of establishing an isomorphism between theoretical nonobservables and empirical observables. The more similar the model and the empirical case are in all respects covered by the theoretical statement, the better the theory will explain or predict. The point of indifference is reached when the error in prediction is within the limits of measurement error.
> (Willer and Willer 1973:26).

Explanatory models derived by abstraction are logically correct, but they are not absolute nor are they beyond alteration or displacement by other models with different structures that provide more sufficient or efficient explanations. Thus fewer cautionary tales should result. While discussed here as possible alternatives, argument by elimination and use of theoretical causal relations or logical model building operate most effectively when used together, as in the examples reviewed above.

Gifford (1981) implies suspected causal linkages can come from attendant bodies of theory for each stage in a taphonomic history (Figure 3.2). Gifford-Gonzalez (1989b:46) emphasizes the necessity of causal linkages in analogically based arguments when she notes "the goal of actualistic research should be to distinguish causal/functional relations" between taphonomic processes or causes, and their attendant effects (Figure 3.3). But she also argues that we must begin to expand our analogical horizons from one line of evidence and analogy to embrace "clusters" of relational analogies; this is simply a restatement of the notion that multiple lines of evidence are better than single lines (see also Gifford-Gonzalez 1991). This is especially so due to the generally hypothetical nature of causal relations between taphonomic processes and their effects. If all lines of evidence and analogically founded results point toward the same conclusion regarding the taphonomic history of a bone assemblage, we can have greater confidence in the truthfulness of that particular conclusion than in one founded on a single line of evidence or single analog.

Analogy

> Neotaphonomic analogs are based on direct observation of cause and effect, through actualistic or experimental data.
> (E. Johnson 1985:159)

The actualistic method involves argument by analogy, so it is appropriate to review the structure of strong analogical arguments. In archaeology such things as middle-range theory building (Binford 1977, 1981b) and the identification of site formation processes (e.g., Schiffer 1987) include major aspects of actualistic research and thus often take the form of analogical arguments. Such arguments have been described in detail by a philosopher with interests in archaeological science.

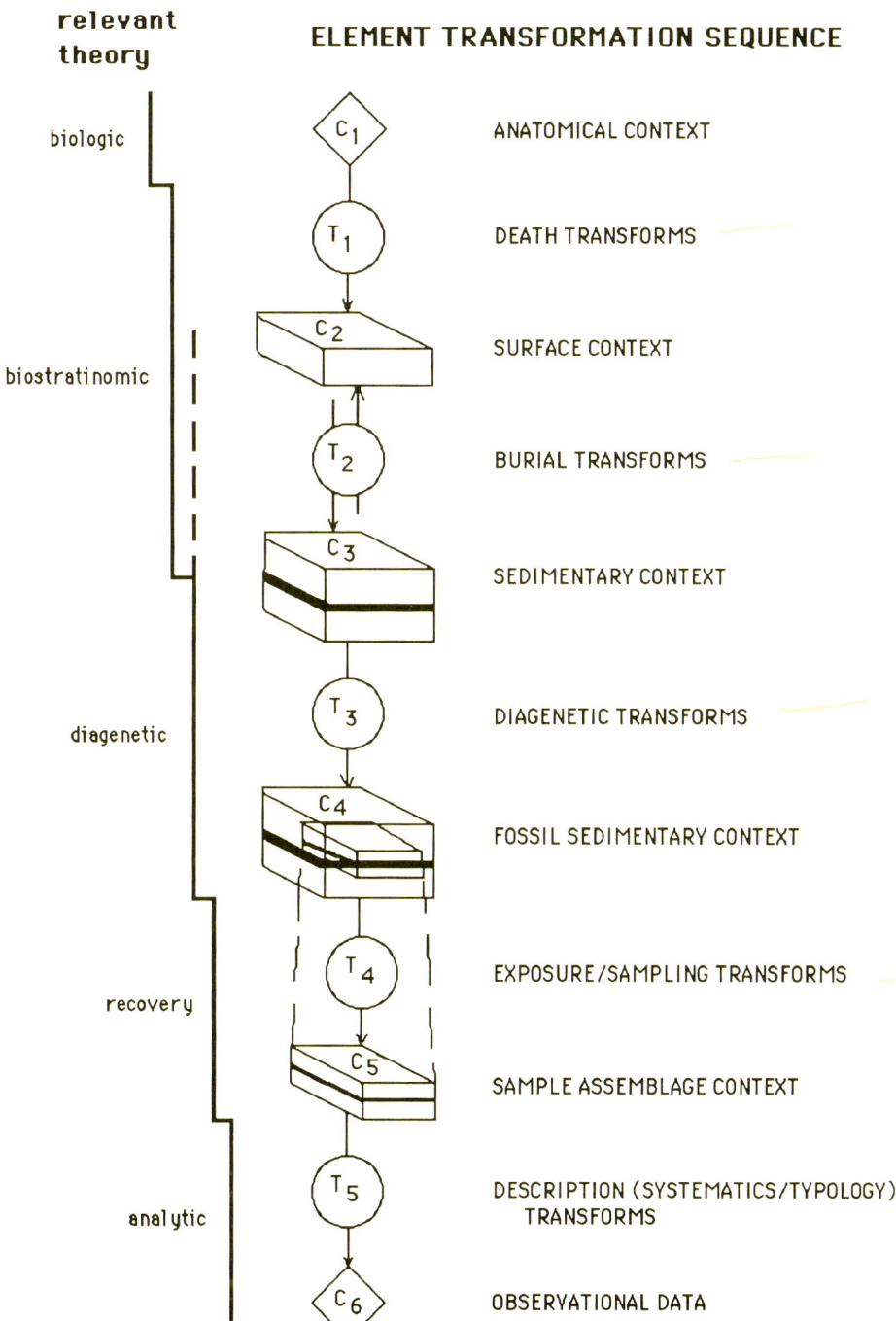

Figure 3.2. Schematic representation of the transformation of an animal from being a living organism to being a fossil showing where particular bodies of theory are relevant, and general categories of transforms and contexts (after Gifford 1981:387, Figure 8.1; courtesy of the author and Academic Press).

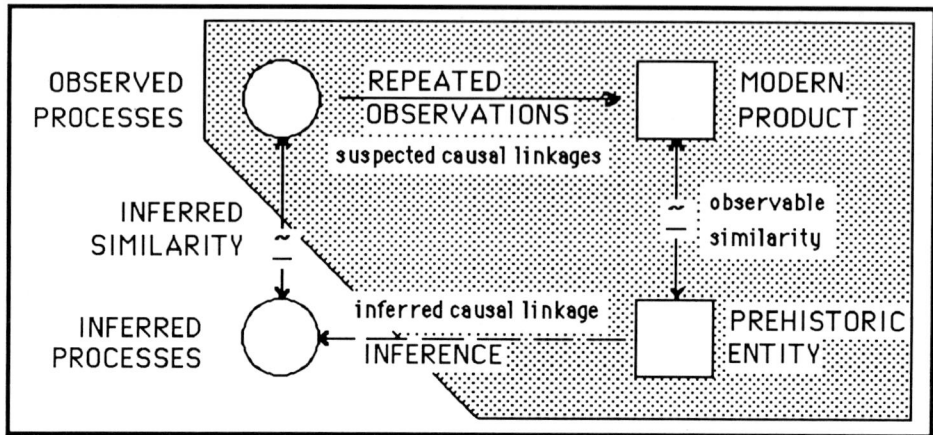

Figure 3.3. A model of (relational) analogical reasoning. Shaded area is where observations are made in the present or where actualistic research takes place (after Gifford-Gonzalez 1989b:44, Figure 1; courtesy of the author and The Center for the Study of the First Americans).

Wylie (1982a, 1982b, 1985, 1988, 1989a, 1989b) explores the structure of analogical arguments, and notes first such arguments are inductive, and they are ampliative; they result in conclusions that generally contain more information than the initial data and premises. This sometimes leads to scepticism that any reliable knowledge of the past will be derived, but that scepticism is "misplaced" (Gifford-Gonzalez 1989b) and maintains a misconception of science. Wylie (1982b:42) characterizes a realist view of science as one that acknowledges and emphasizes the ampliative aspect of scientific knowledge and seeks to improve it. Analogical arguments are not inherently faulty, but they can be categorized as involving either weakly or strongly argued formal or relational analogies (e.g., Hodder 1982, above). Weakly argued analogies are those which fail to "specify the (usually limited) points on which an analogy holds (i.e., to specify the positive, negative and neutral aspects of an analogical comparison of items or contexts) and an indiscriminate carrying over of all features of the [actualistic] source to the [prehistoric] subject" (Wylie 1982b:43). Formal analogies are those which only specify points of similarity (and less often, points of dissimilarity) between the modern source and the prehistoric subject phenomena, and conclude similar processes created both phenomena without consideration of the possible causal and/or structural linkages between the observed phenomena and the processes (after Hodder 1982; Wylie 1982a, 1988). "Analogical argument is formally valid; the difficulty is just that the relevant major premises are insecure," but it is not just formal similarities of the source and subject but rather that the observed (effects) properties are somehow nonaccidentally related to, or more than simply correlated with, the inferred (causal) properties that serves as the basis for analogical arguments (Wylie 1988:136).

As Kelley and Hanen (1988:170) note, "does saying that A caused B mean anything more than saying that whenever A occurs B occurs?" That is, "the concept of cause involves at least correlation" (Kelley and Hanen 1988:249), but does it involve more than this empirically? They suggest being able to *predict* successfully the result of a process indicates more than a simple correlation (Kelley and Hanen 1988:170), but conclude that:

> relevance is the chief desideratum, and in order to determine relevance, it is necessary to explore the negative and neutral aspects of analogies as well as the positive ones. Causal, functional, and structural analogies are more likely to be theoretically relevant than "mere" similarities.
> (Kelley and Hanen 1988:269)

The relational properties (causal and/or structural linkages) not only warrant the analogical conclusion, but guide inquiry directed toward improving them (Wylie 1988:144). Murray and Walker's (1988) refutationist strategy is simply part of building strong inductive or confirmationist analogical arguments. Thus Wylie (1988, 1989a) can report that archaeologists working in the Great Plains of North America were able to refute the analogically based notion that late prehistoric people there were nomadic horsemen and conclude that this adaptation appeared very late in time while previous cultures involved more sedentary horticulturalists. Detailed work on the modern or source side of the analogical equation "can serve not just to widen the range of analogs on which interpretation can draw, but also to sharply limit them, providing an important basis for critical assessment of their credibility" (Wylie 1989a:13). One manner of working on the subject or prehistoric side of the equation involves calling upon multiple, independent lines of evidence (Wylie 1989a:15). Simply put, working back and forth or horizontally between subject and source sides of the equation, and up and down or vertically within each side of the equation by studying multiple attributes, should ultimately produce strong, analogically based inferences. To use Wylie's (1989a) metaphor, such tacking back and forth, and building of suspension cables, respectively, will simultaneously reinforce and constrain our inferences. Building strong, relational analogies, then, involves clearly specifying the "number and extent of similarities between source and subject, the number and diversity of sources cited in the premises in which known and inferred similarities co-occur as postulated for the subject, and the expansiveness of the conclusions relative to the premises" (Wylie 1985:98).

Gifford-Gonzalez (1991:226) argues that as zooarchaeologists first and taphonomists second, we generally wish to study "the life relationships – ecological, social, and cultural – of prehistoric hominid species," and such study is analogical in nature. She suggests that during the 1980s taphonomists became very adept at identifying taphonomic agents; that is, we now know what kinds of visible traces a hungry carnivore will leave on a bone and how those differ from the visible traces produced on a bone that was butchered by a stone tool-wielding hominid. The relational analogies used to produce such identifications are strong and we tend to have much confidence in the identifications.

To build confidence in the inferences built upon our identifications of the creators of traces, Gifford-Gonzalez (1991:228–229) suggests use of a hierarchy or "nested system of analytical categories" described by the following set of terms:

> TRACE: visible attribute displayed by a bone that has undergone a taphonomic process (e.g, a scratch on a bone);
>
> CAUSAL AGENCY: the immediate physical causes producing a trace; the immediate interaction of materials producing a trace (e.g., the grating of a sand grain against a bone);
>
> EFFECTOR: the item or material that effects the modification of a bone (e.g., a sand grain grating against a bone);
>
> ACTOR: the source of force or energy that creates traces (e.g., an ungulate with sand grains adhering to its hoof trampling a bone);
>
> BEHAVIORAL CONTEXT: the prehistoric systemic environment in which the taphonomic process took place (e.g., a herd of ungulates milling about a waterhole);
>
> ECOLOGICAL CONTEXT: the type of ecosystem and environment in which the actors lived (e.g., an African savanna).

Identification of traces involves description of the formal attributes (size, shape, location, orientation, frequency, etc.) of those traces, and the use of relational analogy to identify the causal agency that created them. Using aggregates of different traces in conjunction with aggregates of identifications, we derive analogy-based identifications of effectors and actors. Because of the complexities of such multi-faceted reasoning and decreasing certainty of the cause–effect linkages between adjacent levels within the hierarchy, there is a "dilution of certainty" (Gifford-Gonzalez 1991:231) as we move from description of the traces through the identification of actors towards inferring the behavioral context. While some of this decrease in confidence can be overcome as we perform more actualistic research focused on the kinds of aggregates of traces created in different behavioral and ecological contexts, we may never have strong confidence in inferences of such contexts based on prehistoric materials given the complex nature of the reasoning involved. As one progresses from casual agency to context, the number of necessary relational analogies increases, and botanical, sedimentological, artifactual, site structural, and other kinds of data come increasingly into play. This is an important topic that will reappear in later chapters, and to which I return for detailed consideration in the final chapter.

Summary

> It seems that archaeology must rely on analogy in the formulation of hypotheses and theories to a greater degree than must other social sciences.
> (J. H. Kelley and M. P. Hanen 1988:378)

Actualistic research is presently perceived as the basis for most taphonomic and archaeological analysis and interpretation. Data derived from actualistic research are, however, commonly used as a source of empirical generalizations or formal analogies rather than to build relational analogies and postulate diagnostic criteria. While the derivation of diagnostic signature patterns or criteria of the biconditional "if, and only if" sort are sometimes suggested to be the desired end product of actualistic research, because of the difficulties in establishing such relations and criteria, the result is usually a series of empirical generalizations or formal analogies because the relevant linkages (necessary and sufficient causal/structural relations) are not always clearly specified, even in hypothetical form. Fortunately, this situation seems to be improving as taphonomists gain a clearer understanding of the actualistic method of learning about the past.

In following chapters I describe various kinds of data derived from actualistic research and analytic techniques for contending with those kinds of data in prehistoric contexts. The explicit reasoning behind the relevance of these kinds of data and analytic techniques to taphonomic analysis is epitomized in Figure 3.3. Typically, the desire of zooarchaeologists performing taphonomic analyses is to identify the agents, implements, and processes which resulted in the attributes displayed by a bone assemblage, where agents and processes, loosely defined, are the sources of the forces that result in modifications to carcasses and bones, and implements are the objects that serve as the mechanism of force transmission to carcasses and bones. The distinction of agents, processes, and implements is for heuristic purposes and may be unnecessary to some forms of analysis. But now it is time to turn to the materials of taphonomic analysis prior to consideration of its techniques.

4

STRUCTURE AND QUANTIFICATION
OF VERTEBRATE SKELETONS

The fossil record is composed almost entirely of the preserved hard parts of organisms.
(D. K. Meinke 1979:122)

Introduction

I presume the reader possesses some basic knowledge of archaeological, paleontological, biological, ecological, and anatomical principles and concepts, but it is important to review some basics of vertebrate skeletal anatomy. In this chapter I present a general discussion of what appear to be taphonomically significant properties of bones, teeth, and related materials. Both microscopic and macroscopic features are reviewed, as well as the principles of ontogeny and allometry. While only superficially covered here, these topics all warrant careful consideration in many taphonomic analyses, and the serious student will find the references cited a good place to start learning more about them.

I also consider some basic issues regarding the quantification of vertebrate remains. In this chapter I review the quantitative units commonly used in vertebrate taphonomy; additional details are provided in other chapters. As with the structure of vertebrate skeletons and skeletal tissues, an extensive literature concerning the quantification of vertebrate remains exists, and the interested reader is encouraged to inspect that literature.

Ontogeny and allometry

Ontogeny and allometry are two interrelated phenomena that often play an influential role in controlling the kind of skeletal tissue upon which taphonomic processes might operate and upon the possible effects of those processes. *Ontogeny* involves the growth and development of an organism from its conception to its death (Figure 4.1). Embryonic development, fetal development, and postnatal development and growth are all included in vertebrate ontogeny (Shea 1988b:401). *Allometry* is the study of relations between the sizes of various body parts, and it may involve the scaling of individual body components relative to body size (Alexander 1985; Shea 1988a:20). For example, large mammals have not only absolutely larger skeletons than small mammals, but due to the larger body mass (weight) of the former they also have

70

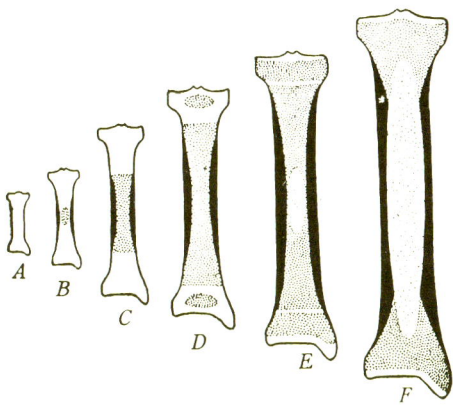

Figure 4.1. Schematic illustration of ossification and growth of endochondral long bone (tibia) of a mammal. White, cartilage; heavy stipling, endochondral bone; black, perichondral bone; light stipling, medulary cavity. A, cartilaginous stage. B, C, initial ossification of diaphysis. D, growth of diaphysis and ossification of epiphyses. E, near fusion of epiphyses and diaphysis, and initial formation of medullary cavity. F, bone growth (ontogeny) complete.
Reproduced by permission from: Romer, A. S. and Parsons, T. S. *The vertebrate body*, Figure 108. Philadelphia: W. B. Saunders Company. Copyright 1977 W. B. Saunders Company.

relatively larger skeletons. That is, to build a mouse the size of an elephant would be possible but would involve making the mouse 100 times longer, 100 times wider, 100 times higher and thus increasing its volume by 1 million, but the cross-sectional areas of the bones would be increased by only 10,000 (Alexander 1985:26; see also LaBarbera 1989; Shea 1992; Smith 1984).

There are at least three kinds of allometric relations. "An organ or structure can grow more quickly than the body as a whole (positive allometry), more slowly (negative allometry), or with the same speed (isometry)" (Rensch 1960:133). One obvious negative allometric relation is the decreasing size of the human head relative to the body during growth (Shea 1988a:21). The relative growth rates, or allometric relations, of two organs or structures may vary (e.g., from positive to negative) during the ontogeny of an organism. Growth rates can be measured for an organ or structure, a part of an organ or structure, or the complete organism (Rensch 1960).

Allometric relations can be determined for a trait across multiple taxa, as with the large mouse endowed with an elephant skeleton, within a taxon as a function of the ontogenic age of individuals, or as a function of different-sized adults. Of interest to a taphonomist is the process of bone formation in a growing vertebrate, as when the cartilaginous model of a limb bone is replaced by bone tissue (Figure 4.1). The relevance of such allometric relations is well illustrated by the numerous zooarchaeologists who have suggested the reason that the remains of ontogenically young mammals are rare in many archaeological collections is that their bones are generally of low structural density. These bones are not fully ossified and thus are readily consumed by scavenging carnivores or removed by other taphonomic agents such as fluvial action due to their relatively low bulk density and high porosity.

Skeletal tissues

> Because skeletons resist decay after death, they have become the awe-inspiring epitome of death itself in every human culture.
> (H. Francillon-Vieillot *et al.* 1990:473)

Vertebrate bodies consist of various kinds of soft tissues, and various kinds of hard tissues. The former consist of muscles, ligaments, tendons, and hide; these may be found in archaeological contexts where preservational conditions are exceptional, but typically they do not preserve. Hard tissues are those that make up the skeleton, and include bone, tooth, cartilage, and horn and antler. It is the hard tissues that typically preserve in archaeological contexts and thus it is the hard tissues, the skeletons, with which nearly all taphonomic analyses are concerned. The reader should understand that in the following, unless otherwise noted, the words bone, cartilage, tooth, antler, and horn refer to the tissue, and not some specific skeletal element.

Bone

Bone is a tissue that has evolved as a structural material (Meinke 1979). It thus displays rather remarkable mechanical properties in an engineering sense (Currey 1984; MacGregor 1985). As well, because it is a living tissue, it can not only respond mechanically to stresses, it can also respond by altering its properties. Anyone who has suffered an accident resulting in one or more broken bones readily appreciates these facts. Bone is often referred to as a "two-phase" or "compound" material (Meinke 1979). This is so because bone consists of about 70% inorganic material, usually *hydroxyapatite*, a calcium phosphate mineral with the general chemical composition $Ca_{10}(PO_4)_6.2OH$, and 30% organic matter, mostly the structural protein *collagen*, which can also vary in chemical composition (Francillon-Vieillot *et al.* 1990:515). The former material is resistant to compression forces and the latter to tension forces (Currey 1984; Johnson 1985). The hydroxyapatite crystals are supported in an extensive system of collagen fibers, being found both within and around those fibers; "one of the long axes of the mineral plates is always fairly well aligned with the collagen fibrils" (Currey 1984:26). Collagen fibers are infinitely long compared to the discrete apatite crystals. The minerals confer rigidity and hardness, and the organic matter confers toughness, resiliency, and elasticity to bones (Hildebrand 1974:96; Romer and Parsons 1977:150).

The term "apatite" refers to a diverse group of calcium phosphate minerals, with other minor included elements varying from tissue to tissue (the word "apatite" derives from the Greek word *apati* which means "to deceive"); thus there is a diversity of apatite minerals (Carlson 1990:531). "Studies of inorganic apatite have demonstrated that a high fluorine content increases the stability (decreases solubility) of the mineral ... Increased carbonate content raises apatite solubility" (Carlson 1990:531). The chemical composition of skeletal tissue, thus, can influence diagenesis (see Chapter 11).

Bone tissue is alive while the animal is alive. It serves as a metabolic reservoir for various minerals, especially calcium and phosphates (de Rousseau 1988:95). Bone tissue is "in a state of constant metabolic exchange with the rest of the body" (MacGregor 1985:7) and thus if the body is not taking in appropriate nutrients in the proper amounts, bones will react accordingly. This creates some skeletally visible signatures of dietary deficiencies and disease that are of great use to zooarchaeologists (e.g., Baker and Brothwell 1980) and perhaps to taphonomists as well.

There are two basic modes of bone tissue formation. *Dermal bone* forms directly in the mesenchyme or near body surface (skin) cells. Bones of the human cranial vault are dermal bones. *Endochondral bone* is "preformed" by a cartilaginous model which is replaced by bone tissue (Figure 4.1). Long bones in humans are endochondral. During the ontogeny of endochondral bone, a long bone consists of three distinct parts. The *diaphysis* is the shaft portion, and the *epiphyses* (pl.; singular form is epiphysis) are the two ends. Each is a center of ossification during ontogenic development (Figure 4.1). Long-bone growth occurs at the ends of the diaphysis in the zone known as the *metaphysis* where the cartilage is replaced by bone. Before the epiphyses and diaphysis begin to grow together they are separated by a disc of cartilage known as the *epiphyseal plate* and, upon its disappearance, the epiphyses and diaphysis grow together and fuse into one discrete object, the adult bone. Long bones grow in circumference or girth as they grow in length, with successive layers of compact bone (sometimes referred to as *periosteal bone*) deposited around the outside of the diaphysis.

Cells that deposit bone material are called *osteoblasts*, and as the osteoblasts become embedded in the bone matrix they become osteocytes or, literally, bone cells. Bone growth occurs until adult size is reached, but bone remodeling occurs throughout the life of an organism. Remodelling of bone is accomplished by *osteoclasts* which resorb bone, creating in part incomplete Haversian systems (see below and "interstitial systems" in Figure 4.3). *Haversian systems* are series of small canals containing blood vessels and nerves which branch through the bone; bones receive nutriment from these. The *periosteum* is the membranous sheath found on the external surface of a bone that can be stimulated to produce new bone, as when an organism breaks a bone. The articular surfaces of bones are not covered with periosteum, but a thin layer of cartilage. Interior (marrow or medullary) cavities of some endochondral bones, especially the long bones of the limb, are lined with a membrane known as the *endosteum* (bone here is sometimes referred to as *endosteal bone*).

The major structural element of bone tissue is the *osteon*. An osteon is simply a roughly cylindrical structure of successive lamellae surrounding a centrally located Haversian canal. The alignment or orientation of collagen fibers and apatite crystals varies from one lamella to another within the osteon, whereas the long axis of the osteon tends to be parallel to the long axis of the bone (Figure 4.2f). Individual Haversian canals are linked to one another between osteons by radiating blood vessels that occupy *Volkman's canals*. Individual osteons are joined to one another by a substance called *cement*.

There are several forms of bone that can be distinguished, basically at different scales. At a fine scale, for mammals, woven bone, lamellar bone, and parallel-fibered bone can be distinguished (Currey 1984:26–27). *Woven bone* forms quickly and has randomly oriented fine collagen fibers. Mineral crystals in woven bone are also randomly oriented (MacGregor 1985:4). Woven bone has irregular trabeculae and is transient, being the initial kind of bone deposited in the fetus and around bone fractures. Spaces around blood vessels in woven bone are more extensive than those found in lamellar bone. *Lamellar* or laminated bone forms more slowly and has an organized structure with the collagen and bone fibers arranged in layers called *lamellae* (Currey 1984; Meinke 1979). Within each lamella the collagen fibers form groups; the individual groups of fibers may display distinct orientations, and these can vary between adjacent lamellae (Figure 4.2). "*Parallel-fibered bone* is structurally intermediate between woven bone and lamellar bone" (Currey 1984:27), and seems to be more rare than lamellar and woven bone. Small cavities or *lacunae* permeate all three kinds of bone, and blood vessels in all types are found in *canaliculi*.

At a more general structural scale four kinds of bone can be distinguished: woven bone, lamellar bone, Haversian system lamellar bone, and fibrolamellar bone (Currey 1984:28–29). Woven bone at this scale occurs in areas several millimeters in all directions in young bone and fracture calluses. *Lamellar bone* at this scale also extends over relatively large areas, such as around the outside surface of mammalian long bones ("circumferential lamellae" [Currey 1984:28]) (Figure 4.3). *Haversian system lamellar bone* is formed when the bone material around a blood vessel is eroded by osteoclasts, and new bone is deposited in concentric layers on the (inner) surface of the resulting cavity (Figure 4.3). The blood vessel(s) remains at the center of the concentric lamellae. These Haversian systems can vary considerably in how they are laid out. Their outer limit consists of a "cement sheath" and very few canaliculi cross it so that cells outside the sheath are "cut off metabolically from the blood vessel in the middle of the Haversian system" (Currey 1984:29), although individual Haversian systems are connected by Volkmann's canals (MacGregor 1985:5). *Fibrolamellar* (or *laminar*) bone is "found particularly in large

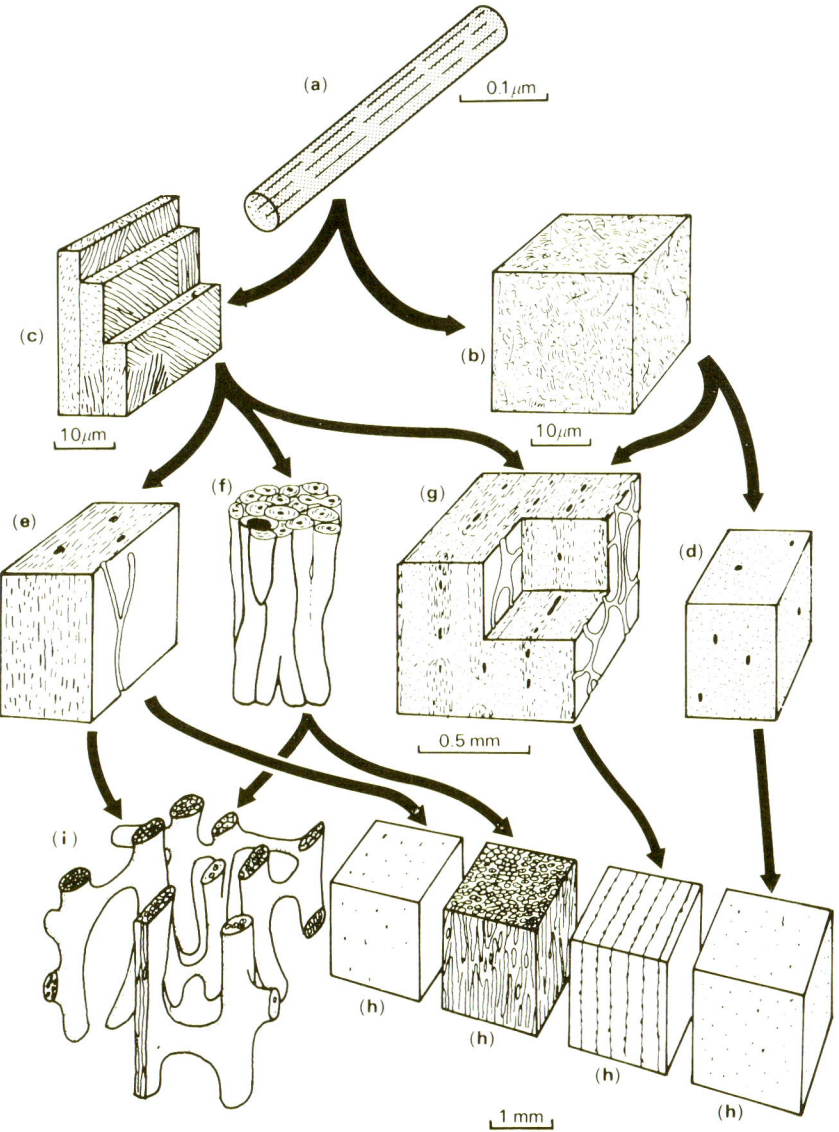

Figure 4.2. Structure of mammalian bone at different scales and levels of organization. a, collagen fibril with associated mineral crystals; b, woven bone, collagen fibrils randomly arranged; c, lamellar bone, showing separate lamellae with collagen fibrils oriented in particular domains; d, woven bone, with blood channels as black spots; e, primary lamellar bone with lamellae shown as light dashes; f, Haversian bone, each Haversian system with concentric lamellae around a central blood channel; g, laminar bone with alternating layers of woven and lamellar bone; h, compact bone types from lower levels; i, cancellous (trabecular) bone. Reproduced by permission from: Wainwright, S. A. *et al.* 1976, Figure 5.14. London: Edward Arnold Ltd. Copyright 1976 by S. A. Wainwright, W. D. Biggs, J. D. Currey and J. M. Gosline.

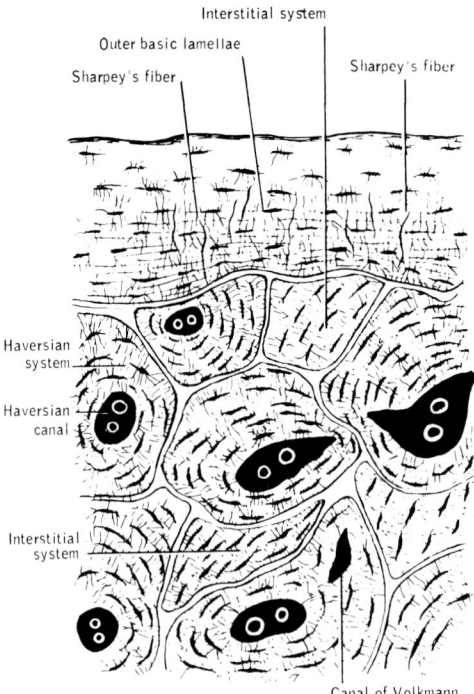

Figure 4.3. Microstructure of mammalian bone showing Haversian and lamellar bone. Reproduced by permission from: Romer, A. S. and Parsons, T. S. *The vertebrate body*, Figure 105. Philadelphia: W. B. Saunders Company. Copyright 1977 by W. B. Saunders Company.

mammals" the bone of which must grow quickly in diameter (Currey 1984:29). This is a combination of woven or parallel-fibered and lamellar bone, with woven bone first quickly forming a kind of scaffolding, and lamellar bone forming later on the outside of the structure and eventually filling in the cavities of the originally formed woven bone. This creates alternating layers of woven or parallel-fibered and lamellar bone (Currey 1984:29; MacGregor 1985:5).

At the final and highest scale, *compact* (dense) *bone* is relatively solid whereas *cancellous* (*trabeculated*, spongy) *bone* has large spaces. Porosity of the former is limited largely to the Haversian canals and lacunae; cancellous bone is quite porous given its structure of plates and struts. The amount and distribution of compact and spongy bone tissues is dependent on each individual bone's particular structure and function (Romer and Parsons 1977:151) (Figure 4.4). Long bones typically consist of mostly cancellous bone at their epiphyseal ends and compact (lamellar) bone makes up the diaphysis. Cancellous bone can take several forms, but basically consists of bone struts and plates that are variously interconnected and of various orientations, largely depending on the function of the particular skeletal element. These struts and plates are called *trabeculae*

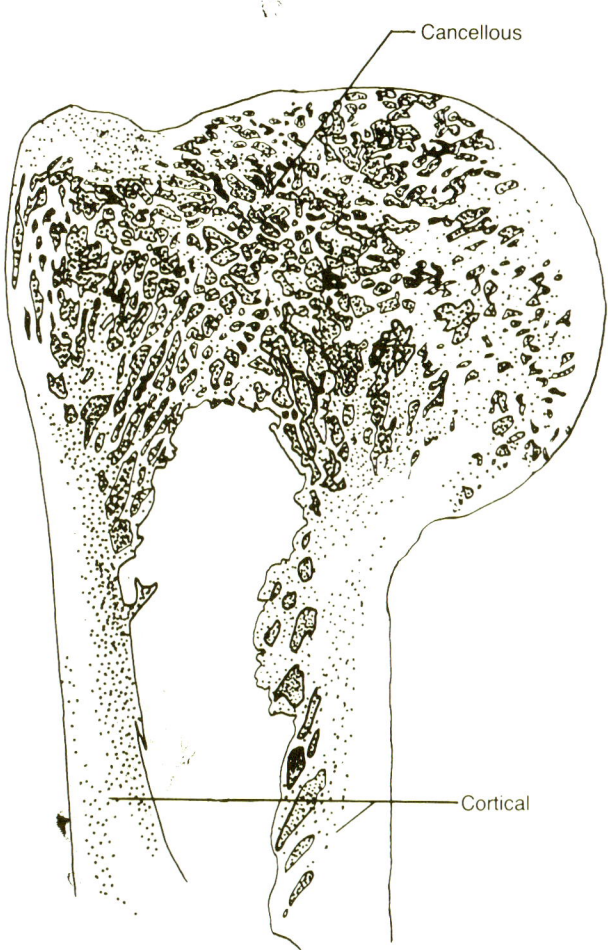

Figure 4.4. The appearance and distribution of trabecular or cancellous and compact or cortical bone in a typical mammalian long bone, a proximal humerus. Reproduced by permission from: Hesse, B. and Wapnish, P. *Animal bone archeology*, Figure 25. Washington, D.C.: Taraxacum Press. Copyright 1985 by B. Hesse and P. Wapnish.

(trabecula means strut). In mammals, the spaces or pores of cancellous bone are usually filled with marrow, and most cancellous bone occurs in the ends of long bones (Currey 1984). Laminated or lamellar and Haversian bone make up compact bone.

The bone tissue of birds is similar to that of mammals, but the thickness of the walls of bird bones tends to be less, relative to a bone's diameter, than in mammals. Bird bones are not marrow filled, but they fill with calcium as a reservoir for egg-shell production (MacGregor 1985:8). Reptilian bone is cellular and "in all but one major reptilian group the bone tissues are vascular

[but] in virtually all adult reptiles, some localized areas can be non-vascular in structure" (Enlow 1969:45, 47). The bone of amphibians is avascular. Haversian systems are lacking in the bone of many reptiles including lizards and snakes (Enlow 1969:47). Compact bone of lizards and snakes is "virtually non-vascular" and lizard long bones have a "relatively limited extent of cancellous trabeculae in the mid-diaphysis" (Enlow 1969:62–63). Bone of teleost fishes has no bone cells and lacks osteocytes (Currey 1984; Enlow 1969).

Turtle shells are perhaps the most intriguing vertebrate skeletal parts. These shells consist of the dish-shaped dorsal *carapace* and the ventral flat *plastron*; the two are connected along the sides by an area called the *bridge*. The shell consists of an outer epidermal, horny surface cover and an inner dermal, bony armor (Zangerl 1969:312). In adult turtles, the outer and inner layers are separated by a "spongy middle region containing large numbers of spherical cavities of different sizes. On both sides of this there are zones of compact lamellar bone, containing moderately numerous radial vascular canals" (Zangerl 1969:313). The inner layer of lamellar bone is more vascular than the outer layer. The shell does not serve as a mineral reservoir whereas the limb bones do (Zangerl 1969:313).

Currey (1984:36) considers individual bones to consist of three basic shapes: tubular, tabular, and short bones. The former are elongated in one direction and in cross-section are approximately circular. They are expanded at the ends, and include the long bones of the limbs and the ribs. Tabular bones are those that are partially flattened, such as the pelvis, scapula, and some bones of the brain case. Short bones are roughly the same dimension in all directions, and include carpals, some tarsals, and some phalanges. Davis (1987:47) distinguishes cylindrically shaped bones (e.g., long bones of the limbs), flat bones (e.g., skull, scapula, innominate, rib), and irregularly shaped bones (e.g., vertebrae). Micozzi (1991:54) distinguishes four "morphological types of [human] bones." Long bones include the mandible, clavicle, humerus, radius, ulna, femur, tibia, and fibula. Short bones include carpals, metacarpals, tarsals, metatarsals, and phalanges; phalanges are the "true short bones" (Micozzi 1991:54). Flat bones include the frontal, parietal, occipital, temporal, sternum, scapula, sacrum, ilium, ischium, pubis, and ribs. Irregular bones include the vertebrae, patella, hyoid, sphenoid, maxilla, nasal, ethmoid, lacrimal, palate, and vomer. All three arrangements are simple morphological typologies meant to underscore some of the similarities and differences in bone shapes. A more geometric and less anatomical approach to categorizing bones according to their shape is described in Chapter 6 (Figure 6.6). As indicated there, that system is quite relevant to some taphonomic problems.

Cartilage

Cartilage is sometimes found in archaeological contexts. There are several basic types of cartilage, but all are somewhat elastic and are harder than most soft tissues such as muscle but softer than bone (Hildebrand 1974:126; Romer and Parsons 1977:150). Cartilage may take the form of a firm gel through which is spread a network of connective tissue fibers. Cartilage cells are called *chondrocytes*, and these are isolated within the matrix (gel) they have secreted. Cartilage typically does not contain blood vessels. Cartilage may become calcified via the deposition of calcium salts in the matrix, which creates a hard, brittle material similar to bone. Such calcification precedes the replacement of cartilage by bone during the ossification of bones that are growing (Francillon-Vieillot *et al.* 1990:520-523; Romer and Parsons 1977:149–150).

Cartilage mainly occupies internal areas of the body (Romer and Parsons 1977:150). That is, with few exceptions (such as the mammalian ear and nose) cartilage is never in the skin and seldom near the surface of the body. Cartilage is more abundant in young animals, and less abundant in adults due to the replacement of much cartilage with bone during ontogeny. Because cartilage generally has a lower structural density than bone, it tends to mediate the effects of taphonomic processes less well than bone.

Tooth

The major function of teeth is mastication (Kay 1988). Mammalian teeth are largely composed of enamel and dentine, with a minor amount of cementum (Figure 4.5). *Enamel* is the hardest skeletal tissue, and is composed of elongate crystals of hydroxyapatite; the apatite crystals in enamel are larger than they are in dentine or in bone (Carlson 1990:535). Only about 1–3% of enamel tissue is organic (a non-collagenous form); the remainder is mineral which is organized into prisms (Hildebrand 1974:95; Romer and Parsons 1977:301). Enamel is harder, denser, and less soluble than dentine because of its low porosity and low organic content. It is a stiff, nonelastic material that is quite resistant to diagenetic chemical change. Enamel is, however, "more brittle and has less compressive strength than dentine" (Carlson 1990:537) because it is so highly mineralized and so non-porous. *Dentine* is harder than compact bone but softer than enamel. It is also composed of hydroxyapatite but the tissue itself is about 30% organic and 70% mineral by weight (Carlson 1990:533). Dentine is "permeated by sinuous sub-parallel canals called dentinal tubules," and it is thus a "fairly porous material" (Carlson 1990:533, 534). But like bone, "the structural intermixture of collagen fibers and apatite crystals in dentine creates a composite material with a high degree of elasticity and strength" (Carlson 1990:534). *Cementum* is a type of bone but is in part acellular (Meinke 1979). It is a mineralized connective tissue that "is similar to bone both ultrastructurally and biomechanically" (Carlson 1990:534). It is 70% apatite, 25% organic, and 5% water by weight (Carlson 1990:534).

Mammal teeth have a crown, neck, and root (Figure 4.5). Ontogenically, the enamel crown forms first, then the body of the tooth (dentine) forms, and finally as the tooth erupts the root forms. The tooth is set in an *alveolus* or

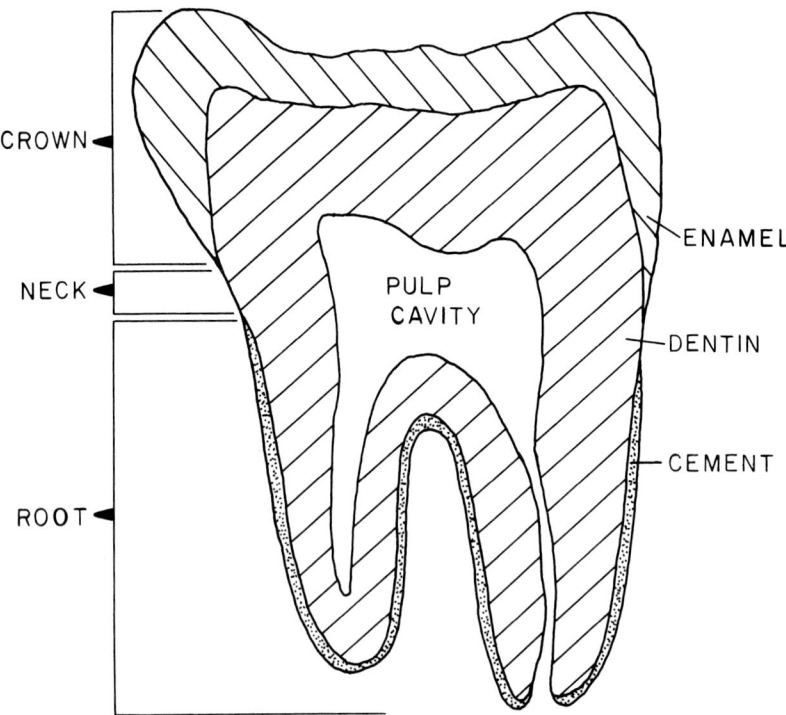

Figure 4.5. Cross section of a typical mammal tooth showing major components and regions.

socket in the jaw. Mammalian dentitions are *heterodont*; that is, different kinds of teeth are found in each jaw. These typically are called, from front to back, incisor, canine, premolar, and molar teeth. The number and morphometry of teeth is taxonomically distinctive in most animals. The tusks of elephants, walrus (*Odobenus rosmarus*), and narwhal (*Monodon monoceros*) are specialized teeth (see Carlson 1990:546–553 for a summary of the microstructural intertaxonomic variation in teeth). Elephant tusks are, in fact, a specialized version of dentine (Francillon-Vieillot *et al.* 1990:474).

Given that the chemical, macro-structural, and shape properties of teeth are different from bones, teeth tend to respond differently to taphonomic processes than bones. Weathering (Chapter 11) of teeth, for instance, is quite different from weathering of bones. While bone tends to split apart and exfoliate in what might be thought of as a time-transgressive process leading eventually to bone dust, teeth simply fall to pieces.

Antler and horn

Horns and antlers are composed of quite different materials. Horn is largely restricted to bovines, and is mainly keratin, a protein that also makes up hair, hoof-sheaths, and feathers. It is, in effect, functionally part of the exoskeleton (MacGregor 1985:19). As a tissue horn tends to grow slowly in an epidermal layer around a vascular bony core referred to as the "horn core" and arising from the frontal bone of the skull. The *horn core*, or "*os cornu*, originates in the subcutaneous connective tissue and fuses secondarily with the underlying skull . . . cartilage is not known to be present in the development stages of horn bone [which] develops as an independent center of ossification" (Goss 1983:67). "Horny lamellae are added [at the base] internally, in the nature of a cone within a cone. This results in the distal displacement of material produced earlier" (Goss 1983:57), and sometimes results in the formation of growth increments. Generally if removed, horn is not regenerated, although the pronghorn (*Antilocapra americana*) of North America annually casts the old horn and grows a new horn (O'Gara and Matson 1975; Romer and Parsons 1977:133–134; Solounias 1988).

Antler resembles bone, and grows at the tip within a blood-rich skin called *velvet* from a protruding portion of the frontal bone called a *pedicle* (Goss 1983:57). Antler grows seasonally, ossifies, and is shed annually (Modell 1969). Individual antlers of modern cervids are long cylinders that variously bend and branch, and which are filled with cancellous bony tissue (Goss 1983). It is generally thought that antlers grow by ossification at the growing tips with additional antler material being added to the surface. Blood vessels are internal and external (in the velvet) to the growing antler. These blood vessels shrivel and die when the antler completes growth, and the velvet is shed. Antler tissue "consists primarily of coarsely-bundled woven bone" (MacGregor 1985:12). Individual antlers have a cancellous core encased by compact tissue; the former occupies progressively less cross-sectional area as one progresses out individual antler tines such that the tips of the tines are composed only of the compact tissue. The ratio of mineral to organic matter of antler is similar to that of bone, and antler production may result in reduction of bone mineral content (MacGregor 1985:13). Antler growth and development depends not only on the age of the animal but on the nutritional status and health of the animal as well (Brown 1983).

Other tissues

Thus far I have considered the skeletal tissues with which most vertebrate taphonomists deal. There are several other tissues, however, that warrant mention. Amphibians, like fish, have bony scales, but the scales of reptiles are keratinized structures (Francillon-Vieillot *et al.* 1990:483). Fish scales are complex, polymorphic structures that have a varied but basically dermal origin. "The scales of fishes form a more or less continuous dermal skeleton on the body, which modulates into the specialized dermal elements of the fins (fin rays) on one hand, and the mouth and pharynx (teeth) on the other"

(Francillon-Vieillot *et al.* 1990:477). Fin rays and scales are homologous with mineralized skeletal elements (Francillon-Vieillot *et al.* 1990:474). Related structures found in some fishes are the bony plates known as scutes and spines, both of which are modified scales (Francillon-Vieillot *et al.* 1990:486). Otoliths are "acellular mineralized concretions which occur in the inner ear" of many fishes (Francillon-Vieillot *et al.* 1990:524). They are organs of equilibrium and are "generally composed of calcium carbonate in the form of pure aragonite" (Casteel 1976:20). The shell of the armadillo (Dasypodidae) is composed of scales, each of which consists of a dermal bony plate, an epidermal keratinous covering, and connective tissue between the two.

While it is sometimes found in archaeological contexts (e.g., Keepax 1981), avian egg shell is, apparently, rarely found as it is seldom reported in the literature. Avian egg shell is mainly calcite but has an organic component as well. It is somewhat porous, and apparently preserves well in alkaline sediments (Keepax 1981:317). Information on the microstructure of avian egg shell is comparable to that outlined above for bones and teeth (e.g., Becking 1975; Tyler 1969).

Properties of skeletal tissues and taphonomy

> [The knee] is a structure of such mechanical implausibility that it it must be designed to do something very well. Unfortunately, I do not know what that something is.
> (J. D. Currey 1984:184)

What is the taphonomic significance of whether bone tissue is composed of woven or Haversian bone? Why should we worry that bone tissue consists of both a mineral component and an organic component? These and similar concerns are important because they influence the ultimate effect a particular taphonomic agent or process will have on a bone specimen. Currey (1984:3–4) notes that "bones function mainly by not deforming appreciably under load . . . The stiffness of a bone and its strength depend on two factors: the stiffness and strength of the bone material itself and also the build [morphometry, or size and shape] of the whole bone." Thus, bones have a function in the body of an organism, and that function dictates the *taphonomic strength* of the bone; that is, a bone's anatomical function determines the bone's mechanical properties, and the mechanical properties of a bone in turn mediate the effects of taphonomic processes on the bone. We begin with a review of some of the basics of biomechanics before we turn to mechanical properties of skeletal tissues. Most research on the mechanical properties of skeletal tissues has focused on bone, and that skeletal tissue is therefore the one I focus on in the following, although comments about the other basic kinds of skeletal tissues are offered where possible.

Basic biomechanics

Stress is a measure of the intensity of force, and can be measured as the force per unit of cross-sectional area. *Strain* involves the change in shape of a body under stress and is dimensionless (Currey 1990:11). Stress and strain are related and are initially proportional to one another. The steep portion of the curve in Figure 4.6a shows this initial relationship between the origin of the curve and the *yield point*. Beyond the yield point of the curve, "irreversible changes have taken place in the material" and strain increases at a much more rapid rate than stress (Currey 1990:11). The initial part of the curve is called *Young's modulus of elasticity* and measures the stiffness of the material. The mineral component of skeletal tissue has a higher Young's modulus and is more brittle than the organic component. Eventually the specimen breaks, at which point the *ultimate stress* and *ultimate strain* can be measured. The total area under the curve is a measure of the toughness of the material. Antler tissue, for example, tends to be tougher than bone (Figure 4.6b). A tough material resists cracks traveling through itself (Currey 1984, 1990), and the composite nature of skeletal tissue, being part mineral and part organic, makes it very tough, tougher than either the mineral or organic fraction alone. Note that the rate of loading can influence the shape of the stress–strain curve in Figure 4.6a (Currey 1990:12; see Chapter 8).

Teeth, because they are mostly mineral, have a high Young's modulus relative to compact bone. But enamel is brittle and not very tough (Currey 1990:19). Young's modulus is lower in cancellous bone, which is better able to bend due to its relatively porous structure, than in compact bone (Currey 1990:23). The greater the mineralization of compact bone, the greater the stiffness and the less the toughness. Antlers, which have relatively low mineral content, have a low Young's modulus but are quite tough relative to compact bone with high mineral content.

Biomechanics and skeletal tissues

Currey (1984:58) observed that apatite is the mineral of choice for bones rather than, say, calcium carbonate, because "apatite is very reluctant to form large crystals" and "large crystals are brittle and in general to be avoided in materials that need to be tough." [Martill (1990:272) notes that the "ultra-microscopic size of the bone mineral has made it difficult to study, especially with regard to diagenetic processes that may have affected the original structure."] The inclusion of organic matter in bone increases the toughness and strength of the tissue because "there is a limit to greater strength provided by hydroxyapatite apposition alone, for at a certain level increased inorganic material will make the composite material brittle" (Burr 1980:113). Those studying the mechani-

a.

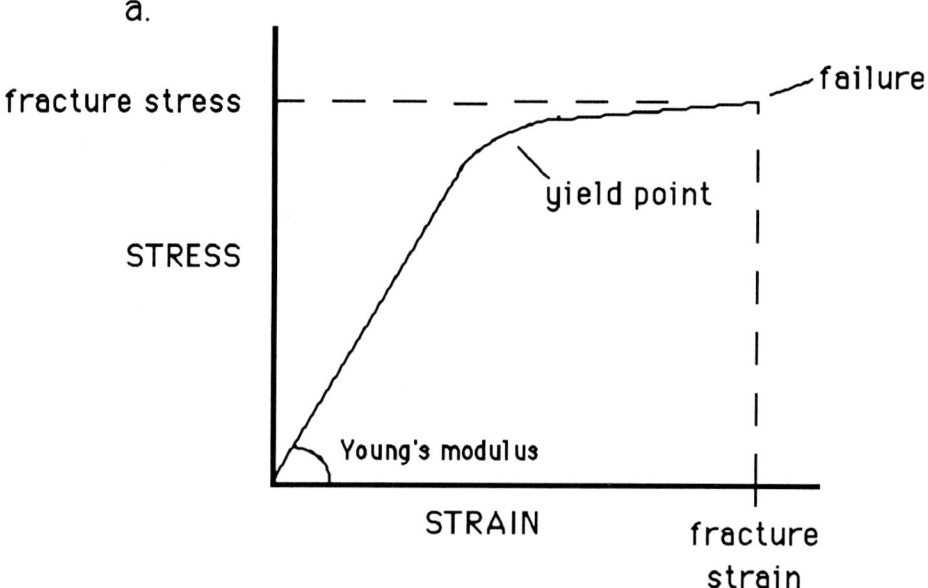

b.

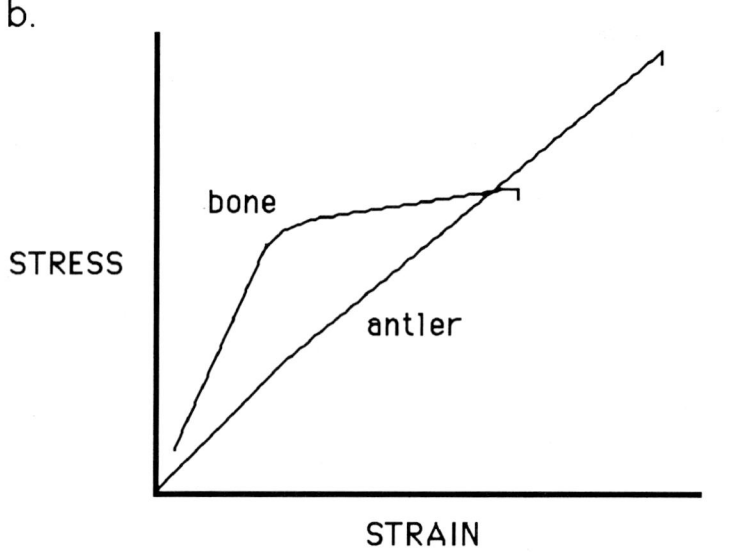

Figure 4.6. a, Modeled relation of stress and strain, showing Young's modulus of elasticity and point of failure (fracture). b, Comparison of stress–strain relation between bone and antler (after Currey 1990).

cal properties of bones have, nonetheless, found that "an understanding of the related effects of porosity and mineralization provides the most adequate picture of the mechanical behavior of bone" (Burr 1980:120).

Burr (1980:110) notes that a bone's mechanical integrity is determined in part by the mineral mass (density) of the bone tissue. "Density is a function of mineralization of the osteon and the porosity per unit volume of bone. Mineral salts account for about 90% of the density of bone" (Burr 1980:110). Because bone mineral content varies across a bone, what I term the *structural density* of the bone (Chapter 7) will vary across the bone. For example, increased weight bearing increases the density of a bone, and the parts of a bone with the greatest density are those parts that tend to undergo the greatest compressive and tensile stresses during the life of the organism (Burr 1980:111). Because the structural density of bone is determined as a weight:volume ratio, the porosity of a bone specimen influences the strength of that specimen: "increasing porosity with age reduces the strength of the bone" (Burr 1980:113), and "increased mineralization intensifies the deleterious effects of porosity in older bone by decreasing the ability of bone to absorb energy by plastic and elastic deformation" (Burr 1980:117).

The relation between bone strength and mineral content holds for both cortical and trabecular bone (Burr 1980:117), as well as other skeletal tissues. But due to the greater macroscopic porosity of trabecular or cancellous bone, it is less stiff and weaker than compact bone (Currey 1984:80). The arrangement and contiguity of trabeculae influence the stiffness of trabecular bone (Burr 1980:118). Porosity is also found in bone tissue at a microscopic level; "the porosity of compact bone is represented by the percentage (volume) of cavities or spaces formed by the Haversian canals" (Johnson 1985:166). Thus, Burr (1980:120) notes that "increasing reconstruction of bone – i.e., a greater percentage of secondary Haversian systems – reduces the mechanical strength of bone by (1) increasing porosity, (2) decreasing mineralization, since younger bone is less calcified than older bone, and (3) increasing the number of osteons per unit area." Further, Burr (1980:120) notes that because the periosteal bone found on the outside of a mammalian long bone is lamellar rather than Haversian in structure, it is denser (because it is more highly mineralized) and less porous than the Haversian bone which is found on the internal surface of the skeletal element forming the walls of the marrow cavity (Figure 4.1). Currey (1984:86) concludes that "in no respect does Haversian bone seem to have mechanical properties superior to those of fibrolamellar bone." Young bones have a higher impact strength (are more able to withstand impacts without breaking) than older bones (Currey 1984:93). This is no doubt because the former are less mineralized, and thus less brittle, than the latter. And, as noted in the previous section, woven bone is more porous than, say, lamellar or Haversian bone.

The number, size, and orientation of collagen fibers, and the degree of their crosslinking all influence the mechanical properties of bone (Burr 1980:120–121). An example with taphonomic implications is provided by Tappen (1969; Tappen and Peske 1970) who shows that the orientation of split-lines and weathering cracks, which are induced during the drying and resulting shrinkage of bone (the initial stages of subaerial weathering; Chapter 9), is related to the orientation of the majority of the collagen fibers in the bone. Thus weathering cracks and split-lines on long bones tend to be parallel to the long axis of the bone. Bones with a spiral orientation to their collagen fibers tend to break spirally (Hill 1976). The implication here is that individual bones or skeletal elements can be conceived of as having a *grain*; "most bone, except woven bone, has a definite grain, produced by the co-oriented cementing together of collagen fibrils and their mineral. This gives bone a microstructure equivalent to that of a fibrous composite with a very high volume fraction of fibers" (Currey 1984:86).

Johnson (1985:167) notes that the fracture of compact bone on a microscopic level is "directly related to the amount and distribution of osteons, the distribution and orientation of collagen fibers, and the combined response to force of osteons and collagen fibers." She (Johnson 1985:167) writes that "osteons whose collagen fibers follow a steeply spiraled course around the longitudinal axis of the osteon exhibit greater tensile but less compressive strengths than those osteons whose fibers have a low angle of spiral (lower tensile but greater compressive strengths)." The point here is that the fracturing of bones is in part related to the grain of the bone tissue (see Chapter 8).

We know very little about how the biomechanical properties of skeletal tissues relate to forces such as abrasion (Currey 1990:11). Geniesse (1982:38) notes that because force applications at different angles to the grain of bone tissue have different deformation and fragmentation effects, abrasion forces will probably have different effects depending on the direction of those forces relative to bone tissue grain. She artificially abraded compact bovid bone under controlled conditions, and collected and weighed the bone tissue removed by several instances of both 300 abrasion strokes and a 30 second period of abrasion. While statistically significant results were not obtained, in every case more bone tissue was removed by abrasive force applied perpendicular to the grain of the bone than by abrasive force applied parallel to the grain. Geniesse (1982:40) suggests that one major factor controlling this difference was the greater chance of tearing off larger pieces of bone tissue when working against the grain whereas abrasion forces working parallel to the grain tend to wear down the bone in smaller pieces.

Summary

The preceding comments have focused on two properties of bone tissue: its structural density (and porosity), and its grain. Both properties owe their manifestations within a particular skeletal element to the microstructure of bone tissue and its variation across the skeletal element. The morphometry of a skeletal element has been discussed in only general terms in this section, but it is clear (Chapter 8) that this variable too is important in terms of whether or not a particular skeletal element can be made into a usable tool. As well, the morphometry of a bone will influence how readily it is fluvially transported and whether or not the bone will move downslope due to gravity (Chapter 6). Bone density in particular is considered in some detail in Chapter 7, and the influence of bone microstructure on how a bone fractures is considered in Chapter 8. The influence of bone microstructure on diagenetic processes is considered in Chapter 11. The discussion in this section has been superficial, but that is in part because, with the exception of fracture mechanics, the influence of bone tissue microstructure in particular on taphonomic processes has not been intensively explored. I have indicated why the microstructure of all skeletal tissues should be so explored.

Vertebrate skeletons

> The skeleton's obvious mechanical function is to serve as the body's scaffolding.
> (H. Francillon-Vieillot *et al.* 1990:473)

A *vertebrate* is, simply, an animal with a backbone. The taxonomic phylum Chordata includes animals that have, either throughout their life or at some stage in their development, an internal supporting rod. That rod is continuous down much of the length of the body along the dorsal midline. If it consists of individual vertebral elements, the animal is classified as a member of the vertebrate subphylum (Hildebrand 1974; Romer and Parsons 1977).

There is a great deal of variation in the skeletons of vertebrates. That variation is found not only in the kinds of tissues making up the skeletons, but in the actual structure of the skeleton. A bird skeleton is not nearly the same as, for instance, the skeleton of a reptile although the two are phylogenetically related. Figures 4.7 through 4.13 present generalized sketches of various vertebrate skeletons. In following chapters I presume much of what is shown in these drawings is well known to the reader. The general directional terms used to describe skeletal parts are illustrated in Figure 4.14.

A *skeletal element* (or bone or tooth) is an "anatomical organ" (Francillon-Vieillot *et al.* 1990:480). Skeletal elements making up vertebrate skeletons can be divided into two basic categories: axial and appendicular. *Axial elements* are found near the midline of the torso, they may be single or paired, and most tend to be bilaterally symmetrical. These include the cranium (single), mandibles (paired), vertebrae (single), ribs (paired), and sternum (single). *Appendicular elements* are often said to be paired; that is, there are discrete left and right, bilaterally symmetrical instances of each. These include the limb bones, the

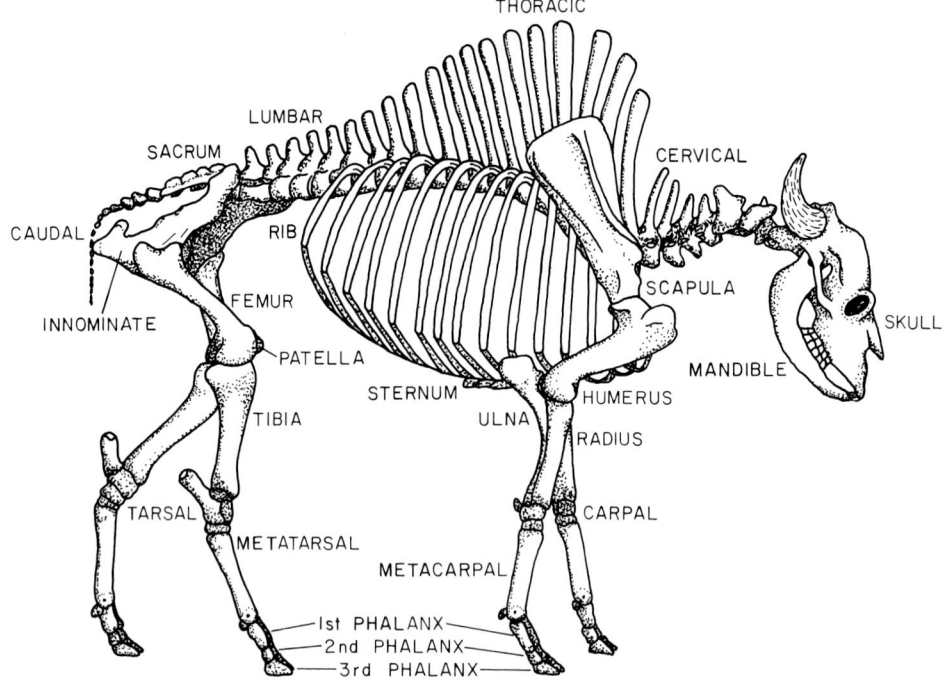

Figure 4.7. North American bison (*Bison bison*) skeleton, showing locations of major skeletal elements.

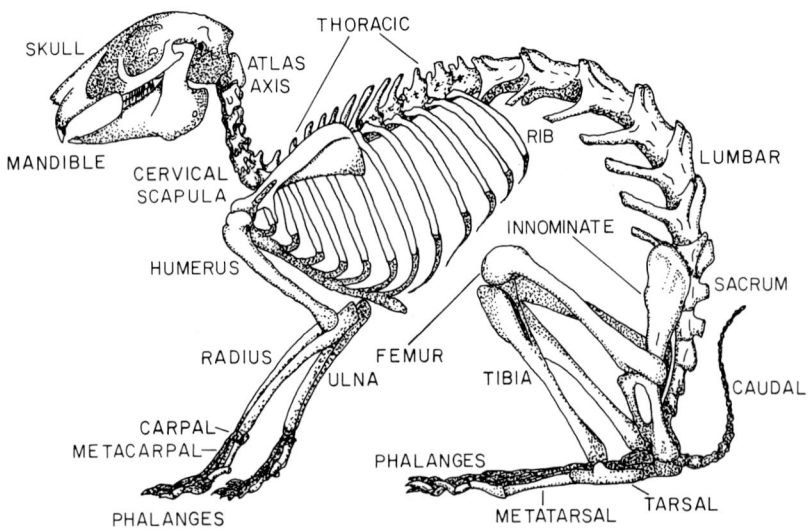

Figure 4.8. Generalized leporid or rabbit skeleton, showing locations of major skeletal elements.

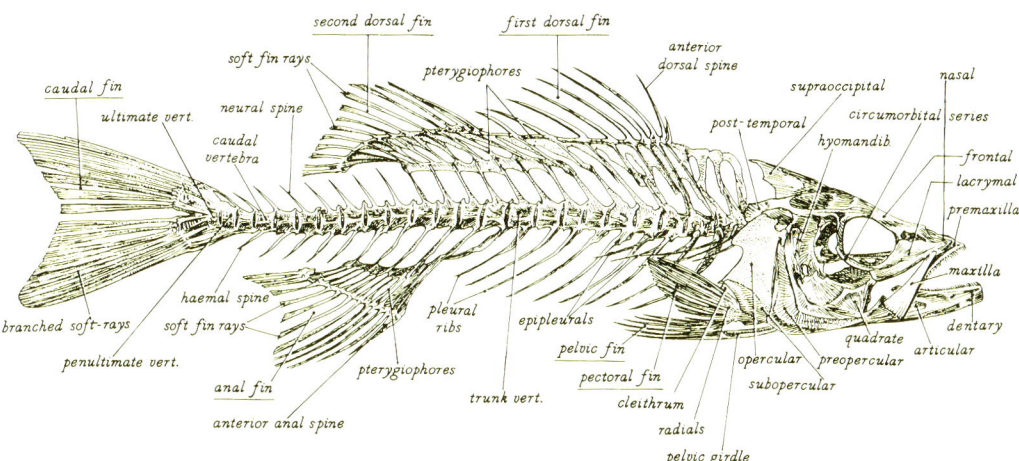

Figure 4.9. Generalized teleost fish skeleton, showing locations of major skeletal elements. Reproduced by permission from: Olsen, S. J. *Fish, amphibian and reptile remains from archaeological sites*. Peabody Museum Papers, vol. 56, no. 2, Figure 3; Copyright 1968 by the President and Fellows of Harvard College.

bones of the pectoral (shoulder) girdle (clavicle if present, scapula, humerus), and the bones of the pelvic girdle (innominate, consisting of the pubis, ilium, ischium). In mammals, "blood-forming red marrow [is] found within the flat bones of the skull, sternum and ribs and in the cancellous extremities of the long bones. Yellow marrow, mostly fat, occupies the interiors [medullary cavity] of the long bone diaphyses" (MacGregor 1985:9; see also Currey 1984:104–105).

Vertebrate skeletons in general have four limbs (Figures 4.7 and 4.8), with the exception of fish which have fins instead of limbs (Figure 4.9). Snakes have no limbs (Figure 4.12) but some species have skeletal vestiges of the limbs their reptilian ancestors once sported. Turtles and tortoises are unique for the shells they carry (Figure 4.11). In birds the forelimbs are modified for flight, bones are thin-walled to decrease weight, and the sternum is enlarged to support the flight muscles (Figure 4.13).

Most taphonomic research has focused on within-skeleton variation in the effects of taphonomic agents and processes. That is, analysts have studied how bovid femora respond to particular taphonomic processes and compare these responses to those of bovid scapulae to the same processes. Much less work has been devoted to comparisons of, say, the responses of a bird humerus and a dog humerus to a particular taphonomic process. Some research along these lines is beginning to appear (e.g., Chambers 1992; Kreutzer 1992; Lyman *et al.* 1992a),

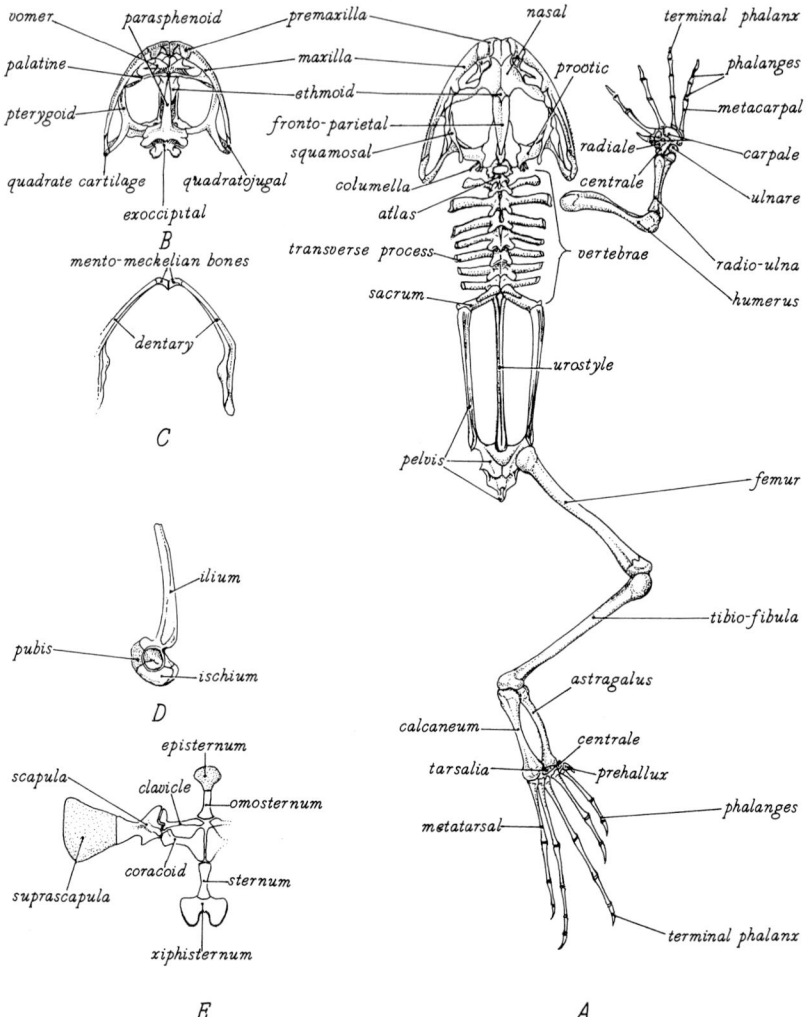

Figure 4.10. Generalized frog (amphibian) skeleton, showing locations of major skeletal elements. Reproduced by permission from: Olsen, S. J. *Fish, amphibian and reptile remains from archaeological sites*. Peabody Museum Papers, vol. 56, no. 2, Figure 10; Copyright 1968 by the President and Fellows of Harvard College.

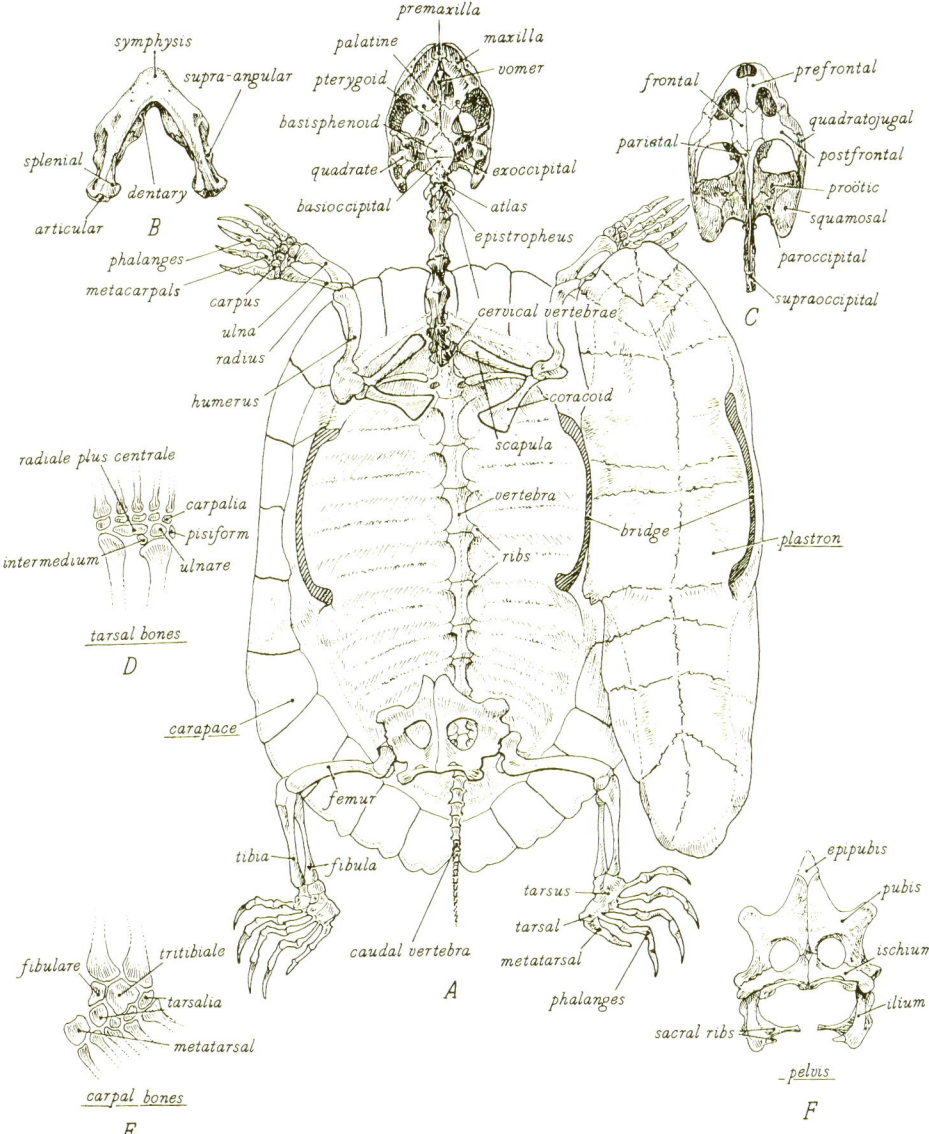

Figure 4.11. Generalized turtle (reptile) skeleton, showing locations of major skeletal elements. Reproduced by permission from: Olsen, S. J. *Fish, amphibian and reptile remains from archaeological sites*. Peabody Museum Papers, vol. 56, no. 2, Figure 11; Copyright 1968 by the President and Fellows of Harvard College.

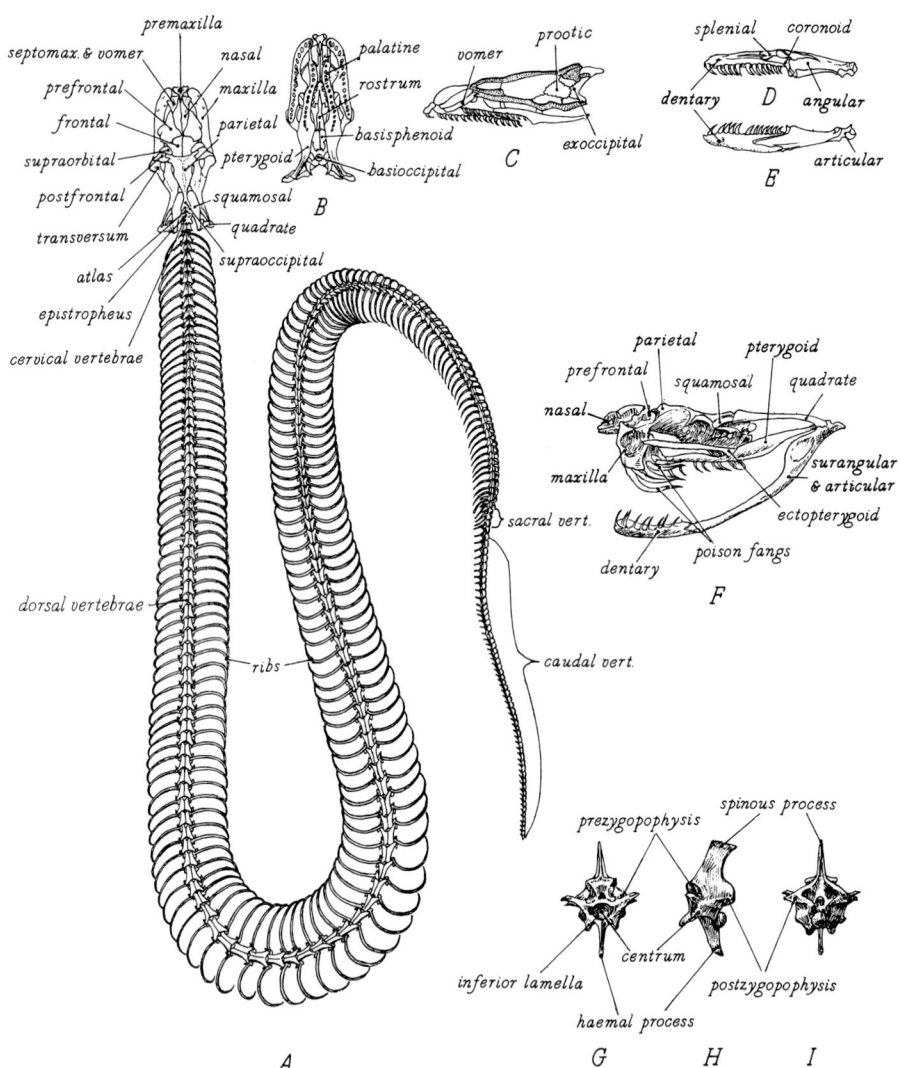

Figure 4.12. Generalized snake (reptile) skeleton, showing locations of major skeletal elements. Reproduced by permission from: Olsen, S. J. *Fish, amphibian and reptile remains from archaeological sites.* Peabody Museum Papers, vol. 56, no. 2, Figure 12; Copyright 1968 by the President and Fellows of Harvard College.

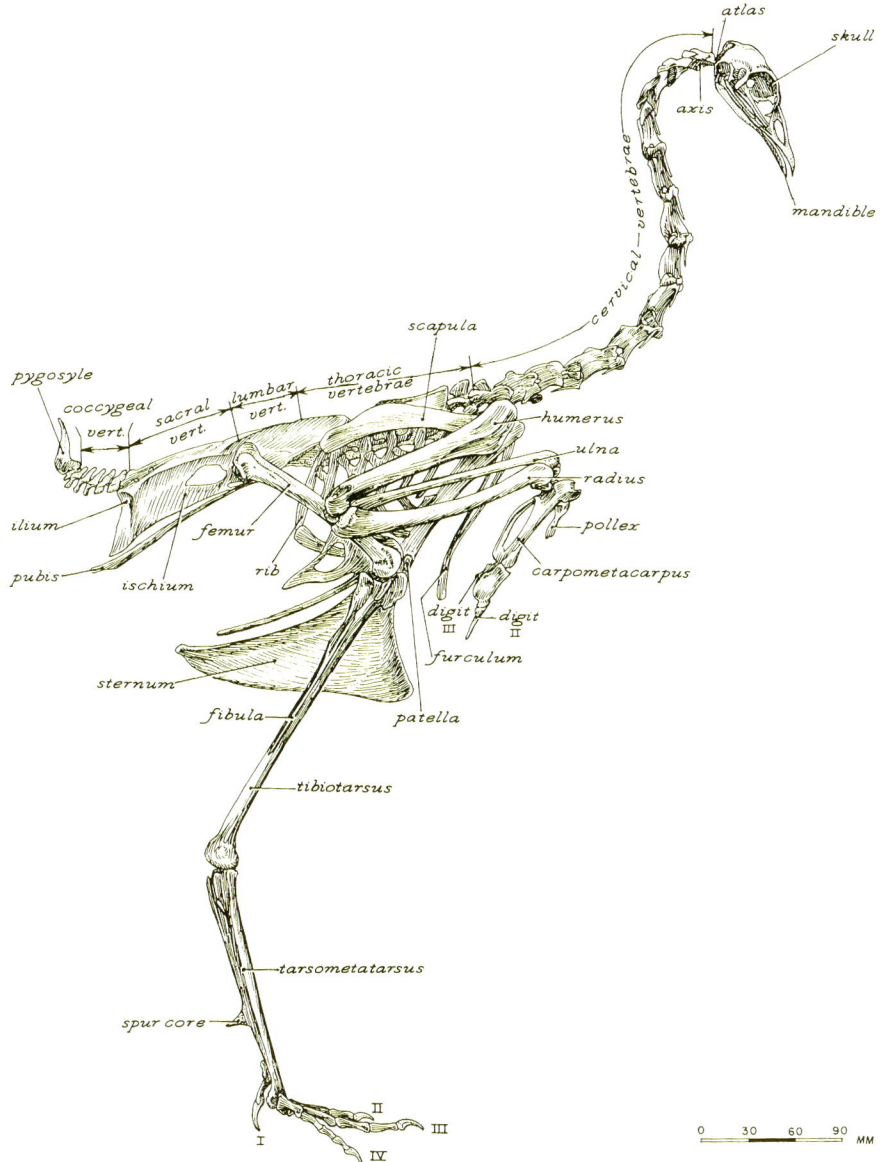

Figure 4.13. Generalized bird skeleton, showing locations of major skeletal elements. Reproduced by permission from: Olsen, S. J. *Fish, amphibian and reptile remains from archaeological sites.* Peabody Museum Papers, vol. 56, no. 2, Appendix Figure 3; Copyright 1968 by the President and Fellows of Harvard College.

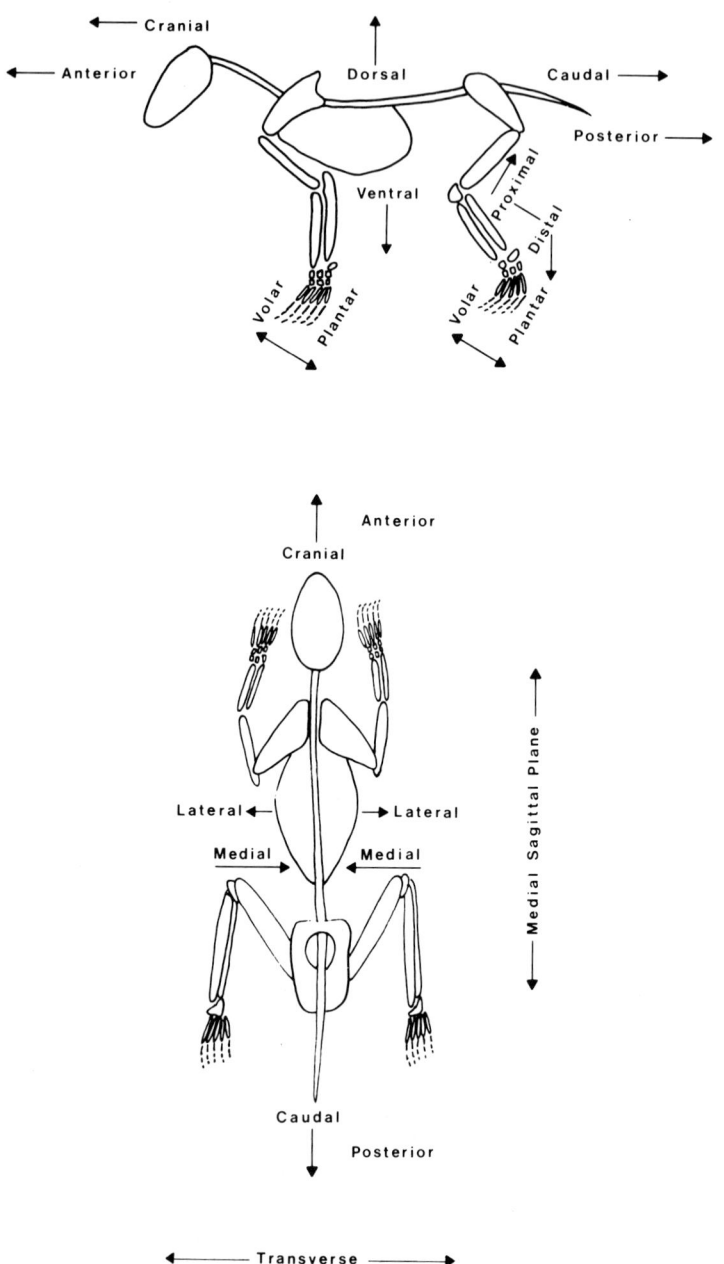

Figure 4.14. Directional terms for vertebrate skeletons. Reproduced by permission from: Hesse, B. and Wapnish, P. *Animal bone archeology*, Figure 30. Washington, D.C.: Taraxacum Press. Copyright 1985 by B. Hesse and P. Wapnish.

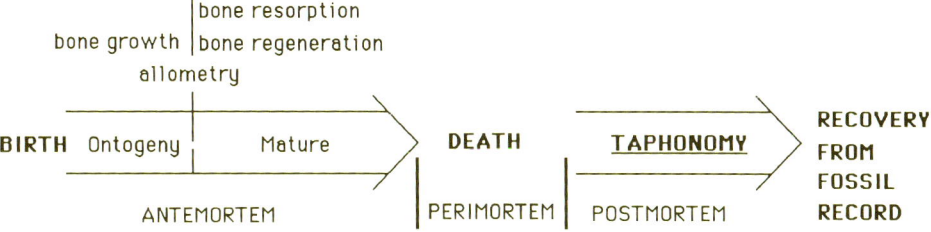

Figure 4.15. Chronological relations of bone ontogeny, bone remodeling, death, and taphonomy. Time passes from left to right.

but is thus far restricted to different mammalian taxa. Given the mediating influences of microstructural properties of skeletal tissues on taphonomic processes, and the major macrostructural and microstructural variation between fish, bird, reptile, amphibian, and mammal skeletons and skeletal tissues, such research seems necessary to the continued growth of our taphonomic knowledge.

Modification of skeletal tissues and time of death

The word *skeleton* is taken from the Greek *skeletos* which means "dried up" or "withered." Taphonomists are concerned with the conditions of, and the modifications to, skeletons, skeletal elements, and skeletal tissues. That is, they identify and study such conditions and modifications based on models of the skeletons of living organisms, individual skeletal elements of living organisms, and skeletal tissues of living organisms. Documenting and explaining differences between the living and the fossil appearance of skeletons, elements, and tissues is the subject of taphonomic inquiry. But in using a model of a specific living skeleton as the comparative baseline, the analyst must consider that the model is based on a norm or average. Skeletal elements and skeletal tissues in particular can be modified from the norm during the lifetime of an organism; such modifications occur before death and are known as *antemortem* modifications. Given the general consensus that taphonomic histories begin with the death of organisms, it must be realized that taphonomists tend to be interested not in antemortem modifications but rather modifications that occur at the time of death or *perimortem* modifications, and modifications that occur after death or *postmortem* modifications (Figure 4.15).

Depending on the research questions being asked, archaeologists may be interested in antemortem damages, such as when congenital defects or injuries to an organism are studied (e.g., Baker and Brothwell 1980:82–95). If human warfare or human hunting practices are the subject of interest, then perimortem modifications may be of interest (e.g., Wilson 1901, and Noe-Nygaard 1974, 1975a, 1975b, respectively). Similarly, study of archaeological traces of human cannibalism tends to focus on perimortem and postmortem modifications of human remains (e.g., Flinn *et al.* 1976; C. G. Turner 1983; White 1992), although here in particular the line separating the time of perimortem from postmortem modifications may be unclear. Modifications to skeletons that result from processing of freshly taken prey for immediate consumption may be difficult to distinguish from modifications that occur, say, a month or more after death, such as when prey are cached or stored for later consumption. Distinction of perimortem from postmortem modification is perhaps less critical than distinguishing these two from antemortem modifications. That is so because taphonomic histories begin with animal death. Of course the cause of death may be an important taphonomic variable (e.g., Weigelt 1927/1989), and we return to it in Chapter 5. Here, we must consider how the analyst identifies antemortem modifications.

Antemortem modifications include readily identified phenomena such as healed fractures. Ortner and Putschar (1981:61) indicate that antemortem "traumatic, fatigue and pathological fractures will vary in the healing process." A traumatic fracture results in the rupture of blood vessels in Haversian canals, the periosteum, and the marrow; released blood forms a hematoma around the fracture. Periosteum is stripped away from the bone surface for a few millimeters adjacent to the fracture site, and this appears to initiate the formation of *callus*. The hematoma (blood clot) is permeated by fibrous connective tissue which may be the source of the fibrous callus that unites the fracture surfaces. The unmineralized fibrous callus provides the matrix for the formation of woven bone and the primary bone callus. The woven bone is eventually replaced with lamellar bone. Healed fractures may be marked by excessive bony tissue at the fracture site, and non-normal alignment of the two pieces (Ortner and Putschar 1981:61–64).

Other antemortem modifications, such as abnormal skeletal development and disease or malnutrition induced modifications, are, like perimortem and postmortem modifications, identified on the basis of models of normal or average skeletons, elements, and tissues. Many antemortem modifications are identified on the basis of the presence of abnormal bone growth or resorption (Baker and Brothwell 1980; Ortner and Putschar 1981). In order for either to occur, the skeletal tissue must be alive, indicating the organism was alive. There is more resorption and less regeneration immediately after a wound to a bone is inflicted, and as the wound heals there is less resorption and more regeneration. *Resorption* often creates pits in bone, bone *regeneration* (initial deposition of woven bone that is eventually replaced by lamellar bone) first occurs as tiny nodules of new bone material. Healed wounds to bones display a relatively smooth surface and extra bone tissue relative to the norm (after Morse 1983).

Garland (1988:324) argues that macroscopic inspection of bones "gives little insight into the interactions which have taken place between bones and the various chemical and physical factors and biological agents within the burial environment." He suggests that microscopic inspection of skeletal tissues, especially histological studies, can be important for distinguishing pathological conditions from taphonomic effects (e.g., Garland *et al.* 1988). That is so because of the microscopically distinct responses of skeletal tissues to disease and to, in particular, post-depositional taphonomic processes (Chapter 9). I suspect Garland is correct, and the research of others tends to confirm this suspicion (e.g., Hackett 1981; Hanson and Buikstra 1987; Piepenbrink 1986). There is no doubt an area of microscopic taphonomic research concerning paleohistology and diagenetic processes that is, as yet, little explored.

The preceding overview focuses largely on mammalian tissues as those are the best known and the most typical in the fossil record. Other hard tissues that might be of concern to a vertebrate taphonomist such as the baleen of whales (a keratinous material) have not been covered. Consultation of the references cited and their respective references should provide a good start towards providing information on the materials not covered as well as further details about the materials that have been reviewed.

Quantification

> [In rational or well-reasoned quantification] the investigator admits to his graphs, so to speak, only items of evidence that are relevant to the particular matter under investigation, and that are as accurate as practicable, with the probable limits of sampling and experimental error expressed graphically.
> (J. H. Mackin 1963:139)

An analyst interested in taphonomic issues has an advantage over, say, a lithic technologist or someone interested in prehistoric ceramics. The taphonomist has a model from which to work. The taphonomist knows, for instance, mammals have four legs, the number of bones and teeth making up a single skeleton of a particular species (Table 4.1), and what each complete bone and tooth of that skeleton looked like when that skeleton was endowed with muscles, a heart, lungs, etc., and was walking around. Documentation of how and why the pile of bones and teeth lying on the laboratory table differ from that model of a living organism is the work of the taphonomist. One of the first important steps in such documentation involves measuring how abundant particular skeletal parts and particular taxa are in a collection of faunal remains (Grayson 1984). In this section, I define several key concepts and terms, describe several of the regularly used quantitative units, and discuss several basic concepts of quantification and measurement.

Concepts

The process of measurement involves "the application of a set of procedural rules for comparing sense impressions with a scale and for assigning symbols to observations" (Gibbon 1984:40). *Measurements* result from comparing observations made on phenomena with a scale according to a set of rules and

Table 4.1 *Frequencies of major kinds of skeletal elements in different mammalian taxa*

Skeletal element	*Homo*	Bovid/Cervid	Equid	Suid	Canid	Felid	*Castor*	Pinniped
cranium	1	1	1	1	1	1	1	1
mandible	1	2	2	2	2	2	2	2
atlas	1	1	1	1	1	1	1	1
axis	1	1	1	1	1	1	1	1
cervical (3–7)	5	5	5	5	5	5	5	5
thoracic	12	13	18	14–15	13	13	14	15
lumbar	5	6–7	6	6–7	7	7	5	5
sacrum[a]	1(4–5)	1(4–5)	1(5)	1(4)	1(3)	1(3)	1–4	1(3)
innominate	2	2	2	2	2	2	2	2
rib	24	26	36	28–30	26	26	28	30
sternum[a]	1(6)	1(6)	1(6–8)	1(6)	1(8)	1(8)	1(5)	1(8–9)
scapula	2	2	2	2	2	2	2	2
clavicle	2	0	0	0	0	2	2	0
humerus	2	2	2	2	2	2	2	2
radius	2	2	2	2	2	2	2	2
ulna	2	2	2	2	2	2	2	2
carpal	16	12	14–16	16	14	14	14	14
metacarpal	10	2	2	8	10	10	10	10
femur	2	2	2	2	2	2	2	2
patella	2	2	2	2	2	2	2	2
tibia	2	2	2	2	2	2	2	2
fibula	2	distal only	proximal only	2	2	2	proximal only	2
astragalus	2	2	2	2	2	2	2	2
calcaneum	2	2	2	2	2	2	2	2
other tarsals	8	6	8	10	10	12	12	10
metatarsal	10	2	2	8	10	8	10	10
first phalanx	20	8	4	16	20	18	20	20
second phalanx	16	8	4	16	16	16	16	16
third phalanx	20	8	4	16	20	18	20	20

Note:

[a] In mature individuals there is one sternum and one sacrum, made up of the number of individual sternabra and sacral vertebra indicated in parentheses.

assigning one or more symbols or values to each observation per phenomenon. Quantitative *units* make up the scale and are of different levels of mathematical power (Shennan 1988:11–12; Stevens 1946), and of different kinds (Gibbon 1984:55). The level of a quantitative unit is determined by rules of ordering and distance, and is critical to determining appropriate statistical tests (see Grayson 1984 for relevant discussion regarding zooarchaeological data). *Nominal-level* measures record differences in kind or category with no ordering or distance between categories. *Ordinal-level* measures record rank orders of magnitude; greater-than and less-than relations are specified, but not how much greater or less than. There is an order to but no specification of distance between measurements. *Interval-level* measures record an order and how much greater-than and how much less-than in terms of fixed and equal units of measurement. Both order and distance between measurements are known. *Ratio-level* measures are the same as interval scale measures, but also have a natural zero point. Nominal-level measures are sometimes described as qualitative, discrete, and/ or discontinuous; ordinal, interval, and ratio-level measures are sometimes characterized as quantitative and/or continuous.

Quantitative units can be observational or analytical. *Observational units* are empirical manifestations that are easily observed properties of phenomena; they are easily experienced with one's senses and can be directly measured. Measuring observational units by tallying the frequency of specimens or determining the length of individual specimens produces *fundamental measurements* of properties of specimens. *Analytical units* consist of observational units that have been modified, often mathematically, to reflect some complex, indirectly observable property of the phenomena under study. In archaeology the property of interest is indirectly observable because the archaeological record is static, and the property of interest tends to include a dynamic process that is believed to be somehow related to that record.

Analytical units may take either of two forms: derived, or interpretive. *Derived units* are more complex than observational units because they are defined by some specified mathematical relation between fundamental measurements. Derived units include such things as ratios of fundamental measurements, and require analytical decisions above and beyond the choices of a scale of comparison and a rule set for assigning symbols. Measurements of derived units produce *derived measurements* (Gibbon 1984:55). Derived units, in the following, tend to have non-explicit, unclear, or only weakly established relations to theoretical or interpretive concepts, although they may play a significant role in comparative analyses. *Interpretive units* are very complex because they are structured to measure some abstract or theoretical concept, and the terms applied to such units often include a name for them. Measurements of interpretive units have been called *fiat* or *proxy measurements* (Gibbon 1984:55).

Derived measurements, in my view, are mathematically generated in the hopes that some hidden pattern within the units measured will be revealed; that pattern may or may not be causally related to the property we wish to measure. Similarly, proxy measurements are mathematically generated but there tends to be some reason to suppose the measured interpretive units are causally related to the property we seek to measure. For example, measures of an individual's socioeconomic status often include fundamental measures of annual income, occupation, and the level of education attained. On one hand, the typically strong correlation between income and status serves as an empirical generalization warranting the conclusion that measures of socioeconomic status can be obtained by mathematically manipulating fundamental measurements of income, occupation, and the like, to produce proxy measures of an interpretive unit we might term a "socioeconomic status index." Derived units, on the other hand, are less closely allied with theoretical concepts, and may simply be used to produce derived measures of different sets of phenomena one wishes to compare. Derived units may attain the status of interpretive units as we learn more about their theoretical connotations, and conversely, interpretive units may change their status and become analytical units if research suggests their suspected causal relation to a dynamic process is less strong than originally believed.

NISP and MNI

NISP and MNI are the quantitative units most commonly encountered in the zooarchaeological literature, and they tend to have generally agreed upon meanings; that is, the units they seek to measure are clear (e.g., Grayson 1984; Klein and Cruz-Uribe 1984; Lyman 1985a). *NISP* is defined as the number of identified specimens per taxon; it is an observational unit. The taxon can be a subspecies, species, genus, family, or higher taxonomic category. *MNI* is defined as the minimum number of individual animals necessary to account for some analytically specified set of identified faunal specimens; it is a derived unit because it may or may not take inter-specimen variation such as age, sex, or size into account. It is important to understand that MNI traditionally means the minimum number of individual animals necessary to account for all the kinds of skeletal elements found in the skeleton of a taxon, the humeri, the scapula, the cervical vertebrae, etc. As we will see, MNI can mean something else if the analytically specified set of identified faunal remains does not include all the kinds of skeletal elements in a skeleton.

In the definitions of NISP and MNI, the word "identified" must be clear. Typically, "identified" means "identified to taxon." It can also mean "identified to skeletal element," such as a humerus, a tibia, a thoracic vertebra, or a vertebra. Because it is typically necessary to identify the skeletal element prior to identifying the taxon represented by a specimen (Lyman 1979c), the meaning of identified as "identified to taxon" usually also (implicitly) entails the meaning "identified to skeletal element." That is, the latter is often necessary to the former. The former is not, however, always necessary to the latter.

It is critical to define explicitly what is meant by "specimens" and "skeletal elements." Grayson (1984:16) defined a *specimen* as "a bone or tooth, or fragment thereof, while an *element* is a single complete bone or tooth in the skeleton of an animal." A specimen is an archaeologically discrete phenomenological unit, such as a complete humerus, a distal half of a tibia, or a mandible with teeth in it. A skeletal element is a discrete, natural anatomical unit of a skeleton, such as a humerus, a tibia, or a tooth. Specimens can, but need not be skeletal elements and are observational units. Skeletal elements are anatomical units that may be represented by fragments or whole bones and are represented, partially or completely, respectively, by specimens. A complete femur recovered from a site is a specimen, an observational unit, and a skeletal element. A fragment of a femur such as the distal end is a specimen, an observational unit, and *represents* but is phenomenologically not, technically, a skeletal element.

Taphonomists have a skeletal model to which the observational units (specimens) they study can be related. That model consists of the individual, anatomically discrete and variously articulated skeletal elements making up the skeleton. For example, it is not unusual to find a table listing the frequencies of forelimbs, femora, proximal tibiae, or thoracic sections of the vertebral column in a zooarchaeological report. These anatomical categories are founded on the model of a skeleton consisting of discrete skeletal elements, and the categories are of varying scales of inclusiveness of the skeletal elements. The anatomical category "distal tibia" is less inclusive than the category "humerus" which in turn is less inclusive than the category "thoracic section of the vertebral column." Only the second category is directly comparable to an anatomically discrete skeletal element. The first category includes some *analytically* specified portion of an anatomically discrete skeletal element whereas the last includes several analytically specified articulated but anatomically discrete skeletal elements. These anatomical categories of varying scales of inclusiveness can and often do serve as the quantitative units within which observational units (specimens) are tallied.

Some authors use the term "bone fragments" or just "fragments" when in fact they mean "specimens" as defined earlier. Thus, when either of the former two terms is used, what often are included in the tallies are both fragments of skeletal elements and complete skeletal elements. "Specimens" is a more satisfactory term as it has no connotation about the kind of the part of the skeleton being tallied (bone, tooth, horn) or about the anatomical completeness of the part. The explicit distinction of elements and specimens is critical to taphonomists concerned with measuring the extent and intensity of bone fragmentation because such measures include NISP:MNE ratios, NISP:MNI ratios, and the like (Chapter 8). If it is not clear what a specimen is and how it might differ from a skeletal element, ratios like these will not be replicable and their taphonomic significance will be obscure.

Because specimens, as defined above, are the fundamental observational units of zooarchaeology, it should be clear that the tenacity and identification skills of the analyst may influence NISP measures (White 1992). The specimens I can identify to skeletal element and taxon may be different from those someone else can identify. While inter-analyst variation in what is identifiable (and thus countable) has not been studied in detail, I suspect that this source of variation between analysts may be minimal in a great many cases.

While MNI (as defined above) and NISP are quantitative units that are basic to much of zooarchaeological analysis, there are other quantitative units that play major roles in modern taphonomic analysis. We turn to several of these now.

MNE

The recent focus in zooarchaeology on taphonomic issues has brought with it a shift from measuring frequencies of taxa (usually accomplished with NISP and MNI) to measuring, among other things, frequencies of portions of skeletons of individual taxa. Many analyses with such a focus use the quantitative unit MNE, or some derivation thereof. The term *MNE* signifies the minimum number of a particular skeletal element or portion of a taxon, such as the minimum number of bovid *proximal* humeri, or the minimum number of caprine thoracic sections of the vertebral column. Analysts typically depend on the form of data presentation to make it clear that the MNE values they publish are not necessarily the minimum number of anatomically complete skeletal elements (analogous to MNI), but rather are often of some portion of a skeletal element or some multi-skeletal element portion of a skeleton.

Because MNE is the minimum number of skeletal portions necessary to account for the specimens representing that portion, the same problems plague the derivation of MNE values as plague the derivation of MNI values (see Grayson 1984 for a discussion of the latter). One may (e.g., Bunn and Kroll 1988; Hesse and Wapnish 1985; Potts 1988), or may not (e.g., Binford 1984b), for instance, take into account age, sex, size, or even taxonomic differences between the specimens for which a minimum number is desired. MNE is therefore an analytical unit rather than an observational unit. The analyst uses some set of criteria by which specimens are considered to be independent (each of two or more specimens represents a separate case) or interdependent (two or more specimens represent the same case). These criteria should be, but seldom are, explicit.

MNE has become an important quantitative unit, reflecting the fact that taphonomists are concerned with how and why archaeological faunal remains differ from the set of skeletal elements making up a complete skeleton. Each mammal, for example, has two humeri, two tibia, and seven cervical vertebrae. The relative abundances of different kinds of skeletal elements one observes in an archaeological collection can be compared to the relative abundances of skeletal elements in a complete skeleton. Explaining why archaeologically observed relative frequencies of skeletal elements differ from or are similar to those in a complete skeleton has proven to be an important and fruitful analytical step (Chapter 7). Further, NISP:MNE ratios are useful for measuring the degree of fragmentation of different skeletal elements (see discussion of Figure 8.11). It is critical, then, to consider how MNE values are derived from a set of specimens.

Table 4.2 *FLK* Zinjanthropus *bovid limb bone data (from Bunn 1986; Bunn and Kroll 1986)*

Skeletal element	NISPends[a]	NISPshafts	MNEends	MNEshafts	MNEcomp
humerus	30	58	19		20
radius	28	57	14		22
metacarpal	21	32	15		16
femur	14	58	6(8[b])	17[b]	22
tibia	20	128	11(15[b])	21[b]	31
metatarsal	24	28	15		16

Notes:
[a] number of identified specimens with one or both ends; the latter are complete
[b] from Bunn and Kroll (1988:142); taxonomic/size differences are accounted for in values in parentheses.

Marean and Spencer (1991:649–650) mention two ways to derive MNE values. One involves measuring the percentage of the complete circumference represented by a long-bone shaft fragment, and then summing those percentages for each measured portion of a skeletal element. This is similar to a method described by Klein and Cruz-Uribe (1984:108) as recording the "fraction by which an identifiable bone is represented [using] common and intuitively obvious fractions (e.g., 0.25, 0.33, 0.5, 0.67) and not attempting great precision." The fractions are summed to produce an MNE value for each skeletal portion. For instance, if the analyst records one complete proximal femur, a fragment representing one half of a proximal femur, and a fragment representing one third of a proximal femur, the sum of the fractions would be $(1.0 + 0.5 + 0.33 =)$ 1.83, for an MNE of two proximal femora. On one hand, Marean and Spencer's (1991) method seems to be relatively accurate, but can result in slight (and probably statistically insignificant) overestimates. Care must be taken with Klein and Cruz-Uribe's (1984) method. That is so because if the three proximal femora pieces noted above all include the greater trochanter, then the MNE is not two, it is in fact three. Klein and Cruz-Uribe's (1984) method can, of course, easily be modified to account for such overlap of specimens. In fact, the second method mentioned by Marean and Spencer (1991:652) "involves using the computer to count the number of portions with *overlapping sections*" (emphasis added), but they do not describe this method.

Bunn and Kroll (1986, 1988) describe three ways to derive MNE values. The analyst may determine (1) the minimum number of *complete* limb bone skeletal elements necessary to account for only the specimens with one or both articular ends, (2) the minimum number of *complete* limb bone skeletal elements necessary to account for only the specimens of limb bone shafts (without an articular end), and, (3) the minimum number of *complete* skeletal elements necessary to account for both the specimens with one or both articular ends and the shaft specimens. These are labeled, respectively, the MNEends, the MNEshafts, and the MNEcomp in Table 4.2 and, as Bunn (1986, 1991) emphasizes, these values can be different. What is perhaps confusing is that all three are labeled MNE in the published record, yet there are significant differences between the three sets of values due to variation in how they are derived. The confusion is exacerbated by the fact that in all but one of Bunn's (1986, 1991; Bunn and Kroll 1986, 1988) published reports on the FLK *Zinjanthropus* faunal assemblage the MNEends for femora is listed as 6 and for tibiae is listed as 11 while in Bunn and Kroll (1988) those values are easily derived as 8 and 15, respectively, from their data tables. The difference here resides in the higher values resulting from *my* distinguishing the size class of taxa represented by the specimens, and thus my introducing yet another kind of MNE value, one that takes into account the size of the organism represented by the bones.

MNE is a derived measure. As with MNI, there are several ways to derive MNE values. The important point here is that, like Marean and Spencer (1991) and Bunn and Kroll (1986, 1988), the analyst must be explicit about the criteria used to derive MNE values when they are presented. This will ensure that comparisons of one analyst's MNE values with another's MNE values is not a comparison of apples and oranges.

MNI and MAU per skeletal portion

Binford (e.g., 1984b) demonstrated and popularized the use of MAU as a quantitative unit. The history of this quantitative unit is interesting. Binford began his zooarchaeological studies with a chapter in a book published in 1977. In that chapter he presented quantitative data as "MNI" values, or "the minimum number of individual animals represented by each anatomical part" (Binford and Bertram 1977:79). While one might presume, given this description and Binford's terminology, that his MNI values are calculated like White's (see the discussion of MNI above, and especially the discussion in following paragraphs), they are not. Binford (and Bertram 1977:146) pointed out he was not interested in the quantitative unit signified by the traditional (e.g., Whitean) meaning of MNI, but rather in the survivorship of different skeletal parts (Binford and Bertram 1977) and how humans differentially dismember and transport carcass portions (Binford 1978). Therefore, Binford (and Bertram 1977:146; Binford 1978:70) divided the observed bone count (MNE) for each anatomical unit (such as proximal femur) by the number of times that anatomical unit occurs in one complete skeleton. He was, in effect, standardizing the observed frequencies of all "anatomical units" according to their frequency in one animal in order to monitor how many of each of the various portions of carcasses were represented. Binford (1984b:50) made it clear that his "MNI" values per skeletal portion were not the same as White's MNI values when he "decided to reduce the ambiguity of language by no longer referring to anatomical frequency counts as MNIs" and introduced the term

MAU for his standardized frequencies of skeletal parts. *MAU* stands for the minimum number of animal units necessary to account for the specimens in a collection.

White (e.g., 1953a) can be credited with popularizing the traditional technique of calculating MNI values in North America. He wrote, "the method I have used is to separate the most abundant element of the species found into right and left components and use the greater number as the unit of calculation" (White 1953a:397). That is, how many individual animals are necessary to account for the *combined* humeri, scapulae, mandibles, cervical vertebrae, etc. in a collection of remains representing a taxon. The analytically specified set of faunal remains consists of all of the kinds of skeletal elements included in the skeleton of a taxon. White is, however, also describing a method for deriving an MNI value for *each* paired skeletal part; that is, how many individual animals are necessary to account for the humeri, for the scapulae, for the mandibles, etc. The MNI determined on the basis of the humeri, then, may not be the same as the MNI determined on the basis of the scapulae. In this case the specified set of faunal remains consists of some analytically defined set of anatomically limited categories of faunal remains, such as proximal femora and cervical vertebrae. The latter MNI values, what I call *MNI per skeletal portion* values, are what are of importance here because of their similarity to MAU.

White sometimes listed both MNI values per skeletal portion and the MNI frequencies of both lefts and rights of paired bones (White 1952b, 1953c, 1955, 1956), although he did not consistently do so (White 1952c, 1953b, 1954). He suggested that to "divide [the total MNE of paired elements] by two would introduce great error because of the possible differential distribution of the kill" (White 1953a:397). That is, White suspected some significant within-site distributional data might be masked by calculating what Binford calls MAU values. For example, White (1953c:59) wrote "in most of the features in the sites from which I have identified the bone the discrepancy between the right and left elements of the limb bones was too great to be accounted for by accident of preservation or sampling. This leads one to believe that studies on the distribution of the kill might be profitable." When comparing two assemblages, the analyst should look "for large discrepancies between the [frequencies of] right and left elements. Small discrepancies are not necessarily significant because they might be due to the accidents of preservation or sampling" (White 1953c:61). White did not suggest how the analyst might distinguish between "great" and "small" discrepancies, nor did he study the spatial distribution of left and right elements in a site. In identifying such analytical avenues, however, White was clearly offering an argument to justify how he calculated MNI per skeletal portion values.

White believed hunters, butchers, and consumers might distinguish between the left and right sides of an animal, and butcher, transport, and distribute the

Table 4.3 *Frequencies of pronghorn antelope skeletal portions from site 39FA83. Columns B and C from White (1952b). Column D is the maximum of either B or C. E = (B + C) ÷ 2*

Skeletal part (A)	MNI Left (B)	MNI Right (C)	MNI (D)	MAU (E)
mandible	18	19	19	18.5
pelvis	13	19	19	16
scapula	24	24	24	24
P humerus	3	0	3	1.5
D humerus	26	30	30	28
P radius	28	25	28	26.5
D radius	23	22	23	22.5
P ulna	23	22	23	22.5
P metacarpal	27	11	27	19
P femur	11	6	11	8.5
D femur	6	10	10	8
P tibia	9	9	9	9
D tibia	19	31	31	25
P metatarsal	22	15	22	18.5

two sides of a large mammal differentially, yet he did not attempt to find evidence for this in any of the bone assemblages he studied. Binford's (1978:70) experience with the Nunamiut suggested to him that "hunters make no such discrimination" between left and right sides of large mammal carcasses. This substantiated his belief that investigating how people differentially butcher, transport, and distribute portions of prey carcasses demanded a counting unit which ignored the distinction of left and right elements, and focused on, say, the number of forelimbs versus the number of hindlimbs versus the number of rib-cages that were transported. Binford thus proposed the MAU quantitative unit to fulfill this analytical function. The difference between how White chose to count carcass portions and how Binford chose to count carcass portions is not a trivial distinction. If White is correct in his suspicion that carcasses may have been differentially distributed based on the side of the carcass, then calculating MNI values of both left and right elements and not dividing their sum by two would be more appropriate than deriving MAU values. MAU masks such variation, and MAU values can be easily derived from frequencies of left and right elements; simply sum the lefts and rights, and divide by two. Note that I am not saying MAU is a poor quantitative unit; it is not.

Binford (1978, 1981b, 1984b; Binford and Bertram 1977) typically norms his MAU per skeletal portion values to what are called %MAU values by dividing all MAU values by the greatest MAU value in the assemblage. White (1952b, 1952c, 1953b) did not norm the MNI per skeletal portion values in his early publications, but did in his later ones (1953c, 1954, 1955, 1956) using a

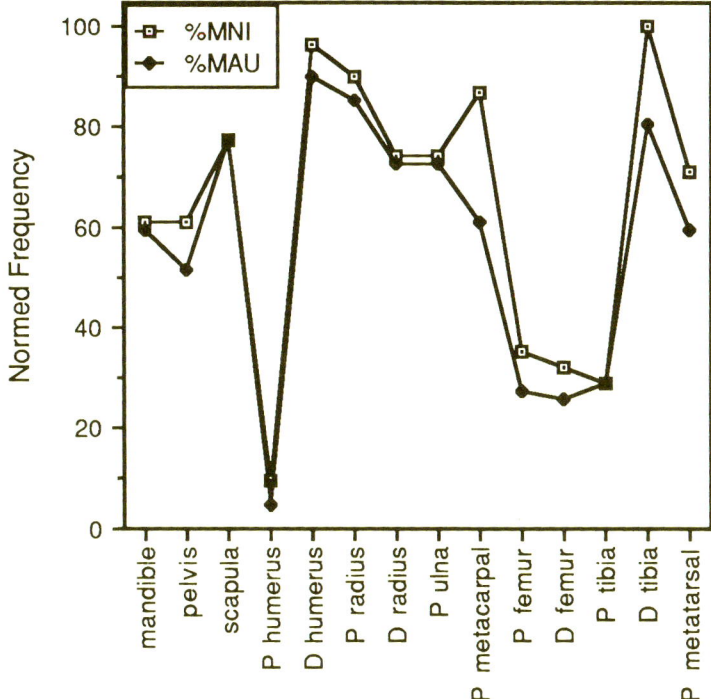

Figure 4.16. Normed MNI per skeletal portion frequencies and normed MAU per skeletal portion frequencies for pronghorn antelope remains from 39FA83 (from Table 4.3).

procedure similar to Binford's. I suspect White normed frequencies to remove the effects of sample size when comparing different sized assemblages. It was not necessary to norm the values when he was describing different assemblages, but in his later publications comparison of assemblages is a major part of his research. White's norming technique was used by other workers for comparative purposes as well (Gilbert 1969; Kehoe and Kehoe 1960; Wood 1962, 1968). Wood (1962) in fact produced graphs of normed Whitean MNI values much like the graphs used by Binford (1978, 1981b, 1984b; Binford and Bertram 1977), except Binford graphed MAU values. Such a graph, using both Whitean MNI values and Binfordian MAU values, is presented in Figure 4.16. Data for that graph are based on the pronghorn antelope (*Antilocapra americana*) remains recovered from site 39FA83 and reported by White (1952b) (Table 4.3).

The graph in Figure 4.16 is of historical interest and, more importantly, it underscores the difference between the MNI and MAU quantitative units when they are used to measure frequencies of skeletal portions. For example, using the data from 39FA83, Figure 4.17 shows that the MAU values for skeletal portions tend to be less than the MNI values for skeletal portions (all

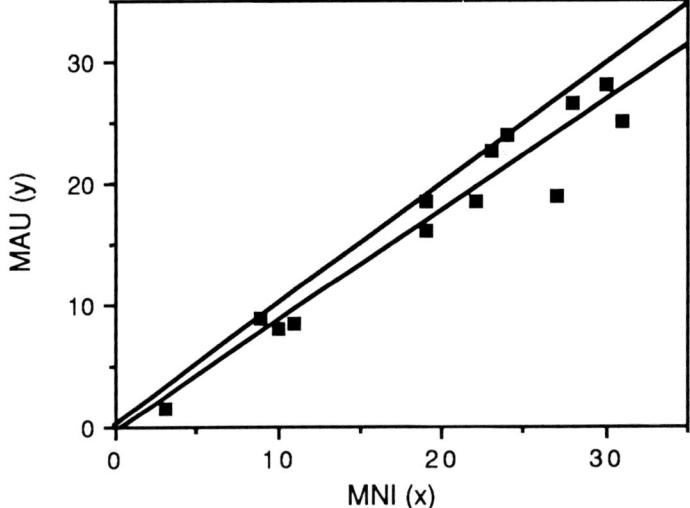

Figure 4.17. Bivariate scatterplot of MNI per skeletal portion frequencies and MAU per skeletal portion frequencies for pronghorn antelope remains from 39FA83 (from Table 4.3). Lower line is the simple, best-fit regression line; upper line is diagonal (origin of 0, slope of 1).

plotted points fall below the diagonal line). This is the predictable result when the frequencies of left and right elements differ. Interestingly, the simple, best-fit regression line through the point scatter ($y = -0.2931 + 0.9018x$; $r = 0.96$, $P = 0.0001$) has a slope < 1 and suggests that as frequencies of skeletal portions increase the difference between MNI and MAU values increases.

Figure 4.18 illustrates the differences between frequencies of left (total $= 252$) and right (total $= 243$) elements of pronghorn antelope in the 39FA83 collection. The simple, best-fit regression line ($y = 2.7837 + 0.8096x$; $r = 0.72$, $P < 0.01$) has a slope < 1, suggesting increasingly greater differences between the frequencies of left and right elements as frequencies of elements increase. If left and right elements were consistently of equal or near-equal frequencies, the best-fit line would be a diagonal line ($y = 0 + 1x$). Herein lies one way to search analytically for what White characterized as "discrepancies" in the frequencies of left and right elements. If points above the diagonal line in Figure 4.18 represent bones from one archaeological context, and points below the diagonal line represent bones from another context, then intra-site differential distribution of the kill possibly occurred. But, are the differences between the frequencies of left and right elements significant, and if so, are such differences found for all paired elements?

To address the preceding question, I calculated adjusted residuals for each category of skeletal part (Table 4.4; see Everitt 1977 for a description of the procedure; adjusted residuals are read as standard normal deviates). They suggest two, or perhaps three of the skeletal parts occur in abundances

Table 4.4 *Observed and expected (in parentheses) MNI frequencies of pronghorn antelope skeletal portions from site 39FA83, and adjusted residuals and probability values for each skeletal portion*

Skeletal part	MNI Left	Adjusted residual	*p*	MNI Right	Adjusted residual	*p*
mandible	18(18.8)	−0.26	0.397	19(18.2)	0.28	0.390
pelvis	13(16.3)	−1.21	0.113	19(15.7)	1.22	0.111
scapula	24(24.4)	−0.12	0.452	24(23.6)	0.12	0.452
P humerus	3(1.5)	1.76	0.039	0(1.5)	−1.76	0.039
D humerus	26(28.5)	−0.71	0.239	30(27.5)	0.73	0.233
P radius	28(27)	0.29	0.386	25(26)	−0.30	0.382
D radius	23(22.9)	0.03	0.492	22(22.1)	−0.03	0.488
P ulna	23(22.9)	0.03	0.492	22(22.1)	−0.03	0.488
P metacarpal	27(19.3)	2.61	0.004	11(18.7)	−2.62	0.004
P femur	11(8.7)	1.13	0.129	6(8.3)	−1.16	0.123
D femur	6(8.1)	−1.07	0.142	10(7.9)	1.09	0.138
P tibia	9(9.2)	−0.10	0.460	9(8.8)	0.10	0.460
D tibia	19(25.5)	−1.95	0.026	31(24.5)	1.98	0.024
P metatarsal	22(18.8)	1.09	0.138	15(18.2)	−1.10	0.136

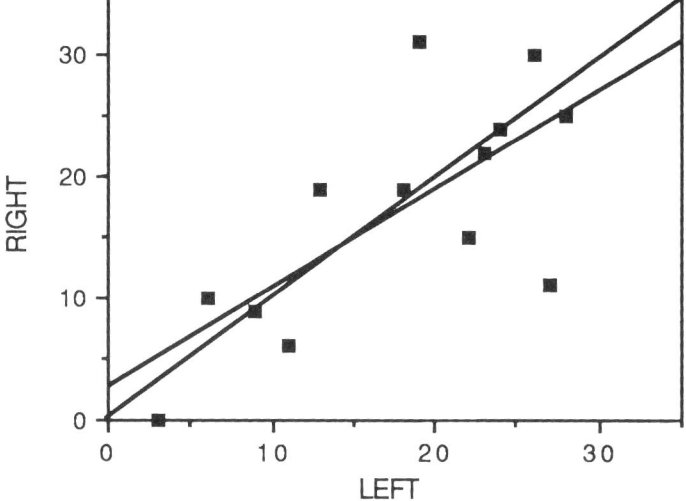

Figure 4.18. Bivariate scatterplot of MNI per skeletal portion frequencies for left and right skeletal portions of pronghorn antelope from 39FA83 (from Table 4.3). Diagonal line has a 0.0 origin on the y axis; simple, best-fit regression line has an origin of 2.78.

significantly different from that expected given random chance. There are more left proximal metacarpals and more right distal tibiae (and fewer right proximal metacarpals and fewer left distal tibiae) than chance alone allows. There may also be more left proximal humeri (and fewer right proximal humeri) than chance alone allows, but there are so few of this skeletal portion that I worry that sample size may be influencing the result. Regardless, it is clear that in this sample the analyst may want to examine the associational contexts of proximal metacarpals and distal tibiae to determine if, as White suggested, there is evidence for differential distribution of the kill. And, analysis of adjusted residuals provides another way to search analytically for White's "large discrepancies" in frequencies of left and right elements.

The importance of comparing frequencies of skeletal portions as measured by Binford's MAU and White's MNI is that the two units measure different properties of a bone collection. Thus, correlations of skeletal part frequencies with either Binford's (1978) economic utility indices or a measure of the survival potential of skeletal parts (e.g., Lyman 1984a) may be significantly influenced by the counting unit used (see Chapter 7). Again turning to the 39FA83 data, we find the following Spearman's rho correlation coefficients:

Quantitative unit	Caribou MGUI	Bone density
MNI $r_s =$	0.73	−0.51
$P =$	0.008	0.06
MAU $r_s =$	0.64	−0.53
$P =$	0.02	0.053

This set of coefficients follows the tradition of using ordinal scale statistics when comparing bone frequencies with a utility index or the structural density of a bone part (see Chapter 7 for more complete discussion). While the coefficients and probability values do not differ greatly whether one uses the MNI values or the MAU values, variation between the coefficients underscores the fact that the quantitative unit used can influence these kinds of statistical results. Clearly, MAU is the unit of choice for both correlations as it more accurately measures the relative frequencies of skeletal parts than MNI, especially when differences between frequencies of left and right elements are great. Two left humeri represent an MAU of 1, but an MNI of 2. When one is interested in determining if frequencies of skeletal parts are the result of differential transport or differential destruction, the number of individual animals is irrelevant; whether more humeri or more tibiae are represented is paramount. And, because not all bones of the skeleton are paired, the frequencies of skeletal parts must be weighted in order to assess accurately which skeletal parts are abundant and which are rare, compared to their relative abundances within a complete skeleton. Weighting is accomplished by dividing the observed frequency of each skeletal part by the expected frequency (the maximum possible frequency if all were present).

Discussion

Both how quantitative units are defined and how they are operationalized must be explicit in order to ensure concordance between the counting units used and the research question addressed with those units. The ultimate concern is that the analyst makes clear what is being counted, how it is being counted, and why specimens are counted that way. Part of the key to producing reliable and valid quantitative measures resides in the accurate definition of target populations and sample populations. Zooarchaeologists interested in determining paleo-environmental conditions from faunal remains require a measure of the fauna that was extant at the time of site occupation whereas zooarchaeologists interested in determining prehistoric human subsistence need to measure the fauna that was killed or harvested by human hunters (Lyman 1982a:337). A. Turner (1983:312–313) made this point when he distinguished between the excavated sample, the killed population, and the living population of animals. The first is the set of faunal remains recovered by the archaeologist from a site; the second is the set of animals procured by the prehistoric occupants of the site; the third is the fauna extant at the time the site was occupied. Turner was concerned with estimating taxonomic abundances within extant faunas (target population) on the basis of excavated samples (sample population), and he made it abundantly clear that the two were not necessarily correlated, in part because the killed population (archaeologically sampled population) need not be a random sample of the extant fauna.

Brewer (1992:207) distinguishes a *target population*, the group of things the analyst wishes to make inferences about, from a *sample population*, what the analyst works with and what serves as the basis of one's inferences; the "sample population must be relevant to the target population, which in turn must be defined by the questions being asked." That is, there must be concordance between the hypothesis being evaluated, the analytic techniques used, and the counting units which are analyzed. It is, for example, the lack of concordance between MNI measures of fossil taxonomic abundances and taxonomic abundances in prehistorically extant faunas that has contributed to this quantitative unit's fall from analytical favor. Likewise, it is the lack of concordance between NISP measures of skeletal part frequencies resulting from differential fragmentation of skeletal elements and actual frequencies of skeletal parts that resulted in the introduction of MNE. We must be explicit about why we have chosen the quantitative unit we have used in our analyses (Lyman 1982a:361). In other words, *we must specify how the quantitative units we use to measure the sample population relate to the quantitative properties of the target population we wish to infer*. Such specification should help us determine if the quantitative unit we have chosen is the appropriate one.

Which counting unit should one use, and how should that unit be operationalized, in a particular situation? Sometimes it is clear (e.g., Badgley 1986a), sometimes it is not. Here is where a major, two-step research effort should be directed, as it may not be clear which of several units is the appropriate one to use. The first step is to specify explicitly the target population one is trying to measure. That is, what is the quantitative property of interest? Such specification should provide clues to an appropriate sample population and quantitative units that give accurate measurement of the target population's properties. This first step may be all that is necessary to a particular analytical problem. If it is not, the second step involves actualistic studies geared toward determining which of the analytical units (and how it is actually counted) most clearly, consistently, and unambiguously reveals archaeologically detectable patterns that reflect the property of interest. Such research will allow some analytical units to be utilized as interpretive units. Brain (1969), for example, detected the important unit of measurement now called %survival (or, %survivorship) during ethnoarchaeological research, and that unit is regularly used today to help explain frequencies of skeletal portions in archaeological assemblages (Chapter 7). Stiner (1991b, 1991e) likewise described several units of measurement that allow the identification of taphonomic processes that have influenced archaeological assemblages. Both Brain and Stiner had particular research questions in mind, questions that suggested the kinds of counting units that logically should be used. Neither operated in a strictly inductive or blindly empirical mode, and, importantly, they both were sufficiently clear about how they defined their counting units and how they operationalized them that others can replicate and test their results, as well as evaluate the appropriateness of those units for measuring specific properties.

Summary

[An] approach to the practice of taphonomy is to enumerate and then explain the differences between fossil collections and living communities of animals.
(A. Hill 1988:563)

A vertebrate skeleton is a complex entity. Its structure, from microscopic to macroscopic levels, can have significant influences on the effects taphonomic processes have on its constituent parts (the skeletal elements). Further, how those constituent parts are counted during analysis is a complex matter. If Andrew Hill's statement quoted above is correct, then clearly quantification is important to taphonomic analysis. How many and which skeletal elements of ontogenically young individuals are present in an assemblage? Are those values different from the values observed for ontogenically old individuals? Have antlers and teeth been modified by taphonomic processes in such a manner as to alter their abundances relative to the abundances of limb bones? Were vertebrae fluvially transported but carpals not so transported? Are femora broken but phalanges not broken? Answers to these and similar questions consist of a large part of the taphonomic data researchers record and analyze as they try to unravel, understand, and write taphonomic histories of assemblages of vertebrate remains. And, they all depend on quantitative data of one form or another. With this chapter as background to the basic issues, it is, then, to describing how one produces answers to such questions that we turn in the remainder of this volume.

5

VERTEBRATE MORTALITY, SKELETONIZATION, DISARTICULATION, AND SCATTERING

> Klähn has made a sharp distinction between the two main groups of causes of death; dying and being killed. By *dying* he means normal death due to old age or sickness. By *being killed* he refers to vigorous individuals that become victims of accident, enemies, or the forces of nature.
> (J. Weigelt 1927/1989:21)

Introduction

Taphonomy is concerned with the differences between what the paleontologist or zooarchaeologist lays out in the laboratory for study, and, variously, the biotic community and/or individual animals represented by that laid-out material. In a way, taphonomic histories begin with the death of an organism. This is not exactly true, although it is precise given most definitions of taphonomy (see Chapters 1–3). It is not exactly true because the behavioral patterns, ecological predilections, and life history of an organism may influence the mode of mortality and the taphonomy of that organism's carcass. As a simple example, terrestrial vertebrates have different taphonomic histories than aquatic vertebrates simply due to the different medium in which they normally die. Knowing something about the behaviors, ecology, and lives of the organisms whose remains are being studied can thus be a great benefit to the taphonomist.

In this chapter we explore the various ways animals die and are killed, how those modes of death might influence subsequent episodes in the taphonomic history of a carcass, and some analytic techniques used to determine prehistoric modes of death. We also explore how mortality influences age and sex demographic parameters indicated by fossils. There are many ways to die. Intimately related to these modes of death are whether the remains being studied represent an active or a passive accumulation (Chapter 6), a synchronic or diachronic accumulation (Chapter 6), the number of organisms represented, and the demographic properties of those dead organisms. In this chapter I introduce the subjects of skeletonization of carcasses and scattering of bones, variables that are intimately related to the mode of death.

114

Modes of death

> An individual animal may die accidentally, of old age, as a victim of parasites or other enemies, from lack of food or as a consequence of external forces.
> (W. Schäfer 1962/1972:9)

Causes of death of organisms are many and varied, especially when compared to the limited number of ways the life of an organism might be initiated. This is because of the accidental factor in mortality. Weigelt (1927/1989) provides the following list of accidental modes of death:

1. death due to volcanic activity;
2. death due to poisonous gases;
3. death due to fire;
4. death by drowning;
5. death by becoming mired in mud, quicksand, oil, or asphalt;
6. death due to flooding;
7. death due to fluctuation in salinity of water;
8. death due to drought;
9. death due to overpopulation (malnutrition);
10. death due to hunting (predation);
11. death due to freezing;
12. death due to falling through ice.

Other accidents include falling from high places and intraspecific activities such as competition between males for breeding privileges leading to the death of one or both of the combatants or the death of a juvenile who got in the way. All of these have a source external to the organism that dies in contrast to disease or old age the source of which is relatively internal to the organism.

Berger (1991:127) suggests that "animals with 'normal' lives rarely make fossils [and thus] preservation of the unlucky is the norm in paleontology." Those unfortunate individuals that die due to an accident are the ones that typically experience a taphonomic history conducive to preservation in contrast to those that die of old age. This "survival of the unlucky" (Berger 1991:127) may be true, but it may also be trivial because it is unusual indeed when all members of an age group die of senility. In the examples described below, typically less than half of an age group die of old age; more than half die of other causes. Thus we have a better chance of finding the remains of an organism that died accidentally than the remains of a senile individual simply because more of the former than of the latter enter the fossil record.

The demography of mortality

> Among the tools developed by neoecologists, that of the life table analysis may at least confer upon paleoecology a welcome addition to exact methodology.
> (B. Kurtén 1953:47)

Modes of death can be categorized as density dependent or independent, age dependent or independent, and occasionally as sex dependent or independent.

The latter two often depend on and have their greatest effect when different age cohorts or sex groups display different behaviors. Mortality factors such as predation, disease transmission, and competition (perhaps leading to malnutrition) are dependent on the density of the population being affected; their effects on mortality increase in magnitude as population density increases. The young are the expendable part of the population, and the rates of their survival and recruitment influence population density. Such population parameters of mortality lead us to consider first the demographics of mortality.

The basics

Study of population parameters of fossil taxa has been undertaken in detailed, systematic fashion for at least four decades (e.g., Kurtén 1953, 1958; Voorhies 1969). Theoretical aspects of the demographic characteristics of vertebrates, especially mammals, are now rather refined (Barlow 1984; Caughley 1966, 1977; Craig and Oertel 1966; Czaplewski *et al.* 1983; Deevey 1947; Polacheck 1985). As well, techniques of mortality analysis are being applied to more and more vertebrate taxa in paleontological and archaeological settings, and mortality in extant populations continues to be studied (Berger 1983; Coe *et al.* 1980; Cribb 1985, 1987; Korth and Evander 1986; Lyman 1989a, 1991a, 1991b; Stiner 1990b, 1991a). In fact, a recent edited volume is devoted specifically to analyzing and interpreting vertebrate mortality patterns as evidenced by archaeofaunas (Stiner 1991c).

A *cohort* is a group of individual organisms that were born simultaneously (Caughley 1977:85). Cohorts can include all individuals born during a particular day, week, month, season, or, typically, a solar year. Because death is continuous throughout the existence of a cohort, regardless of how old the individuals are, there are always fewer individuals in each succeedingly older age class than in the immediately preceding age class; that is, using the typical time unit of one year, with the passing of each year there are fewer individuals still alive in a cohort. Real cohorts (all individuals born in one year) are seldom studied by biologists or paleontologists; rather, techniques of demographic analysis described here are usually applied to a population of individuals of all ages as if that population was made up of the demographic history of a cohort from the birth of that cohort's first individual to the death of its last individual.

In the following, it is presumed that a relatively accurate technique for assessing the ontogenic age of modern and fossil specimens has been used; that is, a technique allowing determination of an individual's age at death within a maximum ± 2 month period has been employed. Ageing techniques are not described here, but many such techniques exist and are described in many of the references cited in this section.

A cohort's mortality pattern is typically and formally presented in the form of a "life table." An example of a life table for the Himalayan thar (*Hemitragus*

Table 5.1 *Life table for female Himalayan thar (from Caughley 1966)*

Age	Frequency	Survival	Mortality	Mortality rate	Survival rate
x	f_x	l_x	d_x	q_x	p_x
0	205	1.000	0.533	0.533	0.467
1	96	0.467	0.006	0.013	0.987
2	94	0.461	0.028	0.061	0.939
3	89	0.433	0.046	0.106	0.894
4	79	0.387	0.056	0.145	0.855
5	68	0.331	0.062	0.187	0.813
6	55	0.269	0.060	0.223	0.777
7	43	0.209	0.054	0.258	0.742
8	32	0.155	0.046	0.297	0.703
9	22	0.109	0.036	0.330	0.670
10	15	0.073	0.026	0.356	0.644
11	10	0.047	0.018	0.382	0.618
12	6	0.029			

jemlahicus) is given in Table 5.1 (from Caughley 1966). The table is constructed as if the individual animals were born on the same day and the numbers surviving to each subsequent birthday were recorded. Table 5.1 follows biological tradition and lists only the females, as they are the source of new individuals in the population; paleontologists typically list all ageable individuals regardless of sex. By column, x is the age of the cohort by one-year intervals. The number surviving in each successive year is listed under f_x. The probability that an individual of a particular age will survive to its next birthday is given under l_x; this value is calculated by dividing the respective f_x value for an age group by the original size of the cohort, in this case 205. The fourth column, d_x, lists the probability of dying during the age interval x to $x + 1$. Thus the probability that a newborn thar in this cohort will die before reaching its first birthday is calculated as the probability that it will survive when born (1.000) minus the probability of survival when it is one year old (0.467), or $1.000 - 0.467 = 0.533$. By convention d_x values are plotted on the row for the beginning of the age interval. The rate of mortality, q_x, is calculated as d_x/l_x and is the proportion of animals alive at age x that die before reaching age $x + 1$. The final column lists survival rates, symbolized as p_x. These are the proportion of animals alive at age x that survive to age $x + 1$ and are the compliments of their respective q_x values, or $1 - q_x$.

Life tables have seldom been published for archaeofaunal assemblages, perhaps because their value to taphonomic analysis is not evident. Koike and Ohtaishi (1985) and Lyman (1987c) independently showed that the data contained in a life table can be useful for detecting predatory pressure on a population and resultant changes in its demographic structure. Knowing that a taxon's behavior is age specific, such as juveniles forming separate groups away from adults, the demographic structure of a fossil population of that taxon might be accounted for by referral to that behavior.

Typically, taphonomists look at age–frequency distributions. In such cases, the number of dead individuals per age class is tallied (usually as MNI values) and all are plotted in a histogram, each vertical bar representing an age class and the height of a bar being scaled to the frequency of individuals per age class. Two basic types of age–frequency distributions, or what are often called *mortality profiles*, are recognized by paleontologists (Hulbert 1982; Kurtén 1983; Voorhies 1969). Zooarchaeologists have further distinguished several variants of the two basic patterns to aid in their interpretation of food-getting practices of prehistoric people (Klein 1982a; Levine 1983; Stiner 1990b, 1991a). One basic mortality type is referred to as "attritional" or "normal" mortality. It is modeled as a frequency distribution of age classes in which very young and very old individuals are overrepresented relative to their live abundances, and reproductively active adults are underrepresented because of varying mortality rates across age classes (Craig and Oertel 1966). Attritional mortality is selective. Those age classes most susceptible to natural ecological mortality, such as juveniles too young to flee predators or escape traps and senile individuals too weak to escape predation or survive less than optimal environmental conditions, are more prone to die whereas healthy adults in their prime reproductive years are less prone to die. This susceptibility to mortality results in a bimodal frequency distribution, with one mode to the far left and the other to the right of center (Figure 5.1b); the frequency distribution is often referred to as "U-shaped" (e.g., Klein 1982b:53). Attritional mortality results from normal or routine ecologically related (accidental) deaths of population members. The model of attritional mortality given in Figure 5.1b assumes that mortality is slow and reflects the rate of biomass turnover, and is said to illustrate "a balanced picture of a fauna as it existed in nature" over time (Voorhies 1969:23).

The second basic mortality type is referred to as a "catastrophic" or "mass" mortality pattern. It is modeled as a frequency distribution of age classes in which successively older age classes are represented by fewer and fewer individuals; in other words, the frequency distribution is unimodal with extreme positive skewing (Figure 5.1a) and is referred to as "L-shaped" (e.g., Klein 1982b:53). The choice of the label "catastrophic" for the L-shaped mortality profile is founded in the fact that mass, or non-selective, mortality produces a synchronic "snapshot" (Voorhies 1969) of a population's age structure at the time of death, but as noted below, this label may be a poor choice. Theoretically, because it is non-selective, catastrophic mortality will result in proportionally more prime-age adults dying than attritional, selective mortality (Voorhies 1969:46–47). Natural catastrophic mortality events include floods (Boaz 1982), droughts (Conybeare and Haynes 1984; Haynes 1984, 1987, 1988a, 1988b), and volcanic eruptions (Lyman 1987c, 1989b);

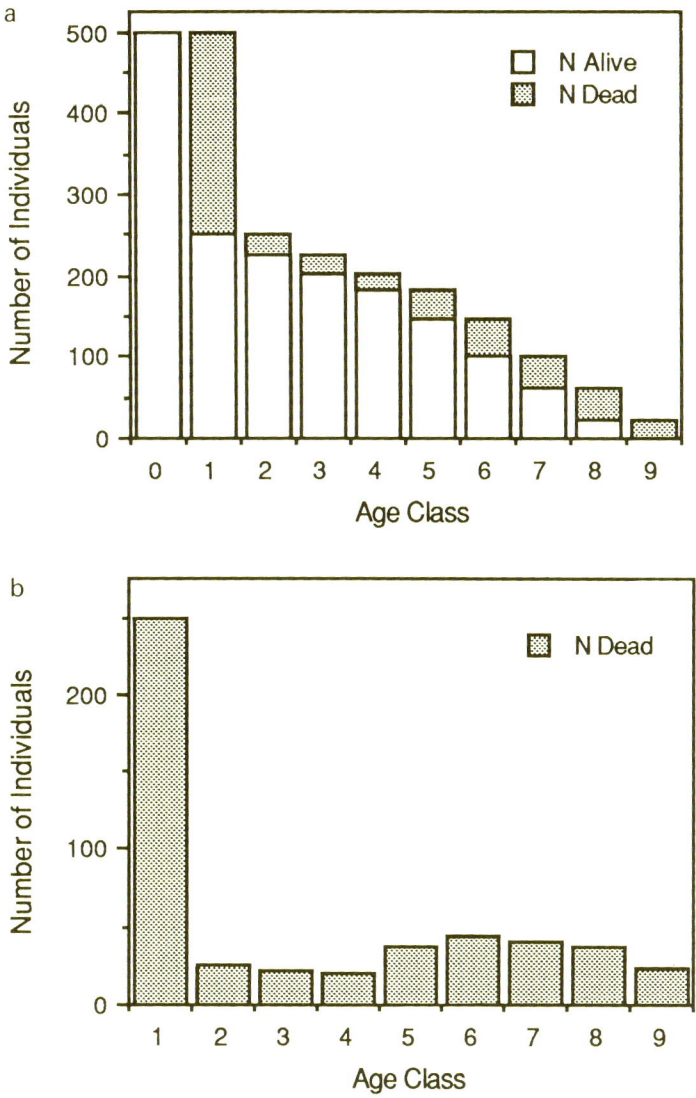

Figure 5.1. Two basic types of age (mortality) profiles (from Table 5.2A). a, blank bars denote the number of individuals alive in a cohort each year during the existence of the cohort, and stippled bars denote the number that must die each year (in each age class); b, the number that die each year (in each age class).

archaeological equivalents of catastrophic mortality can be found in jump sites and traps where large numbers of animals were killed simultaneously (e.g., Levine 1983; Reher 1974).

What are typically called catastrophic and attritional mortality are, from a demographic and an ecological perspective, interrelated. An instantaneous event that causes the synchronous death of all members of a population produces an age–frequency distribution of those individuals who have, to that

Table 5.2 *Life tables for two hypothetical*
populations of mammals. For A, 500 females give
birth once a year; for B, 500 females give birth twice
a year or all have twins (from Klein 1982b)

	Age	Number alive	Number dead	Mortality rate
	x	l_x	d_x	q_x
A.	0	500	250	0.50
	1	250	25	0.10
	2	225	22	0.10
	3	203	20	0.10
	4	183	37	0.20
	5	146	44	0.30
	6	102	41	0.40
	7	61	37	0.60
	8	24	24	1.00
	9	0	0	
B.	0	1000	700	0.70
	1	300	150	0.50
	2	150	75	0.50
	3	75	30	0.40
	4	45	9	0.20
	5	36	7	0.20
	6	29	6	0.20
	7	23	12	0.50
	8	11	11	1.00
	9	0	0	

instant, survived more routine or attritional mortality (Klein 1982b:58). "The perfect mass-mortality population is a 'frozen' living population" (Craig and Oertel 1966:351). Thus the two frequency distributions in Figure 5.1 could derive from the same population of animals. An attritional age–frequency distribution or mortality pattern can be derived from the age–frequency distribution for the living population (or the "catastrophic" mortality profile) "simply by graphing only the loss of individuals for each age group" (Korth and Evander 1986:228). Thus the stippled bars in Figure 5.1 represent deaths in a population over some multi-year time period while the blank bars represent the age structure of the living population at any one point in time during that time interval (presuming, as we shall see, that the population was of stable size and makeup during the time interval).

But there are complicating factors in the relation between the frequency distributions (blank versus stippled bars) in Figure 5.1. In particular, recruitment and mortality rates must be off-setting if a population is to remain stable in size through multi-generational (usually multi-year) time. As well, if the age structure of the population is to remain unchanged through time, mortality must be distributed among the age classes in such a manner as to ensure a stable age structure. Klein (1982b) illustrates how mortality rates must change in the face of rapid recruitment rates (or vice versa) in order to maintain a stable population size and structure. In a population having 500 births per year and maximal longevity of 9 years, 500 deaths per year must occur with the distribution shown in Table 5.2a and Figure 5.1. If that population of 500 individuals has 1000 births per year, such as all females giving birth to twins, then 1000 deaths must occur per year with the distribution shown in Table 5.2b and Figure 5.2. Comparing Figure 5.1 and Figure 5.2 shows how increased recruitment suppresses the mode signifying the death of old individuals past their prime, producing an age–frequency distribution with more of an "L-shape" (such an effect could be enhanced by making the scale on the y-axis larger). While the data show the frequencies of deaths are "U-shaped," this exercise underscores that simple inspection of the shape of an age–frequency distribution like Figure 5.2 can be misleading.

Given the above complicating factors, it is not surprising that paleontologists and zooarchaeologists have used a set of criteria to help determine whether, for example, an L-shaped age–frequency distribution derived from fossil remains actually represents an instance of catastrophic (age-independent) mortality, or an instance of attritional mortality. The shape of the age–frequency distribution is one of those criteria, but it cannot be the only one. Other criteria used to identify the kind of mortality pattern represented include the discreteness of age classes, taphonomic information, and paleoecological data (Klein 1982b:58–59). By discreteness of age classes I mean the observed range of variation in ontogenic ages around age class mid-points. The discreteness of age classes criterion applies only to those taxa with seasonally restricted birthing seasons (Craig and Oertel 1966:351). This is so because if births occur throughout the year, then a cohort would consist of individuals of all ages (any month) within a year. If births are temporally restricted to a month or season, then all individuals in a cohort will be within a month or two of exactly the same age. For a seasonally birthing taxon and age classes that each represent one year, a range of variation in time of death of ±5 months per age class suggests mortality was attritional as individuals were dying virtually all year long. A range of variation in time of death of ±1 month suggests a rather limited time during the year when animals died and that mortality was catastrophic.

As with the shape of the age–frequency distribution, age–class discreteness should not be used alone. That is so because individuals in a population might be dying throughout the year, or attritionally, yet the locality sampled may represent a seasonally restricted accumulation of those remains, such as around water-holes used only during the dry season. Some examples will make the importance of this clear.

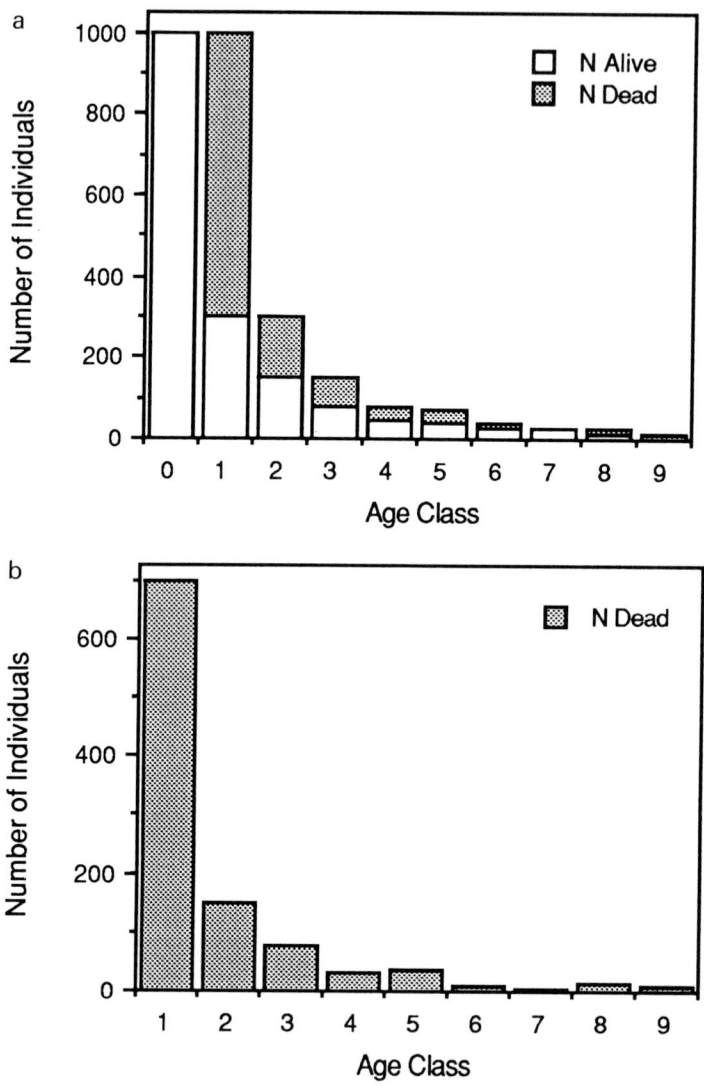

Figure 5.2. Age (mortality) profiles for a population with high mortality and recruitment (from Table 5.2B). a and b as in Figure 5.1.

Hulbert (1982) describes a set of fossil three-toed horse (*Neohipparion* cf. *leptode*) remains the age–frequency data for which indicate mortality was attritional (Figure 5.3) and the age classes for which are ±0.35 years. The lack of relatively non-discrete age classes and the U-shaped mortality profile, once the apparent under-representation of the youngest age class is accounted for, suggest mortality was attritional. Kurtén (1983) describes a set of fossil antelope (*Pachytragus solignaci*) remains from Tunisia the age–frequency distribution of which suggests mortality was catastrophic (Figure 5.4) yet the

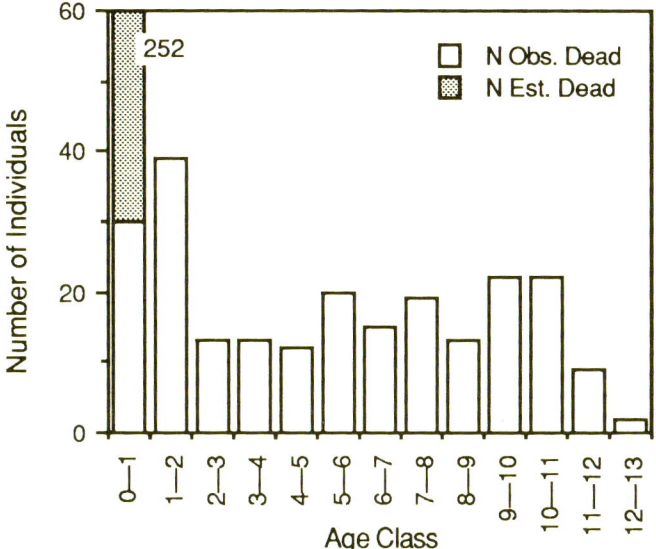

Figure 5.3. Mortality profile for fossil horses (data from Hulbert 1982). Age classes are one year each. Blank bars are frequencies observed in fossil record; stippled bar is estimated.

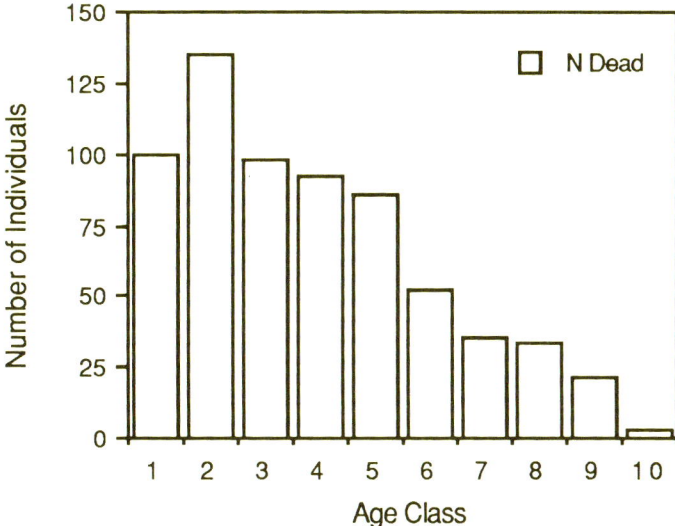

Figure 5.4. Mortality profile for fossil antelope (data from Kurtén 1983). Age classes are one year each.

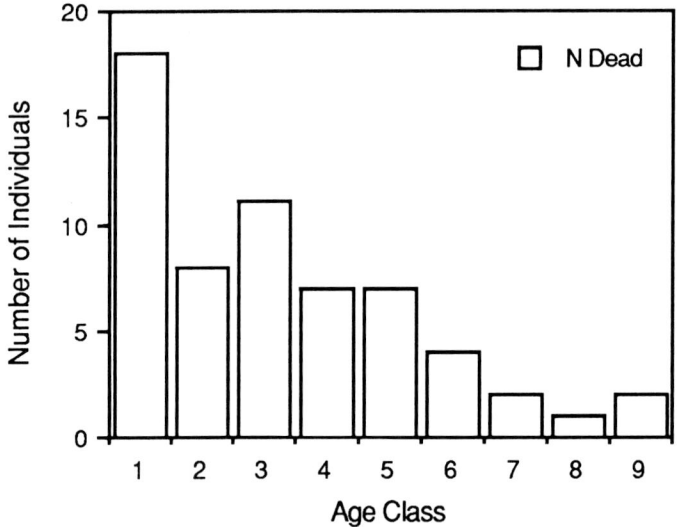

Figure 5.5. Mortality profile for archaeological deer remains (data from Simpson 1984). Age classes are one year each.

lack of discrete age classes and geological data indicate the sample resulted from seasonal sampling of a population undergoing attritional mortality. Similarly, in her study of mule deer (*Odocoileus hemionus*) remains from an archaeological site, Simpson (1984) found an L-shaped age–frequency distribution (Figure 5.5) but argued that because the age classes were not discrete (were ± 3 months), the remains accumulated during one season over multiple years. Nimmo (1971) derived a U-shaped age–frequency distribution (Figure 5.6) for an assemblage of pronghorn antelope (*Antilocapra americana*) remains, but argued that because the age classes were relatively discrete (± 1 month), the taxon had seasonally-restricted birthing, and the fossil remains were recovered from an archaeological kill site, mortality was catastrophic.

More than the shape of the age–frequency distribution derived for a set of fossils must be considered if one wishes to infer whether mortality was "attritional" or "catastrophic." And the latter term may well be a misnomer if it is allowed to denote whether mortality was diachronic or synchronic rather than just the shape of the mortality profile described by an age–frequency distribution. But the shape of such frequency distributions is a logical place to start one's analysis. Klein (e.g., 1982a; Klein and Cruz-Uribe 1984:57–60) suggests using the Kolmogorov–Smirnov two-sample D statistic as a way to determine whether the shape of an age–frequency distribution approximates an L-shaped or a U-shaped frequency distribution. The procedure is to take an age–frequency distribution for a known mortality event, whether attritional or catastrophic, and compare the cumulative percentage distribution of age classes in that distribution with that of the population of unknown mortality. An example will make this clear.

Table 5.3 *Observed and expected frequencies of wapiti from catastrophic mortality resulting from volcanic eruption of Mount St. Helens (from Lyman 1987c)*

Age Class	Number of Dead Observed (Cum. %)	Number of Dead Expected (Cum. %)	Difference (D)
0	13 (0.157)	19 (0.275)	0.118[a]
1	20 (0.389)	14 (0.478)	0.089
2	11 (0.517)	10 (0.622)	0.105
3	14 (0.68)	8 (0.738)	0.058
4	18 (0.889)	6 (0.825)	0.064
5	3 (0.924)	4 (0.883)	0.041
6	1 (0.934)	3 (0.926)	0.008
7	2 (0.957)	2 (0.955)	0.002
8	1 (0.967)	2 (0.984)	0.017
9	1 (0.977)	1 (1.00)	0.023
≥10	2 (1.00)	0 (1.00)	0.000
Total:	86	69	

Note:
[a] greatest D.

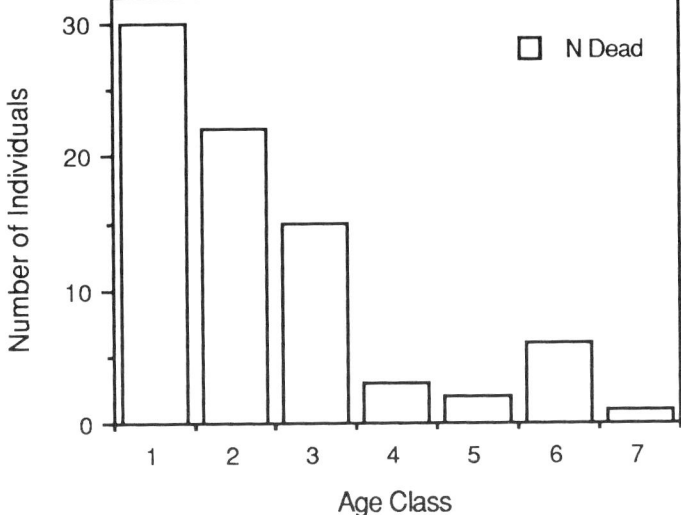

Figure 5.6. Mortality profile for archaeological pronghorn antelope remains (data from Nimmo 1971). Age classes are one year each.

The mortality of wapiti (*Cervus elaphus*) resulting from the volcanic eruption of Mount St. Helens in the state of Washington in May of 1980 produced the known catastrophically-generated age–frequency distribution described in Table 5.3 (Lyman 1984b, 1987c, 1989b). Because the sample is not representative of the extant wapiti population due to behavioral variation between different age–sex groups (Lyman 1987c, 1989b), the observed number of

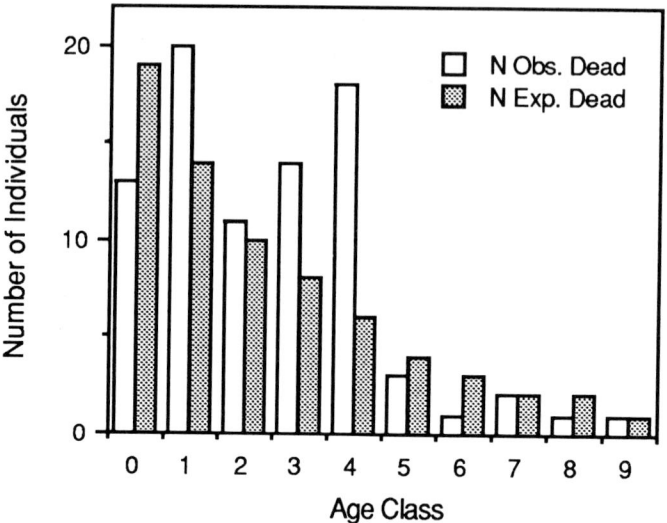

Figure 5.7. Expected and observed mortality profile for wapiti killed by the volcanic eruption of Mount St. Helens (from Table 5.3). Age classes are one year each.

individuals described per age class was smoothed using polynomial regression (Caughley 1977) to produce the expected number of individuals per age class given in Table 5.3. The frequency distribution of the expected number of dead individuals per age class is clearly catastrophic and differs visually from that for the observed number of individuals. But is that difference significant? Cumulative percentages were calculated for each age–frequency distribution and the differences between those percentages by age class or row were derived (Table 5.3). The greatest difference, or *D* statistic, is 0.118 and falls in the youngest age class. That observed *D* is less than the tabled *D* of 0.22 when *P* = 0.05. Thus, the Kolmogorov–Smirnov test suggests there is no statistically significant difference between the two frequency distributions in Figure 5.7. If, for example, these were both fossil assemblages, one would conclude, on the basis of the Kolmogorov–Smirnov two-sample *D* statistic, that they both represented catastrophic mortality events (on the basis of mortality profile shape and the discrete age classes) even though the frequency distributions appear different.

Interpreting the demographics of mortality

Thus far the discussion has concerned identifying two basic types of mortality profiles originally identified in paleontology. As noted above, zooarchaeologists have expanded the number of mortality profile types (e.g., Levine 1983), in part during attempts to distinguish wild prey animals from domestic prey animals (e.g., Collier and White 1976; Wilkinson 1976). We still have much left to learn in this particular regard (e.g., Cribb 1985, 1987), yet the demographics of mortality remain an important taphonomic clue. This is in spite of the fact that "the cause of death does not necessarily indicate whether the pattern of mortality is attritional or catastrophic" (Shipman 1981b:18). Similarly, as we have seen in conjunction with the discussion of Figures 5.3 through 5.6, whether the pattern of mortality is attritional or catastrophic does not necessarily indicate the cause of death or whether mortality was diachronic or synchronic.

Stiner (1990b, 1991d) emphasizes, correctly I think, that until we know and understand the modern range of variation in mortality patterns, we will have difficulty understanding what various mortality patterns signify when identi- fied in prehistoric contexts. She suggests that we therefore must study that modern variation in conjunction with the ecologies and behaviors of both predator and prey species. For example, Lyman (1989b, 1991a, 1991b) reviews the modern behaviors of pinniped taxa during the breeding and birthing season and how escape behaviors, or lack thereof, resulted, in the case of one taxon, in the death of many adult breeding-age males that defend their breeding territories against all comers and few deaths of other age–sex classes. In the case of another pinniped taxon, the adult males and females of which tend to flee at signs of danger, mostly helpless and relatively naive and immobile newborns were killed by human hunters. Hudson (1991) shows how procurement technology exerts similar influences on the age–sex structure of the killed population of a small cervid. Based on neotaphonomic data, Hockett (1991) argues that assemblages of leporid (rabbits and hares) bones accumulated by raptors will be dominated by remains of juveniles. Human hunters, in contrast, create leporid bone assemblages with relatively high frequencies of remains of adults. This difference exists because of an upper size limit of individual prey that raptors can effectively prey upon (Hockett 1991). Similar selection of younger age classes of prey has been documented for some mammalian carnivores (e.g., Andrew and Evans 1983; Haynes 1980b). In fact, in a unique study of the demography of prey mortality, Smith (1974) suggests wolves (*Canis lupus*) focus their predation of deer (*Odocoileus virginianus*) on old adults and juveniles whereas the prehistoric human group he was studying took mostly near-adults and prime adults. Smith concludes there was, then, no direct competition for deer between the humans and wolves, and that the two predators kept the deer population stable because the net effect of predation by the two predators was that all age classes were exploited.

Study of mortality profile shape, age–class discreteness, and context of the fossils grants insights to whether mortality was synchronic (mass, or literally catastrophic) or diachronic (attritional). Once that is determined, study of other variables such as behavioral variation between distinct age–sex groups and topographical, geological, and archaeological contextual data may reveal why mortality was synchronic or diachronic. Why does an instance of prehistoric mortality have the demographic expression it does? We next turn to how this question is answered.

Analyzing the demography of mortality

Klein (1982a) suggests that inspection of a mortality profile in conjunction with knowledge of the behavior of the taxon involved and study of the geological context of the fossil sample allows the distinction of animal populations that were generated by hunting from those that were generated by scavenging. In archaeological contexts, taxa that can be taken in large numbers simultaneously, such as in drives or traps, will produce catastrophic profiles (L-shaped age–frequency distributions) whereas taxa that are not behaviorally amenable to being taken in large numbers at one time must be hunted individually and will tend to produce attritional mortality profiles (U-shaped age–frequency distributions). In the latter, very young individuals are more rare than in paleontological assemblages, suggesting that in the latter scavenging by carnivores has selectively removed the remains of the youngsters. Thus, in archaeological contexts low proportions of youngsters in a population displaying attritional mortality suggests scavenging by humans of prey animals that had been killed and consumed by non-human carnivores. High proportions of youngsters, however, suggest active hunting by humans. If a catastrophic mortality profile is found in an archaeological context, hunting and scavenging are best distinguished by geological data. In these cases, scavenging by humans of a population of animals that died due to a natural catastrophe is, Klein (1982a:153) supposes, a rare event. Therefore, if a catastrophic mortality profile has been derived from fossils that clearly accumulated over a long time span (hundreds or thousands of years given stratigraphic and contextual data), then it is likely that mortality represents artificial catastrophes and hunting whereas a relatively short-term accumulation (e.g., a single event) may represent hunting or scavenging by human predators. Geological data must be called upon to determine if the remains are in a sedimentary context suggestive of natural mortality (such as a flash-flood deposit) or artificial mortality such as at the base of a cliff.

Stiner (1990b, 1991a, 1991d) employs a three-pole graphing technique to assess mortality patterns. Each of the three axes represents a different age class: juveniles (approximately the first 20% of natural ecological longevity, but defined by Stiner [1990b:311] as from birth to the age at which a particular, taxon-dependent deciduous tooth is shed), prime-age adults (breeding age), and old adults (past their prime, approximately the last 30% of natural ecological longevity) (Figure 5.8). Based on populations with known mortality profiles, Stiner defines areas on the three-pole graph that corresponded to the traditional U-shaped and L-shaped age–frequency distributions as well as three additional types of mortality that she labels juvenile-dominated, prime-dominated, and old-dominated. Her graphing technique gives the analyst another way to visualize variation in mortality across age classes. Based on demographic data compiled for prey taxa, Stiner uses this graphing technique

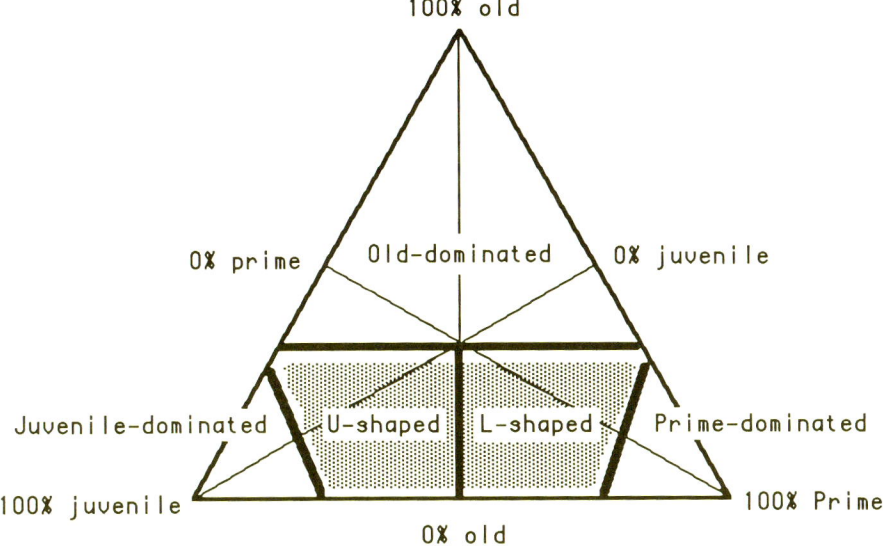

Figure 5.8. Three-pole graphing technique for assessing demographic (mortality) data (after Stiner 1990b:318, Figure 6).

to distinguish, in a general way, the kinds of hunting techniques that produce a particular kind of mortality. Cursorial predators tend to "engage in long chases of their quarry" and generate U-shaped mortality profiles (Stiner 1990b:322). Predators which ambush their prey tend to produce L-shaped mortality profiles because the encounter of prey is determined by chance. Predators that depend on scavenging tend to produce old-dominated age–frequency distributions, whether they are cursorial or ambush predators, probably because prime adults are less susceptible to attritional death and the remains of juveniles are less likely to survive gnawing by carnivores.

A data set described by Klein (1982a) can be used to illustrate Klein's and Stiner's techniques. That data set concerns two collections (Table 5.4) of remains of the extinct giant African buffalo *Pelorovis antiquus*, a large bovid that Klein (1982a) characterizes as rather immune to predation due to its size and suspected social cohesion in defensive situations. Both collections apparently represent palimpsest accumulations; that is, multiple stratigraphic units and/or depositional events are represented by each collection. Plotting the age–frequency distributions for the two assemblages, it is obvious that the mortality profile for Klasies River Mouth contains many more juveniles than the profile for Elandsfontein (Figure 5.9). These two mortality profiles are statistically different ($D = 0.557$, $P < 0.01$). Klein (1982a) believes the Klasies River Mouth data represent active hunting of the animals due to the significantly greater number of individuals in the first age class relative to the low frequency of young individuals in the Elandsfontein assemblage, which seems to represent a

Table 5.4 *Mortality data for two fossil assemblages of* Pelorovis antiquus *(from Klein 1982a). Each age class represents 10% of the natural ecological longevity of the taxon*

Age class	Klasies River Mouth	Elandsfontein
1	41	8
2	1	12
3	1	2
4	4	11
5	4	10
6	7	12
7	3	19
8	3	15
9	0	6
10	0	0

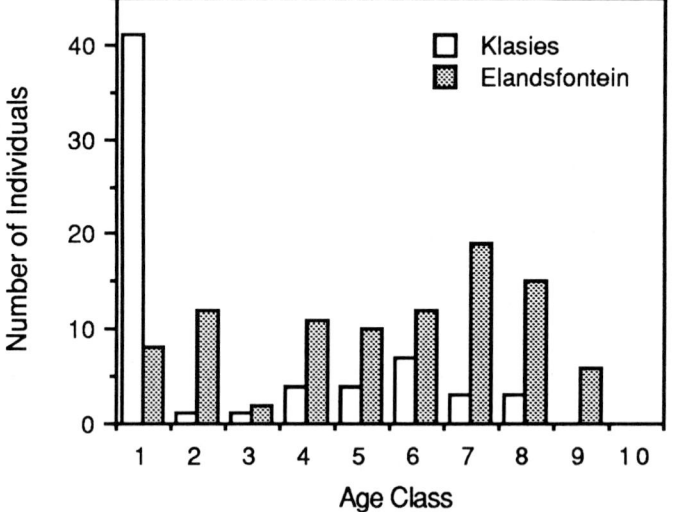

Figure 5.9. Mortality profiles for African bovid remains from Klasies River Mouth and from Elandsfontein (from Table 5.4). Age classes are each 10% of natural ecological longevity.

natural accumulation. The greatest difference in the cumulative frequency distribution in fact occurs in the youngest age group.

Following Stiner's (1990b) procedure and lumping the first two age classes into the juvenile age group, the third through the seventh age classes into the prime age group, and the eighth through the tenth age classes into the old age group, the plot of points on the three-pole graph (Figure 5.10) indicates the two assemblages are different, and that the Klasies River Mouth assemblage

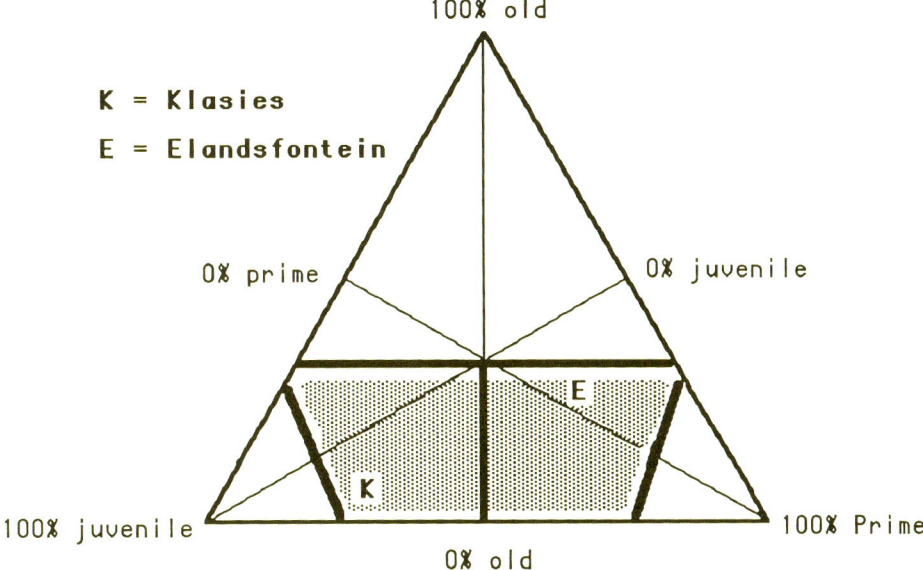

Figure 5.10. Three-pole graph of mortality data from Klasies River Mouth and Elandsfontein (from Table 5.4; see Figure 5.8).

approximates a U-shaped profile whereas the Elandsfontein assemblage approximates an L-shaped mortality profile. The latter is perhaps a function of the graph as other data described by Klein (1982a) indicate both profiles are attritional. Cumulative frequencies of the Klasies River Mouth profile do not differ significantly from an attritional profile modeled from the number of dead individuals in Table 5.2a ($D = 0.141$, $P > 0.05$) but cumulative frequencies of the Elandsfontein profile do differ from that modeled attritional profile ($D = 0.416$, $P < 0.01$). Both the Klasies River Mouth and Elandsfontein assemblages differ significantly from the catastrophic age–frequency distribution modeled by the number of live individuals in Table 5.2a ($P < 0.01$ for both). Klasies River Mouth has proportionately more juveniles than the modeled catastrophic profile and Elandsfontein has proportionately more prime-aged individuals. Stiner's graph (Figure 5.10) suggests just such a difference between the two collections.

If Klein (1982a) is correct that the Elandsfontein assemblage represents a natural accumulation, what does the weakly prime-dominated mortality profile for that assemblage signify? Perhaps this assemblage consists mostly of prime-age males that died due to malnutrition after the stress of the breeding season. Klein (1982a:155) reports that many of the remains occur "near what [at the time the remains accumulated] were probably perennial sources of water." Did this species of bovid respond in a manner similar to Barnosky's (1986) Irish elk (see Chapter 3)? Alternatively, the Elandsfontein remains are weakly old-dominated (Figure 5.10), so perhaps it was individuals of both sexes that were nearing the end of their prime years that were dying; these age–sex cohorts would have been more susceptible to the vicissitudes of dry-season environments. Whether these bovids were mostly males or individuals of both sexes almost past their prime, this analysis of the demography of their mortality directs us toward other kinds of data that may contribute to a taphonomic explanation for their death and occurrence as fossils. A bimodal distribution of sizes of remains would suggest both sexes were represented. If season of death could be determined, that might provide corroborating evidence for dry-season stresses. Regardless of the outcome of such analyses, this possibility leads us to another topic, the seasonality of mortality.

The seasons of mortality

As should be clear from the preceding, the season of mortality may be an important taphonomic variable in explanations of why a bone assemblage appears the way it does, especially in terms of the demography of mortality. Methods for determining the season of animal death are many. Monks (1981) provides a detailed review of most of them, and I do not reiterate that discussion here. It suffices to note that two properties allow the season of death to be determined. First, many taxa have seasonally restricted breeding and birthing seasons. Second, due to patterned ontogenic development, it is possible to assign a season of death to organisms with a restricted season of birth based on their observed stage of ontogenic development. The age of an individual organism at death can be assigned to it based on the second property, but season of death can only be determined for individuals of those taxa with the first property.

For cervids such as deer (*Odocoileus* spp.) in the northwestern United States, the seasonally restricted birth season occurs between May 1 and June 30. An individual that dies when it was exactly one year old died June 1 ± 30 days. Using that knowledge of modern deer, plus knowledge of the ontogeny of modern deer, one can estimate the season of death of deer represented by prehistoric remains. Figures 5.11a and 5.12a illustrate examples taken from two archaeological sites in eastern Washington state. In both cases, the seasonality–frequency distributions, or *seasonality profiles*, suggest most deer were dying in late summer through early winter months. Because the presence of butchery marks (see Chapter 8) on many of the bones indicates humans played a role in their accumulation and deposition, the analyst might interpret the seasonality profiles as indicating that deer hunting was most intensive in the fall, and call on evidence such as the condition of ungulates in temperate latitudes as indicating the fall months are the time when deer are of the most nutritional value to humans (e.g., Speth and Spielmann 1983).

The preceding interpretation of Figures 5.11a and 5.12a suggests that the mortality profiles for the two assemblages should approximate a catastrophic

a

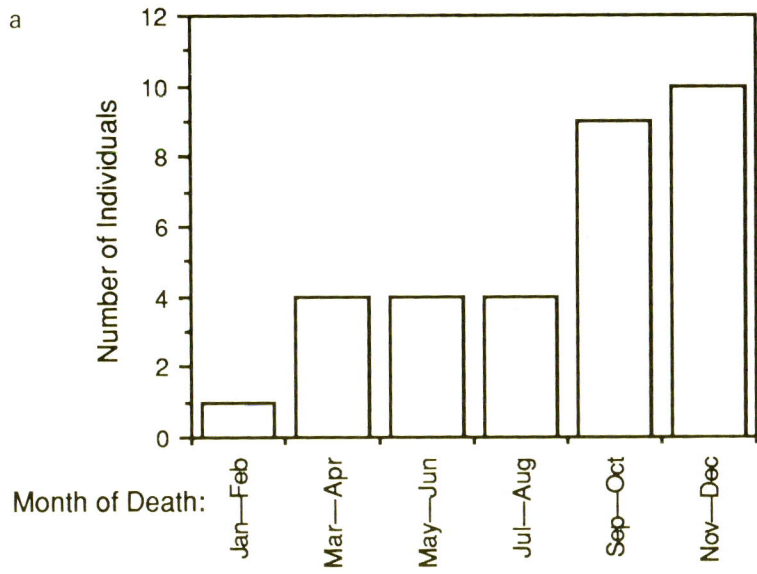

Month of Death:

b

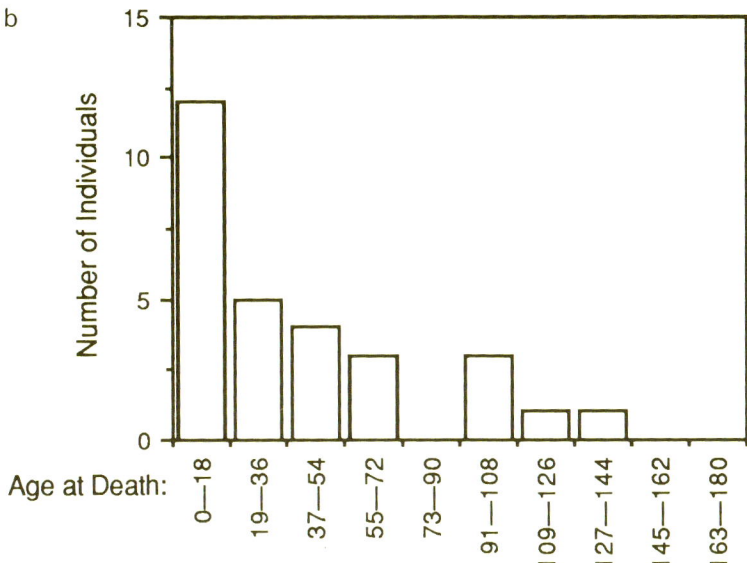

Age at Death:

Figure 5.11. Seasonality (a) and mortality (b) profiles for deer (*Odocoileus* spp.) remains from archaeological site 45DO189 in eastern Washington (from Lyman 1988b). Age at death in (b) is in months.

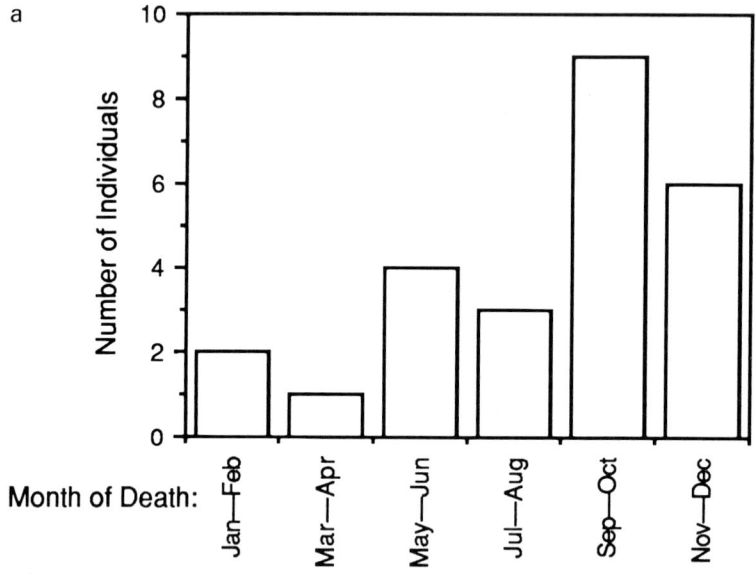

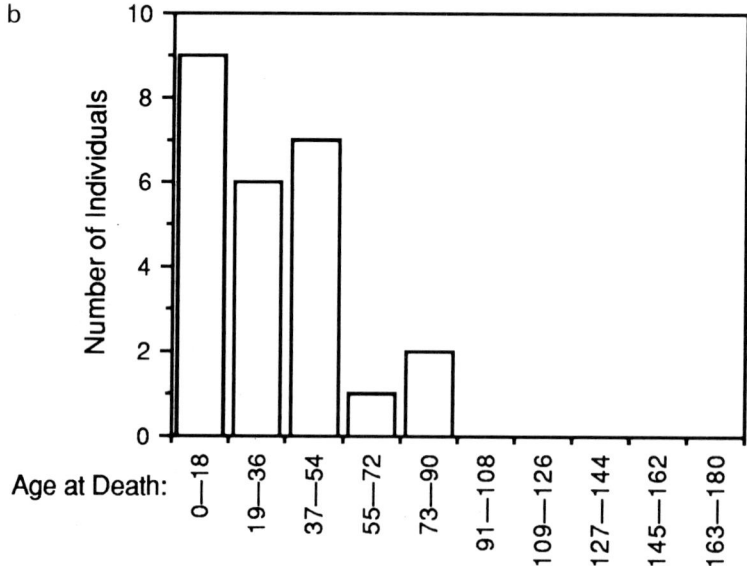

Figure 5.12. Seasonality (a) and mortality (b) profiles for deer (*Odocoileus* spp.) remains from archaeological site 45DO176 in eastern Washington (from Lyman 1985b). Age at death in (b) is in months.

or L-shaped age–frequency distribution. They in fact do (Figures 5.11b and 5.12b). That this is the case is due to the seasonally-restricted birthing season for the taxon involved, and, like the mortality profiles of Kurtén (1983; Figure 5.4) and Simpson (1984; Figure 5.5), the more or less seasonally-restricted accumulation of deer remains. In these cases, the seasonally restricted mortality of deer helps account for the mortality profile, and the mortality profile helps explain the seasonality profile.

A less frequently used technique for assessing seasonality is the presence of chitinous exuvia of fly pupae. Gilbert and Bass (1967) determine the season when humans were buried in a site in South Dakota by the presence of such insect remains and the limited season (during warm months) when such organisms are active today. Chomko and Gilbert (1991) infer a June through September season of site formation and deposition of bone pieces in a pit based on the association of fly pupae with the bones. Guthrie (1990) similarly infers the season of death for an individual of a late Pleistocene bison (*Bison* sp.) in Alaska based on the presence of fly pupae. These insects tend to attack carcasses quickly, and their presence suggests some delay between death and burial. In fact, Guthrie (1990:83) cites evidence from South Africa indicating mammal carcasses there may be completely skeletonized by flies within five days of death in the summer, and within 14 days in the winter. Similarly, Payne (1965) reports that immature pig (*Sus scrofa*) carcasses about 1.0 to 1.4 kg in size placed outside in South Carolina were completely skeletonized (bits of hide remained) by feeding insects after eight days. Thus the impact and utility of insects for understanding taphonomic processes seems great. Not only might their remains, when associated with vertebrate skeletons, suggest delayed burial and season of death, but their presence may help explain why some carcasses are found in a more or less articulated state while others are disarticulated. That is, as we see in the next section, the appearance of death in the fossil record can be mightily influenced by mere insects as well as larger taphonomic agents.

Skeletonization and disarticulation

> If air is cut off, malignant putrefaction sets in, causing the formation of noxious substances.
> (J. Weigelt 1927/1989:5)

Natural traps such as chimney caves, bogs, and tar pits are generally evident from geological data, and the cause of death is rather more evident in such contexts than in other kinds. For the more typical assemblages of bones recovered from open sites, geological data may provide clues as to the mode of death, but the appearance of the dead animals may also provide important data on how they died. For example, Barnosky's (1985, 1986) Irish elk fossils were recovered from peat deposits suggesting they died in a bog, and that is the

Figure 5.13. A partial, articulated wapiti (*Cervus elaphus*) skeleton *in situ*.

source of the miring hypothesis he tests. Lyman (1989b) and Voorhies (1981) describe the effects of volcanic eruptions on the appearance of mortality. Individual mammals that were shielded from the explosive force of the eruption simply collapsed and were quickly buried by volcanic ash from the eruption, thus their skeletons were still largely articulated when recovered (e.g., Figure 5.13). The fact that these skeletons were stratigraphically within a thick deposit of volcanic ash suggested the individuals died of suffocation. But their rapid burial resulted in mostly complete skeletons with individual bones in their proper anatomical positions. What I am touching on here are those immediate postmortem processes of soft tissue decomposition, skeletonization, and disarticulation, and it is these processes to which I now turn.

I was eating my lunch the first time I read Weigelt's (1927/1989) graphic descriptions of immediately postmortem alterations to animal carcasses; it was all I could do to finish my peanut-butter and jelly sandwich. But Weigelt's and others' descriptions of what happens in the first few days or weeks after death are important because they illustrate how the mode of death can influence the taphonomic history of an animal carcass. Schäefer (1962/1972:10) was quite aware of this when he wrote "usually enough time has passed [between death and burial] for the skeletal parts to have lost their original [soft tissue] connections or for external agents to have separated them. Hence there are only two ways in which complete skeletons of vertebrates can be preserved: either quick burial in areas of rapid sedimentation or burial in places sheltered from transportation forces." While paleontologists and zooarchaeologists need not be greatly concerned with immediately postmortem changes in organisms due to the temporal remoteness of the mortality event represented by a fossil (e.g., Shipman 1981b hardly mentions such changes), we do need to have a basic grasp of the processes involved and their potential effects on a carcass. This is especially so when one subscribes, as I do, to the position that taphonomic effects are cumulative, and what happens early in a taphonomic history can influence what happens later in that history.

In the following, it is important to realize that *skeletonization* (soft tissue removal) occurs due to micro-organisms such as bacteria and fungi, to small organisms such as insects, and to medium and large organisms, such as vultures, hyenas, and other scavenging carnivores. The first two categories often are discussed under the term decomposition, even though the bacteria and insects are scavenging food if that is taken to mean eating the flesh of dead animals which the consumer has not killed. But here I follow the literary tradition of considering bacteria and insects under the general category of decomposition agents and processes. I begin with a description of the kinds of settings and processes resulting in those rare cases when soft tissues are preserved before turning to the more common cases of soft tissue decomposition, skeletonization, and disarticulation (see Allison and Briggs 1991a for more extensive treatment).

Soft tissue preservation

> Postmortem transformation by the action of micro-organisms has not been well studied.
> (M. S. Micozzi 1991:38)

Some of the best-known cases in which mammalian soft tissues have preserved over several centuries or millennia concern the human mummies found in Egypt. There are multiple ways for a mummified carcass to result from taphonomic processes. The term "*mummification* refers to all natural and artificial processes that bring about preservation of the body or its parts" (Micozzi 1991:17) This definition of the process may be too inclusive as it can be perceived to include the preservation of just bones. It is my impression that the term *mummy*, when applied to an entity, typically refers to a carcass with at least some preserved soft tissue (often a major portion of the soft tissue found on the living organism) in addition to some bones. It is in this last sense that I use the term mummification; to denote the processes that bring about the preservation of at least some soft tissues, as well as at least some of the bones, of an animal. I do not consider the slippery question of "how much soft tissue must be preserved for a partially skeletonized carcass to be considered a mummy rather than simply a skeleton?"

Micozzi (1991:17) suggests there are three major categories of mummification processes. *Natural mummification* involves natural processes, either singly or several, such as desiccation and freezing. *Intentional mummification* involves human processes which deliberately exploit natural processes, such as freezing a carcass to preserve it. *Artificial mummification* is also intentional and involves humans but the processes used involve ones not available naturally. These include the use of preservative substances such as resins and oils, and processes such as smoke curing. Because the latter two kinds of mummification tend to have been little explored by taphonomists, I do not consider them further here (see Micozzi 1991 for a useful discussion of such processes). In the remainder of this section, then, I review only natural mummification.

It is the extremes of environmental conditions that tend to favor natural mummification. For example, extreme cold such as freezing temperatures tend to favor preservation of soft tissues (see the review in Guthrie 1990). Low temperatures inhibit bacterial activity which results in soft tissue decomposition. Freezing also results in sublimation, or freeze-drying (Micozzi 1991:9), which inhibits putrefaction (see below), a process which requires moisture. Hot, dry environmental conditions are also conducive to the preservation of soft tissues due to desiccation which, as noted, in turn prevents putrefaction. Carcasses deposited in mineral salts sometimes have preserved soft tissues due to the desiccatory effects of the salts. Aquatic environments that are acidic and anaerobic also lead to soft tissue preservation due to inhibition of bacterial activity (Allison 1990), as evidenced by Danish "bog people" preserved by the tannic acids created in peat bogs (Micozzi 1991). Many plant products have anti-bacterial and/or insecticidal effects which might inhibit decomposition of animal soft tissues, but whether these properties accrue to carcasses of animals deposited near plants producing such products is not clear.

The duration of soft tissue preservation, even if it does not preserve down through the centuries for the taphonomist, is important. Bones can take on different postmortem positions relative to one another given changes in attached soft tissues. Weigelt (1927/1989) distinguishes what he calls the *passive position* of vertebrate skeletons of carcasses that decomposed in water from what he calls the *contorted position* of skeletons of carcasses that decomposed in subaerial contexts. He writes "when a carcass dries out, shrinking mechanisms often bring about unnatural, contorted positions not usually found in living animals. The most conspicuous is the often observed [dorsally concave] curvature of the cervical part of the spinal column" (Weigelt 1989:103). Weigelt (1989:88) describes the passive position as "similar to the natural structure of the body." I take "the natural structure of the body" to mean the positions of anatomical parts relative to one another when the organism is alive. Dodson (1973:15) has noted that "after *rigor mortis* is lost, a carcass carried by flowing water to its place of burial will show a passive position, rather than the rigid position of an animal buried in a state of *rigor* or after desiccation on land." The orientation of carcasses (see Chapter 6) may, thus, reveal details about deposition.

Distinction of the passive and contorted positions of skeletons underscores the importance of detailed spatial and contextual data for taphonomic analysis. That kind of data cannot be derived from a paper bag full of bones even though all might have been recovered from the same excavation unit. Later in this discussion I describe procedures for the analytical rearticulation of skeletons that can be applied when contextual data are lacking. But, even if the skeleton of an animal has been arranged in a passive or contorted (or some other) position, soft tissues seldom preserve; more often than not, they decompose.

Schäfer (1962/1972:20) notes that "carcasses of marine mammals that die from natural causes drift for weeks on the surface of the sea," and as soft tissues deteriorate, bones become disarticulated and "fall to the [sea bottom] one by one as from a drifting sack [and thus are] spread over miles of the sea floor." Carcasses float due to gas production during decay, and the phenomenon Schäfer describes has been termed "bloat and float" (Allison and Briggs 1991a:27). Allison and Briggs (1991a:27) indicate that the propensity for a carcass to float during decay of soft tissues is controlled by the "strength" of soft tissue, the rate of gas production, and hydrostatic pressure. When the tissues are sufficiently weakened by decay the increasing volume of gas breaks the carcass apart, the gas escapes, and the carcass sinks. Rather unusual carcass orientations, degrees of skeletal completeness, and states of articulation may result.

Soft tissue decomposition

> Dead organisms are a valuable food source in any environment. If this food source is utilized by macro-organisms it is termed scavenging; if it is utilized by microbes, such as fungi and bacteria, it is termed decay.
> (P. A. Allison 1990:213)

The removal of soft tissues from vertebrate skeletons involves organisms of many sizes, from bacteria to large mammalian carnivores. In the following, I begin with discussion of the small organisms that remove soft tissues, the process and effect of which are usually referred to as decomposition or decay, and work up through progressively larger organisms.

Micro-organisms

Subsequent to the death of a vertebrate, soft tissues generally decompose due to the action of bacteria and enzymes. Following the forensic literature, where much research concerning soft tissue decomposition has been reported, *decay* involves the decomposition of protein under aerobic conditions; *autolysis* involves enzymatic breakdown of tissue by the enzymes in the (once living) organism, enzymes that assisted in metabolic functions; *putrefaction* involves the bacterial breakdown of protein under anaerobic conditions, with the source of the bacteria being either internal to the organism or external to it (Haglund 1991:25). Soft tissue decomposition usually proceeds successively "from within the carcass due to the action of enteric micro-organisms [indigenous bacterial microflora, some of which are anaerobic], and from [outside the carcass] by colonization [of the carcass] with soil micro-organisms and decay organisms" (Micozzi 1991:37, 39, 42). Putrefaction occurs only in the presence of moisture and in moderate temperatures; desiccation and temperatures less than 4°C prohibit putrefaction because these conditions inhibit bacterial growth (Janaway 1990; Micozzi 1991:37, 38, 40). "Decomposition due to bacterial action is rapid in environments characterized by temperatures between 15°C and 37°C" (Micozzi 1991:41). The rate of decomposition slows when a carcass is buried, in part due to the low temperature of the enveloping sediment inhibiting bacterial growth, but also due to the decreased access of the soft tissues to carrion insects (Micozzi 1991:37). Fungi which may contribute to the decay process tend to be aerobic and "are restricted to the surface of the cadaver" (Janaway 1990:147).

Haglund (1991) follows Payne (1965; see also Coe 1978, 1980), and lists six chronological stages of decomposition: fresh, bloated, active decay, advanced decay, dry, and skeletal remains. "Small amounts of decaying tissue remain throughout the dry stage [and] the skeletal stage may retain some ligamentous tissue, articular cartilage, and other cartilage" (Haglund 1991:26). Micozzi (1991:43) suggests that soft tissue decomposition which is free from the effects of carrion insects is slower than when such insects are present, and is characterized by five stages: fresh, bloated and decomposition, flaccidity and dehydration, mummification, and desiccation and disintegration. Disarticulation sets in by the final stage. Johnson (1975) describes four stages of decomposition in small mammals: fresh, bloat, decay, and dry. Bloating results from gases produced by anaerobic protein decomposition caused by putrefaction; the bloating stage is brief (two to five days) in warm environments, long (several weeks) in cool environments. The decay stage involves mostly aerobic protein decomposition, and carrion insects abandon the carcasses during the dry stage (Johnson 1975). Different authors have broken the decomposition *process* into different stages, perhaps because differing environmental conditions exacerbate or inhibit various decompositional processes, or because decomposition is a process, it is a continuum, the stages of which are artifacts of our observations. Whatever the case, the important point is that decomposition is a process.

The rates of decomposition and skeletonization are dependent on the environment of carcass deposition, the cause of death, the condition of the carcass at death, and other factors. Emaciated bodies decompose more rapidly than healthy bodies. Given the rule of thumb that a chemical reaction doubles in rate for every 10°C rise in temperature (Haglund 1991), carcasses decompose more rapidly in warmer environments (e.g., Coe 1978). Frozen bodies preserve quite well as chemical reactions are negligible (e.g., Guthrie 1990). Decomposition is most rapid in carcasses located on the ground surface or in air, of moderate rate for carcasses in water, and slowest for buried carcasses (Haglund 1991:29). *Saponification*, a chemical reaction in which fat is hydrolyzed and converted to adipocere, occurs in cool, moist environments, and "adipocere formation serves to retard other forms of decomposition" (Haglund 1991:32).

Insects as agents of soft tissue removal
Moving up the size scale of agents and processes that remove soft tissues from animal carcasses, the category above bacteria generally discussed by forensic scientists involves insects. Three kinds of insects are typically observed on soft tissue remains (Micozzi 1991:44). *Necrophagus insects* are primary consumers of carrion and mainly include larvae of insects that are not normally necrophagus themselves. *Omnivorous insects* consume carrion and necrophagus insect larvae. *Predator* and *parasite insects* prey on necrophagus insects alone. Other categories of insects include those simply living in or on a carcass or which by chance are found there (Micozzi 1991:44; see also Johnson 1975; Payne 1965). Only the necrophagus and omnivorous insects are involved in soft tissue removal.

Actualistic research by Johnson (1975), Payne (1965), and others (e.g., Rodriguez and Bass 1983) indicates different insect taxa tend to characterize different stages of soft tissue decomposition, although insects are most abundant during the decay stage (between the bloat and dry stages). Thus, a taphonomist might gain insights into when during decomposition a carcass was buried if remains of particular insect species are found associated with the remains. The taphonomist might also gain some idea of how long the carcass was exposed as most insects tend not to exploit buried carrion. Environmental factors such as temperature and moisture influence insect activity, and thus different carcass depositional habitats may influence whether and, if so, which insect taxa exploit animal carrion.

Finally, it is important to note that some insects eat bone tissue (e.g., Behrensmeyer 1978; see Chapter 9), and some insects move bones (Shipman and Walker 1980). The former may not be mediated by the size of the bone relative to the size of the insect, although the latter probably is (Shipman and Walker 1980). These two effects as generated by insects have, however, not been studied in detail in actualistic settings, and perhaps for that reason these effects have not been frequently inferred in prehistoric settings.

Joint anatomy

Because large-scale agents that remove soft tissues have tended to be discussed in terms of their influence on the disarticulation of vertebrate skeletons, it is difficult to separate skeletonization and disarticulation for discussion purposes. The difficulty arises as well because the size of large-scale tissue removers allows them to remove soft tissue and disarticulate the included bones simultaneously. Small-scale tissue removers such as insects and bacteria, due to their small size, typically cannot separate a bone and soft tissue package from a carcass and then move that package. In the absence of large-scale soft tissue removers vertebrate carcasses appear simply to fall apart and remain semi-articulated (Coe 1978, 1980; Haynes 1991; Johnson 1975; Payne 1965; Payne *et al.* 1968). For both small- and large-scale soft tissue removers, one factor seems consistently to mediate disarticulation, or the anatomical disassociation of skeletal parts, and that is joint anatomy. Thus, it is important to review joint anatomy before discussing disarticulation and large-scale soft tissue removers.

Syndesmology, the study of ligaments, joints, and articulations, and *arthrology*, the study of joints, have resulted in a wealth of information on joint anatomy. Much of that information concerns the mechanics of various joints, but there are also data relevant to taphonomic studies of disarticulation. In particular, it is reasonable to suppose that the kind of articulation between two skeletal elements, and the kind(s) of soft tissues holding those two skeletal elements together, influence the postmortem interval between death and disarticulation of the two bones.

Some joints or articulations result in two bones being relatively immobile relative to one another, such as the sutural joints between bones of the skull. In other joints the articulated bones are only slightly movable relative to one another, such as is found with the intervertebral joints and pubic symphysis of mammals. Finally, the articulated bones of some joints are quite mobile relative to one another, such as in the elbow or knee joints of humans. These types of joints vary in terms of the degree to which the articulated bones interlock with one another and in terms of the amounts and kinds of connective tissues holding them together (Hildebrand 1974:450–455; Micozzi 1991:49–50).

Joints can be classified according to the type of material holding them together. The following is abstracted from Moore's (1985:33–36) discussion of such a classification, in which major joint categories are indicated by numbers and joint types within a category are indicated by letters.

1) Fibrous joints: bones are united by fibrous tissue;
   A) Sutures: bones united by a thin layer of fibrous tissue; little to no movement of bones; occur only in the skull;
   B) Syndesmosis: bones joined by a sheet of fibrous tissue, either a ligament or interosseous fibrous membrane (as between radius and ulna); slight to considerable movement of bones possible;
2) Cartilaginous joints: bones united by cartilage;
   A) Primary cartilaginous joints: bones united by hyaline cartilage which permits slight bending (e.g., epiphyseal cartilage plate between epiphysis and diaphysis of a long bone, rib to sternum connection);
   B) Secondary cartilaginous joints: articular surfaces of articulated bones covered with hyaline cartilage and cartilage surfaces joined by fibrous tissue and/or fibrocartilage; strong and slightly moveable (e.g., inter-vertebral centra, pubic symphysis);
3) Synovial joints: normally provide free movement between joined bones; have four distinguishing features, (1) joint cavity, (2) articular cartilage, (3) synovial membrane, and (4) fibrous capsule; accessory ligaments strengthen the capsule and limit movements of the joint in undesirable directions;
   A) Plane joints: permit sliding, as between two carpal bones;
   B) Hinge joints: permit movement in one axis at right angles to bones involved, as human elbow;
   C) Pivot joints: permit rotation around a longitudinal axis through a bone, as the atlas–axis joint;
   D) Condyloid joints: allow movement in two directions, as the metacarpal–phalange joint of humans;
   E) Saddle joints: allow movement in two directions as articular surfaces are saddle-shaped, as the carpal–metacarpal joint of the human thumb;
   F) Ball and socket joints: highly movable, multi-directional movement, as the human hip and shoulder joints.

This classification is somewhat taxonomically structured, as the type of connective tissues seems most important, followed by finer-scale details of the articulation, such as the joint's morphology and mobility.

Disarticulation involving only micro-organisms (bacteria and insects) seems to begin with mobile joints, and progresses to the slightly mobile joints many of which have intraosseous ligaments holding them together. Immobile suture joints are the last to disarticulate because they are held together by ligaments and the bones have, in a way, interwoven with each other (Micozzi 1991). These general natural sequences may vary slightly depending on the kinds of organisms involved in soft tissue removal.

Disarticulation

From a taphonomic perspective, the importance of soft tissue removal resides in the function of such tissues to hold the bones together in the form of a skeleton. As the soft tissues are removed, a skeleton will fall apart; that is, bones that were articulated in life will become disarticulated, and perhaps eventually spatially disassociated or scattered (Hill 1979a, 1979b). The access of sca-vengers, whether mammals, insects, or bacteria, to carcasses influences the rate of skeletonization. Burial effectively removes carcasses from many large scavengers, and not only retards the rate of soft tissue decomposition, but also the rate of bone disarticulation and scattering. Abler (1985), for example, reports on a domestic sheep (*Ovis aries*) that became mired in a muddy swamp. Only the dorsal part of the back and the head were not buried, and only these exposed portions were "most vulnerable to scavenging, decay, and weather-ing" (Abler 1985:250). Lyman (1989b) reports similar results for carcasses of North American wapiti (*Cervus elaphus*) killed by a volcanic eruption; only portions of carcasses not buried by volcanic ash were exposed to scavenging and weathering, buried portions of carcasses were still articulated and unweathered a year and a half after the eruption. Micozzi (1991:49, 51) suggests "decomposition of soft tissue occurs from the top (head) downward" and thus "the mandible and skull generally have the first opportunity to become disarticulated from the remainder of the skeleton." Abler's (1985) suggestion may thus simply concern a particularistic factor that exacerbates the generality suggested by Micozzi (1991).

Hill (1979b) describes a disarticulation sequence for the topi (*Damaliscus korrigum*), and Hill and Behrensmeyer (1984) describe similar sequences for the wildebeest (*Connochaetes taurinus*), domestic cow (*Bos* sp.), Burchell's zebra (*Equus burchelli*), and Grant's gazelle (*Gazella granti*). All are summarized in Table 5.5. Hill and Behrensmeyer (1984) found great overall similarity between the five disarticulation sequences. While they document some differences between them, they found those differences could not be explained by reference to taxonomic variation or by variation in carcass size. Importantly, Micozzi (1991:50) correctly reports that Hill's studies did not identify and distinguish the activity of decomposition organisms (bacteria and insects) from large-scale scavengers (e.g., hyenas).

Table 5.5 *Rank order of joint disarticulation in five mammalian taxa (from Hill 1979b, Hill and Behrensmeyer 1984). Numbers denote order of disarticulation; 1 = first joint to disarticulate, 30 = last joint to disarticulate. Joint type (after Moore 1985) given in parentheses; S, synovial joint; C, cartilaginous joint; F, fibrous joint*

Joint	Topi	Wildebeest	Domestic Cow	Burchell's Zebra	Grant's Gazelle
cranium–mandible (S – hinge)	5	4.5	2	2	6
cranium–atlas (S – hinge)	20	4.5	6	8.5	6
atlas–axis (S – pivot)	7	20	13	22.5	6
axis–third cervical (C – secondary)	29	30	24	26	30
cervical–cervical (C – secondary)	30	29	25	29	29
seventh cervical–first thoracic (C – secondary)	27.5	25	22.5	22.5	23
thoracic–thoracic (C – secondary)	22	24	26	20	17
thoracic–rib (S – condyloid)	23	21	21	11	11
thirteenth thoracic–first lumbar (C – secondary)	25.5	22.5	28	27	24.5
lumbar–lumbar (C – secondary)	25.5	25	29	28	27
seventh lumbar–sacrum (C – secondary)	27.5	27	30	22.5	28
sacrum–first caudal (C – secondary)	2.5	9	20	8.5	2.5
caudal–caudal (C – secondary)	2.5	3	9	4	2.5
sacrum–innominate (S – plane)	24	28	22.5	22.5	24.5
innominate–femur (S – ball and socket)	11	9	8	6.5	13
femur–tibia (S – condyloid)	19	13.5	15	12.5	20.5
tibia–tarsals (S – plane)	14.5	13.5	15	12.5	20.5
tarsals–metatarsal (S – plane)	21	15.5	19	25	20.5
metatarsal–first phalanx (S – hinge)	17	15.5	11	17	18.5
first phalanx–second phalanx (hind) (S – hinge)	17	17.5	10	17	15.5
second phalanx–third phalanx (hind) (S – hinge)	17	19	4	5	15.5
forelimb–body (none)	1	1	1	1	1
scapula–humerus (S – hinge)	4	2	5	3	4
humerus–radius, ulna (S – hinge)	11	17.5	16	17	18.5
radius–ulna (F – syndesmosis)	14.5	22.5	27	30	26
radius–carpals (S – plane)	6	11	14	14	13
carpals–metacarpal (S – plane)	8.5	12	18	17	13
metacarpal–first phalanx (S – hinge)	8.5	7	12	10	9
first phalanx–second phalanx (fore) (S – hinge)	13	9	7	17	9
second phalanx–third phalanx (fore) (S – hinge)	11	6	3	6.5	9

To determine if joint anatomy, as characterized by Moore's (1985) typology, influenced the order of joint disarticulation as determined by Hill, I assigned each joint in Table 5.5 to one of Moore's types. Hill (1979b) notes that most of the joints he considered are synovial joints. In fact, 19 of the 30 joints in Table 5.5 are synovial joints (five kinds of synovial joints are represented), 10 are secondary cartilaginous joints, and one is a fibrous syndesmosis joint. I calculated the average rank of disarticulation for all joints of three types for the topi; other joint types were only represented by one or two joints. The average rank of disarticulation for the 10 secondary cartilaginous joints is 21.5 ($\pm$ 10.3), for the five synovial plane joints the average rank is 14.8 ($\pm$ 7.8), and for the 10 synovial hinge joints the average rank is 12.4 ($\pm$ 5.4). These crude data suggest joint anatomy does influence the order of joint disarticulation documented by Hill.

Hill's (1979a, 1979b, 1980; Hill and Behrensmeyer 1984, 1985) research focused on spatially isolated carcasses and joints. After studying several loci with multiple carcasses of domestic cows in close proximity, Todd (1983a:40–41) reports that "as the number of carcasses increases and as the distance between carcasses decreases, the greater the potential for [a disarticulation sequence] different from that observed by Hill (1979b)." Further, if a carcass is on the periphery of a cluster of carcasses, it will tend to disarticulate more rapidly than a carcass near the center of the cluster due to the former's greater accessibility to scavenging carnivores and trampling agents (Todd 1983a). Todd (1983b:54) suggests some of the differences between the decomposition and disarticulation of an isolated carcass from a group of carcasses results from the creation by the latter of a "distinctly different taphonomic micro-climate than that which would operate on a single carcass in the same setting." This is similar to, but at a different (multi- or between-carcass) scale than that (single- or within-carcass) used by Micozzi (1991:42) who notes that the temperature of decomposing soft tissue may differ from the ambient sediment or air temperature "due to the creation of an internal microenvironment by the action of bacteria."

Given the importance of moisture for skeletonization, the season of death in temperate climates may exert strong influences on rates of skeletonization and thus on disarticulation. In temperate North America, Todd (1983b:71) suggests "we could expect a greater degree of skeletal disarticulation and dispersal where animals died in the late spring or summer than in those where death occurred in the fall or winter" (see also Galloway *et al.* 1989). The presence of snow drifts exacerbates the effects of moisture as the drifts melt (Todd 1983a, 1983b:72). With the exception of the effects of snow drifts, Haynes (1991) has come to conclusions similar to Todd's, even though Haynes studied carcasses of African elephants, animals several times larger than the domestic cows studied by Todd. Finally, Micozzi (1986) found that previously frozen carcasses tend to disarticulate more rapidly than fresh carcasses.

Numerous actualistic studies indicate that as the frequency of predators increases relative to the frequency of prey, the intensity of exploitation of a given prey carcass by predators will increase, and thus the rate of skeletonization and extent of bone dispersal will also both increase (D'Andrea and Gotthardt 1984; Haynes 1980a, 1980b, 1982; Hewson and Kolb 1986; Todd 1983a). In his study of North American wolves (*Canis lupus*), Haynes (1980b, 1982) found that as the number and hunger of wolves increased, the intensity and extent of carcass exploitation increased, and, the wolves tended to more fully exploit carcasses of animals they had killed than carcasses of dead animals they found. He described the sequence of prey carcass utilization and the sequence of bone destruction in order to approximate stages of carcass exploitation (Table 5.6).

A sequence of exploitation of prey carcasses by African carnivores was developed by Blumenschine (e.g., 1986a, 1986b, 1987), who monitored numerous carcasses of animals both killed and consumed by the carnivores, and scavenged carcasses the original predator for which was unknown. That sequence (Table 5.7) is similar to the one Haynes documented for North American wolves, although detailed comparison between the two is not possible because Haynes' sequence is not as detailed as Blumenschine's. Blumenschine's general carcass consumption sequence indicates that tissue external to bones, or flesh, is eaten in a particular sequence by mammalian scavengers first, and then that same sequence is followed when scavengers exploit tissues such as marrow and "pulp" that are inside bones (Table 5.7). The soft tissue consumption sequence tends to begin with the skeletal parts with large amounts (by weight) of tissue (Blumenschine and Caro 1986), and ends with the skeletal parts with low amounts of tissue (Figure 5.14; $r_s = 0.673$, $P < 0.05$). Omitting the rib-cage and cervical vertebrae, which Blumenschine (1986a:644) suggests are anomalous because these parts are "difficult to gain complete access to without first eating parts posterior or (less frequently) anterior to them," significantly strengthens the relationship between the consumption sequence and flesh weight ($r_s = 0.833$, $P = 0.02$).

Haynes' (Table 5.6) and Blumenschine's (Table 5.7, Figure 5.14) data are important because they help explain Hill's (Table 5.5) sequence of disarticulation. Hill's sequence was apparently influenced by mammalian scavengers, but to an unknown degree, and thus it is not surprising that it tends to indicate the appendicular skeletal elements become disarticulated prior to the axial skeleton. This is similar to the sequence of soft tissue consumption found by Haynes and Blumenschine. More detailed comparisons cannot be performed with the data at hand.

Hill (1979a, 1979b, 1980; Hill and Behrensmeyer 1984, 1985) suggests the natural disarticulation sequences he describes can be compared to disarticulation sequences produced by humans butchering similar taxa as a way to monitor how joint anatomy influences butchery. Hill (1979a; Hill and Behrensmeyer 1985) finds great similarity between his natural sequences and one North American Paleoindian bison (*Bison* sp.) kill site assemblage, and also finds

Table 5.6 *Sequence of damage to bones of ungulates exploited by North American wolves (from Haynes 1982; with modifications and additions from Haynes 1980a)*

Skeletal element	Light to moderate utilization	Full utilization	Heavy utilization
mandibles	still articulated, angular process lightly gnawed, hyoids present	one hyoid present, angular process gnawed	disarticulated, ventral border broken off, medial (lingual) surface gnawed
skull	no damage to bones	nasals tooth-scratched	nasals ragged at ends, anterior premaxillaries may be broken
vertebrae	some processes gnawed		most vertebral processes gnawed and removed
pelvis	edges of ilia and ischia gnawed, trabecular bone exposed	ilia and ischia partly gone	only stumps of ilia and ischia remain
scapula	vertebral border gnawed and ragged	vertebral border splintered and jagged, disarticulated from humerus	blade crunched and splintered, only glenoid fossa portion may remain
humerus	greater tuberosities gone or furrowed	tuberosities gone, tooth scoring on shaft	proximal end gone, $\frac{1}{3}$ of proximal shaft gone, distal condyles gnawed
femur	trochanteric stump left, greater trochlear rim scored at right angle to long axis, minor damage to medial condyle	medial condyle gouged, surface of lateral condyle gone, trochlea well opened, trochanteric stump gone, tooth marks undercut head	distal end gone, head nearly gone, shaft breaking up
tibia	lateral proximal end grooved or beveled, some furrowing and gouging	crest open or gone, medullary cavity exposed at lateral proximal end, medial edges	proximal end gone, fracture edges sharp with localized rounding, tarsals still articulated, shaft broken last
GENERAL OBSERVATIONS	two or three limbs still articulated with body, three or four matapodials and all phalanges present	three or four limbs disarticulated from body	few metapodials and phalanges present

Table 5.7 *Ranked general consumption sequence,
from first to last eaten (from Blumenschine 1986a)*

Carcass unit	Rank	Carcass unit	Rank
Hindquarter flesh		Hindlimb marrow	
pelvis	1	femur	13
lumbar	3	tibia	14
femur	2	metatarsal	16
tibia	7	phalanges	15
Forequarter flesh		Forelimb marrow	
rib-cage	4	humerus	17
scapula	6	radius	19
humerus	5	metacarpal	20
radius–ulna	9	phalanges	18
cervical	8	Head contents	
Head flesh		maxilla pulp	21
tongue	10	mandible pulp	23
mandible	11	brain	22
maxilla	12	frontal pulp	24

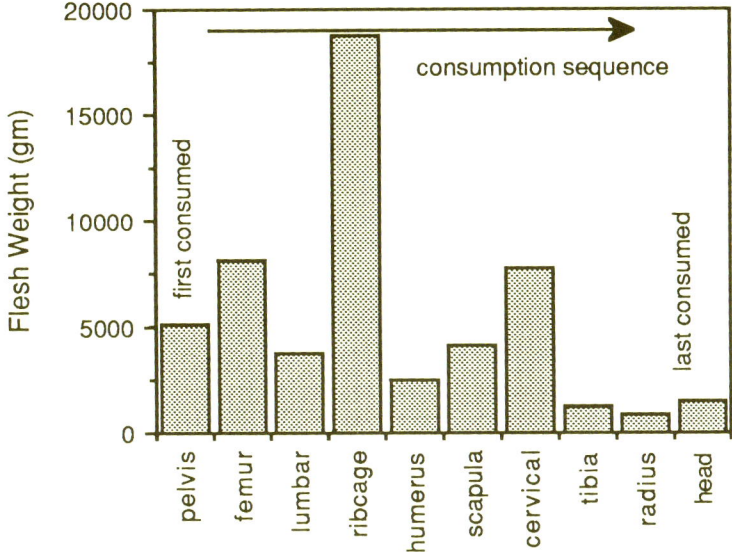

Figure 5.14. Blumenschine's (1986a, 1986b) consumption sequence plotted against flesh weight. See text for discussion.

consistent differences between the taxon-specific natural sequences and that bison assemblage. He thus concludes "disarticulation is extremely uniform in different large herbivore species, whatever the agent of dismemberment, human or non-human" (Hill and Behrensmeyer 1985:144).

Humans, like large quadrupedal carnivores, have a size advantage relative to many prey carcasses such that they can move carcass parts rather than simply cause a carcass to fall apart as bacteria and carrion insects do. That advantage is exacerbated by the use of tools by human butchers to aid in tasks of skinning, defleshing, and dismemberment. Tool use, in turn, often creates traces in the form of butchering marks which allow the analyst to identify humans as agents of disarticulation (Chapter 8).

It is important to note that, thus far, discussion of disarticulation has been limited to separation of bones at their points of articulation, or at joints. Most authors suggest that joints are the easiest places for non-industrial human butchers to disarticulate bones of a carcass (e.g., Binford 1978; Gilbert 1979; Read 1971). But bones can also become anatomically disassociated by breaking them during butchery (or by consumption by large quadrupedal carnivores), as noted by Binford (1978:63) when he writes "under certain conditions (e.g., frozen carcasses) bones may be broken during [dismemberment]" (see Chapter 8).

Analysis of disarticulation and scattering

> It is well worth while considering fragments in their context, as found. Several undistinguished chips and flakes of bone, found together, may have joins preserved which will enable something significant to be rebuilt.
> (I. W. Cornwall 1956:197)

If *articulation* is defined as two or more skeletal elements being in their proper anatomical positions relative to one another, and within a centimeter of each other if not in fact touching, and if *scattering* is defined as increasing the spatial distance between anatomically related bones (see the Glossary), then clearly both are most readily measured in the field during the collection of faunal remains. Recording which bones were found in contact with one another and recording the distance between bones lying close to, but not in contact with, one another are preferable to simply noting which bones came from the same or adjacent excavation units, and then turning the bones and those notes over to the taphonomist for study. Yet the latter is commonplace.

Abundant ethnoarchaeological data indicate Thomas (1971:367) was correct twenty years ago when he wrote "the dietary practices of man tend to destroy and disperse the bones of his prey-species." It is the rare case when bones that were articulated in life are found articulated in archaeological (or ethnoarchaeological) contexts. (Although articulated skeletal elements may be more common in paleontological contexts than archaeological ones, there are

Table 5.8 *Joint articulation data for bison bones from the Casper site and the Horner II site. See text for discussion*

Joint	Casper			Horner II		
	Hill's n	CJF	Potential[a] (% articulated)	Hill's n	CJF	Potential[a] (% articulated)
scapula–humerus	1	0.11	70 (1.4)	17	0.70	77 (22.1)
humerus–radius/ulna	38	4.28	65 (58.5)	45	1.84	122 (36.9)
radius–carpals	22	2.48		37	1.52	
carpals–metacarpal	21	2.36		61	2.50	
metacarpal–1st phalanx	26	2.93		57	2.33	
1st phalanx–2nd phalanx	25	2.82		55	2.25	
2nd phalanx–3rd phalanx	25	2.82		52	2.13	
femur–tibia	15	1.69	53 (28.3)	30	1.23	95 (31.6)
tibia–tarsals	19	2.14	45 (42.2)	50	2.05	110 (45.4)
tarsals–metatarsal	24	2.70	52 (46.2)	89	3.64	126 (70.6)
metatarsal–1st phalanx	26	2.93		96	3.93	
1st phalanx–2nd phalanx	25	2.82		94	3.85	
2nd phalanx–3rd phalanx	25	2.82		92	3.77	
Hill's N	= 444			= 1221		

Note:
[a] Potential number of articulations (% of potential that are actually articulated).

no data I am aware of to substantiate such a claim.) In the following, I review several techniques for measuring disarticulation and several others for measuring scattering. It is thus important to note that the two processes and their results are not mutually exclusive. The analytical techniques described below are good ones, irrespective of whether they are used to measure disarticulation or dispersal. The sub-section headings are, then, heuristic devices rather than logical ones.

Articulation and disarticulation

Hill (1979a, 1979b) measures the order of joint disarticulation with the following equation:

$$100n \div NR = CJF \qquad [5.1]$$

where N = the total number of all intact or articulated joints in a bone assemblage, n = the total number of a particular skeletal joint in a bone assemblage, R = the number of times a particular joint occurs in a single skeleton, and CJF is the corrected joint frequency. As an example, I calculated CJF values for the front and rear limbs of North American bison skeletons as documented at the Casper Site (Frison 1974) and the Horner II site (Todd 1987b). Those frequencies (Table 5.8) indicate the scapula–humerus (shoulder) joint and the femur–tibia (knee) joint were the "first" to be disarticulated

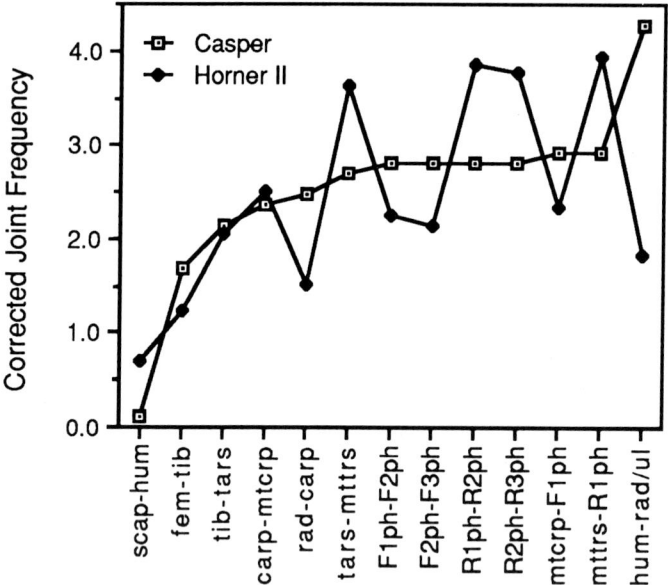

Figure 5.15. Order of joint disarticulation at Casper and Horner II sites as determined by Hill's (1979a, 1979b) method. Abbreviations: scap, scapula; hum, humerus; rad, radius/ulna; carp, carpals; mtcrp, metacarpal; F1ph, forelimb first phalanx; F2ph, forelimb second phalanx; F3ph, forelimb third phalanx; fem, femur; tib, tibia; tars, tarsals; mttrs, metatarsal; R1ph, rear limb first phalanx; R2ph, rear limb second phalanx; R3ph, rear limb third phalanx.

according to Hill's model, but then the "order" of disarticulation differs between the two assemblages (Figure 5.15). Spearman's rho between the two sets of CJF values is weak ($r_s = 0.52$, $P = 0.07$), suggesting that bison limbs at these two sites were disarticulated in rather different orders.

Hill's (1979a, 1979b) method is founded on calculating the proportional frequency of articulated joints for each type of joint represented, weighted by how many times a joint type occurs in a skeleton and the total number of all types of articulated joints. Thus the analyst may find one distal humerus and one proximal radius that are not articulated, or the analyst may find only the distal humerus or the proximal radius, or the analyst may find neither the distal humerus or the proximal radius, and the CJF value for the elbow joint would be the same in all three cases because it only tallies the articulated joints. That is, Hill's method and equation [5.1] may be useful for comparative purposes (e.g., Figure 5.15), but this does not consider how many joints are possible given the bones recovered, and it does not consider how many of those possible joints are not articulated.

Todd (1987b:142–146) suggests the analyst count the number of articulated joints that are possible for a bone assemblage, and then calculate the percentage of those possible joints that are actually articulated. Burgett (1990:163)

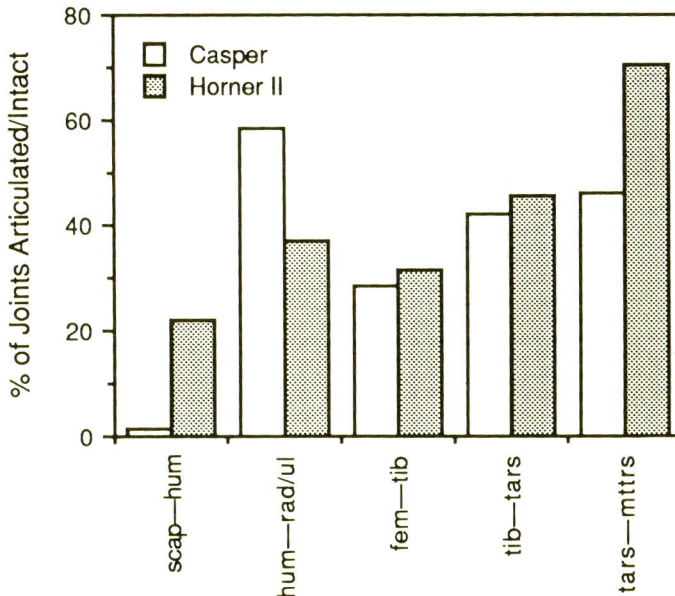

Figure 5.16. Proportion of articulated joints at Casper and Horner II sites as determined by Todd's (1987b) method. Abbreviation: scap, scapula; hum, humerus; rad/ul, radius-ulna; fem, femur; tib, tibia; tars, tarsals; mttrs, metatarsal.

labels this the "percentage of potential articulations," or PPA. For example, an assemblage of 15 distal left humeri and 20 proximal left radii has the potential to have 15 articulated left elbow joints. Note that the potential number is the lower of the two values because both bones are required to make an articulating joint. The procedure assumes that all specimens are from individuals of the same sex and ontogenic age. If 10 of those 15 potential elbow joints are articulated or intact, then 67% (10 of 15 = PPA) are articulated. Such calculations for selected joints in the Casper and Horner II bison assemblages (Table 5.8) indicate, for example, about the same proportion of knee and ankle joints are articulated (and disarticulated) at both sites, but more shoulder and ankle joints at Horner II are articulated than at Casper, and more elbow joints are articulated at Casper than at Horner II (Figure 5.16).

Burgett (1990:163) labels another statistic the "percentage of surviving articulations," or PSA. The PSA is "the frequency of articulations recorded for each joint observed, divided by the frequency in which each joint occurs in a complete skeleton, [and multiplying the result by 100 to provide] a percentage value" (Burgett 1990:155). This statistic was designed specifically for actualistic research wherein the number of animals (and thus the frequency of potential joints) is known. In Burgett's case, the number of animals was always one. In archaeological cases, the analyst could use the MNI as an approximation of the number of animals in an assemblage of faunal remains, and multiply the number of joints per skeleton by the MNI to derive a denominator for calculating the PSA. Finally, Burgett (1990) suggests comparing PSA values for the total vertebral column (e.g., 27 joints in one bison [*Bison bison*] would be the denominator) with the PSA values for the complete forelimb (denominator = 16 when MNI is 1) and PSA values for the complete hindlimb (denominator = 14). Forelimb joints are: thoracic–scapula, scapula–humerus, humerus–radius-ulna, radius-ulna–carpals, carpals–metacarpal, metacarpal–first phalanges, first phalanges–second phalanges, second phalanges–third phalanges, times two for both limbs. Hindlimb joints are: innominate–femur, femur–tibia, tibia–tarsals, tarsals–metatarsal, metatarsal–first phalanges, first phalanges–second phalanges, second phalanges–third phalanges, times two for both limbs.

Burgett (1990) found little relation between disarticulation of bison and wapiti (*Cervus elaphus*) skeletons scavenged by coyotes (*Canis latrans*) and the live weight of the animals. Similarly, there was no relation between the ontogenic age of the prey carcasses and the PSA. He did find, as Hill (1979b) did, that the axial skeleton remained articulated longer than bones of the limbs. Once the state of articulation and disarticulation are recorded, the analyst can begin to study other phenomena, such as frequencies of butchery marks, seasonal variation in animal death, post-depositional disturbance, bone weathering, and the sedimentary context in seeking an explanation for variation in the observed state of articulation and disarticulation.

Dispersal and scattering

The usual terms for analytical re-articulation are *refitting studies* or studies of *conjoining pieces*. Such studies had their origin in the archaeological analysis of stone artifacts. Schild (1976:96), for example, reports that the "demanding study technique involving matching stray flint flakes and blades, and even completed tools, with the flint 'cores' from which the blanks used for tool manufacture were struck, generates what we call *articulation nets*, which define the boundaries of a single occupation unit." Cahen and Moeyersons (1977:813) report that their

> reassembly of worked stones (that is, joining together the tools, flakes, core and fragments struck off from the same block) [produced] fitting artefacts at several different depths from which different radiocarbon dates have been obtained (the vertical distance between joining pieces sometimes exceeds 1 m) ... the vertical distribution of the elements of the reconstructed cores does not follow the order in which they have been struck off. This excludes the hypothesis of a later reutilisation of older artifacts.

These observations led them to conclude that subsurface movement of the stone artifacts had resulted in their differing vertical proveniences.

Study of the horizontal and vertical proveniences of refitting or conjoining pieces has become a valuable analytical tool for understanding site formational processes (Hofman 1981, 1986). Such mechanical refitting is not unlike working through a complex jigsaw puzzle wherein the sample of pieces one is trying to fit together actually consists of several variously incomplete and complete puzzles. It was attempted with faunal remains soon after intriguing results began to emerge from the study of refitting stone artifacts. For example, Johnson (e.g., 1982, 1987) refit bone pieces distributed across a single horizontal provenience to identify "butchering articulation nets." Conjoining pieces of skeletal elements from different vertical proveniences were reported by Villa *et al.* (1986:433) in their attempt to identify clusters of bones that had been "processed and discarded at the same time." Such clusters, when disturbed by post-depositional (and by implication post-butchering) processes, were identified by Villa *et al.* (1986:433) as clusters of bone "pieces that can be refitted and by a higher (than average) density of pieces within a restricted area as shown by horizontal and vertical plots of their observed positions."

Villa *et al.* (1986) examined the pieces of particular skeletal elements that fit together. Johnson (1982, 1987) examined both pieces of skeletal elements that mechanically fit together, and skeletal elements that anatomically seemed to represent the same individual animal. That is, Johnson performed both *mechanical refitting* and *anatomical refitting* analyses. The former is analogous to refitting stone artifacts and involves the mechanical process of assessing whether two pieces fit together (just like with a jigsaw puzzle); these may be fragments of the same bone, such as a shaft fragment and an end fragment, or it may be two articular ends that were articulated in life. Anatomical refitting may involve the use of biological observations such as age and sex variation between the individuals represented in the faunal collection; if only one adult and one subadult organism are represented in the collection, then it is a relatively straightforward matter to assign specimens the ontogenic age of which can be determined to one or the other individual. This kind of anatomical refitting becomes more difficult as the number of individuals within particular age, sex, and/or size categories increases. For large samples with many individuals in each age/sex/size category Todd (1987b) suggests the principle of bilateral symmetry may allow the anatomical refitting of paired bones. Such bilateral refits require much comparative data to establish the range of variation between paired bones. Similarly, Todd (1987b:179) suggests such comparative data may also permit "intermemberal refitting" of skeletal elements because for many taxa "the length of a femur can be predicted from the length of the humerus of the same animal."

If the assemblage consists of only 10 specimens, each specimen must be checked against the other nine to determine if the two refit. Thus, for 10 specimens there are 45 possible refits of any two. These 45 possible pairs do not, of course, account for the fact that for each two possibly refitting specimens there are multiple ways they might fit together. For example, if each specimen has two ends, then there are four possible ways for an end of one specimen to refit with an end of the other specimen. The 45 possible pairs, in the case of our 10 specimens, becomes 180 possible ways to refit an end of one specimen to the end of another specimen. The magnitude of the refitting task is exacerbated by the fact that bone specimens not only have ends, but they also often have two or more sides or edges, such as when a specimen consists of only the lateral portion of a long bone diaphysis.

There are, then, several ways to refit bone specimens, or to detect which specimens conjoin. The important questions, once the refitting specimens have been identified (perhaps regardless of the method of refitting, but see below), concern how to summarize those data, and determining what the kind and degree of refitting means in terms of taphonomic processes. One might simply count the number of bone specimens, including non-identifiable ones, in an assemblage, and also count the number of refit pieces to derive a proportion of specimens with refitting pieces (if two specimens refit, tally both as having a specimen that conjoins with it). Given the labor-intensive nature of refitting, however, the unclear taphonomic meaning of a statistic such as the proportion of specimens that refit with at least one other specimen may be pointless to calculate.

Studies of conjoining bone specimens focus not just on the fact that some specimens can be refit with one another. Those studies also examine the horizontal and vertical provenience of the refit specimens. Thus it is the relative spatial locations of refitting specimens that are important to refitting studies. Villa *et al.*'s (1986) map shows that the dense clusters of bones thought on the basis of contextual and stratigraphic data to represent synchronous depositional events had refitting pieces somewhat removed from them. Johnson's (1982, 1987) "butchering articulation nets" indicate refitting pieces were found dispersed between several distinct clusters of bones, suggesting the clusters were contemporary and probably represent subsets of the same butchering event. If it is remembered that *disarticulation* is the simple anatomical disassociation of bones and involves their spatial separation some minimal distance, and that *scattering* is the further spatial disassociation of the bones, then the distance between conjoining specimens can be used as a measure of disarticulation and scattering.

Todd (1987b:189, 193) presents two methods for measuring the degree of scattering. For paired bones such as femora, mandibles, and the like, which can be anatomically refit, he uses the *index of skeletal disjunction* (ISD). This index is individually calculated for each kind of paired skeletal element (for femora, for mandibles, etc.) by first measuring the minimum distance between each anatomically refit pair. Then, the average minimum distance is calculated for each kind of paired skeletal element. The number of conjoined pairs is multiplied by two, and added to the number of unpaired specimens in the

Table 5.9 *Index of skeletal disjunction (ISD) and index of fragment disjunction (IFD) for the Horner II bison remains (from Todd 1987b)*

Index of skeletal disjunction

Element	N of pairs	MD	Total N	% Pairs	(MD ÷ % Pairs) 100	Standardized ISD
Humerus	9	0.99	43	41.86	2.365	48.58
Radius	10	1.10	43	46.51	2.365	48.58
Femur	8	2.05	38	42.11	4.868	100.00
Tibia	11	0.85	44	50.00	1.700	34.92

Index of fragment disjunction

Element (1)	N of refit fragments (2)	Minimum N of elements with refits (3)	MD (4)	Col. 3 ÷ Col. 2 (5)	Col. 4 ÷ Col. 5 (6)	Standardized IFD (7)
Humerus	14	7	0.39	0.50	0.78	29.78
Radius	4	2	0.35	0.50	0.70	26.73
Femur	31	13	1.10	0.42	2.62	100.00
Tibia	12	4	0.41	0.33	1.24	47.44

collection to produce the total number of bones in a category. The number of bones with a conjoined mate is divided by the total number of bones in a category to derive the percentage of bones with mates. The mean minimum distance between anatomically refit specimens (in meters) is then divided by the percentage of bones with mates and multiplied by 100. The resulting numbers are then scaled from 1 to 100 to derive the ISD. Equation [5.2] summarizes the calculations of the ISD, and the method for calculating Todd's (1987b) *index of fragment disjunction* (IFD) is summarized in equation [5.3]. In these equations, MD is the mean minimum distance between refit specimens, and $_i$ is the skeletal element under consideration (e.g., femur, or mandible, or humerus). In [5.2] relatively complete, anatomically refitting skeletal elements are considered, and in [5.3] mechanically refitting fragments of skeletal elements are considered.

$$\frac{(MD_i \div \% \text{ of }_i \text{ that anatomically refit) } 100}{\text{maximum value in the numerator for the assemblage}} \quad [5.2]$$

$$\frac{MD_i \div (\text{minimum number of }_i \text{ with refit fragments} \div \text{total number of refitted fragments of }_i)}{\text{maximum value in the numerator for the assemblage}} \quad [5.3]$$

An example will help bring the indices into focus. Data presented by Todd (1987b:189, 193) for the Horner II bison remains are given in Table 5.9. Both the ISD values and IFD values are greatest for the femur, and are relatively low for the other limb bones. These examples illustrate how to calculate ISD and IFD values, but what is the taphonomic significance of such values? For one thing, greater index values indicate greater scattering (spatial disassociation),

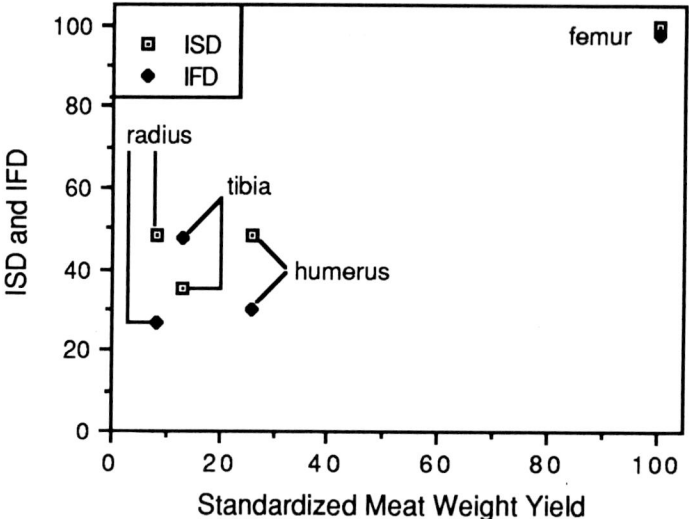

Figure 5.17. Bivariate scatterplot of index of skeletal disjunction (ISD) and index of fragment disjunction (IFD) against standardized meat weight yield for Horner II bison.

and thus one might calculate either index to determine which kind(s) of skeletal elements are greatly scattered and which are not. The indices might also be calculated for different taxa to determine if there is taxonomic variation in scattering. But again, the question will arise as to why some categories of skeletal portions are more scattered and other categories are less scattered.

As Thomas (1971) observed, humans tend to scatter the remains of animals they butcher. Thus one might predict skeletal portions with high economic utility will be more scattered than skeletal portions with low economic utility (see Chapter 7). I followed Todd's (1987b) lead and plotted the ISD and IFD values in Table 5.9 against the average meat yield and against the average marrow yield for each kind of bone as measured on three adult bison (one male, two females) by Emerson (1990). The results suggest femora display the greatest ISD due to their high meat yield as the other skeletal elements are neither as greatly scattered (lower ISD) or as meaty (lower meat yield) (Figure 5.17). The scattering of these complete, non-fractured bones may result from dismemberment to enhance transport (lighten the weight) of skeletal portions or to enhance removal of meat from the bones. The IFD does not seem to be strongly related to the marrow yield per category of skeletal portion (Figure 5.18). This is perhaps so because breaking bones for marrow extraction may not result in the disassociation of fragments of the broken bone as once it is broken and the marrow extracted there is little economic reason to move the fragments. For example, Figure 5.19 indicates that the ISD and IFD are not greatly different for each kind of skeletal element in this sample. That is, the

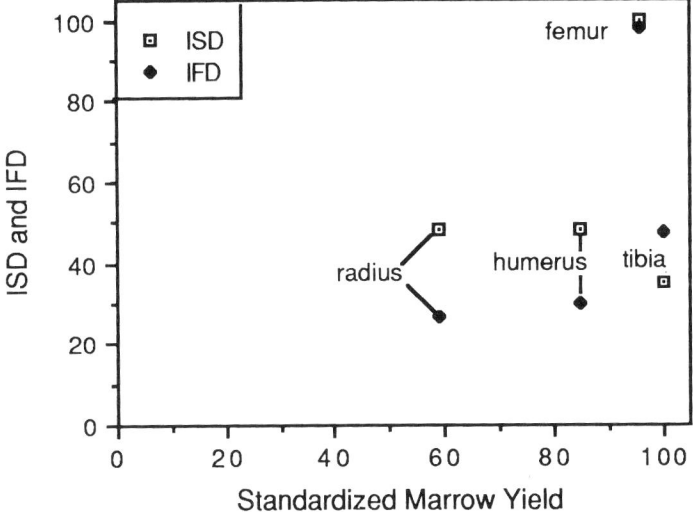

Figure 5.18. Bivariate scatterplot of index of skeletal disjunction (ISD) and index of fragment disjunction (IFD) against standardized marrow yield for Horner II bison.

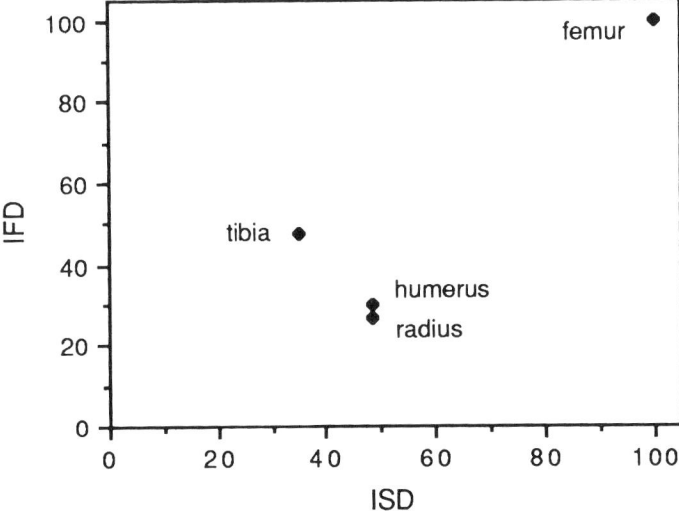

Figure 5.19. Bivariate scatterplot of index of skeletal disjunction (ISD) against index of fragment disjunction (IFD) for Horner II bison.

anatomically refitted bones (measured by the ISD) were dispersed to a degree similar to the mechanically refitted fragments (measured by the IFD).

What I have sought to do with the above example is not to solve the taphonomic problem of why skeletal parts have the distributions they do at the Horner II site. Taphonomic questions are seldom so easily answered. Rather, I have simply illustrated how two measures of scattering might be applied to a bone collection when the analyst seeks to explain the distribution of skeletal parts in a particular site.

Summary

Taphonomic processes, by definition, can only begin after an animal dies. The cause of mortality often has significant taphonomic consequences. Catastrophic or non-selective mortality of multiple individuals in a small geographic area can result in an abrupt and major input to the fossil record. Normal attritional mortality provides a relatively constant input of a few carcasses each year, generally scattered over a large spatial area. Study of the demography of the dead population of organisms represented by a fossil collection helps the analyst to ascertain the mode, and perhaps the cause, of mortality. When analysis of the demography of mortality is done in conjunction with determination of the season of mortality, additional insights to causes of mortality may be granted.

Seldom are soft tissues of animal carcasses preserved. When they are, such preservation often involves what might be thought of as mummification processes. More typically, soft tissues are removed and the skeleton progressively looses its anatomical integrity; that is, the bones become disarticulated and eventually spatially disassociated. There are many ways for an animal carcass to be skeletonized, and these usually involve the consumption of soft tissues by various microscopic and macroscopic organisms. The order of joint disarticulation in a skeleton seems to be a function of the structure of a joint and its associated soft tissue. Analysts interested in skeletonization and disarticulation typically calculate the proportion of articulated joints represented in a collection of bones, and may perform various kinds of analyses of conjoining bones and bone fragments to estimate the degree of disarticulation and scattering.

Sometimes complete carcasses and sometimes only parts of carcasses are dispersed from the site of animal death and original carcass deposition. These carcasses and carcass parts may eventually be accumulated in a particular geographic place by any of a myriad of bone accumulating processes and agents. It is the dispersal and accumulation of faunal remains to which we turn in the next chapter.

6

ACCUMULATION AND DISPERSAL OF VERTEBRATE REMAINS

What are all these bones *doing* here?
(P. Shipman 1979:42)

Introduction

Paleontologists and zooarchaeologists tend to be optimal foragers when choosing a place to collect faunal remains. For the former, it saves time and money. For the latter, it not only saves time and money, but the places chosen are usually selected by an archaeologist because they contain a dense concentration of artifacts; if bones are spatially associated with the artifacts, then they too are usually collected (because they are there and) regardless of their frequency per unit volume of sediment. Typically, few archaeological sites are excavated or collected simply because they contain animal remains. Determining how the animal remains came to be in the locations from which they are collected, regardless of their geographical and geological position when collected, is one of the most fundamental aspects of taphonomic research.

Why are bones densely concentrated in a particular location, but not in surrounding areas? Why are some kinds of bones present or abundant and other kinds absent or rare? Why are some skeletons completely articulated, some partially articulated, and others totally disarticulated? Why are some bones spatially close and others spatially distant from one another? Why are some bones oriented one way and others oriented another way? Why do some assemblages have lots of carnivore remains and others have few relative to the frequency of herbivore remains? This sampling of questions implies a number of variables that might be measured in an analysis of accumulation. Some of these are discussed at length in other chapters. In this chapter the variables that seem to be most directly related to determining how bones have been accumulated and/or dispersed are considered. We begin with some general concepts before delving into details of variable measurement and analysis.

Dispersal, scattering, and accumulation

> Concentration of remains from multiple individuals occurs either by active transport of prey remains to a focal area or by passive accumulation at a site where predation occurs repeatedly.
> (C. Badgley 1986a:336)

Animal carcasses begin their taphonomic history as articulated skeletons. Skeletonization leads to disarticulation and scattering of individual segments

161

of carcasses and skeletons. Sometimes those segments are sets of articulated skeletal elements held together by connective tissue (such as a complete limb) and other times they are individual, discrete skeletal elements or fragments thereof. Hill (1979b:268–269) defines *scattering* as "the increase in dispersion of the parts of a skeleton," and implies that skeletal parts become more dispersed as they become more spatially distant from the skeleton that was their origin; "scattering can be seen to have occurred locally around the point of death [and] leads to the dispersion of remains over a considerable area." *Dispersal* is synonymous with scattering. The degree of dispersal of skeletal parts of individual carcasses depends in part on the length of time between skeletonization and burial; bones of carcasses that are buried prior to skeletonization tend to stay articulated and not be dispersed.

The degree of scattering of skeletal parts of a carcass also depends on the taphonomic processes and agents of scattering or bone transport that act upon the parts. Such processes and agents include scavengers, carnivores, humans, fluvial action, gravity, and trampling. The first three focus on exploitation of soft tissues and bones (see Chapter 7), the latter three tend to act on the physical and morphometric attributes of bones regardless of their soft tissues or economic anatomy (see Chapter 7). The kind and degree of scattering of bones, then, is often dependent on skeletonization processes (which particular individual bones or subsets of articulated bones are initially detached from a carcass), the depositional environment of the carcass, and the processes of scattering.

Scattering has as its starting point the complete set of articulated bones making up a skeleton. Behrensmeyer (1983:98) notes, dispersal of bones by predators and scavengers results in "limbs and sometimes even the head being dragged away from the vertebral column, which generally acts as a kind of 'anchor' from which other parts are removed." Thus, scattering, as a taphonomic process, is at least implicitly monitored or measured from the original depositional position of the complete skeleton. Accumulation, as a taphonomic process, is usually monitored from a different position on the landscape relative to the location of the originally deposited complete skeleton.

Two basic types of bone accumulating processes can be distinguished: active and passive (Badgley 1986a, 1986b). *Active accumulation* processes are those which, via transport or movement of skeletal parts (whether or not as complete carcasses/skeletons) significant distances from the location of animal death, result in relatively dense concentrations of bones and teeth in a spatially limited area. Such processes are labeled "spatially focused processes" by Behrensmeyer (1983:94). Active accumulation involves forces and energy external to the animal(s) whose bones are accumulated. *Passive accumulation* processes are those which do not involve transport of skeletal parts significant distances from the location of animal death; such processes are not spatially focused, and have been considered to represent normal attritional mortality and deposition of animal remains close to the place of death (Behrensmeyer 1983). Passive accumulation involves forces and energy internal to the animal(s) – its behavior – whose bones are accumulated. A good example of an active accumulation is an animal den (including human habitation sites) to which adult meat eaters bring portions of prey carcasses for provisioning themselves and/or their young. Passive accumulations can involve loci where multiple individual animals have died or were killed, such as fluvially created paleontological quarry sites (e.g., Voorhies 1969), volcanically created catastrophic mortality of gregarious animals (e.g., Lyman 1989b), or humanly generated bison kill sites (e.g., Frison 1974); in these instances animal remains have not been moved far from the location of animal death. Often, but not always, the form of mortality in these cases is attritional (see Chapter 5), and when mortality is attritional, passive accumulation across a large section of landscape results in the slow but continuous addition of animal carcasses to the (future) fossil record (e.g., Behrensmeyer 1982, 1983; Behrensmeyer and Boaz 1980; Behrensmeyer *et al.* 1979).

Unlike dispersal, accumulation need not begin with an entire carcass or skeleton. That is so because whereas measures of dispersal typically monitor the *spatial disassociation* of parts of individual skeletons, measures of accumulation typically monitor the *spatial concentration* of bones regardless of whether those bones derive from the same individual carcass or multiple, independent carcasses, although determination of the number of carcasses that contributed bones (e.g., MNI) may be important. Basically, when taphonomists consider accumulation, they are seeking to identify the taphonomic agents and processes that collected and deposited the faunal remains. Dispersal involves movement of bones away from the location of animal death and carcass deposition, whereas accumulation involves movement and deposition of bones in a location away from the location of animal death and carcass deposition. Both involve movement of animal parts, but for the former the monitoring perspective of the analyst is located at the site of animal death and original carcass deposition whereas for the latter the analyst's monitoring perspective originates at the site of bone deposition (typically spatially removed from the site of carcass deposition).

Behrensmeyer (1987:430) outlines a typology of the "major types of bone occurence, for animals larger than 15 kg, according to causes [of mortality and accumulation] and how these relate to biological and physical agencies, transport and time" (Figure 6.1). That typological system concerns what can be considered *interpretive types*, as the "types of occurence" include such categories as "caches" and "predation arenas." Behrensmeyer (1987:431) outlines "general taphonomic characteristics" of each type; that is, she provides a list of attributes the analyst can study to determine if a particular bone assemblage represents a "den" or a "trap." The typological system is instructive because it identifies four variables that play a significant role in the accumulation history

TYPE OF OCCURRENCE

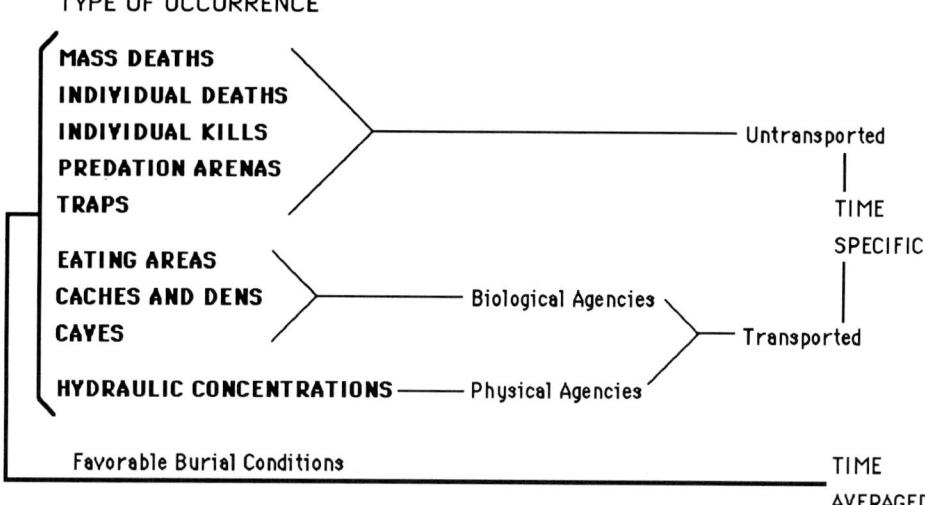

Figure 6.1. Types of bone occurrence based on mortality type (individual, mass), bone accumulation agencies, transport, and duration of accumulation (after Behrensmeyer 1987:430, Figure 1, courtesy of the author and Plenum Press).

of a fossil assemblage. First, the number of individuals that die (mass or multiple versus single) establishes the frequency of remains that may eventually be incorporated into the fossil record. Second, bone assemblages may be untransported (passive accumulation) or transported (active accumulation), and third, agents of transport may be biological (e.g., carnivores) or physical (e.g., fluvial action). Finally, assemblages may represent "time specific" occurrences consisting of faunal remains accumulated over a relatively short time period, or "time averaged" occurrences that consist of faunal remains accumulated over a long time period.

Each of the four variables – mortality, passive or active accumulation, accumulation agent, and accumulation duration – in Behrensmeyer's (1987) scheme has two variable states (Table 6.1). Arranging these four dimensions of variability so that they intersect one another to form all possible combinations of the four variables and eight variable states results in 16 possible classes of bone accumulations (Table 6.2, Figure 6.2). These classes are not interpretive, but rather simply describe all possible combinations of the variable states identified by Behrensmeyer (1987), something her typology (Figure 6.1) does not do. Further, the classes do not differentially weight the dimensions of variability, as Behrensmeyer's typology does, and thus each resulting class is defined on the basis of the same suite of dimensions. Behrensmeyer's (Figure 6.1) "mass deaths" type is different than her "eating areas" type due to differential weighting of variables; the former is distinguished on the basis of

Table 6.1 *Dimensions of variability in the process of bone accumulation*

VARIABLE
Variable State
comments

I.   MORTALITY
      A. Single Individual
            1. attritional mortality (see Chapter 5)
      B. Multiple Individuals
            1. mass or catastrophic mortality (see Chapter 5)
II. ACCUMULATION ACTION
      A. Untransported
            1. passive accumulation
      B. Transported
            1. active accumulation
III. ACCUMULATION TYPE
      A. Biological
      B. Physical
IV. ACCUMULATION DURATION
      A. Time Specific
            1. short duration, fine grained
            2. one mortality event (either one or multiple individuals)
            3. one accumulation action
            4. one accumulation type
      B. Time Averaged
            1. long duration, coarse grained
            2. more than one mortality event (multiple instances of either or both one or
               multiple individuals)
            3. more than one accumulation event
            4. perhaps more than one accumulation action
            5. perhaps more than one accumulation type

the number of dead individuals whereas the latter is distinguished on the basis of one possible outcome for one or more dead individuals (they were eaten).

The classification system described by Table 6.2 and Figure 6.2 can subsume all of the interpretive types described by Behrensmeyer (1987), and can accommodate other interpretive types as well (Table 6.3). The classification system does not, however, result in the same separation of some of the types as Behrensmeyer's typology. Part of the reason for that is the taxonomic structure of the typology in Figure 6.1 and the differential weighting of the variable states in that typology. All variable states are equally weighted in the classification in Table 6.2 and Figure 6.2, thus all variable states make an equal contribution to defining the classes of bone accumulation.

My intent in presenting the typology in Figure 6.1 and the classification in Figure 6.2 is to highlight some of the important variables in both schemes. For example, whether or not a bone occurrence is the result of passive (untrans-

Table 6.2 *Classes of variation in bone accumulation (from Table 6.1)*

Mortality type	Accumulation type	Accumulation action	Accumulation duration	Cell in Figure 6.2
single	biological	passive	short	1
single	biological	active	short	2
multiple	biological	passive	short	3
multiple	biological	active	short	4
single	physical	passive	short	5
single	physical	active	short	6[a]
multiple	physical	passive	short	7
multiple	physical	active	short	8
single	biological	passive	long	9
single	biological	active	long	10
multiple	biological	passive	long	11
multiple	biological	active	long	12
single	physical	passive	long	13[a]
single	physical	active	long	14
multiple	physical	passive	long	15
multiple	physical	active	long	16

Note:
[a] Cell not visible in Figure 6.2.

ported) or active (transported) accumulation is significant because in the case of the former there often, but not always, tend to be fewer fossils per unit area than in cases of the latter (see below). But even passive accumulation, if it is spatially focused and of sufficient temporal duration, can result in rather dense concentrations of fossils. Further, as was emphasized in Chapter 5, the duration of accumulation can result in a set of spatially juxtaposed but temporally separate individual deaths which may have the appearance, in some respects, of a short-duration mass death event. Biological and physical accumulation agents can (a) exploit individual or mass deaths, and (b) operate over short or long time periods. It is those fossil assemblages that have formed due to multiple accumulation events that are termed "time averaged" by Behrensmeyer (e.g., 1982).

Taphonomists often begin their study of a fossil assemblage with a question like "Why are *these* bones *here*?" They seek to identify the processes which accumulated the fossils in order to be able to make inferences about the fossil assemblage and/or the fauna it represents, inferences that are not based on an assemblage that is somehow biased, with regards to those inferences, by the processes which resulted in the accumulation. Knowing the agents and processes of bone accumulation, one can allow, say, for the fact that a hydraulic concentration will not contain many bones with low bulk densities because such bones tend to float more readily than bones with high bulk

Table 6.3 *Alignment of types of bone occurrence (Figure 6.1) with bone accumulation classes (Table 6.2, Figure 6.2)*

Class	Type(s) of bone occurrence (references with examples of type)
1	individual deaths, individual kill sites, traps (Behrensmeyer *et al.* 1979; Burgett 1990)
2	eating areas, caches, dens, caves (Binford 1981b; Brain 1981; Blumenschine 1986b)
3	mass deaths, predation arenas, traps (Frison 1974; Haynes 1991; Olsen 1989b)
4	eating areas, caches, dens, caves (Binford 1981b; Brain 1981; Lam 1992)
5	individual deaths, traps (Lyman 1989b; Oliver 1989)
6	hydraulic concentrations (Boaz 1982; Stewart 1989)
7	mass deaths (Butler 1987; Graham and Oliver 1986; Lyman 1989b)
8	hydraulic concentrations (Boaz 1982; Stewart 1989)

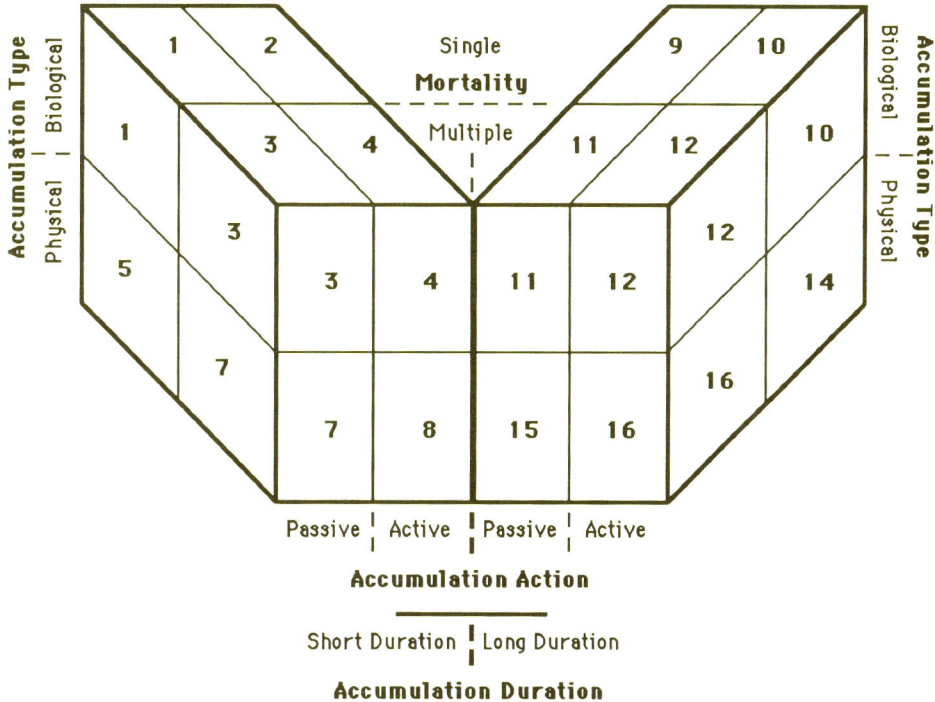

Figure 6.2 Classes of bone occurrence defined by dimensions of variability in accumulation agent (physical, biological), mortality (single individual, multiple individuals), accumulation action (passive, active), and duration of accumulation (short, long) (from Table 6.2). Each numbered block is a distinct class; classes 6 and 13 are not visible.

densities and thus the former are winnowed out whereas the latter, while they might be moved by fluvial action, tend to remain as a lag deposit (e.g., Behrensmeyer 1975b; Korth 1979; Voorhies 1969). It is to the techniques used for identifying both dispersal (which is where accumulation, in a way, begins) and accumulation that we now turn.

Analyzing dispersal

> The primary assumption for the archaeologist to evaluate is that the dietary practices of man tend to destroy and disperse the bones of his prey-species.
> (D. H. Thomas 1971:367)

Dispersal and scattering of skeletal parts has not been as extensively or intensively studied as the accumulation of skeletal parts. Dispersal and scattering begin with disarticulation (see Chapter 5) and end at some (spatial and temporal) point of bone accumulation and deposition. The transition from disarticulation to accumulation is not clear, so distinguishing a stage of dispersal between the two is not easy. However, as noted earlier, dispersal, by definition, can be (spatially) measured from the original depositional location of an individual carcass. Thus, the refitting techniques and analyses of disarticulation described in Chapter 5 are both applicable to studying dispersal. The indices of skeletal disjunction (ISD) and fragment disjunction (IFD) (Chapter 5) can be used to measure the degree of dispersal of the parts of individual skeletons if other evidence suggests the location of animal death is included in or near the sampled space. Large ISD or IFD values denote significant dispersal, but do not identify the agent(s) or process(es) responsible for that dispersal. If the monitoring perspective of the analyst must originate from the location of animal death and carcass deposition in order to measure dispersal, how is that locus identified from the zooarchaeological record? Such an identification would seem to be the first step in the analysis of dispersal, and may involve techniques for assessing how bones were accumulated.

Non-zooarchaeological as well as zooarchaeological criteria indicating that a site is a kill site seem to be the ones often used to identify a location as the place of animal death and original carcass deposition. Similarly, non-faunal archaeological criteria are often used in conjunction with zooarchaeological criteria to infer a site is a habitation, consumption, or occupational site, that is, something other than a kill site. In the case of the former, archaeological evidence of a trap or jump is often used, although other lines of evidence should also play an important role. Olsen (1989b), for example, shows that the traditional interpretation of the Upper Paleolithic site of Solutré in France as a horse jump is wrong for a number of reasons. Zooarchaeological and taphonomic evidence suggests the wild horse (*Equus ferus*) remains recovered from this site represent a catastrophic mortality event (Figure 6.3), but the topographic setting of the remains, the inferred migratory patterns of the prehistoric

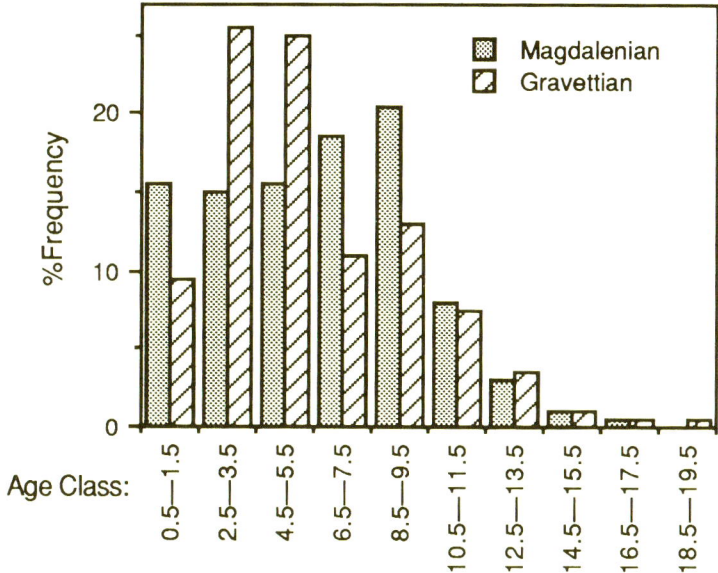

Figure 6.3. Equid mortality profiles for Magdalenian and Gravettian levels at Solutré, France (after Levine 1983). Age classes are in years and are given as class mid-points.

horses, and the behavior of modern wild horses all point to prehistoric hunters intercepting and directing migrating herds into a cul-de-sac at the base of a cliff where horses were corralled and killed.

Faunal data suggestive of mass kill sites include catastrophic mortality profiles, thick, dense concentrations of bones of multiple individuals with some bones still articulated, and, the complete or near-complete representation of animal skeletons. Levine's (1983) ontogenic data for the horse remains from two cultural horizons at Solutré are suggestive of catastrophic mortality (Figure 6.3; see Chapter 5 for discussion of the analysis of ontogenic data), with the exception that young horses are rare (but see also Hulbert 1982). The latter may be attributable to poor preservation of the remains of young horses due to their low structural density (see Chapters 7 and 9). Olsen (1989b) reports 4,483 bone and tooth specimens of horse were recovered from 36.3 m² of excavation, for an average of 123.5 specimens per m². She refers to the stratigraphic occurrence of the horse remains as the "dense 'horse magma'" (Olsen 1989b:298). Of the skeletal elements Olsen (1989b) describes, all major portions of the skeleton are present (she does not describe vertebrae or ribs), although in varying frequencies (Figure 6.4; see Chapter 7 for techniques of dealing with such frequencies), and many individual animals are represented. That people had something to do with this accumulation of bones is indicated by the fact that of the 2,484 bone specimens represented in Figure 6.4, seven have butchering marks on them (see Chapter 8 for discussion of butchering marks).

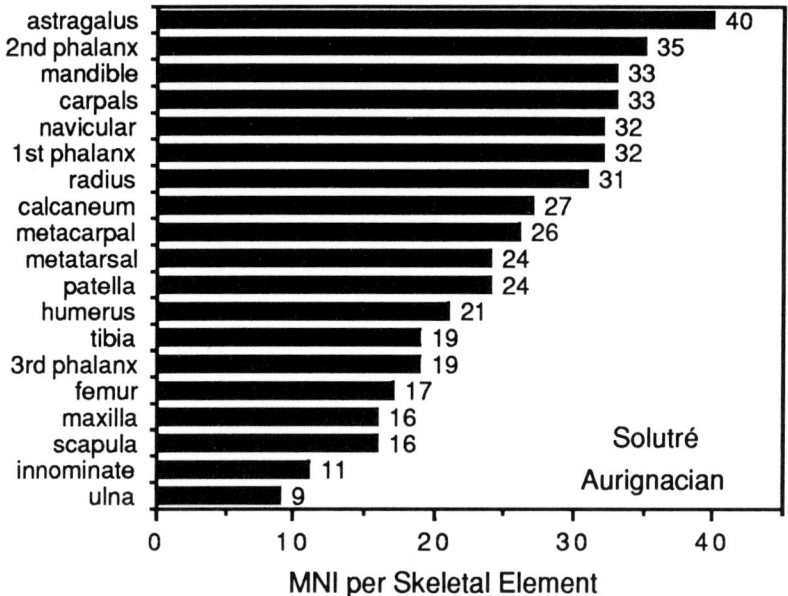

Figure 6.4. Frequencies of equid skeletal parts in the Aurignacian level of Solutré, France (after Olsen 1989b).

Wheat (1979, see also Wheat 1972) suggests the set of criteria summarized in Table 6.4 allows the distinction of kill sites, processing sites, and habitation or consumption sites. The values given there are not absolutes but rather serve to illustrate the relative frequencies of the listed variables for a given site type. For example, as one reads down the list from kill to processing to consumption sites, the proportion of bones still articulated with other bones decreases, the proportion of bones not articulated with other bones (that are disarticulated) increases, and the average number of bones making up a set of articulated bones decreases. This is only logical as it measures disarticulation and dispersal of bones, two major effects of butchering (see Chapter 8). But we must not lose sight of the fact that other processes also result in the disarticulation and dispersal of bones; the attributes in Table 6.4 do, nonetheless, provide one clue to whether or not the site represents the original location of animal death and carcass deposition.

Evidence of occupational features such as houses in a site, abundant evidence of stone tool manufacturing, and/or numerous plant processing tools may prompt the investigator to explore the possibility that the site was a habitation, where such activities are expected, rather than a kill site, where such activities probably rarely took place. As Table 6.4 indicates, it seems that few bones will be articulated, and the number of bones in a set of articulated bones will be small. But the proportions of articulated and disarticulated bones tell us little about dispersal; rather, they measure the degrees of articulation and disarticulation and only indirectly monitor dispersal. To measure dispersal, as

Table 6.4 *Criteria proposed by Wheat (1979) for distinguishing kill sites, processing sites, and consumption sites. Note that the frequencies presented are for several specific sites, and are meant as guides to relative frequencies rather than as absolutes*

Site type	% of bones articulated	% of bones not articulated	Average number of bones articulated	Make-up of tool assemblage
Kill	49	51	10	many stone and few bone butchering tools
Processing	28	72	6	few stone and many bone butchering tools
Camp or consumption	2.5	97.5	3	?

defined above, we need to know the original death location of the animals represented by the bones in a habitation site, but because the bones in a habitation site were accumulated in that site, we probably could not determine the location where the represented animals were killed (see O'Connell *et al.* 1992 for an interesting discussion of kill sites created by humans). We would, perhaps, be able, via analytical refitting, to determine how far the bones were dispersed from one another within the consumption site once they had been accumulated there, and thus we could measure the relative degree of dispersal of bones within a site. But we would not know how or how far the bones were dispersed from the kill site. For example, Lyman (1989b) used anatomical refitting to identify the dispersal of bones from known locations of carcass deposition. Some bones had moved downslope, apparently as a result of fluvial transport. Bones found upslope of the original carcass location had been gnawed by carnivores, leading to the inference that these specimens had been dispersed by carnivores. But Lyman was dealing with a limited number of carcasses representing a known catastrophic mortality event, and he sampled a horizontally large area (ca. 100 m × 100 m) in which most bones were found on the surface. Without these additional data, and when bones have been prehistorically accumulated from kill sites a kilometer or more from the archaeologically sampled site of bone deposition, such inferences and measures of dispersal may not be possible.

The study of dispersal of skeletal parts from the original site of animal death and carcass deposition may be too demanding; that is, we may be unable analytically to determine the original location of carcass deposition. Dispersal of bones can be, and in fact often is, studied without meeting this demand. In the following I review several ways dispersal has been and can be studied without clear knowledge of the location of original carcass deposition.

Fluvial dispersal

Thus far dispersal by humans has been considered in the sense that the fossil contexts have been described as archaeological. What, then, about natural

Table 6.5 *Mammalian skeletal elements grouped by their susceptibility to fluvial transport (from Voorhies 1969:69). Columns labeled "I & II" and "II & III" are transitional between main groups*

Group I immediately moved, may float or bounce along bottom	I & II	Group II gradually removed, stay in contact with bottom	II & III	Group III lag deposit
rib vertebra sacrum sternum	scapula phalange ulna	femur tibia humerus metapodial pelvis radius	ramus of mandible	skull mandible

agents of bone dispersal? One of the most-studied mechanical dispersal processes is fluvial transport (e.g., Behrensmeyer 1975b, 1982; Boaz 1982; Boaz and Behrensmeyer 1976; Dodson 1973; Frostick and Reid 1983; Hanson 1980; Korth 1979; Voorhies 1969; Wolff 1973). Voorhies (1969) pioneered these studies when he performed experiments with disarticulated bones of domestic sheep (*Ovis aries*) and coyote (*Canis latrans*) in a flume. His results indicate some skeletal elements are more likely to be moved by fluvial processes than others (Table 6.5). Behrensmeyer (1975b) subsequently elaborated on Voorhies' scheme, noting that the structural density of bones as well as their size and shape influences the probability that a particular bone will be fluvially transported. In fact, what came to be called "Voorhies Groups" of bones (sets of bone displaying varying probabilities of being moved by fluvial processes) are weakly correlated with the structural density of bones as the latter was measured by Behrensmeyer ($r_s = 0.495$, $P = 0.06$). Further, Behrensmeyer's expanded Voorhies Groups are strongly correlated with the structural density of bones ($r_s = 0.775$, $P = 0.001$) (see Chapter 7 for discussion of the structural density of bones). Wolff (1973) referred to these processes and results as *sorting* to denote the bones of a skeleton, if disarticulated, would be sorted by hydrodynamic processes into groups of readily moved and not readily moved skeletal elements.

Behrensmeyer (1975b:489) discusses the "dispersal potential" of bones in a fluvial medium, and notes that "since Voorhies Group I [Table 6.5] is the most easily affected by fluvial transport, its presence or absence in fossil assemblages can provide specific information on the sedimentary history of bone assemblages." The absence of Group I bones suggests the studied assemblage is a lag assemblage (Group I bones having been winnowed out); the presence of Group I bones suggests a non-fluvially winnowed assemblage. Further, presuming that the fluvial transport of the bones began at the site of animal death, then "the proportions of different Voorhies Groups in fossil assemblages should

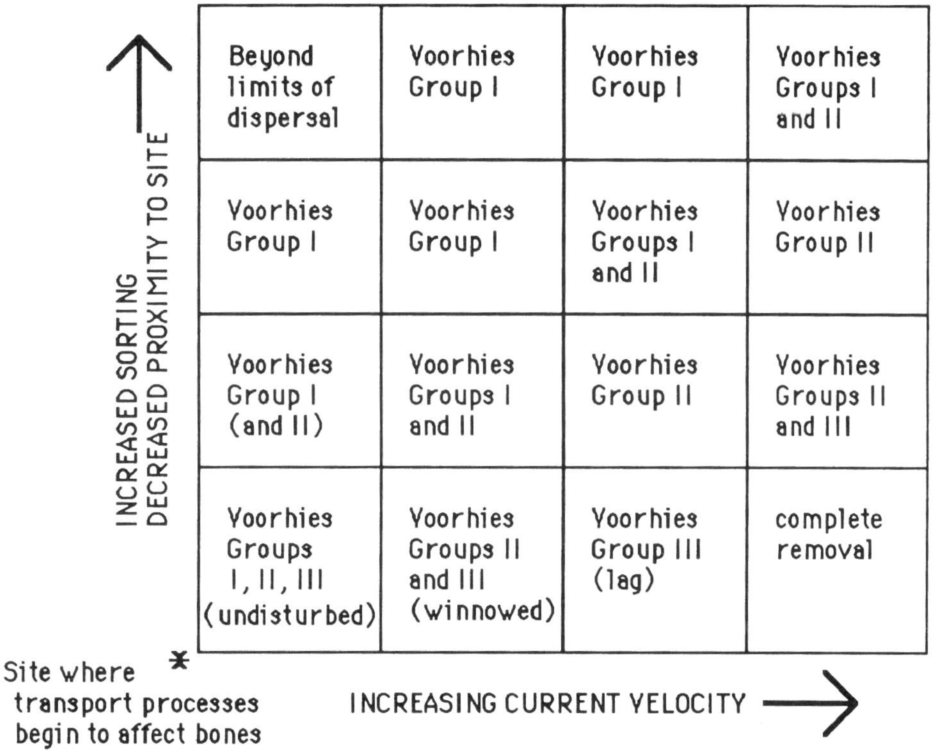

Beyond limits of dispersal	Voorhies Group I	Voorhies Group I	Voorhies Groups I and II
Voorhies Group I	Voorhies Group I	Voorhies Groups I and II	Voorhies Group II
Voorhies Group I (and II)	Voorhies Groups I and II	Voorhies Group II	Voorhies Groups II and III
Voorhies Groups I, II, III (undisturbed)	Voorhies Groups II and III (winnowed)	Voorhies Group III (lag)	complete removal

INCREASED SORTING → (vertical axis)
DECREASED PROXIMITY TO SITE (vertical axis)

Site where ✱ transport processes begin to affect bones

INCREASING CURRENT VELOCITY ⟶

Figure 6.5. Classification of bone dispersal groups according to current velocity and proximity to the site where bones begin transport by fluvial action; see Table 6.5 for skeletal elements in each Voorhies Group (after Behrensmeyer 1975b:491, Figure 5; courtesy of the author and Museum of Comparative Zoology, Harvard University).

provide evidence for the proximity of fossils to the original thanatocoenose and the habitats of the living animals" (Behrensmeyer 1975b:490, 491). This relationship is modeled in Figure 6.5.

Behrensmeyer (1975b) measured the influence of the shape of bones on their potential for fluvial transport by studying their settling velocities, a theme picked up on by Korth (1979) in his study of the taphonomy of microvertebrates (see also Dodson 1973). Behrensmeyer (1975b) and Korth (1979) argue that the settling velocity of a bone is related to the potential that the bone will be fluvially transported. Korth's (1979) experimental results on settling velocities of bones of small mammals and Dodson's (1973) experimentally derived transport groups of mouse bones align fairly well with Voorhies' Groups (Table 6.6). Korth (1979) notes that sedimentary analysis of the matrix surrounding the recovered fossils can provide important corroborating evidence that a particular bone assemblage was, or was not, subjected to fluvial transport. If, for example, the bones and the sediment particles have equivalent

Table 6.6 *Korth's (1979) settling groups aligned with Voorhies' (1969) Groups (Table 6.5). All are mammal bones*

Animal live weight	I	I & II	Voorhies Groups II	II & III	III
Settling group: Taxa < 500 g					
	rib	atlas	calcaneum	molar	mandible
	phalanx	lumbar	astragalus	incisor	tibia
		metapodial	femur	maxilla	
		radius	humerus	skull	
		scapula			
Settling group: Taxa > 10 kg					
	rib	atlas	skull	mandible	mandible
			scapula		
			molar		
			femur		
			astragalus		

settling velocities, then it is likely the bones were fluvially transported. Behrensmeyer (1975b:499) indicates "the most likely places for final burial [of fluvially transported bones] are in the actively aggrading parts of a channel such as point bars and sand or gravel bars [where the coarse fraction of the sediment load is deposited] ... Bones will move along a channel, suffering progressive abrasion, until they encounter such a situation."

Boaz (1982) studied modern sets of bones in the Mara River of Tanzania. Her study of a set of wildebeest (*Connochaetes taurinus*) carcasses and bones indicates that when this migrating species crosses large rivers, many individuals drown and the carcasses then float downstream where they are accumulated in areas of shallow water. She notes that "herd size and crowding may be more important than flood levels and current velocities as causes for mass drownings during river crossings" (Boaz 1982:173). The abundance of carcasses in such accumulations results in low levels of damage by carnivores, and, depending on the sediment load of the fluvial system, the carcasses and bones may be quickly buried only to be re-exposed and abraded by the sediment load. Some bones may be winnowed out by fluvial action, but the horizontal and vertical distribution of bones seems to produce concentrated, dense accumulations of faunal remains.

Frison and Todd (1986) measured the distance each of several individual skeletal elements of an Indian elephant (*Elephas maximus*) was transported by fluvial action over multiple experiments. From those measurements they derived a fluvial transport index (FTI) with the equation:

$$\log_{10}[(MTD_i)\,100] \qquad\qquad [6.1]$$

in which MTD is the mean transport distance of skeletal element i over multiple transport episodes. The values from equation [6.1] are scaled from 1 to 100 (all

Table 6.7 *Fluvial transport index (FTI) values and saturated weight index (SWI) values for various taxa. Indian elephant after Frison and Todd (1986); other taxa derived from Behrensmeyer (1975b). Blanks indicate no data*

Skeletal element	Indian elephant		SWI			
	SWI	FTI	Domestic sheep	Reedbuck	Topi	Zebra
cranium	100.00		100.0	100.0		
mandible	69.35	34.56	19.1	23.4		
atlas	2.76	41.97	10.4	11.1	21.4	20.6
axis			11.0	10.9	21.2	22.3
cervical	2.08	96.64	8.4	12.1	16.4	19.2
thoracic	1.66	76.43	5.0	5.1	7.1	8.2
lumbar	1.10	76.21	4.7	11.9	10.7	5.6
sacrum	6.97	80.11	11.3	17.8	31.9	
caudal	0.37	92.43				
rib	1.74	53.98	3.8	5.6	8.7	4.6
scapula	16.50	62.95	14.5	22.6	40.0	
humerus	26.40	57.77	27.7	54.7	81.0	63.5
radius–ulna	18.70	49.95	21.9	43.5	73.8	50.8
metacarpal	1.05	68.83	9.4	36.4	40.7	20.3
pelvis	57.20	0.00	42.9	43.9		
femur	26.40	24.26	31.9	92.4	96.9	100.0
patella	0.98	90.19	1.0	3.5	6.2	3.3
tibia	8.56	72.84	27.7	92.4	100.0	68.9
fibula	1.22	60.19				
astragalus	1.08	96.83	2.3	7.7	8.1	8.2
calcaneum	1.66	100.00	3.0	11.6	12.9	10.2
metatarsal			10.2	41.4	45.2	28.2
first phalanx			1.4	3.7	5.2	5.7
second phalanx			0.5	4.2	2.7	5.4
third phalanx			0.6		1.8	2.4

values divided by the maximum value and multiplied by 100) to produce the FTI values (Table 6.7). Frison and Todd (1986) found that skeletal elements with FTI values ≥ 75 (sacrum, patella, astragalus, calcaneum, cervical, thoracic, lumbar, sacrum) are basically the same as those found in Voorhies Group I (Table 6.5); skeletal elements with FTI values of 50 to 74 (ribs, scapulae, humeri, tibiae, and metacarpals) are similar to Voorhies Group II; and skeletal elements with low FTI values (< 50; atlas, mandible, pelvis, radius–ulna, femur) are nearly the same as those in Voorhies Group III. In bivariate scatterplots of the FTI values against the frequencies of Columbian mammoth (*Mammuthus columbi*) bones in two clusters at the Clovis-aged Colby site in Wyoming, Frison and Todd (1986) found little clear relation between the two, and concluded that fluvial transport had not created the bone clusters.

Frison and Todd (1986:68) also recorded the weight of the skeletal elements of the Indian elephant after they were saturated, and normed the wet weights to a scale of 1 to 100 to produce a saturated weight index (SWI). The SWI is inversely but imperfectly correlated with the FTI ($r = -0.67$, $P = 0.001$), underscoring "the degree to which factors other than weight influence [fluvial] transport potential" (Frison and Todd 1986:68). Frison and Todd found little clear relation between the two clusters of mammoth bone at the Colby site and the SWI, and concluded that the weight of the bones had exerted little influence on whether or not bones ended up in the clusters. They did, however, find significant relations between what they called the "dispersed bones" not in the two bone clusters at the site, and both the FTI and the SWI, suggesting to them that fluvial transport had moved these "dispersed bones."

Whether or not Frison and Todd's (1986) conclusions regarding the Colby site prove to hold up under further study, it is clear that the FTI and SWI indices they developed are of value in helping us understand the dispersal of bones. Similar indices for other taxa should be developed. Frison and Todd's (1986:66) FTI and SWI values for Indian elephant are listed in Table 6.7. Also listed in that table are SWI values I derived from wet weight data described by Behrensmeyer (1975b:572) for four mammalian species (domestic sheep, *Ovis aries*; reedbuck, *Redunca* sp.; topi, *Damaliscus* sp.; and zebra, *Equus* sp.). Such indices can be used in bivariate scatterplots and/or with correlation coefficients to determine if bone frequencies and the indices are correlated (see Chapter 7 for examples of how such indices are used). Significant correlations would suggest fluvial transport has played a role in distributing (dispersing) the bones.

Shape of fossils

Most researchers studying fluvial transport suggest that the shape of a bone exerts a strong influence on whether it is transported. Frostick and Reid (1983) describe a simple way to measure and classify the shape of bones. They suggest "the three mutually perpendicular axes of the fossils provide a measure of both particle shape and sphericity . . . Axial ratios (c/b, b/a) subdivide specimens into spheres, rods, discs, and blades" (Frostick and Reid 1983:159). The axes they refer to are: a, maximum dimension; b, mid dimension; and c, minimum dimension. These are potentially referred to as length, width, and thickness, respectively, if it is realized that length is not necessarily related to some anatomical orientation of the measured bone. Plotting the axial ratios (c/b, b/a) against one another allows specimens to be classified according to their basic shape, as shown in Figure 6.6. Note that as one moves from the upper right of the graph to the lower left, sphericity decreases. Sphericity is defined by the equation

$$(bc/a^2)^{0.33} \tag{6.2}$$

where a, b, and c are defined as above. Figure 6.6 shows four fictional bones, the dimensions of which are given, plotted on the graph. Frostick and Reid's (1983:163) experimental data indicate that bones classified on the graph as blades and discs are less likely to move downslope than bones with shapes approximating rods and spheres, or at least the former will move more slowly

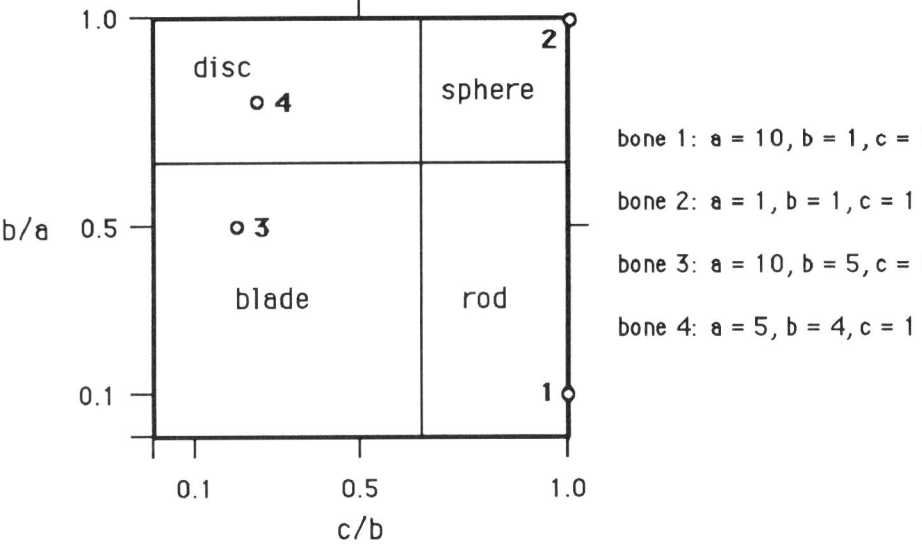

Figure 6.6. Classification of bone shape based on axial ratios (after Frostick and Reid 1983). Bone dimensions can be any unit of linear measurement (e.g., cm). See text for discussion.

than the latter. Dodson (1973:18), for example, reports that vertebrae (sphere-like shape) were more easily dispersed by fluvial action than mandibles (blade-like shape).

Faced with an assemblage of variously complete bones and fragments of bones, the taphonomist could measure the three perpendicular axes of each specimen and plot them on the graph in Figure 6.6. If the majority of the specimens classified as rods and spheres occupy lower vertical positions than those classified as discs and blades, the analyst could then suggest that fluvial transport had dispersed the bones. This presumes, of course, that the relative vertical positions of the bones are not due to faulting or tectonic tilting of the stratigraphic unit in which the fossils are found, and thus that the taphonomist has detailed provenience data.

Orientation of bones
Fluvial action will not only preferentially sort and transport skeletal elements, it will also result in patterned orientations of long bones (Toots 1965a, 1965b; Voorhies 1969). There are two basic ways in which orientation data are typically presented: *rose diagrams*, and *stereographic projections* (Fiorillo 1988; Toots 1965a). The former incorporates two-dimensional orientation data based on the azimuth of the long axis of specimens within a horizontal plane whereas the latter incorporates both the azimuth and the vertical plunge or dip of the long axis relative to a horizontal plane. I first describe the former, and then turn to the latter.

Toots (1965a:220) suggests "two-dimensional thinking accounts for the tendency to depict orientation data graphically in a rose diagram." He argues that a rose diagram is "perfectly correct when studying two-dimensional phenomena" (Toots 1965a:220), and I believe he would agree that a rose diagram is appropriate if there is minimal dip or plunge in the long axis of fossils with a definite long axis. Presuming such is the case, one can measure what is typically called the "preferred orientation" (e.g., Kreutzer 1988) or the azimuth of a long bone with a compass if the bone is *in situ*, or with a protractor if maps of bone distributions are sufficiently detailed. A long bone actually has two orientations, such as 160° and 340°, if measured from (say) the proximal end and from the distal end. That is, ignoring whether the proximal to distal or distal to proximal orientation is measured, then the possible range of variation in orientation spans only 0° to 180°. If a rose diagram of orientation data is produced as a complete 360°, then, for instance, the data between 0° and 180° are the actual measurements, and the remainder of the graph (from 181° to 360°) is a mirror image of the former. The orientation of a long bone may be measured to the nearest degree, but orientation data are typically graphed in a rose diagram by some > 1° interval, such as by each 10° (e.g., Fiorillo 1988; Kreutzer 1988; Shipman 1981b).

"Rose diagrams are useful as indicators of random or nonrandom patterning in the orientation of long axes of bones ... Asymmetry of a diagram is assumed to be an indicator of nonrandom positioning of bones" (Frison and Todd 1986:53). To interpret rose diagrams the analyst must first assume that the skeletal elements whose orientations are plotted were "free to move in response to directional forces" and must further assume that the bones moved across an essentially flat surface that did not influence their final orientation (Frison and Todd 1986:54). (Regarding the former, Dodson's [1973:17] experiments with mouse bones suggest to him that the original orientation of bones prior to fluvial action "has some effect on the susceptibility of bones to movement," an observation also noted by Frison and Todd [1986].) Sets of articulated bones will be oriented differently than their disarticulated constituent parts, and if sufficient soft tissue is still attached such sets may weigh so much that they cannot be moved by fluvial action. Frison and Todd's experiments with elephant bones, and experiments by others (e.g., Hanson 1980; Voorhies 1969), indicate that microtopographic features, channel width relative to bone length and shape, and other features can result in bones having orientations that are not related to current direction. Perhaps for that reason Shipman (1981b:71) recommends that only samples with at least 72 bones for which orientation can be measured be plotted in rose diagrams. Smaller samples may contain too many bones the orientations of which are not a function of fluvial action, but rather are the result of the influence of other factors (e.g., trampling); larger samples should more clearly display orientation patterns. Also, if geomorphic features such as a channel or slope are apparent in the stratigraphic record, bone orientation should be related to that feature and not just the compass azimuth (e.g., Frostick and Reid 1983).

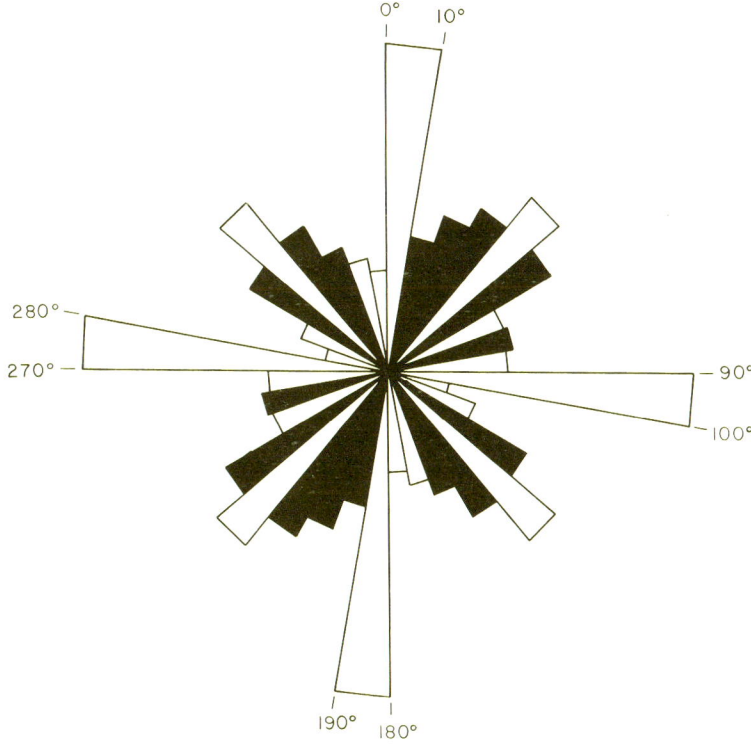

Figure 6.7. A mirror-image rose diagram showing azimuths of long axis of long bone (after Kreutzer 1988:226, Figure 4; courtesy of the author and Academic Press). Black wedges are non-significant values; white wedges are significant values (see Table 6.8).

The mirror-image rose diagram in Figure 6.7 presents orientation data for 1,084 bones compiled by Kreutzer (1988) for two late Pleistocene/early Holocene clusters of bison bones at the Lubbock Lake site in Texas. Kreutzer (1988) compared the observed and expected frequencies of specimens per 10° orientation class by calculating adjusted residuals for each class, and finding the probability of observing such adjusted residuals if orientation was random (Everitt 1977 describes the statistical technique). The results (Table 6.8) indicate that the black wedges (10° classes) in Figure 6.7 are not significantly different from expected values whereas the white wedges are significantly different from expected values. Kreutzer (1988) concludes that four classes have more bones than expected given random chance, and seven classes have fewer bones than expected given random chance. In light of experimental data (Voorhies 1969) indicating that long bones with one heavy end will have their long axes oriented parallel to the current direction and bones with ends of approximately equal weight will have their long axes oriented perpendicular to current direction, Kreutzer (1988:227) concludes that the specimens from

Table 6.8 *Observed and expected frequencies of 1084 bone specimens per 10°
orientation class at Lubbock Lake, Texas (from Kreutzer 1988)*

Degree class	Expected frequency	Observed frequency	Adjusted residual	Probability	Conclusion
0–10°	60.22	119	8.03	<0.05	more than exp.
11–20°	60.22	50	−1.35	>0.05	
21–30°	60.22	61	0.10	>0.05	
31–40°	60.22	69	1.17	>0.05	
41–50°	60.22	83	3.05	<0.05	more than exp.
51–60°	60.22	70	1.30	>0.05	
61–70°	60.22	45	−2.00	<0.05	fewer than exp.
71–80°	60.22	47	−1.74	<0.05	fewer than exp.
81–90°	60.22	44	−2.13	<0.05	fewer than exp.
91–100°	60.22	112	7.05	<0.05	more than exp.
101–110°	60.22	23	−4.85	<0.05	fewer than exp.
111–120°	60.22	34	−3.43	<0.05	fewer than exp.
121–130°	60.22	59	−0.16	>0.05	
131–140°	60.22	80	2.65	<0.05	more than exp.
141–150°	60.22	61	0.10	>0.05	
151–160°	60.22	48	−1.61	>0.05	
161–170°	60.22	42	−2.39	<0.05	fewer than exp.
171–180°	60.22	37	−3.04	<0.05	fewer than exp.

Lubbock Lake were fluvially oriented because two major groups of orientations are detectable, and they are at right angles to one another.

Behrensmeyer's (1990:234) suggestion that "in strong currents elongate bones generally orient parallel to the flow direction, with the heavier end upstream [whereas] in shallow water or weak currents elongate bones may orient perpendicular to the current" does not invalidate Kreutzer's (1988) conclusion. It suggests other lines of evidence one might inspect. Do the skeletal elements plotted in Figure 6.7 have ends of equal weight? Is there sedimentological evidence for two distinct fluvial events, one with a strong current (coarse sediment), one with a weak current (fine sediment)? Kreutzer (1988) identified a significant pattern in the bone orientation data. Other analyses can refine and elaborate her conclusions.

Voorhies' (1969:66) experimental data indicate that long bones tend to orient parallel to flow direction if they are submerged, but when the bones are only partially submerged they tend to orient perpendicular to flow direction. Shipman (1981b:69) suggests the term *preferred orientation* denotes a particular, dominant compass orientation; that is, one of the two ends of a long bone is typically upstream, or downstream. This has been called *polarity*, or the tendency to have a particular orientation given inherent properties in the symmetry of the fossils (Toots 1965a:220). As noted, Voorhies (1969:66) indicates that the "heavier end" of long bones tends to be upstream once a long bone has been subjected to fluvial action. If either end is likely to be upstream (or downstream), Shipman (1981b:71–72) suggests the term *preferred axis* of orientation be used in order to underscore the absence of polarity. Mirror-image rose diagrams result when polarity is absent or not recorded. Rose diagrams may, of course, be drawn when polarity is noted.

Toots (1965a:220) notes that "in geology most phenomena are three-dimensional [and] projecting three-dimensional data into one plane ... may lead to the loss of critical information." Thus, rose diagrams may oversimplify some situations, and Toots (1965a, 1965b) and Fiorillo (1988; see also Rapson 1990) suggest using stereographic projections to illustrate the three-dimensional orientation of fossils. Such projections graph three-dimensional orientations as "lines within a unit sphere. The points where the lines intersect the sphere, projected onto the horizontal plane, form the stereographic projection" (Fiorillo 1988:1). Each point around the periphery of the projection (circle) represents a specimen the long axis of which is horizontal whereas each point near the center of the projection represents a specimen the long axis of which is vertical. The closer a point lies to the outside edge of the projection, the nearer the represented specimen is to being horizontal; the closer a point lies to the center, the nearer the represented specimen is to being vertical (90° from horizontal). Figure 6.8e shows how the vertical plunge or dip of a bone is plotted on the horizontal plane of a stereographic projection. Each point's position around the circle of the projection indicates its orientation in the horizontal plane, plotted to the nearest degree (similar to a rose diagram).

Four models of possible stereographic projections are shown in Figure 6.8. These models show both the plunge or dip and the orientation of each plotted bone. The model in Figure 6.8a describes a case in which all specimens are vertical or nearly so and thus preferred orientations are not detectable. Figure 6.8b illustrates a case in which all specimens are nearly horizontal and display no preferred orientation. Figure 6.8c illustrates a case in which all specimens are essentially horizontal and display preferred orientations. Results like that in Figure 6.8d indicate no preferred orientation and random plunge among specimens. Note that a fifth model, such as would be represented by Kreutzer's (1988) data shown in Figure 6.7, would produce point scatters every 90°, such that Figure 6.8c would have two additional point scatters, one on each side of the circle as well as at the top and bottom.

Stereographic projections are easy to compile and generate. For example, Table 6.9 lists fictional data for five long bones. The plunge or dip of each bone is measured as the degrees from horizontal (0°) of the long axis of each bone (perfectly vertical is 90°). As well, the azimuth of the long axis of each bone is measured from the proximal to the distal end to account for polarity. That is, the azimuth is measured as the angle between the 0° line and the line defined from the proximal end to the distal end of the bone (in Figure 6.9, the distal end is toward the center of the circle and the proximal end is pointed to the outside

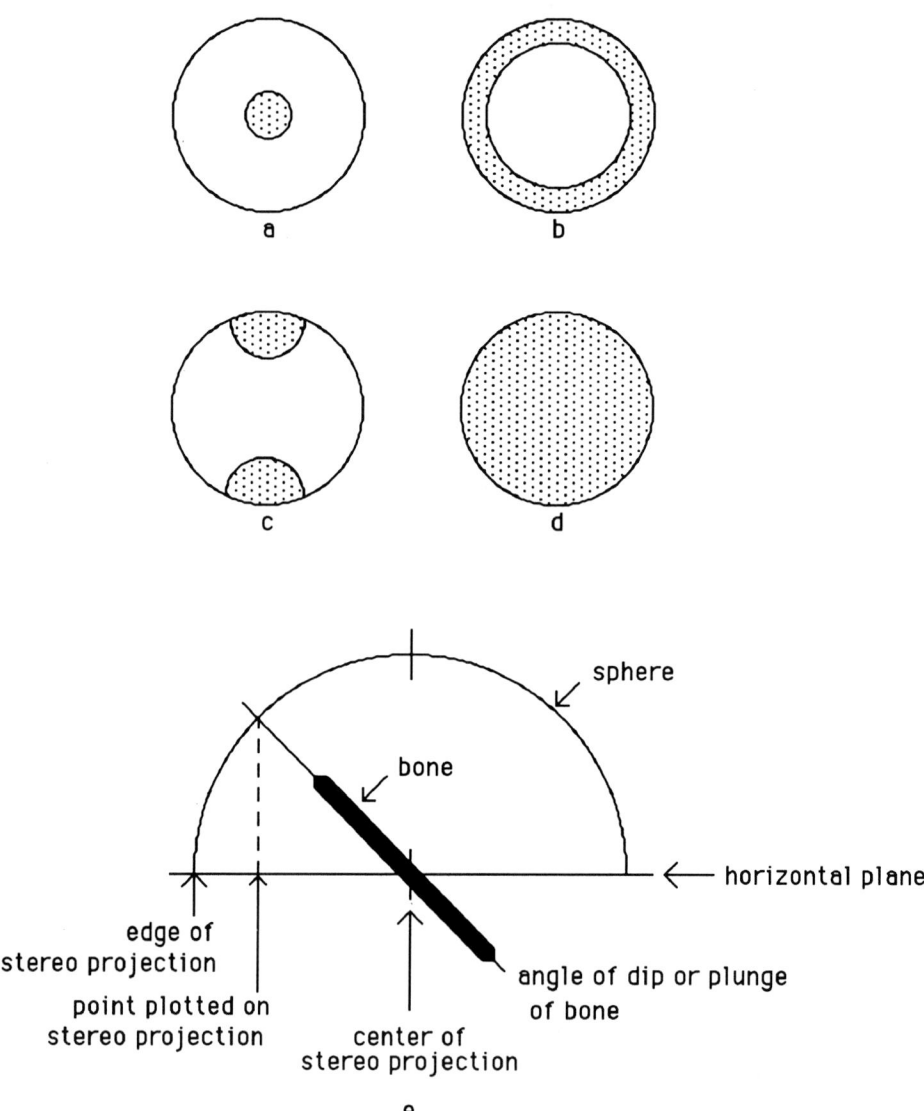

Figure 6.8. Idealized stereographic projections of four possible distributions of long bone orientation and plunge or dip. Modified from Toots (1965a) and Fiorillo (1988). Each individual bone is represented by a dot; borders added to emphasize groups. a, long axes of bones have a near vertical orientation with no preferred azimuth. b, long bones have an approximately horizontal orientation with no preferred azimuth. c, long bones have an approximately horizontal orientation with one preferred azimuth. d, long bones have a random orientation horizontally and vertically. e, a stereographic projection from the side illustrating how plunge or dip is plotted.

Table 6.9 *Three-dimensional orientation data for five fictional long bones (see Figure 6.9)*

Bone	Vertical dip or plunge	Horizontal azimuth
A	30	10
B	5	50
C	0	75
D	85	310
E	45	200

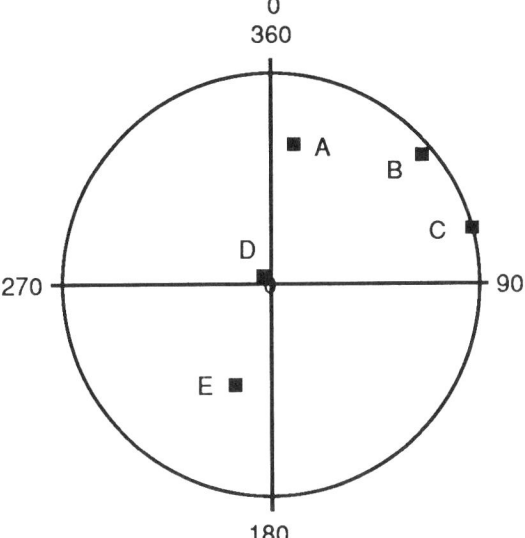

Figure 6.9. A stereographic projection of the horizontal and vertical orientation of five bones (from Table 6.9).

of the circle). Polarity need not be plotted in stereographic projections, in which case a mirror-image stereographic projection would result. The graph in Figure 6.9 of the data in Table 6.9 might be taken to indicate a preferred orientation of most bones to have their long axes more or less parallel to a 45° or 225° line, and three of the five bones are horizontal or nearly so.

Orientation of carcasses

Individual animal carcasses may have a particular orientation just as individual bones may. Carcass orientation data may reveal aspects of their transport and depositional history (Dodson 1973; Schäfer 1962/1972; Weigelt 1927/1989). For example, Lyman (1989b) measured the orientation of carcasses of wapiti (*Cervus elaphus*) killed as a result of a volcanic eruption. Orientation was

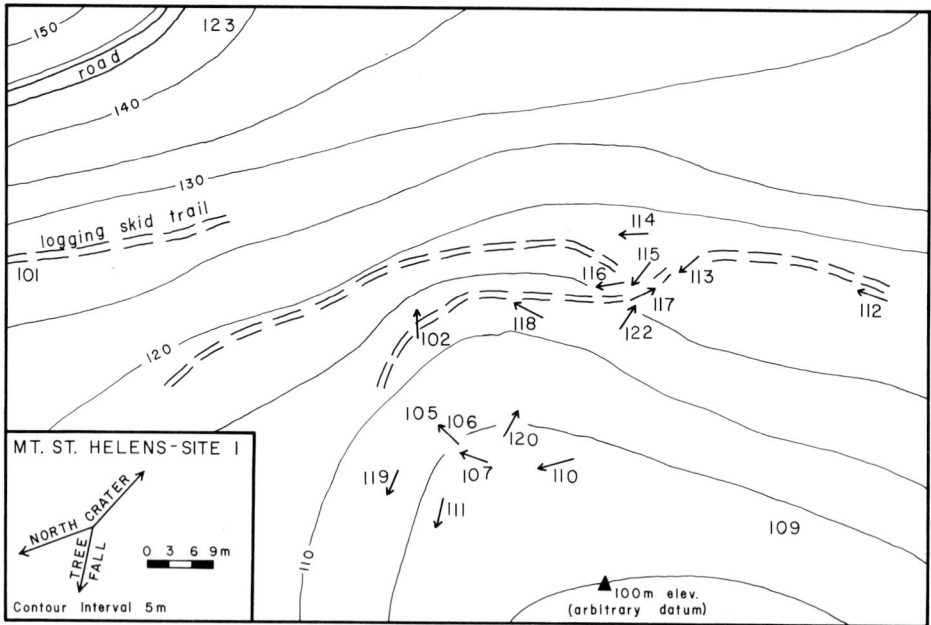

Figure 6.10. Distribution and orientation of wapiti carcasses killed by the volcanic eruption of Mount St. Helens. Three-digit numbers denote individual carcass locations; arrows point from tail to head of carcasses for which orientation could be determined (from Lyman 1989b:161, Figure 12; courtesy of The Center for the Study of the First Americans).

measured as the azimuth of the articulated axial skeleton. Polarity was accounted for by measuring the azimuth of the line defined from the tail to the head of each carcass (Figure 6.10). Lyman (1989b:161) noted the compass bearing from the carcasses to the volcano and concluded that "postmortem orientation of these cervids seems to be unrelated to the location of the volcano responsible for their deaths" based on Figure 6.10. A plot of the carcass orientation data on a polar graph (Figure 6.11), however, suggests 9 of 13 wapiti turned away from the cloud of volcanic ash (away from the volcano) that swept over them and resulted in their suffocation (all carcasses appear to represent animals that simply dropped dead where they stood). Chi² analysis indicates, however, that this value (9 of 13) is not statistically different from a random pattern (chi² = 1.92, 0.25 > P > 0.1).

Regardless of the taphonomic significance of the wapiti carcass orientations, such data may be important. Wheat (1972:29), for instance, concluded that the "orientations" of 14 complete bison skeletons at the Olsen-Chubbuck site in Colorado were due to their being packed into the bottom of a narrow arroyo. Two of the carcasses had, in fact, "nearly vertical" orientations. An additional 27 nearly complete skeletons displayed orientations similar to the 14 complete skeletons, and their orientations also seemed to be related to their position in

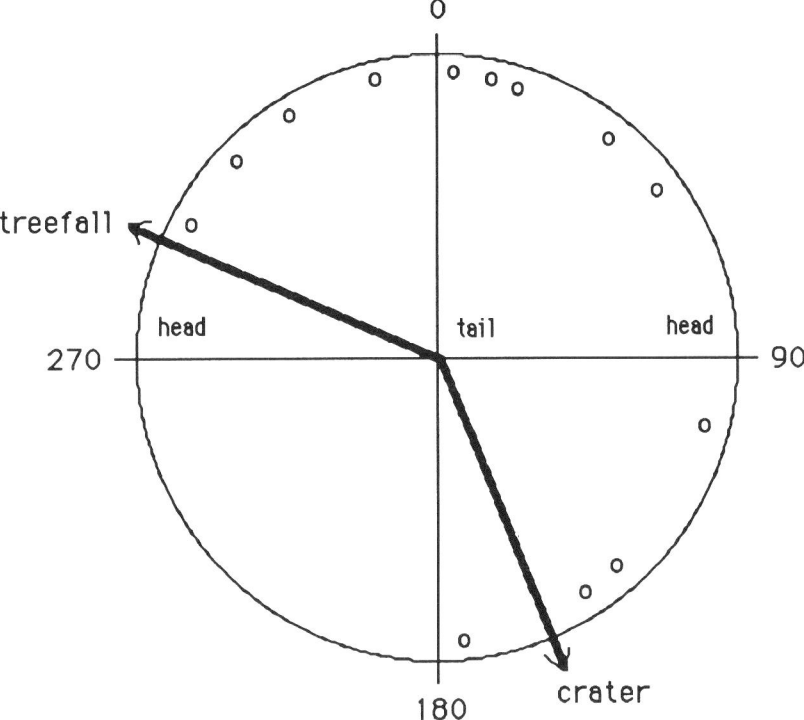

Figure 6.11. Azimuth of wapiti carcasses killed by the volcanic eruption of Mount St. Helens. Each small circle represents a single carcass; polarity is indicated with tail at center of large circle and head away from the center. Note that many carcasses are facing away from the crater (see text for further discussion). Compare with Figure 6.10.

the arroyo. These few examples underscore not only the importance of carcass orientation data for taphonomic concerns (why are the skeletons not in life-positions?), but the significance of the geological context of the carcasses. The latter, of course, applies with equal force to the discussion of bone orientation above. It should also be obvious that if one can measure the orientation of a carcass, one has probably found the location where the carcass of the dead animal was deposited, if not where the animal in fact died, and thus dispersal of individual bones from that location could be measured if individual bones can be anatomically or mechanically refit to the skeleton (see Chapter 5 on refitting).

Abrasion
Behrensmeyer (1975b) and Korth (1979) suggest that fluvial transport might result in the abrasion of bones (see also Bromage 1984). Abrasion (see the Glossary) results from the tumbling of bones in a liquid that contains sediment. The physical erosion of bone results from its constant but shifting contact with

the sediment grains. For example, Boaz (1982:142) reported that in a sample of 236 wildebeest (*Connochaetes taurinus*) bones she collected from fluvial environments in Africa, 59 displayed no evidence of any kind of damage, 48 had been broken by carnivores, 61 had been broken and then abraded, and 67 showed abrasion damage only. She noted that the effects of abrasion by the sandy sediment were "characterized by a 'wearing-away' of the outer table of bone and exposure of the inner cancellous portion" (Boaz 1982:147). Korth (1979:263–265) examined two mouse skeletons (one each of *Peromyscus* and *Microtus*) in tumbling barrels containing water and quartz grains averaging 2–4 mm in diameter. The barrels were rotated so as to approximate a linear velocity of 24 cm/sec, and the bones were examined every 10 hours for 80 hours. The mouse skeletons first disarticulated, and the hypsodont (rootless) teeth fell from the maxillae and mandibles, but the rooted teeth stayed in their alveoli. Bones of the skull then disarticulated, and their edges became rounded. Sharp edges of all bones showed evidence of rounding early in the process, and bone surfaces became progressively thinner. Eventually, some bones broke due to the thinning.

Shipman and Rose (1983a:79, see also Shipman and Rose 1988) caution that "bones in a tumbling barrel are exposed to the impact of sedimentary particles more continuously than are bones in natural stream conditions [and] tumbled bones are also subjected to more constant velocity than is likely to be realistic." Their tumbling barrel experiments with bones of domestic sheep (*Ovis aries*) indicate that microscopic features on bone surfaces are quickly and often completely removed by abrasion, but "grossly apparent changes in bone surfaces occur after about 35 hours of abrasion" (Shipman and Rose 1983a:79). In their experiments "sedimentary abrasion rarely produced scratches or other elongate grooves, regardless of the sediment size, the inclusion or exclusion of water [in the tumbling barrels], the condition of the bones (fresh, weathered, fossilized, whole, or broken), and the duration of tumbling" (Shipman and Rose 1983a:79).

Shipman and Rose (1988:317) list six factors that influence the rate and nature of abrasion of bone by sediments:

1. the grain size of the sedimentary particles with which the bones are transported;
2. the composition of the sedimentary particles;
3. the presence or absence of soft tissue on the bone;
4. the condition of the bone at the onset of transport (fresh or weathered; broken or whole; mineralized or unmineralized);
5. the presence or absence of water in the sedimentary system; and
6. the duration or distance involved in transport.

They observed that only a few bones showed macroscopically or grossly visible abrasion features, such as exposed cancelli (as reported by Boaz 1982), after 20 to 35 hours of tumbling. Shape changes in bones were extremely subtle early in the tumbling process. "The smaller the grain size, the faster the rate of abrasion, all other things being equal" (Shipman and Rose 1988:323). More angular ("sharper") sediment abraded bones faster than more rounded sediment particles. Bone edges abraded faster than surfaces, probably because of greater exposed surface area relative to volume (see also Andrews 1990). Finally, convex surfaces of bones tended to abrade faster and prior to concave surfaces.

Recognizing traces of abrasion is only one aspect of the problem, as several processes can abrade bone, processes such as trampling (Brain 1967a), eolian activity (Shipman and Rose 1988), and fluvial transport. The former two processes are discussed in Chapter 9, where it is noted that the distribution of abrasion damage across individual bone specimens often can help the analyst determine which process was responsible for the abrasion. It suffices here to note that fluvial transport of bones abrades the entire surface of the specimen whereas abrasion by eolian activity abrades only the exposed or top surface(s) of specimens. Trampling may abrade bones, but also creates deep scratches in bone surfaces, something fluvial abrasion does not produce. Finally, sedimentological analysis and determination of whether the matrix enclosing the bones is eolian or fluvial in origin may help the analyst sort out the responsible taphonomic process. But, while it seems that coarser and more angular sediments may abrade bones more extensively and rapidly than fine textured, subangular or rounded sediment particles, Graham and Kay (1988:237) caution that "redeposition and recycling of bone specimens may make breakage, polish, and abrasion appear incongruous with sedimentary particle size." They found heavily abraded bone specimens spatially associated with unabraded bone specimens, and all were in rather fine sediment. Thus while the sedimentary contexts of the bones are important taphonomic variables (see Chapters 10 and 11), the analyst should not be mislead by that context.

Dispersal by hominids and scavenging carnivores

Other intensively studied dispersal mechanisms include animals that move or collect bones. Most have been studied from the vantage of the site to which these dispersal mechanisms bring bones and deposit them, such as lairs, nests, or dens. Thus I consider them in more detail in the discussion of accumulation below. However, there is one interesting way that the analyst might study dispersal of animal bones by mammalian carnivores and scavengers from a kill site. Blumenschine's (1986a) general sequence of consumption of carcass parts (Table 5.7) indicates which carcass parts generally are the first to be removed (consumed or spatially moved away) from the site of animal death. Bones of the hindquarter and lumbar region are the first to go, bones of the forequarter are second, and bones of the head are the last. Similar sequences of transport of skeletal parts away from a kill site by humans have also been developed (see Chapter 7). The point here is, the bones present offer a clue as to whether the

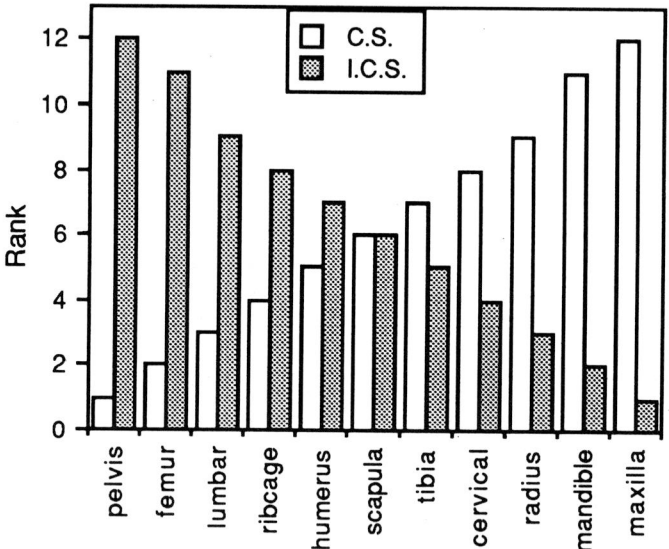

Figure 6.12. Blumenschine's (1986a) consumption sequence. The rank is the order in which flesh associated with the indicated skeletal element is consumed (from Table 5.7). C.S., consumption sequence; I.C.S., inverse consumption sequence.

assemblage represents a kill site from which bones were removed or a site to which bones were brought.

Blumenschine's (1986a) consumption sequence can be modeled as a bar graph, the height of the bars being based on the rank order of consumption of each skeletal part (Figure 6.12; see Blumenschine and Cavallo [1992:93] for a less schematic illustration). The bar height is equivalent to the relative frequency of each particular skeletal part one would expect to find at a kill site where carnivores variously consumed and removed bones following the consumption sequence. The inverse consumption sequence (Blumenschine 1986a) can similarly be modeled (Figure 6.12), and the bars represent the relative frequencies of skeletal parts one would expect at a den or lair to which bones had been transported. For example, Binford (1981b:214–216) summarized the frequencies of bones from carnivore kills of waterbuck (*Kobus ellipsiprymnus*) in East Africa (as reported by A. P. Hill) and frequencies of reedbuck (*Redunca* sp.) bones from a carnivore den (as reported by R. G. Klein). Plotting those two sets of frequencies against Blumenschine's consumption sequence suggests there is little relationship between the two variables (Figure 6.13). In fact, there is no significant correlation between the consumption sequence and either bone assemblage (for the kills $r_s = 0.06$, $P = 0.82$; for the den $r_s = 0.30$, $P = 0.34$). Other evidence suggests the frequencies of bones have been significantly influenced by the destruction of many of them due to carnivore consumption. Thus, in this case, the analyst would want to pursue

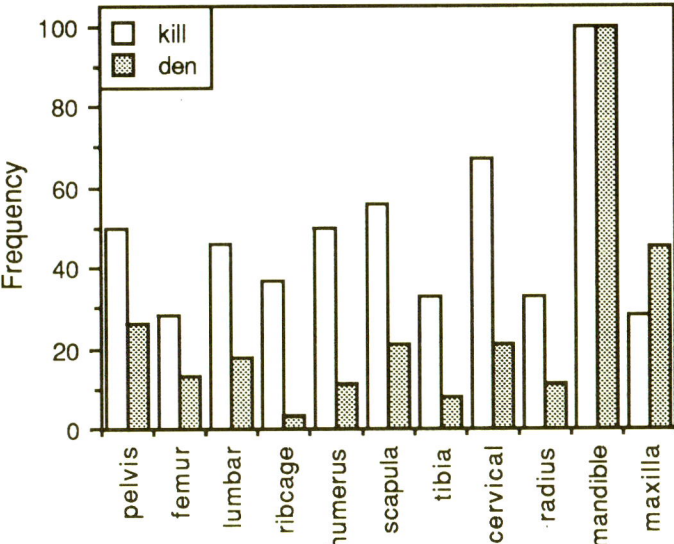

Figure 6.13. Frequencies of skeletal elements from carnivore kills and from a carnivore den plotted against Blumenschine's (1986a) consumption sequence (see Figure 6.12).

other lines of evidence when studying the dispersal of bones, including all factors that might influence bone frequencies (see Chapter 7 for extensive discussion). Nonetheless, Blumenschine's (1986a) consumption sequence is a valuable analytical tool that can be used as part of initial analyses examining the dispersal of bones from the location of carcass deposition. Finally, the inverse consumption sequence can be used to help study bone accumulation, our next topic of discussion.

Analyzing accumulation

> I arbitrarily define *cultural* bone as those fragments of non-human tooth and osseous material deposited as the result of human activity. Bone deposited by other mechanisms is termed *natural* bone.
> (D. H. Thomas 1971:366)

Background scatters

Zooarchaeologists and taphonomists typically deal with bone assemblages that were collected from spatial loci displaying higher than average concentrations of animal remains. An obvious taphonomic problem thus concerns identifying the agent(s) or process(es) responsible for creating those dense concentrations of bones. One way to begin to approach this problem is to examine what the average or normal bone accumulation mechanisms are on the landscape, and/or to determine what has been referred to as the "background" scatter or normal density of faunal remains across the landscape (e.g., Behrensmeyer 1983).

Behrensmeyer (1982, 1983, 1987; Behrensmeyer and Boaz 1980; Behrensmeyer *et al.* 1979) has gone far towards identifying and describing what typical background scatters of bones look like in eastern Africa. In studying the passive accumulation of bones across the landscape, researchers begin with the observation that "the number of bones of a species which accumulate in an environment depends initially on the population size and annual death rate of that species" (Behrensmeyer and Boaz 1980:75). They also point out that the remains of smaller taxa are less likely to (a) preserve and (b) be recovered (Behrensmeyer *et al.* 1979; see Chapter 9 for detailed discussion). Behrensmeyer (1983) notes that there are differences in background scatters that tend to correlate with the depositional microenvironment. For example, the average density of faunal remains (taxonomically identifiable specimens only; ribs, sternabrae, and long bone diaphysis fragments not included) in the swamp habitat (0.0083 specimens/m^2) is four times higher than it is in the bush (0.002 specimens/m^2). These densities represent an accumulation period of 10–20 years, and projecting these frequencies across greater time spans and correcting for the fact that not all of the specimens will be buried and thus preserved for future recovery, Behrensmeyer (1983) suggests that only about 4.4 specimens/m^2 will be deposited and preserved over 10,000 years.

Frequencies of skeletal parts differ not only with depositional habitat, they also vary with whether a bone assemblage represents a passive accumulation across the landscape or an active, spatially focused accumulation. Data (from Behrensmeyer 1983; Behrensmeyer and Boaz 1980) indicating this variation are summarized graphically in Figure 6.14. There, it is clear that the assemblage labeled "dispersed bones" has approximately equal numbers of skulls, vertebrae, forelimb bones, and hindlimb bones; this assemblage represents the passively accumulated attritional background scatter of bones across the landscape. The assemblage labeled "predation patch" is also a passive accumulation, but it is one representing a geographic location where multiple animals were killed, one at a time, over time by predators. The predation patch assemblage tends to have relative bone frequencies opposite those of the "hyena den" assemblage, reflecting the fact that hyenas have selectively removed bones from animal death loci to their dens. The "buried bones" assemblage tends to reflect fairly accurately the "average skeleton" assemblage, suggesting that bones of the four major categories of skeletal elements plotted all have about the same chance of being buried and thus incorporated into the (future) fossil record. There are notable differences, however, between the "average skeleton" and other assemblages, all probably reflecting differential dispersal and destruction of bones within the four categories of skeletal parts.

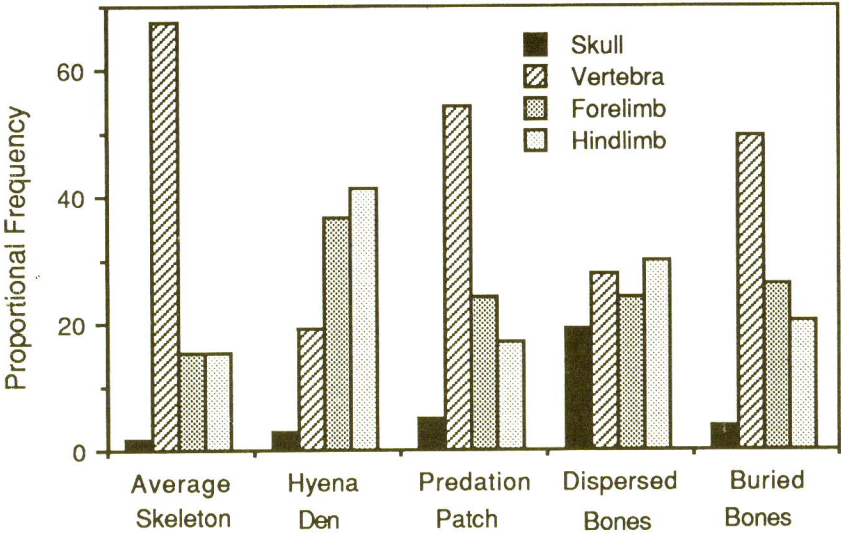

Figure 6.14. Relative frequencies of skeletal portions in different types of bone accumulations (after Behrensmeyer 1983; Behrensmeyer and Boaz 1980).

Haynes (1988b) presents data on natural background scatters of bones for both Africa and Canada. He compared five assemblages of bones from "mass death sites" with two "cumulative assemblages" of bones. A mass death site is defined as a "relatively circumscribed locus where a herd or group of animals (of one or more taxa) died over a brief time span due to a single agency of death," and a cumulative site is defined as a place "where bones of large mammals have amassed over time due to numerous different mortality events such as serial predation, the regular and habitual killing of prey animals in the same loci" (Haynes 1988b:219). Haynes found that the ratio of bones per individual (MNE:MNI) did not differ between the two kinds of assemblages, but the density of bones (frequency per m²) was greater at the mass death sites (avg. = 0.14) than at the cumulative sites (avg. = 0.09). He notes that sites with high densities of bones but only one taxon represented "are not always the result of mass deaths" (Haynes 1988b:230). Haynes also notes, as Behrensmeyer did, that bones of small animals have less chance of survival and incorporation into the fossil record than bones of large animals (see Chapter 9). Finally, he reports that carnivores and scavengers tend to remove bones from carcasses in mass death sites but feed heavily on bones at death sites when carcasses are scarce.

Passive mass accumulations

The world-famous La Brea tar pits contain a well-known passively accumu-lated, multi-taxon, multi-individual fossil assemblages (e.g., Stock 1956). The frequency of bones per unit area tends to be quite high in such settings. There are several ways for such mass accumulations to be generated passively. Haynes (1988b, 1991), for example, has found that loci such as waterholes where multiple individuals die, even though it may be one individual dying at a time, tend to have large, dense accumulations of bones. Individuals may die from thirst such as during times of drought, from starvation, or from multiple events of predation by carnivores. In a way, the waterholes serve as magnets, drawing in nutritionally stressed individuals, or potential prey.

Natural traps may or may not serve as magnets for animals. For example, Oliver (1989) describes a natural pit-fall trap, Shield Trap Cave, in Montana. His analysis of a portion of the excavated material revealed over 5,000 identifiable bones representing at least 52 individual mammals of 10 species. Most of these remains, Oliver (1989) believes, represent animals that accidentally fell into the cave and either died as a result of the fall or could not escape from the cave and died from starvation. Oliver (1989:81) suggests that "carnivores were attracted to the trap by the presence of dead and/or dying animals on the cavern floor" but few were able to survive the 14 m fall into the cave; those that did soon succumbed to injuries resulting from the fall. The notion that some natural traps served as magnets for carnivores is employed as an explanatory device by White *et al.* (1984). They found that some caves which served as natural traps contained much higher proportions of carnivores relative to herbivores (average of 2 caves = 32.8% of the MNI) than (a) caves which did not serve as natural traps (average of 18 caves = 4.7% of the MNI), and (b) a living mammalian fauna (2% carnivores). They explained this difference as resulting from "the peculiar configuration of the two [carnivore-trap] caves and the behavior of the carnivores" (White *et al.* 1984:250). In particular, the two caves they believe to represent carnivore traps were difficult to get out of, and the carnivore taxa represented by the faunal remains tend to be opportunistic feeders of carrion. Animals that died in the carnivore trap caves (including carnivores) would have served as "bait for the trap" (White *et al.* 1984:246).

In brief, passive mass accumulations of bones have three distinguishing attributes. First, there must be some factor which results in animals being attracted to a locus year after year. Second, the probability that at least some of these animals will die in that attractive (?) locus must be greater than in other loci. Third, the animals effectively accumulate themselves; that is, there is no agent of carcass or bone accumulation that is external to the accumulated carcasses or bones, but rather the process of accumulation involves the behaviors of the accumulated animals themselves. In the preceding few paragraphs, several such loci have been identified. These include natural traps where the attractive feature seems to be carrion, the "attractive" feature is that the trap is hidden, or the attractive feature may be a relatively isolated watering hole during dry seasons.

Active mass accumulations

In contrast to passive mass accumulations, active mass accumulations are created by bone-accumulating agents and processes that are external to the accumulated carcasses and bones. These agents and processes include geological processes such as fluvial action, and biological processes such as humans and denning carnivores. I only briefly consider geological processes in the following as they have been dealt with in earlier portions of this chapter. I then consider some of the natural biological processes, and then turn to humans as bone accumulators.

Fluvial accumulations

Frequencies of skeletal parts can be attributed to fluvial accumulation by reference to Voorhies Groups or the like (Figures 6.5 and 6.6; Table 6.5). Andrews (1990:18–19) attempted to simulate the effects of fluvial transport on small mammal bones by placing a sample of rodent bones in a rotary tumbler along with different sediment grades (from relatively fine [silt and sand] to coarse [gravel], and several large clasts added to the previous two). He found that after about 200 hours of tumbling there was "a slight degree of rounding of the bones [and] after 300 hours rounding was pronounced" for all sediment grades. Bones were broken and chipped more when large clasts were included. He concludes that "the nature of the breakage is largely controlled by the structural properties of the bones" (Andrews 1990:19).

Harvester ants as bone accumulators

In a unique study Shipman and Walker (1980) document the bone-collecting activities of harvester ants in Africa. The assemblage of faunal remains created by these insects is characterized by dense concentrations of bones and teeth around the entrance tunnels of the insects' burrows. The taxa represented are all small and of local derivation. There is a wide diversity of mammalian taxa represented, including diurnal and nocturnal, and fossorial and non-fossorial forms; this may be different from attributes of a fauna represented by remains collected by, say, a nocturnal mammalian predator or raptor (see below). The ant-generated bone assemblage contained abundant "robust" skeletal elements (Shipman and Walker 1980:499), a theme we return to below.

Rodents as bone accumulators

Hoffman and Hays (1987) conducted an experiment in which they placed a number of mammal bones in a cave in North America occupied by the eastern wood rat (*Neotoma floridana*). They found that this small rodent moved and redistributed many of the bones. The wood rats did not seem to select bones for gnawing or on the basis of bone morphology or texture, but rather moved all bones < 100 g in weight and > 0.3 g. This species of wood rat weighs between

170 and 340 g. Varying proportions of mammal, fish, and turtle bones were moved by the wood rats; overall, 55% of all bones were moved. Horizontal movement was, on average, 1 to 2 m, with a maximum distance of 5 m. Many bones were also moved 1 to 2 m vertically. Hoffman and Hays (1987) do not indicate how a taphonomist might analytically determine if wood rats have redistributed bones in a site.

Hockett (1989b) indicates the frequencies of skeletal parts accumulated by wood rats may be a function of what was available for them to accumulate. "Vertebrae and ribs may be common simply as a reflection of the fact that they are abundant elements in an ungulate carcass" and rare elements may be rare because many of them were consumed or removed from a carcass by carnivores (Hockett 1989b:31). Regarding the latter, Hockett (1989b) reports that 51% of the non-rodent bones he collected from wood rat nests display evidence of carnivore gnawing. Few of them, however, displayed signs of having been gnawed by the wood rats. Hockett suggests that the size of a bone may limit whether it is transported by a wood rat. Hockett (1989b:33) suggests bones > 1.1 cm wide, > 29.5 cm long, and > 54.5 g can not be transported by the bushy-tailed wood rat (*Neotoma cinerea*), and if "none of the three individual [sizes is exceeded by a fossil bone], then the bone in question may have been deposited at the site by wood rats;" larger bones were probably deposited by other agents. I would add that the presence of wood rat remains in the bone assemblage, evidence of wood rat nests in the site, and the presence of rodent gnawing marks on some bones in the assemblage might be taken as circumstantial evidence that this small rodent is perhaps responsible for the distribution of some bones (e.g., Grayson 1988; Lyman 1988a). Figure 6.15a shows three domestic sheep (*Ovis aries*) astragali collected from a wood rat nest in New Mexico. All show typical extensive rodent gnawing, and while the precise identity of the gnawing agent is not known, the context of the bones suggests wood rats are responsible. Figure 6.15b shows two distal tibiae of deer (*Odocoileus* sp.). One of them was collected from a rockshelter in Missouri and displays extensive rodent gnawing. Again, the precise identity of the gnawing agent is not known, but marmot (*Marmota* sp.) remains are common in the rockshelter deposits and this large sciurid lives in and around the shelter today.

Brain (1980, 1981) reports that the African porcupine (*Hystrix africaeaustralis*) transports and accumulates mammal remains (bone and horn) and tortoise shell. This rodent does not seem to destroy bones of low structural density (see Chapter 7). The African porcupine collects a wide range of sizes of bone specimens; Brain's (1980) sample from a lair occupied by this species contained specimens ranging between 1 and 750 g, with the majority (70%) weighing between 1 and 50 g. The maximum length of the specimens is 90 cm, but the majority fall between 2 and 15 cm. Of the 1708 total bone specimens collected from the lair, 1,043 (61%) display evidence of having been gnawed by porcupines. Most of the bones collected by the porcupines are "bleached, defatted bones" (Brain 1980:123; 1981:116). Brain (1981:116) concludes that "the most reliable indication that a bone accumulation in a cave has been built up by porcupines is the presence of typical gnawing marks on defatted and frequently weathered bones." Importantly, Brain (1981:117) notes that "the incidence of gnawed bones in the whole collection from any site will depend on the abundance of bones available to the porcupines at the time;" if bones are readily available, few will be gnawed; if bones are scarce, many will be gnawed.

Maguire *et al.* (1980:91) indicate that the African porcupine creates two types of damage to ungulate bones. First, gnawing marks take the form of "broad, contiguous shallow scrape marks." Second, cancellous bone is often scooped out of the ends of long bones, creating "tubular shafts" (Maguire *et al.* 1980:93). The scooped or hollowed out cancellous bone may resemble carnivore damage, but Maguire *et al.* (1980:91) report that distinctive porcupine gnawing marks are "invariably also present." They report that "porcupines are incapable of splitting or cracking the shafts of limb bones, so that bone flakes are exceedingly rare in porcupine-collected bone accumulations" (Maguire *et al.* 1980:93). The gnawing they illustrate is similar to that shown in Figure 6.15.

Dixon (1984) suggests the North American porcupine (*Erethizon dorsatum*) may accumulate bones in caves and dens, but thus far clear documentation of such behavior is not available. It has been documented that this species gnaws bones (Dixon 1984). It seems both African and North American porcupines collect bones for gnawing in order to keep their incisors in good condition, thus it is not surprising that most of the bones they collect are dry and at least slightly weathered because dry bones are easier to gnaw (are less slippery from grease; see Chapter 9). As well, this probably accounts for the fact that bones of small vertebrates do not seem to be collected by porcupines (Andrews 1990:7).

Birds as bone accumulators
Limited research indicates various species of (African) vulture (Accipitridae) accumulate bones at their nesting sites (Mundy and Ledger 1976; Plug 1978). This behavior apparently provides a source of calcium for growing chicks (Mundy and Ledger 1976; Richardson *et al.* 1986). There are no rodent or small mammal bones in some of these collections (Plug 1978). Many of the broken bones show evidence of having been broken by carnivores, and some bones are still articulated, suggesting these specimens may have been introduced to the nest as sources of meat or other soft tissue for the chicks (Mundy and Ledger 1976; Plug 1978). Bone specimens of bovids are virtually all broken and range in length from 1 to 40 cm, with a modal value of 3–6 cm. Some of the bones that were regurgitated by vulture chicks had been partially digested and have a corroded appearance (Richardson *et al.* 1986:38). "Artifacts" are also accumulated by vultures. Thus when a nest deteriorates and collapses, the bone and artifact 'fall-out' could create a small 'pseudo-site' on the ground.

Solomon *et al.* (1986) report that the Australian great bower bird (*Chlamy-*

a

b

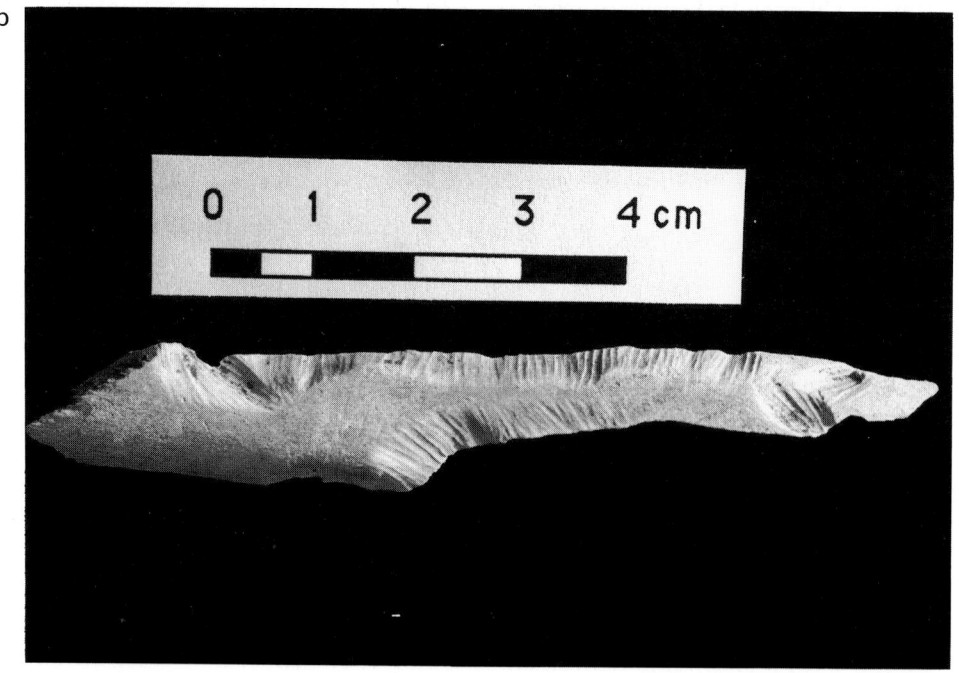

c

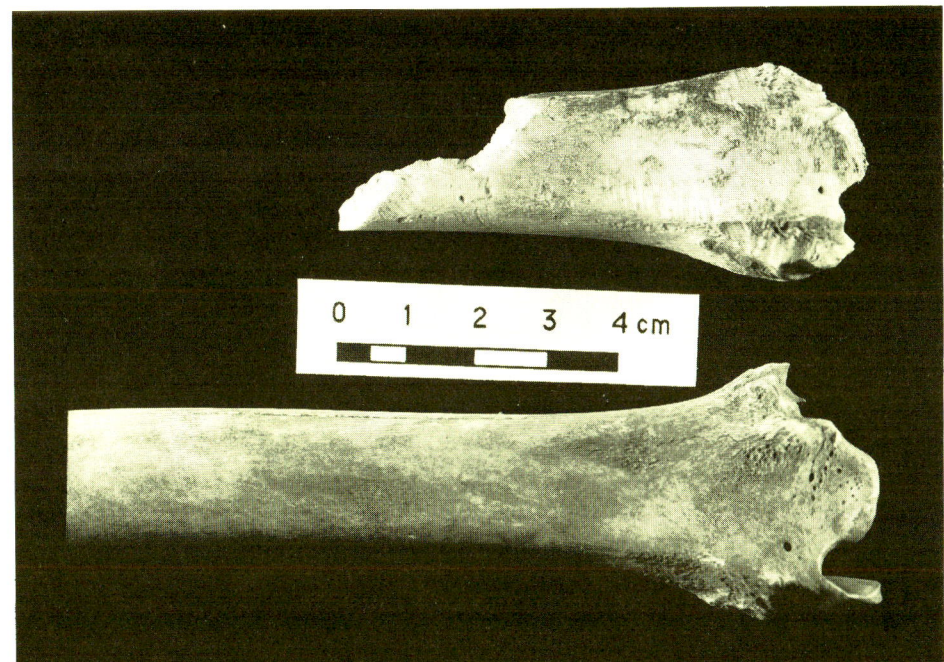

Figure 6.15. Rodent gnawed bones. a, distal view of three domestic sheep (*Ovis aries*) astragali that have been extensively gnawed by wood rats (*Neotoma* sp.) recovered from a wood rat nest in New Mexico; note the exposed cancellous bone tissue in the two left specimens and the incisor grooves on the left margin of all three; b, a long bone shaft fragment gnawed by marmots (*Marmota* sp.) recovered from a rockshelter in Missouri; c, distal tibiae of deer (*Odocoileus* sp.), upper specimen gnawed by marmots and recovered from a rockshelter in Missouri, lower specimen is an ungnawed comparative specimen placed in the same orientation as the upper specimen.

dera nuchalis) collects bones as well as artifacts. They suggest "an abandoned, decomposed and dispersed collection of bower objects could be misinterpreted as an open archaeological site" (Solomon *et al.* 1986:308). They suggest bower birds may remove bones from existing archaeological sites or a decomposed bower near a site could contaminate that site. They collected over 300 vertebrate specimens, all of which represent mammals, from two bowers. All specimens are < 30 g, and 84% of them are < 10 g. All specimens are < 19.0 cm long, and 81% are 0.2 to 0.9 cm in maximum dimension. About 80% of the bones have length:weight ratios less than 30 mm/g. The proportion of specimens identifiable to skeletal element and the proportion identifiable to taxon are both high relative to zooarchaeological collections. Nearly 80% of the specimens have a glossy sheen or polish "on the ends of long bones and in the middle of short, chunky bones" (Solomon *et al.* 1986:314). A few specimens have punctures and beak marks. Fragmentary specimens seem to have been broken when dry, and some appear to have been broken by the bower birds. Finally, many of the specimens have relatively high structural densities (see Chapter 7).

Many species of owls tend to roost in caves and rockshelters. They cast (regurgitate) pellets which are composed of matted fur, hair, bones, teeth, and other undigested material. One or two pellets are cast per day and when a roost is occupied for long time periods, owls have the potential to create major accumulations of bones of their prey (Kusmer 1990). Actualistic research on a number of species of owls indicates that some attributes of such owl-generated bone accumulations may be diagnostic of this bone accumulator. For example, bone assemblages resulting from accumulation by owls tend to have abundant bones of all elements of the skeleton, but mandibles and femora may be the most abundant elements (Dodson and Wexlar 1979; Kusmer 1990). These bones may represent a limited number of the species present in the area (e.g., only the remains of small mammals [no large mammal skeletal parts] are present in owl pellets), many bones may be from immature prey animals, and many of the represented species will probably be nocturnal and/or crepuscular (Kusmer 1990). Brain (1981) summarizes data indicating at least some species of African raptors and owls exploit mammalian prey rather opportunistically, and thus the created fossil assemblage may be a relatively accurate reflection of the small mammal population extant in the predator's foraging area. It seems that the frequencies of different skeletal elements will not allow the analyst to distinguish the particular raptor species responsible for a particular bone accumulation (Hoffman 1988; see below).

Owls tend to damage the bones they accumulate. There is limited digestive corrosion (Kusmer 1990), although bones accumulated by hawks may show more such corrosion than those accumulated by owls. Skulls of prey accumulated by owls in summer months may be more fragmented than those accumulated in winter months (Lowe 1980). Hawks and the screech owl (*Otus asio*) seem to break the bones of their prey more than most owl taxa thus far studied (Hoffman 1988). Most scapulae and innominates will be broken by many owl species, and femora, radii, mandibles, and humeri will tend to be complete and not broken (Dodson and Wexlar 1979; Kusmer 1990). Hoffman (1988) defined nine categories of fragmentation, recording to the nearest 25% the portion of a skeletal element represented by a specimen: complete, 25% of the proximal portion present, 50% of the proximal portion present (proximal half present), 75% of the proximal portion (distal 25% missing), 25% of the shaft present, 50% of the shaft present, 25% of the distal portion present, 50% of the distal portion present (distal half present), and 75% of the distal portion (proximal 25% missing). He found that hawks and the screech owl tend to create more fragmented bones and more kinds of fragmented bones than owls in general.

Hockett (1989a, 1991) found that North American raptors accumulate bones in open sites as well as caves and rockshelters, and tend to accumulate more juvenile leporid bones than bones of adult leporids. Innominates are frequently damaged, the tibia is often broken into a diaphysis cylinder (especially for small leporids or *Sylvilagus* spp.), the transverse processes of vertebrae and the greater trochanter of the femur are often damaged, and forelimb bones tend to outnumber hindlimb bones in raptor-accumulated assemblages. The skull is usually disassociated (dispersed) from the rest of the carcass, the occipital is broken, and there are beak and/or talon punctures behind the eye sockets. The ascending ramus of the mandible often has punctures or is broken. Forelimbs and hindlimbs are often dispersed as articulated units, and long bones may have one or more punctures on one side. Human accumulators of leporid bones tend to create more tibia diaphysis cylinders for large leporids (*Lepus* spp.), accumulate more bones of mature leporids, and variously burn and butcher the carcasses (Hockett 1991).

In the most intensive and extensive study to date, Andrews (1990) describes in some detail the effects a number of avian predators have on mammalian bones. He also compares the modifications made to bones by these avian predators to those made by some mammalian carnivores (summarized in the following subsection). Andrews (1990) indicates there are four basic categories of evidence the analyst can examine in attempts to identify the predator responsible for a bone accumulation. The presence of pellets or scats, some of which can be identified to the taxon that created and deposited them, is one category of evidence, but these seldom preserve (Andrews 1990:28). Second, the size range of prey taxa may provide an indication of the predator's identity, but the size of the predator and the size of the prey are not tightly correlated. Third, the taxa of prey may also provide some hints as to the identity of the predator, but again these two variables do not seem to be tightly correlated as many predators are far too opportunistic in their foraging habits (Andrews 1990:29). It is the fourth category of evidence, modification of bones, that Andrews (1990) finds to be the most accurate signature of particular predatory taxa. He distinguishes three kinds of bone modification – skeletal part frequencies or bone loss, bone breakage, and digestive corrosion – and reviews each in some detail for a number of predators. I summarize that discussion in the following paragraphs.

Bone loss: Noting that owls, hawks, and mammalian carnivores often destroy some bones of their prey, Andrews (1990:45) suggests that calculation of the relative proportions of skeletal elements represented in an accumulation may provide clues to the identify of the bone accumulator. These proportions are calculated with the equation

$$R_i = N_i/(MNI)E_i \qquad [6.3]$$

where R_i is the relative proportion of skeletal element i, N_i is the observed frequency of element i in the assemblage, MNI is the minimum number of individuals, and E_i is the frequency of skeletal element i in one prey skeleton.

This equation accounts for the fact that there are, for example, more humeri than skulls in a single mammal skeleton, and is mathematically identical to equations used by others for similar purposes (e.g., Dodson and Wexlar 1979; Korth 1979; Kusmer 1990; Hoffman 1988; Shipman and Walker 1980).

The purpose of equation [6.3] is to produce a frequency distribution of different skeletal elements so that different assemblages may be compared. For example, if it can be demonstrated using actualistic data that hawks deposit many femora and few mandibles whereas owls deposit few femora and many mandibles, then we will have a valuable analytical technique for identifying the agent of bone accumulation. In fact, previous researchers have suggested such distinctions do not appear to be possible. Hoffman (1988:85) compared the frequency distributions of skeletal elements across seven species of raptor (4 owl and 3 hawk species) and concluded "differential element representation as measured by ordinal ranking is insufficient for distinguishing clearly between raptor species." Similarly, after study of the frequencies of skeletal parts created by 10 species of owls, two species of hawks, and seven species of mammalian carnivore, Andrews (1990:49) concluded that "at this level of analysis it is not possible to distinguish adequately between a diurnal raptor, some of the owls, or mammalian carnivores."

The problem is readily indicated using the R_i frequencies of skeletal parts reported by Andrews (1990). Those frequencies for the barn owl (*Tyto alba*) and the kestrel (*Falco tinnunculus*) are correlated ($r = 0.572$, $P = 0.02$), those frequencies for the barn owl and the coyote (*Canis latrans*) are correlated ($r = 0.63$, $P = 0.007$), and those frequencies for the kestrel and coyote are correlated ($r = 0.726$, $P = 0.03$). These statistics indicate there is perhaps insufficient variation in skeletal part frequencies generated by these bone-accumulating agents to allow them to be distinguished on this basis alone. (Potential inter-analyst variation in how bones were quantified and/or in identification skill precludes comparison of, say, barn owl skeletal part frequencies reported by Dodson and Wexlar [1979], Hoffman [1988], Kusmer [1990], and Andrews [1990].) Thus as Andrews (1990:49) notes, "a rather more detailed form of analysis becomes necessary" if we wish to identify and distinguish these kinds of bone accumulators. We turn to that other form of analysis in later sections of this chapter.

Andrews (1990:45) suggests one could also "express the distribution of skeletal elements against the total number of bones in the sample," although this results in an "exaggeration of the abundances of such elements as ribs and vertebrae." The advantage to this latter technique, Andrews (1990:45) suggests, is that it does not depend on the calculation of MNI, "at best an unreliable estimate of the true numbers of individuals in a sample" (see also Grayson 1984). Andrews (1990:45–46) uses chi² to compare skeletal part frequencies calculated this way from different assemblages. While this analytical procedure may be a reasonable one, it must be emphasized that the analyst needs to be clear whether the frequency distribution of "skeletal elements" is founded on NISP or MNE values (see Figure 8.11 and associated discussion).

Because the average R_i value for an assemblage (calculated from R_i for all included skeletal elements) "is commonly used to characterize assemblages and it is often suggested that depositional agents may be identified through average R_i values" (Kusmer 1990:630; e.g., Korth 1979), if average R_i values are at least partially a function of sample size or NISP (as Kusmer's 1990 research indicates) as well as a measure of differences in the treatment of bones by depositional agents, then yet another problem exists with using these values as analytical tools for identifying agents of bone accumulation. This is so because we as yet do not know what portion of the variation in average R_i values can be attributed to sample size differences and what portion of that variation can be attributed to differences in the identity of the depositional agents.

While one of the most intensively and extensively documented variables available, the relative frequencies of skeletal parts generated by various raptors and mammalian carnivores do not seem to allow the analyst to identify clearly a bone-accumulating agent. This may also be true of large mammal bones accumulated by large mammalian predators, including hominids (see Chapter 7). It is in part for this reason that most analysts have turned to other kinds of data in attempts to identify bone-accumulating agents.

One kind of data involves calculating the ratio of cranial to post-cranial remains, and another involves calculating the ratio of proximal limb elements to distal limb elements. Andrews (1990:49) calculates the ratio of cranial to post-cranial remains two ways. One technique involves calculating the ratio of the sum of the frequencies of femora, tibiae, humeri, radii, and ulnae, and the sum of the frequencies of mandibles (lefts and rights counted separately), maxillae (lefts and rights counted separately), and isolated molars, multiplied by 100 to derive a percentage. Because in a single skeleton the ratio of these two categories of skeletal parts is 10 post-cranial to 16 cranial parts, Andrews multiples the observed ratio by 5/8 to weight the ratio according to the frequency of these bones in one skeleton. The other technique of calculating a ratio of post-cranial skeletal parts to cranial skeletal parts described by Andrews (1990:49) involves calculating the ratio of the sum of the frequencies of femora and humeri, and the sum of the frequencies of mandibles and maxillae (for both, lefts and rights are tallied separately), and multiplying that ratio by 100 to produce a percentage. Because the ratio of these skeletal parts is 4:4 in a single skeleton, no weighting of the ratio is necessary. A graph of the first described ratio of post-cranial to cranial skeletal parts using Andrews' (1990:49) data suggests owls tend to produce the highest ratios, mammalian carnivores produce mid-level ratios, and hawks produce low ratios (Figure 6.16).

Andrews (1990:49–50) also suggests calculating the ratio of major distal limb parts (tibiae plus radii) to major proximal limb parts (femora plus humeri)

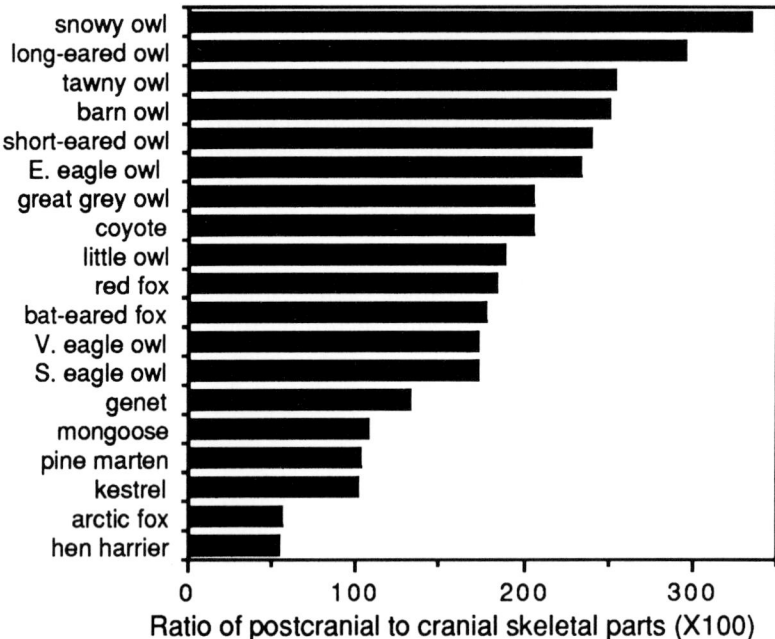

Figure 6.16. Ratio of postcranial to cranial skeletal parts accumulated and deposited by 19 species of raptors and mammals; see text for discussion (from Andrews 1990). Snowy owl (*Nyctea scandiaca*); long-eared owl (*Asio otus*); tawny owl (*Strix aluco*); barn owl (*Tyto alba*); short-eared owl (*Asio flammeus*); E. eagle owl = European eagle owl (*Bubo bubo*); great grey owl (*Strix nebulosa*); coyote (*Canis latrans*); little owl (*Athene noctua*); red fox (*Vulpes vulpes*); bat-eared fox (*Otocyon megalotis*); V. eagle owl = Verreaux eagle owl (*Bubo lacteus*); S. eagle owl = spotted eagle owl (*Bubo africanus*); genet = small-spotted genet (*Genetta genetta*); mongoose = white-tailed mongoose (*Ichneumia albicauda*); pine marten (*Martes martes*); kestrel (*Falco tinnunculus*); arctic fox (*Alopex lagopus*); hen harrier (*Circus cyaneus*).

multiplied by 100. A graph of these data for 19 taxa of bone accumulators indicates owls deposit about equal numbers of proximal and distal parts of limbs (Figure 6.17). Hawks tend to deposit fewer distal limb parts than owls, and mammalian carnivores deposit the fewest distal limb parts relative to proximal limb parts.

Bone breakage: I have already mentioned Hoffman's (1988) technique of tallying the proportion of a bone fragment represented by a specimen, and using that data to calculate the number of kinds of fragments and the diversity (frequency of representation of each) of the kinds of fragments. He concludes that hawks produce more broken bones than most owls (Hoffman 1988:87). Andrews (1990:52) found similar results, plus he noted that small mammalian

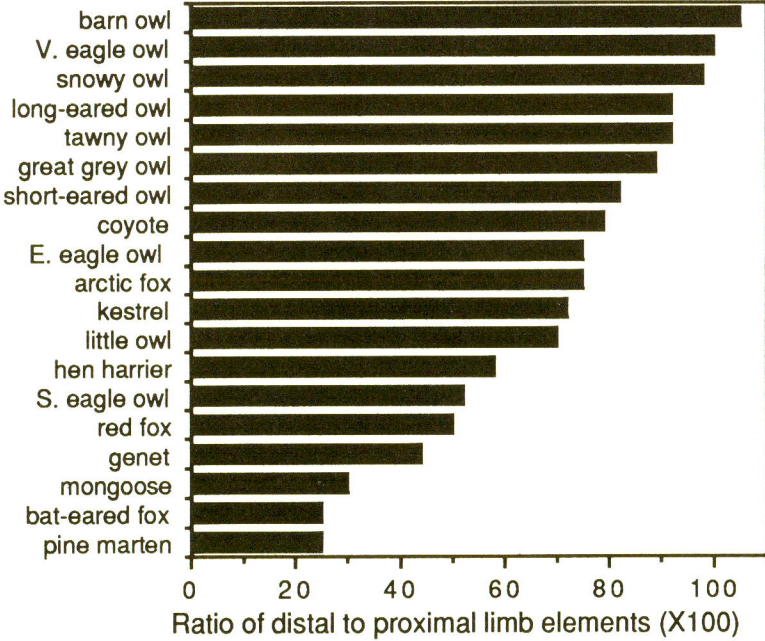

Figure 6.17. Ratio of distal to proximal limb elements accumulated and deposited by 19 species of raptors and mammals; see text for discussion (after Andrews 1990). Bone accumulating taxa as in Figure 6.16.

carnivores produce more broken bones (fewer complete bones) than hawks, and some owls produce very few broken bones.

I calculated the proportion of the MNE (minimum number of elements) that were complete from various assemblages reported by Andrews and Evans (1983), Dodson and Wexlar (1979), Hoffman (1988), and Kusmer (1986). Only data for the humerus, ulna, femur, and tibia were compiled. As graphed in Figure 6.18, these data corroborate the observations of others. Most owls break fewer than 25% of the included skeletal elements. Mammalian carnivores tend to break 50% or more of the included limb bones, and hawks essentially break all of the included skeletal elements (see Chapter 8 for further discussion of the quantification of fragmented bones).

Fragmentation of cranial elements shows similar patterns. Most owls studied by Andrews (1990:53–64) tend to leave skulls with maxillae still attached whereas hawks and carnivores produce more isolated maxillae. Owl pellets contain more incisors (both upper and lower) than hawk pellets and mammalian carnivore scats. Most owls break few mandibles and virtually no molars or incisors whereas hawks and mammalian carnivores break many mandibles. Hawks break a few molars and incisors, but mammalian carnivores seem to produce the most broken molars and incisors. Thus, if the analyst can

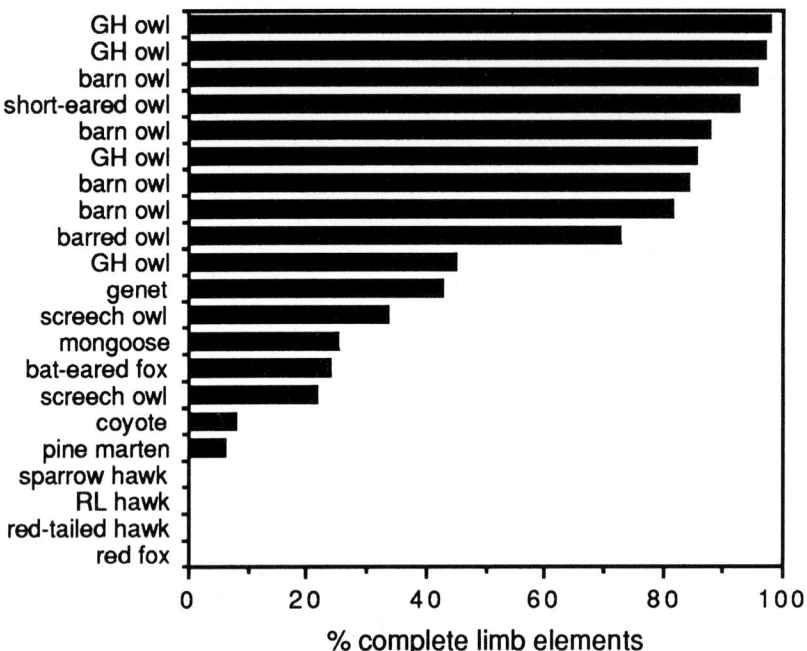

Figure 6.18. Proportion of complete limb elements (humerus, ulna, femur, tibia) in assemblages accumulated by selected raptors and mammalian carnivores. GH owl = great horned owl (*Bubo virginianus*); barn owl (*Tyto alba*); short-eared owl (*Asio flammeus*); barred owl (*Strix varia*); screech owl (*Otus asio*); sparrow hawk (*Falco sparverius*); RL hawk = rough-legged hawk (*Buteo lagopus*); red-tailed hawk (*Buteo jamaicensis*); mammals as in Figure 6.16.

measure the degree to which the specimens making up a fossil collection are fragmented, they will be closer to identifying the agent of bone accumulation.

Digestive corrosion: Andrews (1990:32) found that digestive corrosion was greatest in mammalian carnivore accumulations; while such corrosion was present in raptor accumulations, it was less pronounced. He suggested that this was the case because the former taxa digest food in the stomach and intestine whereas raptors only digest in the stomach, thus food in mammalian carnivores is subjected to longer digestion times. Andrews (1990:32) found that tooth enamel most readily displayed the effects of digestive corrosion, probably because it is the most highly mineralized skeletal tissue (Chapter 4); dentine and bone also displayed digestive corrosion but less extensively than enamel (see also Fisher 1981).

More molars were digested by mammalian carnivores and hawks than by owls in Andrews' (1990:65) samples. Digestion seemed also to be more intensive and extensive for the former than for the latter predators. The occlusal corners of salient angles progress from slightly rounded, to flattened, to complete loss of enamel and exposure of dentine, as one shifts from owls to diurnal raptors (hawks) to mammalian carnivores (Andrews 1990:67). Skeletal elements of the limb were also more intensively and extensively corroded from digestion by hawks and mammalian carnivores and less corroded by owls (Andrews 1990:79; Fernandez-Jalvo and Andrews 1992:410).

Rensberger and Krentz (1988) report that corrosion features on rodent bones and teeth created by the digestive action of coyote (*Canis latrans*) and great horned owl (*Bubo virginianus*) tend to be similar. Many features visible under scanning electron microscopy were not evident under standard light microscopy. These include deep solution fissures with rounded edges and pits. The fissures and pits seem to originate in canals beneath the bone surface.

Fernandez-Jalvo and Andrews (1992) provide a useful synopsis of digestive corrosion stages displayed by rodent cheek teeth and incisors first reported by Andrews (1990). *Light digestion* is restricted to the occlusal corners of the salient angles of cheek teeth and does not penetrate below the alveolar margin; the entire enamel surface of incisors is slightly pitted and the distal tip of the enamel surface may be completely removed, indicating digestion while the incisor was still in place in the jaw. *Moderate digestion* of cheek teeth is signified by the removal of enamel along the entire edge of the salient angle, perhaps with some pitting; the entire enamel surface of incisors is more completely affected and the dentine has a wavy surface. Corners and salient angles of cheek teeth are heavily rounded and much enamel is removed and dentine exposed in *heavy digestion* of cheek teeth; incisors have isolated islands of enamel and wavy dentine surfaces. *Extreme digestion* results in cheek teeth that are rarely identifiable and with quite damaged dentine; incisors have very little or no enamel remaining and dentine is quite damaged.

Mammalian carnivores as bone accumulators

There has probably been more research and more written on this category of bone-accumulating agents than any other, no doubt because many of its members occupy caves and rockshelters just as hominids do. The zooarchaeologist and taphonomist is thus regularly faced with determining if the bones in such a site are present there due to the action of humans or the action of carnivores. Most actualistic studies have focused on relatively large carnivores. Andrews and Evans (1983), however, provide an important study on the bone collecting activities of relatively small mammalian carnivores. They examine a number of attributes of bone assemblages accumulated by small carnivores, some of which are compared above with bone assemblages accumulated by owls and hawks. While in this section we focus on large mammalian predators and bone accumulators, it is important to summarize observations made on the modifications to bones created by small, dog-sized or fox-sized carnivores.

Stallibrass (1984, 1990:153–155) lists several attributes of bone modification produced by fox-sized canids. Bones in scats are highly fragmented, even more so than is found in some owl pellets. Fragments rarely exceed 1 cm in maximum dimension, and bones of domestic sheep (*Ovis aries*)-sized prey are small and unidentifiable. Isolated teeth are common and most mandibles of small mammals are broken. Some bones have puncture marks from gnawing and some display digestive corrosion. Diaphysis fragments are splinters rather than tubes. As we will see, these are rather similar to the modification attributes documented for larger predators.

The African leopard (*Panthera pardus*) carries prey carcasses into trees for uninterrupted consumption (Brain 1980, 1981; Cavallo and Blumenschine 1989). Bones fall out of the tree as the leopard consumes the carcass. If those bones remain around the base of the tree, a small concentration of bones may build up. One of the most interesting attributes of the skeletal remains (other than gnawing damage) is the destruction of the brain case and postero-dorsal portion of the eye sockets (Brain 1981:93).

In the Old World, the most studied large mammalian carnivore is the hyena. There are three extant species of hyena: the brown hyena (*Hyaena brunnea*), the striped hyena (*Hyaena hyaena*), and the spotted hyena (*Crocuta crocuta*). The bone-collecting habits and bone-modifying behaviors and results of those behaviors for each species have been studied in some detail (Brain 1981; Horwitz and Smith 1988; Kerbis-Peterhans and Horwitz 1992; Lam 1992; Richardson *et al.* 1986; Skinner and van Aarde 1991; and references therein). The spotted hyena has been characterized as "the most effective extant bone-cracking carnivore" (Marean and Spencer 1991:648). In the New World, much research has been done on wolves (*Canis latrans*), coyotes (*Canis latrans*), and domestic dogs (*Canis familiaris*) (Binford 1981b; Burgett 1990; Haynes 1980, 1983a; Kent 1981; Klippel *et al.* 1987). Anecdotal data have also been collected for the African lion (*Panthera leo*) and North American bears (*Ursus* spp.) (Haynes 1983a). Many of the attributes of bone modification produced by these large mammalian carnivores can be summarized as follows.

Ragged-edged chewing (Maguire *et al.* 1980:79–80) is typically seen on thick bone such as the ends of limb bone shafts (Figure 6.19). Also known as *crenulated edges* (Binford 1981b:51), this type of damage may occur on very thin bone when the tooth penetrates and removes part of the edge of the bone. *Shallow pitting* (Maguire *et al.* 1980:79–80) or *pitting* (Binford 1981b:44–48) often has a restricted distribution and is produced when the bone is sufficiently strong or dense to withstand the pressures of teeth and not be punctured (Figure 6.20). *Punctures* (Binford 1981b:44–48), *punctate depressions*, or *perforations* (Maguire *et al.* 1980:79–80) result when the bone collapses under the pressures of teeth, leaving a clear, more or less oval depression in the bone, often with flakes of the outer wall of the bone pressed into the puncture (Figure 6.21). Shipman (e.g., 1981a:366) suggests punctures are produced by the pressure of a single tooth cusp or canine applied at an angle approximately perpendicular to the bone surface. Punctures decrease in diameter as depth from the bone surface increases.

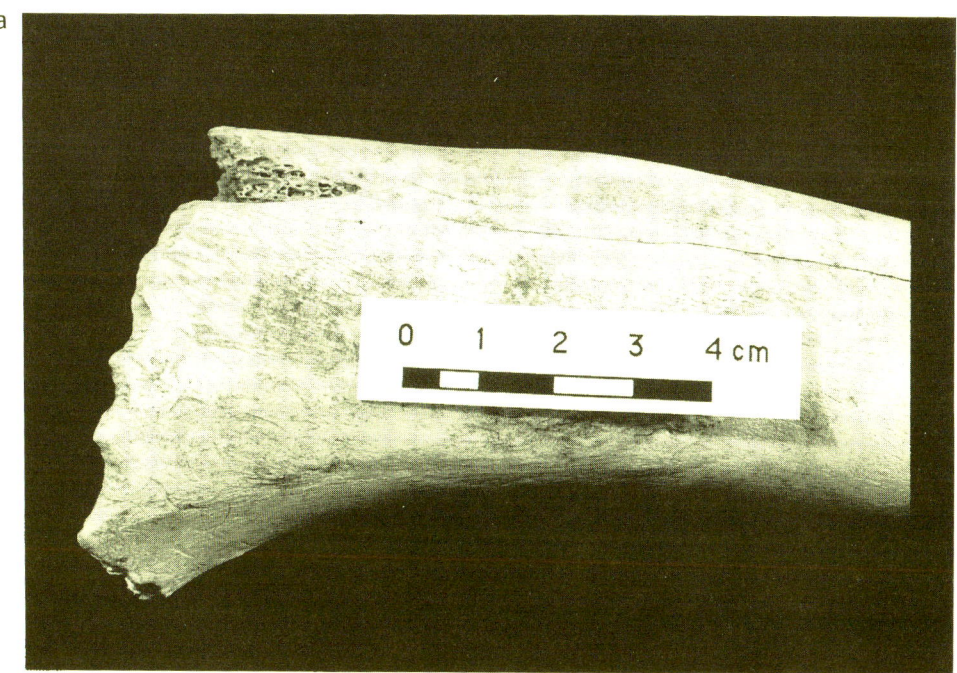

Figure 6.19. a, ragged and crenulated proximal end of a modern wapiti (*Cervus elaphus*) humerus shaft resulting from gnawing by carnivores, probably coyote (*Canis latrans*), collected from Washington state; b, detail of the crenulated edge, note polishing of points of crenulations and furrows.

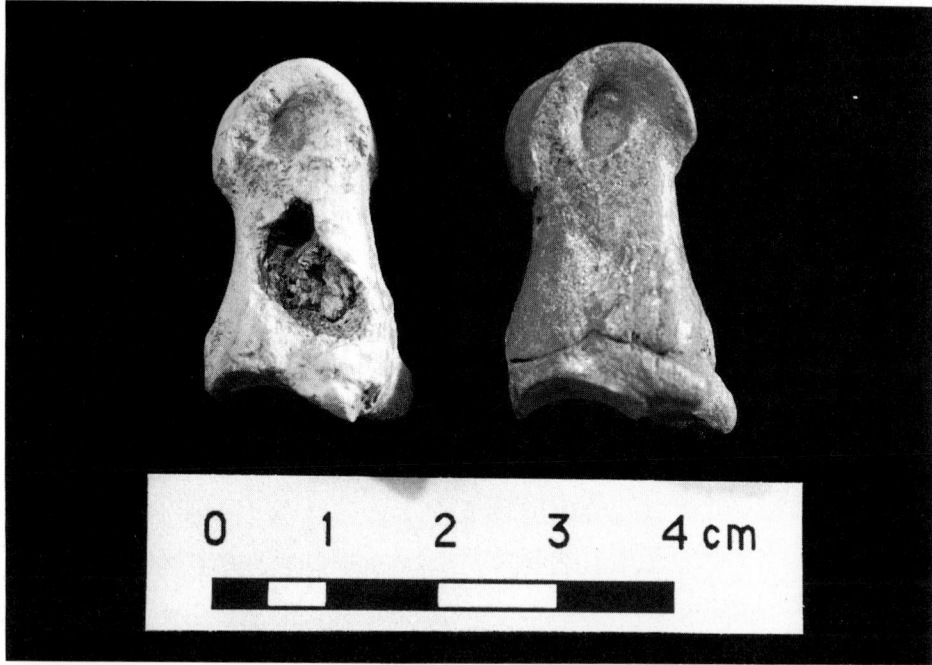

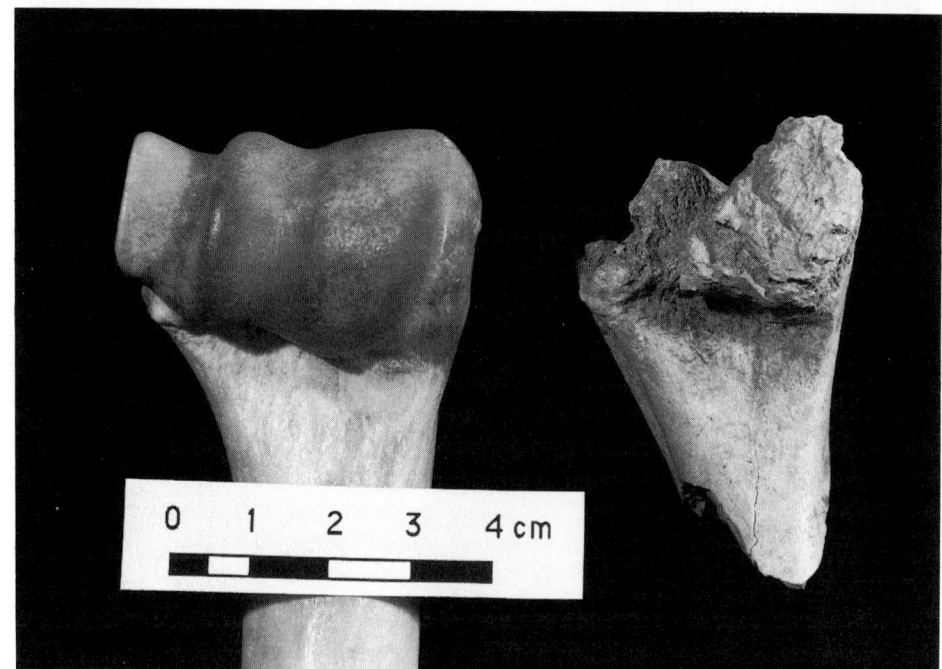

Figure 6.20. Pitting and punctures. a, left, deer (*Odocoileus* sp.) second phalanx collected from a rockshelter in Missouri showing a large puncture (pitting is evident on the reverse side) probably resulting from multiple bites, right, complete ungnawed comparative specimen; b, right, an intensively chewed deer distal humerus showing pitting, collected from a rockshelter in Missouri, left, an ungnawed comparative specimen.

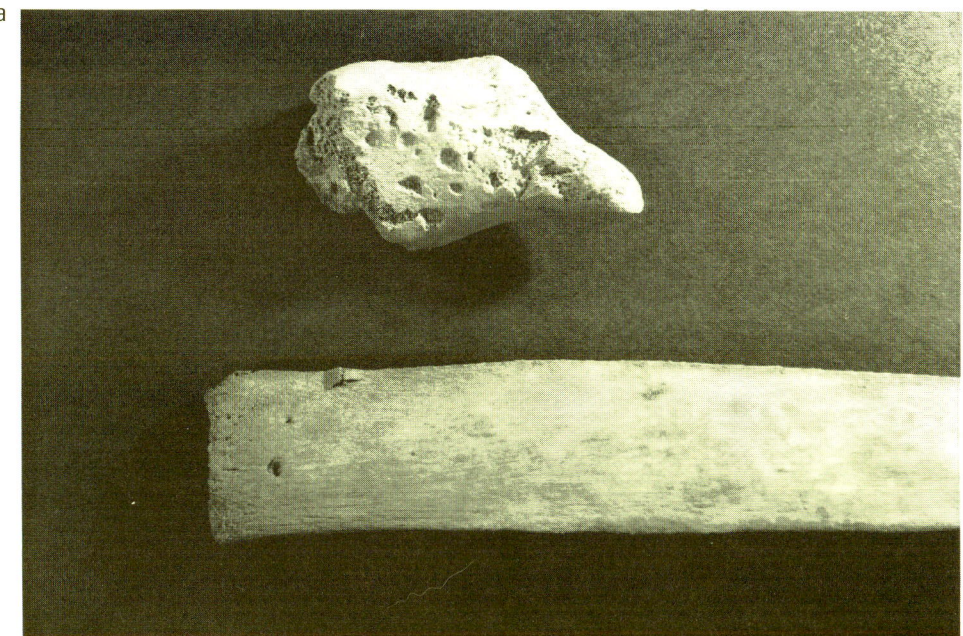

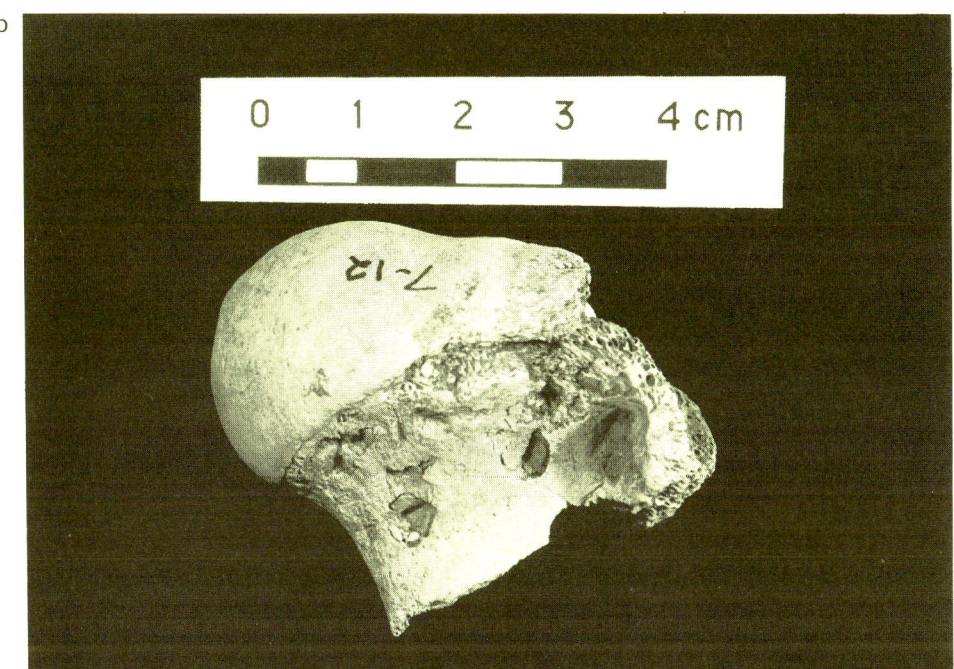

Figure 6.21. Punctures on (a) a patella and a ventral rib of wapiti (*Cervus elaphus*), collected from Washington state (from Lyman 1989b:153, Figure 7; courtesy of The Center for the Study of the First Americans), and (b) a proximal femur of deer (*Odocoileus* sp.) collected from a rockshelter in Missouri.

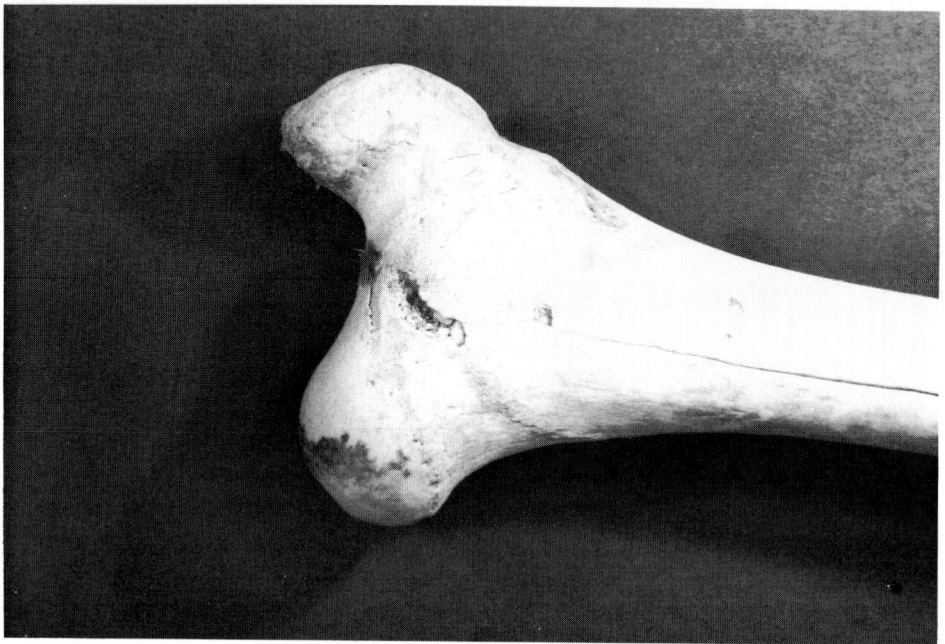

Figure 6.22. Furrow on a modern wapiti (*Cervus elaphus*) proximal femur collected from Washington state (from Lyman 1989b:153, Figure 7; courtesy of The Center for the Study of the First Americans).

Striations, gouge marks (Maguire *et al.* 1980:79–80), or *scoring* (Binford 1981b:44–48) are usually short, parallel, and linear or straight marks that are roughly perpendicular or transverse to the long axis of the bone. They may be quite close together, are usually on the shafts of long limb bones, and tend to follow the surface of the bone. They apparently result from dragging the teeth across the surface of the bone. Shipman (e.g., 1981a:365) labels such marks *tooth scratches* and characterizes them as "elongate grooves that may vary from V-shaped to U-shaped in cross-section [and] the bottom of the groove is smooth." Tooth scratches may occur singly, as sets of parallel or subparallel marks, or as clusters with different orientations. Various contiguous or close, irregular and randomly-oriented grooves (Maguire *et al.* 1980:79–80) or *furrows* (Binford 1981b:44–48; Haynes 1980, 1983a) are generally found on the ends of long limb bones where bone tissue is cancellous (Figure 6.22). *Scooping out* or *hollowing out* (Binford 1981b:44–48; Haynes 1980, 1983a; Maguire *et al.* 1980:79–80) is the result of extreme furrowing, and involves the removal of significant portions of the cancellous bone tissue from epiphyseal ends of limb bones (Figure 6.23). Scooping out results in the production of large, irregular holes in long bone ends.

Both hyenas and North American canids produce *acid-etching* and *corrosion* of bone (Figure 6.24). Such bones must have passed through the digestive tract

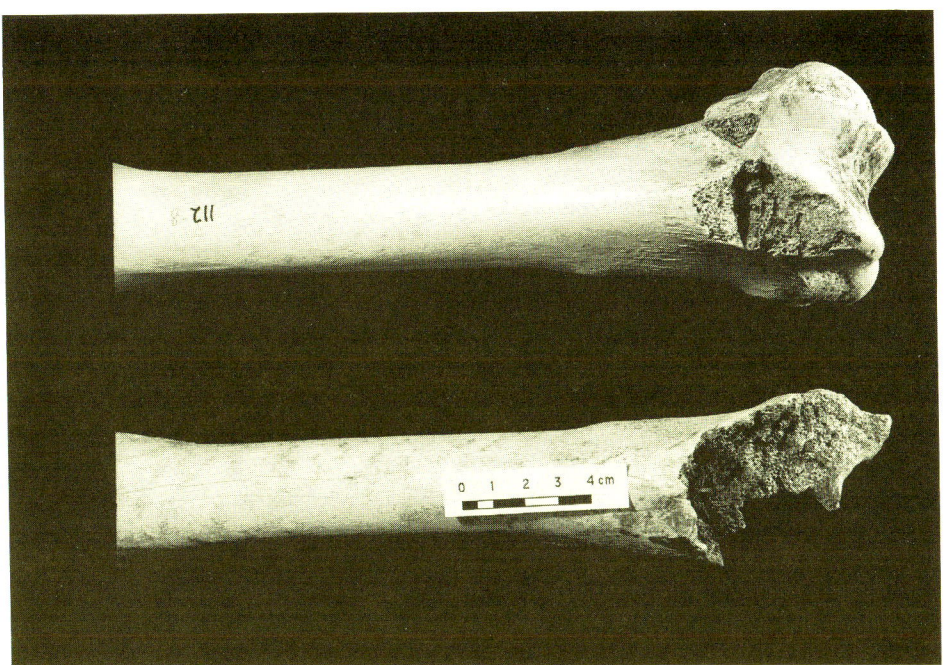

Figure 6.23. Scooping out on two distal femora of modern wapiti (*Cervus elaphus*) collected from Washington state.

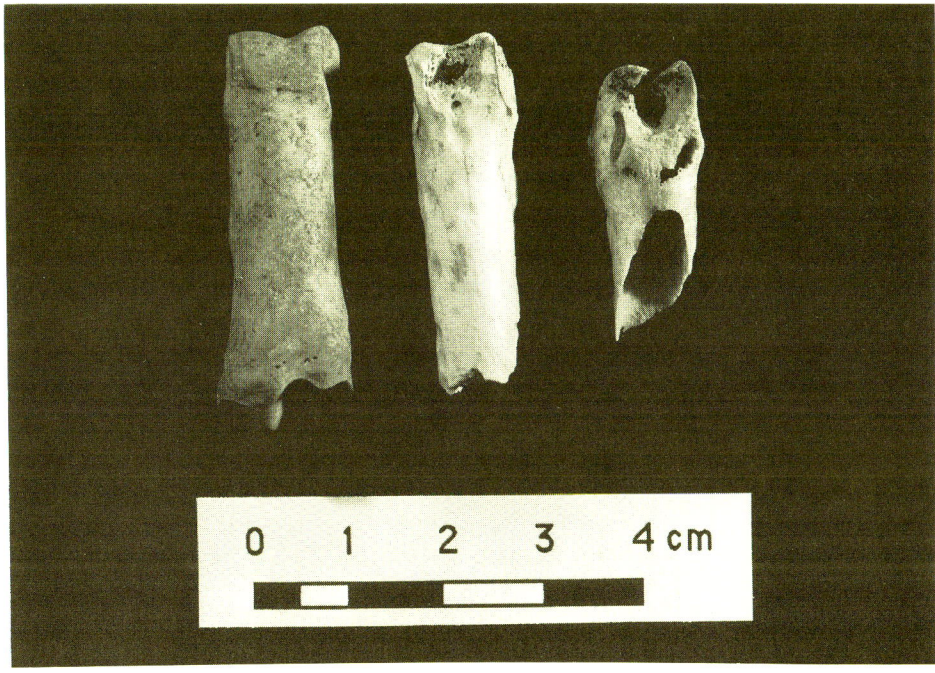

Figure 6.24. Digestive corrosion of first phalanges of domestic sheep (*Ovis aries*) collected from Missouri. Left specimen is a whole comparative specimen; note pitting on center specimen and feathering of fracture edges on right specimen.

or been regurgitated after some time in the stomach. Both groups of carnivores also *splinter and crack* bones (see also Chapter 8). This fracturing produces *lunate* or *crescent-shaped fracture scars*, thought of as "half of a punctate bite mark when the strength of the bite has been sufficient to split the bone so that the fracture line passes through the punctate perforation" by Maguire *et al.* (1980:79–80). There is sometimes an associated bone flake partially detached from the concave surface (medullary cavity side) of the bone at the point of tooth contact. When mammalian carnivores gnaw bones they "attack the ends of long bones first" (Binford 1981b:51). *Channeled bones* are produced by "puncturing the bone back from the transverse edge, leaving a channel running parallel to the longitudinal axis of the bone" (Binford 1981b:51). *Chipping back* results from chewing the edge of a broken long bone; the bone edge is continuously chipped and tooth scoring on the external surface of the bone is frequently associated. Licking chipped edges or ends can produce rounded and polished edges and ends that have the appearance of "use wear" (Binford 1981b; Haynes 1980, 1983a). Often, tooth damage is bipolar due to the vise-like action of the upper and lower tooth rows (e.g., Figure 6.20). Long-bone diaphyses that have had the ends removed can be collapsed by gnawing carnivores, producing many diaphysis splinters. In carnivore-gnawed assemblages, "evidence of gnawing, pressure-flaked edges, incised scarring on the outside of the flake, pitting and abrasion from repeated vise-like mashing of a bone surface, and so on will occur on large splinters (>4 cm long); small splinters will exhibit no such modification. The latter may show signs of having been corroded by stomach acids and be embedded in feces" (Binford 1981b:60).

The mere presence of one or more of these tooth-marking attributes would simply indicate that a large mammalian carnivore had access to the bones, and not directly indicate the carnivores accumulated the bones. These attributes may allow more precise identification of the carnivore taxon responsible if the bones are not too extensively gnawed according to Haynes (1983a) (see Table 5.6 for a scheme of stages of carnivore destruction of bones). His criteria for identifying the general taxonomic family of gnawing carnivore are reproduced in Table 6.10. Haynes (1980, 1983a) and Kent (1981) caution that mammalian carnivores can gnaw and move bones and yet leave no traces of tooth marks on the bones.

Other authors have argued that the size of a tooth puncture or lunate flake scar provides a good indication of the identity of the carnivore taxon. For example, Morlan (1983:256) suggests that the loading points of teeth on bones "will have diameters slightly larger than the contact areas of the teeth, and for most carnivores such diameters will be relatively small, no more than a few millimeters. Larger notches resembling loading points may be preserved on a carnivore-induced fracture, but they are composites of many smaller flake scars and do not represent discrete point loading." That is, multiple bites might be detected as multiple discrete loading points. But where do we measure the

Table 6.10 *Gnawing damage to bones typical of four taxonomic groups of mammalian carnivores (from Haynes 1983a)*

Damage type	Canids	Hyenids	Ursids	Felids
Tooth marking on compact bone:	3	5	2–1	1–0
5 = most expected, 1 = least expected				
Grinding off prominences vs. biting through:	2	2–3	5	1
5 = mostly grinding (leaves smooth stumps);				
1 = mostly biting through (leaves a rough stump,				
usually an irregular rim of compact bone)				
Tooth impression shape in trabecular bone:	3	3	5	1
5 = square or rectangular;				
3 = cone or truncated cone;				
1 = "axe-edge" or elongated V-shape				

"contact area" or diameter of a tooth? This is a sticky problem because teeth can penetrate bony material, thereby increasing the cross-sectional area of contact between the tooth and the bone. This is especially so for trabecular or cancellous bone which is easily and readily penetrated by teeth. And, what about assemblages gnawed by more than one taxon of carnivore? These difficulties can be illustrated with an example.

While analyzing a collection of small mammal remains (approximately 7,000 specimens identified to genus or species) recovered from a rockshelter in the state of Washington, I measured the maximum and minimum diameter of punctures. Measurements were taken at the surface of the bone. Because the assemblage is late Holocene in age, I measured the diameters of a set of modern mammalian carnivore canines approximately 2–3 mm from the distal tip of the canine. The 2–3 mm value was chosen because most of the punctures were about that deep. Plotting the frequencies of tooth puncture diameters against the range of measured canine diameters produces an ambiguous indication of the taxonomic identity of the bone-gnawing carnivore despite the fact that the measured carnivores are the major ones historically recorded in the site area and remains of each of them were found in the rockshelter (Figure 6.25). I suspect that these results were derived because the gnawing marks were produced by several carnivore taxa with different-sized canine teeth.

Blumenschine and Selvaggio (1988, 1991) inspected both pits (Figure 6.20) and hammerstone-generated marks; they term the latter *percussion marks*. They found that the two can be distinguished under low (10 ×) magnification. Percussion marks always have dense patches of microstriations in or emanating from the pit defining the mark. Tooth-generated pits rarely have a few, less densely occurring striations that generally require more powerful magnification to be visible. Unlike tooth marks, percussion marks should occur in approximately the same location on specimen after specimen of the same skeletal element.

Haynes (1983a) provides a descriptive guide for distinguishing the gnawing

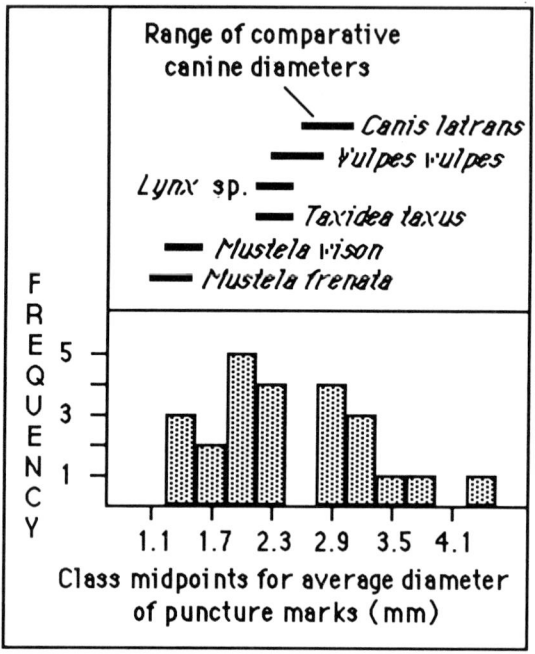

Figure 6.25. Comparison of diameters of puncture marks on small mammal bones collected from a rockshelter in Washington state, and the range of canine diameters of modern carnivores.

damage done to ungulate bones by ursids, canids, and felids that is based on the appearance of the gnawing damage. Richardson (1980; Richardson *et al.* 1986) presents data (summarized in Figure 6.26) indicating hyenas damage and destroy more bones of prey than lions (*Panthera leo*), leopards (*Panthera pardus*), feral dogs (*Canis familiaris*), and black-backed jackals (*Canis mesomelas*). As well, hyenas tend to fragment bones more often than these other carnivores, as indicated by the NISP:MNE ratios in Figure 6.26. Such comparative analyses are useful for establishing carnivore taxon-specific attributes of bone modification but have, to date, been rarely published. Perhaps that is because we are still far from documenting the full range of variability in modifications to animal carcasses and prey bones that can be produced by particular carnivore taxa (e.g., Lam 1992).

What is becoming abundantly clear as research progresses is that it will often be necessary to study multiple attributes of the fossil assemblage in order to identify agents of bone accumulation (e.g., Noe-Nygaard 1989). This is quite clear in a recent synopsis of attributes that might be used to detect the accumulation of bones by hyenas. Cruz-Uribe (1991; see also Klein 1975) describes attributes she believes to be distinctive of a bone assemblage created by hyenas, and notes how these attributes differ from a bone assemblage accumulated by hominids. It is important to note that all of her collections are

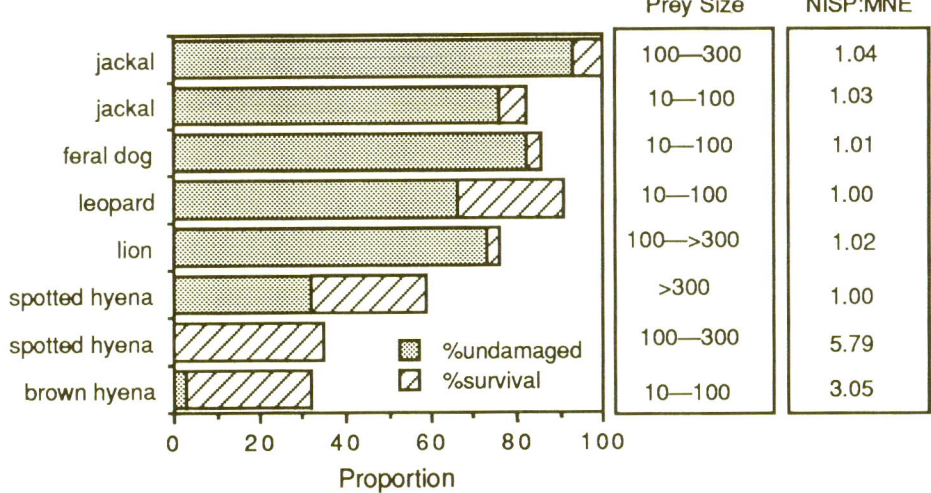

Figure 6.26. Attributes of modification to prey bones created by various African carnivores. Prey size is live weight of animal in kg; NISP:MNE ratio is Richardson's (1980) "fragmentation ratio" divided by 100 (after Richardson 1980:114, Figure 4; and Richardson *et al.* 1986:29, Figure 3).

prehistoric ones; that is, she has no experimental controls over the fossil assemblages she compares because the actual bone-accumulating agents are not known. Many of the attributes she lists do, however, tend to align rather well with attributes documented in actualistic contexts. In her collections, Cruz-Uribe finds that carnivores make up at least 20% of the remains (as measured by MNI) accumulated by hyenas whereas carnivores make up less than 13% of the remains accumulated by hominids. Second, many but not all bones accumulated by hyenas display striations, pitting, grooves, scooping out of cancellous bone, and digestive corrosion. Third, bone assemblages accumulated by hyenas contain many long bone cylinders; that is, the more or less complete diaphysis will be present but will lack the epiphyses. Hominid-generated accumulations tend to have more broken diaphyses and intact epiphyses. Fourth, mortality profiles of prey exploited by hyenas (see Chapter 5) tend to be attritional whereas mortality profiles of prey exploited by hominids may be either attritional or catastrophic. Fifth, in hyena accumulations, small ungulates tend to be better represented by cranial bones and large ungulates by post-cranial bones (as measured by MNI). In hominid accumulations, there is no clear relation between frequencies of cranial and post-cranial bones, and prey size. Finally, small, hard, structurally dense (see Chapter 7) bones such as tarsals, carpals, sesamoids, and phalanges tend to be abundant in well-preserved hominid accumulations, but are rare in hyena-accumulated assemblages because hyenas swallow and digestively destroy these elements.

The criteria outlined by Cruz-Uribe (1991) are similar to ones outlined by Stiner (1990a, 1991b, 1991e) and summarized in Chapter 7. Both authors were interested in distinguishing carnivore-created bone accumulations from those created by hominids. This brings us, then, to the last major bone accumulator zooarchaeologists and archaeological taphonomists tend to consider. I have found this to be a convenient place to make this topical shift, but the perceptive reader will note some overlap between the immediately preceding discussion and what follows.

Hominids as bone accumulators

The problem of recognizing hominids as bone-accumulating agents is one of analytically sorting out what Thomas (1971) calls *cultural bone* from *natural bone* (Lyman 1982a). The former category includes specimens deposited as a result of human activity; the latter category includes specimens (accumulated and) deposited as a result of natural processes. When the analyst is interested in human subsistence practices, a logical second step in the distinction process is the sorting of culturally deposited animal remains into those representing food waste and those representing non-food remains. A buried pet or beast of burden represents culturally deposited bone but may not have been eaten. Several kinds of criteria have been used to distinguish cultural and natural bone, with the implication generally being that these same criteria also help to distinguish culturally deposited food bones from culturally deposited non-food bones. I briefly review each of these criteria in the following.

Burning or charring: Because many humans cook meat before consuming it, there is often a chance that bone embedded in the meat will be burned (criteria for recognizing burned bone are reviewed in Chapter 9). Whether a bone is burned during cooking depends of course on how meat is cooked, and whether, say, bone in meat broiled over a fire will be burned depends on whether the bone is exposed to the heat. Balme (1980) suggests that taxa with much burned bone probably represent human food resources whereas taxa with little burned bone in the same site may not have been used as a food resource. Evidence described by Grayson (1988; see also Lyman 1988a) indicates this distinction may not be a valid one when used alone. Further, Balme (1980) found burned bone in a cave that contained clear evidence of human occupation, but also found burned bone in a cave that apparently had not been occupied by people although in the latter site there was much less burned bone than in the former (see Briuer 1977 for similar results).

Grayson (1988:27–29) examined the proportions of individual burned skeletal elements of jackrabbit (*Lepus* sp.) recovered from a cave in Utah. He found skull and ulna specimens were burned less often, and tibia specimens were burned more often than can be accounted for by random chance in the assemblage as a whole, but when stratigraphically distinct assemblages were studied, these results changed. There was no consistent between-strata pattern in the burning of jackrabbit specimens. For this set of remains, burning was randomly distributed across skeletal elements from stratum to stratum because the burning occurred after the skeletons had been disarticulated and deposited, and the burning was the result of *in situ* natural burning of the dry organic matter in the cave. Grayson (1988:29) emphasized that the lack of precise provenience information for individual specimens precluded addressing the reasons for the differential burning of the faunal remains (were tibiae more often deposited in stratigraphic areas that were to become burned than skulls and ulnae?). Grayson's (1988) study is instructive because it illustrates how the taphonomist might begin to unravel the burning history of a set of animal remains.

Comminution of bone: Boiling bone to extract grease may result in comminuted bone or small bone fragments. The presence of extremely small bone fragments in large quantities might be interpreted as indicative of bone grease production and thus that the taxa represented by the bones were exploited as a food resource. However, small bone fragments are often difficult to identify to skeletal element (Lyman and O'Brien 1987) and thus they are also difficult to identify to taxon. Vehik (1977:172–173) suggests four lines of evidence can be used to indicate bone grease production: (1) the presence of many small pieces of unburned bone; (2) low frequencies to absence of bones with high grease content (see Chapter 7 for measures of grease content of particular skeletal elements); (3) the presence of hammerstones, anvil stones, fire pits, and thermally fractured rock; and (4) the loss of collagen fibers from the bone fragments. The latter, of course, can also result from diagenetic processes (see Chapter 11).

Mineralization, weathering, and staining: The degree of weathering, mineralization, and staining of bones occasionally varies between intrusive or naturally deposited and culturally deposited bone (descriptions of these modifications are given in Chapters 9 and 11). Cultural bone should be more heavily weathered, stained, and/or mineralized than intrusive bone because the latter was deposited in site sediments some time after the former. However, simply because a bone is stained in a manner suggesting it is not intrusive does not mean it was deposited by hominids. A dog, for example, could introduce non-cultural bone to a site simultaneously with the deposition of cultural food bone.

Butchering and technology marks: The presence of butchering marks or modifications to bones resulting from the production of bone tools are good indications that faunal remains were deposited as the result of human activities (these attributes are described in Chapter 8). However, not only is there no detailed study distinguishing these two kinds of modifications (what we might term butchery marks and technology marks, respectively), all culturally deposited faunal remains will not necessarily display one or the other. The usual procedure, however, is to argue that if some of the remains of a taxon display such modifications, then all remains of that taxon were probably accumulated and deposited by hominids (e.g., Noe-Nygaard 1989).

Ethnographic analogy: Ethnographic analogy may be helpful in suggesting which taxa were eaten, but there are at least two potential problems. First, ethnozoological data may be ambiguous. For example, in western North America there are two species of *Lynx*, the bobcat (*L. rufus*) and the lynx (*L. canadensis*), which in some areas are sympatric. Some ethnographic accounts for this region refer to the "wildcat" as an exploited taxon but without listing the taxonomic name. This is the genus *Lynx*, but which species? The second potential problem is that the ethnographic record, as has been pointed out many times, is only one frame of an as yet unfinished feature film. Cultures evolve and change, and what people were eating in, say, 1905 when ethnographic data were recorded may not be what the ancestors of those people were eating a thousand years earlier.

Skeletal completeness: Thomas (1971:367) suggested that because "the dietary practices of man tend to destroy and disperse the bones of his prey-species," the set of taxa with lesser relative skeletal completeness is the one that can be identified as consisting of culturally deposited bone whereas the set of taxa with high relative skeletal completeness is the one that can be identified as consisting of naturally deposited bone. While use of this technique has been advocated by some (e.g., Wing and Brown 1979; Ziegler 1973), others have argued that the computed statistic actually measures sample size (e.g., Grayson 1978b). As well, ethnographic research subsequent to Thomas' (1971) suggestion has shown that humans do not always accumulate and deposit fractions of animal carcasses (see Chapter 7). A measure of relative skeletal completeness cannot distinguish, say, an incomplete rabbit skeleton deposited by natural processes from a complete rabbit skeleton deposited by cultural processes. This does not mean measures of skeletal completeness are not useful analytical tools, and we return to them in later chapters.

Context and associations: Variously burned, comminuted, mineralized, and butchered bone is readily believed to have been modified and deposited by cultural processes when such bones are associated with undisputed evidence of hominids, such as artifacts. That is so because, while a fossorial rodent may die in its burrow within a site and thus its remains may become associated with artifacts, it is doubtful that those rodent remains will also be burned, highly fractured, and variously mineralized or stained, or that they will display butchering marks. The combined attributes of burning, fragmentation, similar mineralization or staining across multiple specimens, butchery marks, and association with artifacts, all point to the same accumulation agent. As more of these attributes fail to be present, the inference that the remains represent culturally accumulated bone progressively weakens.

Accumulation and dispersal as mirror images

One possible effect of accumulation and dispersal that might be considered during analysis is that, beginning with a complete carcass, some bones of that carcass may be dispersed from the location of carcass deposition and accumulated in another location. Thus, dispersal and accumulation of bones may have what might be referred to as *mirror image* effects on the composition of skeletal parts in a carcass from which selected bones have been dispersed and the composition of the dispersed and accumulated skeletal parts. Binford (1981b:222) said it well when he wrote "a transported [actively accumulated] assemblage should look like the 'opposite' of the nontransported [carcasses from which bones have been dispersed] assemblage; that is, it [the transported assemblage] should be the proportional inverse of the parts that were not transported." While this is a logical supposition regarding the effects of accumulation and dispersal of skeletal parts on the relative frequencies of those parts, Binford (1981b:222–229) had some difficulty finding empirical evidence that this predicted relation held in the real world.

Bunn *et al.* (1991) describe and compare the %MAU frequencies of skeletal parts they recorded on the landscape surrounding some settlements occupied by African foragers, and the frequencies of skeletal parts in one of those settlements. They compiled the data within two general size classes of animals; small mammals weigh between 2.2 and 115 kg, large mammals weigh between 115 and 905 kg (Table 6.11). If Binford's (1981b) suspicion is correct, then the frequencies of skeletal parts recorded on the African landscape should be inversely correlated with the frequencies of skeletal parts in the settlement. That is so because at least some of the bones on the landscape display butchering marks and thus prompted Bunn *et al.* (1991) to conclude that these probably derived from animal kill sites created by hominid activities. Scatter-plots of the frequencies of bones on the landscape against the frequencies of bones in the settlement for both size groups of mammals, and their respective statistics, do not meet Binford's (1981b) expectation (Figure 6.27). As Binford (1981b) before them, Bunn *et al.* (1991) suggest the expected relation is not found due, at least in part, to variation in the survival of different skeletal parts. For example, they interpret the bone frequencies as indicating that bones seem to preserve better, and have a greater probability of being incorporated into the (future) fossil record if they are deposited in the settlement than if they are left on the landscape regardless of the size of the represented animal. Further, the remains of larger animals tend to be more abundant on the landscape than the

Table 6.11 *Frequencies (%MAU) of skeletal parts of two sizes of mammals from the landscape and from a hominid settlement. Size classes 1 and 2, animal live-weight size = 2.25 to 115 kg (5 to 250 lbs); size classes 3 and 4 animal live-weight size = 115 to 905 kg (250 to 2000 lbs). Size classes and data from Bunn* et al. *(1991)*

Skeletal part	Size classes 1 and 2		Size classes 3 and 4	
	landscape	settlement	landscape	settlement
cranium	100.0	100.0	100.0	28.6
mandible	29.2	62.5	47.4	57.1
cervical	19.1	50.0	63.9	53.1
thoracic	1.2	21.5	20.2	33.4
lumbar	8.3	70.8	24.6	57.1
sacrum	8.3	25.0	26.3	57.1
rib	0.3	9.6	12.0	18.0
pelvis	25.0	37.5	44.7	14.3
scapula	29.2	100.0	36.8	42.9
humerus	16.7	87.5	26.3	100.0
radius	20.8	25.0	21.1	85.7
femur	16.7	50.0	47.4	85.7
tibia	20.8	62.5	42.1	57.1
metapodial	13.0	25.0	22.4	35.7

remains of the smaller animals regardless of the skeletal part represented. Finally, Bunn *et al.* (1991:52) suggest that the frequencies of limb elements of large animals are higher than the frequencies of limb elements of small animals in the settlement due to differential transport.

While it is doubtful that excavations will often be extensive enough to allow direct comparison of skeletal part frequencies in sites representing locations from which bones were dispersed with skeletal part frequencies in sites representing locations in which bones were accumulated, Binford's (1981b) expectation regarding the inverse relation of these two kinds of bone assemblages is a reasonable one. But as his research and that of Bunn *et al.* (1991) indicate, even in ethnoarchaeological settings the expectation may not be met due to differential preservation and/or differences in the transport of skeletal parts. We return to these topics in Chapter 7.

Summary

> The problem is to determine the identity of the primary accumulator when evidence has been distorted by secondary or even tertiary animal activity on bones ... Interpretation of multi-agency accumulations is difficult, if not impossible, after inadequate excavation or collection.
> (G. Avery 1984:347)

I have in this chapter discussed various of the characteristics of the dispersal and accumulation of animal remains. I have also reviewed many of the

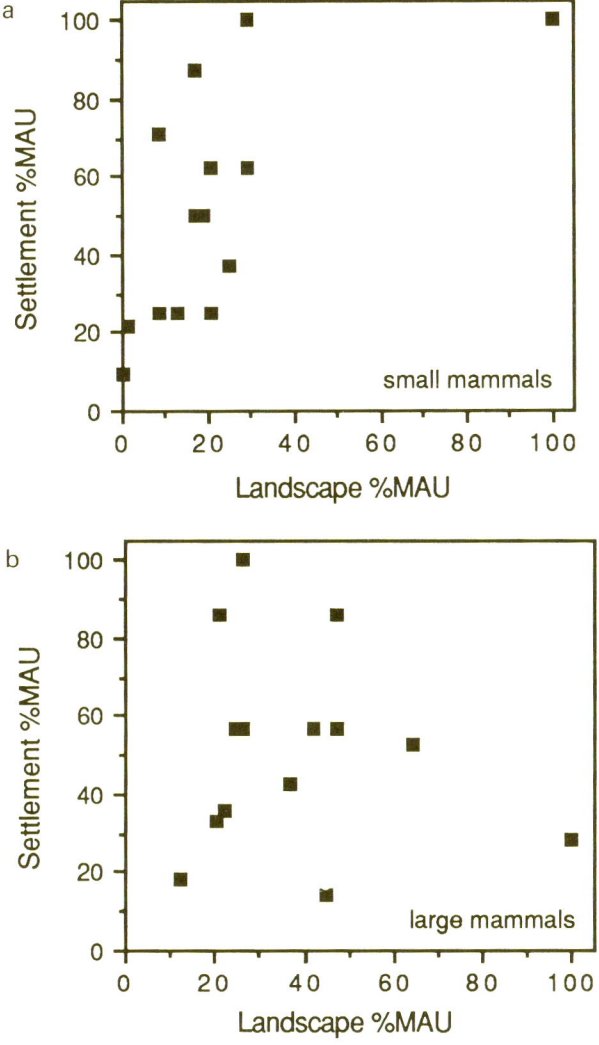

Figure 6.27. Bivariate scatterplots of relative frequencies of bones from (a) small mammals and (b) large mammals on the African landscape against bone frequencies in a hominid settlement. If taphonomic processes other than transport by humans have not altered bone frequencies, the point scatter should define a diagonal line with a negative slope.

attributes of fossil assemblages proposed as diagnostic of particular dispersal and accumulation processes and agents. What warrants emphasizing is that archaeofaunal assemblages may be the result of multiple dispersal and accumulation agents; when that is the case, the analyst is in for some difficult work. In a way, as noted in the previous section, the problem is reduced to a question of whether an assemblage of bones (or which portions of it) was culturally or naturally deposited. The attributes described in this chapter are typically used to help answer this question.

Accumulation and dispersal of vertebrate skeletons are not easily distinguished conceptually. Dispersal tends to be conceived of as taking place relative to the original location of carcass deposition. Accumulation can be of carcasses, in which case the original locus of carcass deposition is examined such as in the case of kill sites. Accumulation is typically conceived of relative to where carcasses and bones are finally deposited. Accumulation can be of individual skeletal specimens, in which case the bones have been transported from the original locations of carcass deposition to another location such as a carnivore's den site or a human habitation site, or it can be of carcasses such as those of animals which are trapped in a bog. Accumulation is characterized as passive when the behaviors of the animals whose bones are being studied dictate how the bones are accumulated. Accumulation is active when it is the action of a taphonomic process external to the animals whose bones are being studied that results in their accumulation. Much of this distinction, as well as additional attributes of accumulation, is summarized in Figure 6.2 and its associated discussion.

Identifying the mechanisms and agents of bone dispersal and accumulation involves the study of many attributes of an assemblage of faunal remains. Knowledge of which faunal remains particular agents tend to disperse and accumulate, and knowledge of how those dispersal and accumulation agents modify bones, are the keys to writing dispersal and accumulation histories of assemblages of faunal remains. Such knowledge helps answer the question "what are all these bones doing here?" Attributes of fossil assemblages other than those discussed in this chapter also help answer this question, but as we see in the next chapter, sometimes the question being asked is more specific than simply identifying an accumulation agent.

7

FREQUENCIES OF SKELETAL PARTS

> After death, vertebrate carcasses are often subjected to the same mechanical laws of transport by geologic agents as any other component of the sediment. Their specific gravity, which changes according to changing buoyancy, and the relation of mass to surface area specify the mechanical arrangement.
> (J. Weigelt 1927/1989:160–161).

Introduction

One of the most obvious and visible properties of a faunal assemblage is the frequencies of each of the particular skeletal elements that make up the collection. A complete mammal skeleton, for example, always consists of two humeri, two scapulae, two mandibles, one skull, etc. From this model of relative frequencies of skeletal parts in an individual, one can predict what should be found in a fossil assemblage that contains, for instance, 10 skulls; here, 20 humeri, 20 scapulae, 20 mandibles, etc., should be found if taphonomic processes have not resulted in the removal of certain kinds of bones, and sampling and recovery processes have not failed to find certain skeletal parts. Analysis of skeletal part frequencies, or what are sometimes called skeletal part profiles, has, in the past 15 years, become a major part of taphonomic research. In fact, the references cited in this chapter show that there has been a major burst of publication on this topic in the late 1980s and early 1990s.

Human utilization and transport of carcass parts

> It is necessary not only to give a reasonable explanation for the presence of the elements found but also for the absence of those not found. From here on out the student is on his own.
> (T. E. White 1953c:61)

Zooarchaeologists have long been interested in explaining the variability with which skeletons of particular taxa are represented. For example, nearly 40 years ago Theodore White, a paleontologist who also did zooarchaeological research in North America, proposed interpretive assumptions for explaining the variation in frequencies of skeletal parts in zooarchaeological collections. These assumptions were founded on White's knowledge of mammalian anatomy and ethnographic data concerning human butchery and subsistence practices. White (1953c:59, 1956:401) phrased the question in the following

223

way: "Which elements were brought into the camp or village and which elements were left at the kill?" The interpretive assumptions were phrased as follows: (1) "While it cannot be expected that all [skeletal] elements will [be equally abundant] it is difficult to escape the inference that some parts were not brought into camp" (White 1952a:337), and (2) "Since the lower limb does not carry any usable meat it is conceivable that it was chopped off and left at the place of the kill in order to reduce the load" (White 1953b:162). Simply put, then, White believed that skeletal elements with large amounts of usable meat were more likely to be transported from the kill to the consumption site than elements with small amounts of usable meat. Basically, elements with large amounts of usable meat were believed by White to consist of the proximal limb bones whereas elements with small amounts of usable meat included the distal limb bones.

Importantly, White also allowed for the fact that not all bone assemblages recovered from suspected habitation, non-kill sites consisted of mostly proximal limb elements. If proximal limb bones were absent from non-kill sites, White believed human butchery practices had destroyed the bones (e.g., reduced them via fragmentation to unidentifiable fragments) and/or that scavenging carnivores such as dogs had consumed them. For any skeletal element that was represented in low abundance, White (1956:402) suggested the analyst consider whether the missing elements were absent due to an "accident of preservation, or an accident of sampling." He even suggested some skeletal elements represented in low abundances may have been made into tools and then been carried away, or if found, were modified or worn beyond recognition.

In the Old World, perhaps the best-known study of skeletal part abundances is that of Perkins and Daly (1968), who proposed the concept of the *schlepp effect* to account for variations in skeletal part abundances (see also Klein 1976). Dart's (e.g., 1957) study of South African Plio-Pleistocene bovid remains prompted him to propose that differential use by early hominids of skeletal elements as tools accounted for variation in frequencies of bones and bone parts, but his proposal was immediately controversial (see Chapter 3). Perkins and Daly's (1968) schlepp effect, on the other hand, did not foster the same level of controversy until a decade after its proposal (e.g., Binford 1978, 1981b). The schlepp effect, from the German verb meaning "to drag," as originally described, "combines a factor related to the size of the game animal with one related to the distance between kill site and home settlement" (Perkins and Daly 1968:104). The schlepp effect can be defined as "the larger the animal and the farther away from the point of consumption it is killed, the fewer of its bones will get 'schlepped' back to the camp, village, or other area" (Daly 1969:149).

Perkins and Daly (1968:104) proposed the schlepp effect as a way to account for low relative abundances of cattle long bones associated with high abundances of cattle foot bones in a prehistoric site occupied by hunters. They suggested that "perhaps because the feet made convenient handles for dragging the meat-filled hide [and because the foot bones] contain useful sinews and have been called 'the hunter's sewing kit'," the archaeologically observed relative abundances of foot bones and long bones could be accounted for by reference to the schlepp effect. According to Perkins and Daly a prehistoric village occupied by pastoralists should produce approximately equal frequencies of cattle long bones and foot bones. Perkins and Daly (1968:104) refer to "New World archaeologists" who, they state, have shown that skeletal parts will be represented in unequal abundances at camp sites to which selected parts of animals butchered elsewhere were brought. Because the only such archaeologist they cite is White (1953c), it is tempting to suggest that their development of the concept of the "schlepp effect" had its roots in his work.

Binford (1978:12) wished to argue that variations in the frequencies of skeletal parts "resulted from variable strategies in the human use of food sources." That is, he wanted to derive the same kinds of conclusions that White, and Perkins and Daly had derived. Binford (1978:11) therefore began with the assumption that "any variability in the relative frequencies of anatomical parts among archaeological sites must derive from the dynamics of their use." He pointed out that people butcher (skin, eviscerate, dismember, disarticulate, deflesh, etc.; see Chapter 8) and scatter segments of individual carcasses during their exploitation. Binford (1978:10) found that "what is clearly lacking from our current understanding [of how people butcher carcasses] is a *specific knowledge* of the particular effects that might be expected to result from activities" such as butchering and schlepping. To provide that specific knowledge he measured the amounts of meat (weight of fat and muscle tissue), marrow (marrow cavity volume multiplied by the percentage of fatty acids present in the marrow), and grease (volume of skeletal part for cancellous parts multiplied by the percentage of fatty acids present in the marrow) associated with skeletal parts of two domestic sheep (*Ovis aries*) and one caribou (*Rangifer tarandus*). The measured values were used to construct indices of the food utility of the individual carcass parts for human consumers. Then, he constructed "a compound index of general utility that monitors quite closely the actual proportions of different usable food components of the animal" based on the amounts of meat, marrow, and grease associated with each skeletal part because it was clear that the three food substances would simultaneously influence choices made by people about how carcasses were to be butchered, stored, and/or transported (Binford 1978:72). Finally, Binford (1978:74) altered the values of the compound, general utility index (GUI) to create a "modified general utility index" (MGUI) that reflected the fact that animals are not always butchered into the discrete skeletal parts for which the GUI values had been measured; rather, some parts with low GUI values may remain attached to (ride with during transport) skeletal parts that have higher GUI

Table 7.1 *Binford's (1978) normed utility indices for domestic sheep and caribou (meat, marrow, grease, GUI, MGUI)*

Skeletal element	Sheep					Caribou				
	Meat	Marrow	Grease	GUI	MGUI	Meat	Marrow	Grease	GUI	MGUI
antler/horn		1.0			1.03		1.0			1.02
skull	12.86	1.0	—	25.74	12.87	9.05	1.0	—	17.49	8.74
mandible										
w/tongue	43.36	10.35	11.75	43.50	43.6	31.1	5.74	12.51	30.26	30.26
w/out tongue	14.12	10.35	11.75	11.65	11.65	11.4	5.74	12.51	13.89	13.89
atlas	18.65	1.0	7.19	18.68	18.68	10.1	1.0	13.11	9.79	9.79
axis	18.65	1.0	9.47	18.68	18.68	10.1	1.0	12.93	9.79	9.79
cervical	55.32	1.0	15.00	55.33	55.33	37.0	1.0	17.46	35.71	35.71
thoracic	46.47	1.0	9.82	46.49	46.49	47.2	1.0	12.26	45.53	45.53
lumbar	38.88	1.0	14.74	38.90	38.90	33.2	1.0	14.82	32.05	32.05
rib	100.0	1.0	9.30	100.0	100.0	51.6	1.0	7.50	49.77	49.77
sternum	90.52	1.0	11.05	90.52	90.52	66.5	1.0	26.00	64.13	64.13
scapula	44.89	6.23	3.85	45.06	45.06	44.7	6.40	7.69	43.47	43.47
P humerus	28.24	28.26	56.67	29.50	37.28	28.9	29.69	75.46	30.23	43.47
D humerus	28.24	41.21	9.38	29.31	32.79	28.9	28.33	27.84	29.58	36.52
P radius	14.01	35.40	32.54	15.18	24.30	14.7	43.64	37.56	16.77	26.64
D radius	14.01	68.98	18.77	15.81	20.06	14.7	66.11	32.70	17.82	22.23
carpals	4.74	1.0	22.98	5.00	13.43	5.2	1.0	36.47	5.51	15.53
P metacarpal	4.74	62.93	13.24	6.37	10.11	5.2	61.68	16.71	8.24	12.18
D metacarpal	4.74	71.85	33.59	6.79	8.45	5.2	67.08	42.47	8.83	10.50
pelvis	81.30	9.57	34.65	81.51	81.50	49.3	7.85	29.26	47.89	47.89
P femur	78.24	38.62	23.68	79.38	80.58	100.0	33.51	26.90	98.32	100.0
D femur	78.24	56.05	100.0	80.58	80.58	100.0	49.41	100.00	100.0	100.0
P tibia	20.76	57.84	56.40	22.71	51.99	25.5	43.78	69.37	27.57	64.73
D tibia	20.76	100.0	26.84	23.41	37.70	25.5	92.90	26.05	29.46	47.09
astragalus	6.37	1.0	24.38	6.64	23.08	11.2	1.0	32.47	11.23	31.66
calcaneum	6.37	23.11	34.38	7.27	23.08	11.2	21.19	46.96	12.40	31.66
P metatarsal	6.37	64.16	12.37	8.02	15.77	11.2	81.74	17.88	15.03	29.93
D metatarsal	6.37	73.52	33.33	8.46	12.11	11.2	100.0	43.13	16.24	23.93
first phalanx	3.37	33.77	15.70	4.33	8.22	1.7	30.00	33.27	3.52	13.72
second phalanx	3.37	25.11	13.33	4.10	8.22	1.7	22.15	24.77	3.03	13.72
third phalanx	3.37	1.0	9.82	3.49	8.22	1.7	1.0	13.59	1.85	13.72

values. Binford (1978:74) reasoned that a skeletal part with a low GUI and attached to a skeletal part with a high GUI should take on a utility value equivalent to the average of the two values (see Binford 1978:15–34, 72–75; Metcalfe and Jones 1988; Jones and Metcalfe 1988; for detailed discussion of how the utility indices were derived). Because Binford (1978) was interested in differential transport of skeletal parts, his MGUI was constructed to account for "riders," or skeletal parts with low GUI values that were transported due to their attachment to parts with high GUI values. All utility values Binford (1978) derived are given in Table 7.1. Note that all of these values have been normed to a scale of 1 to 100 by dividing all derived values by the greatest derived value in a column. Thus, the normed MGUI is typically referred to as the %MGUI.

Binford (1978:81) modeled how variation in the human transport and utilization of carcass parts would be reflected by concomitant variation in skeletal part frequencies and in the %MGUI value for individual skeletal parts. That modeling took the form of a family of curves, each representing a bivariate scatterplot of points derived by plotting the utility of a particular skeletal part on the x-axis against a measurement of the frequency of that skeletal part in an assemblage on the y-axis (Figure 7.1). The unit for measuring the frequency of skeletal parts Binford (1978) chose to use is the *minimal animal unit* (MAU; originally called MNI values, Binford [1984b] changed the term to MAU). MAU values are calculated by first ascertaining how many of a particular skeletal part or element are present; three overlapping fragments of distal humeri indicate a minimum number of three distal humeri, typically called an MNE (minimum number of elements) value (see Chapter 4). After the MNE per skeletal part is determined, those values are then divided by the number of times that part occurs in one skeleton to produce an MAU value. Thus an MNE of three distal humeri would be divided by two to derive an MAU of 1.5 distal humeri. If 26 thoracic vertebrae are present, and the taxon under study has 13 thoracic vertebrae per individual, an MAU of two thoracic vertebrae would be noted. In Binford's procedure, MAU values are normed on a scale of 1 to 100 (by dividing all MAU values for an assemblage by the maximum MAU value in the assemblage) prior to producing the scatterplot from which inferences of differential transport and utility are derived. The normed values are typically referred to as %MAU values.

As an example of how this analytic technique is implemented, the ethnoarchaeologically documented site of Anavik (Binford 1978:78) can be used. This site is a caribou (*Rangifer tarandus*) kill-butchery site occupied in the spring by the Nunamiut of Alaska. Both MNE and MAU frequencies of caribou bones recovered from this site are given in Table 7.2. Note that Binford (1978) did not report MNE values for this site, but for sake of illustration I derived them from the MAU values he presented in order to illustrate the relationship between MNE and MAU values. A bivariate scatterplot (Figure 7.2) of the %MAU

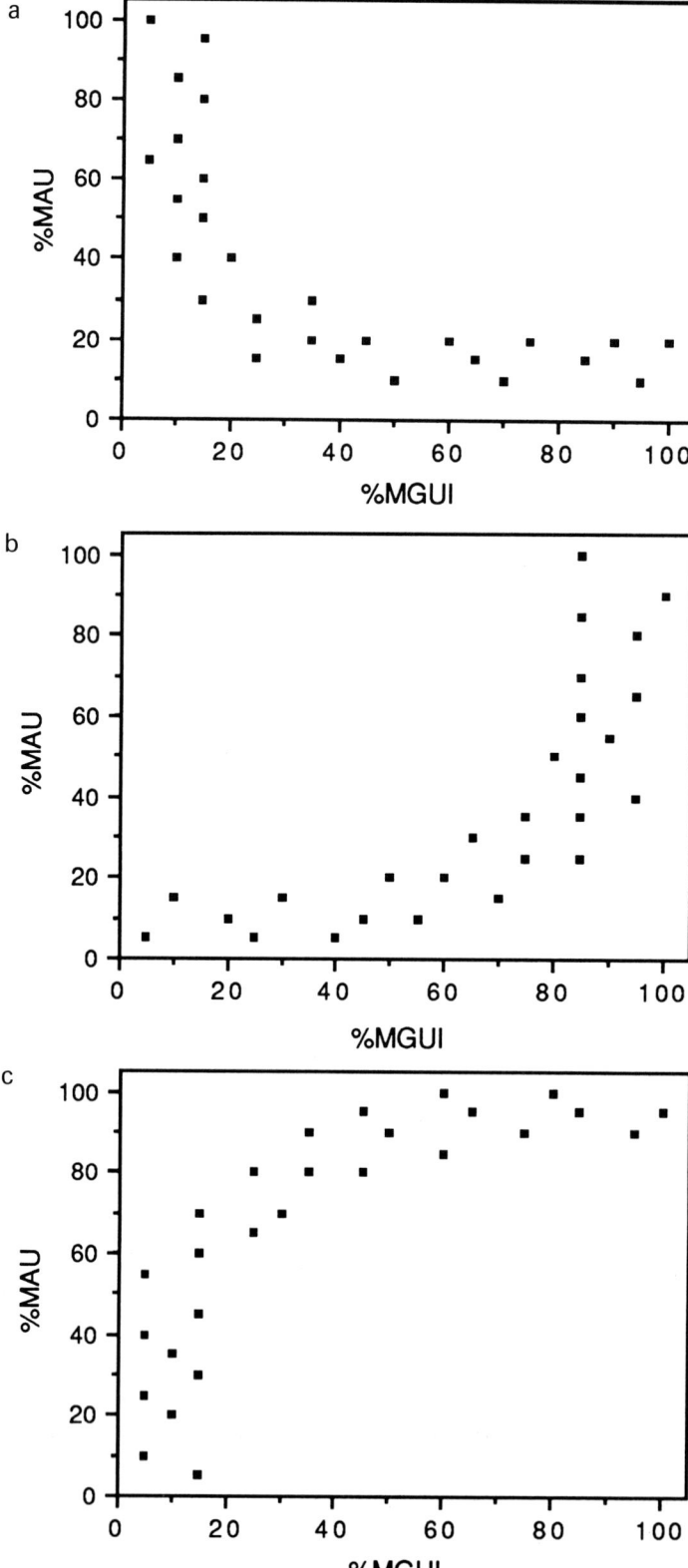

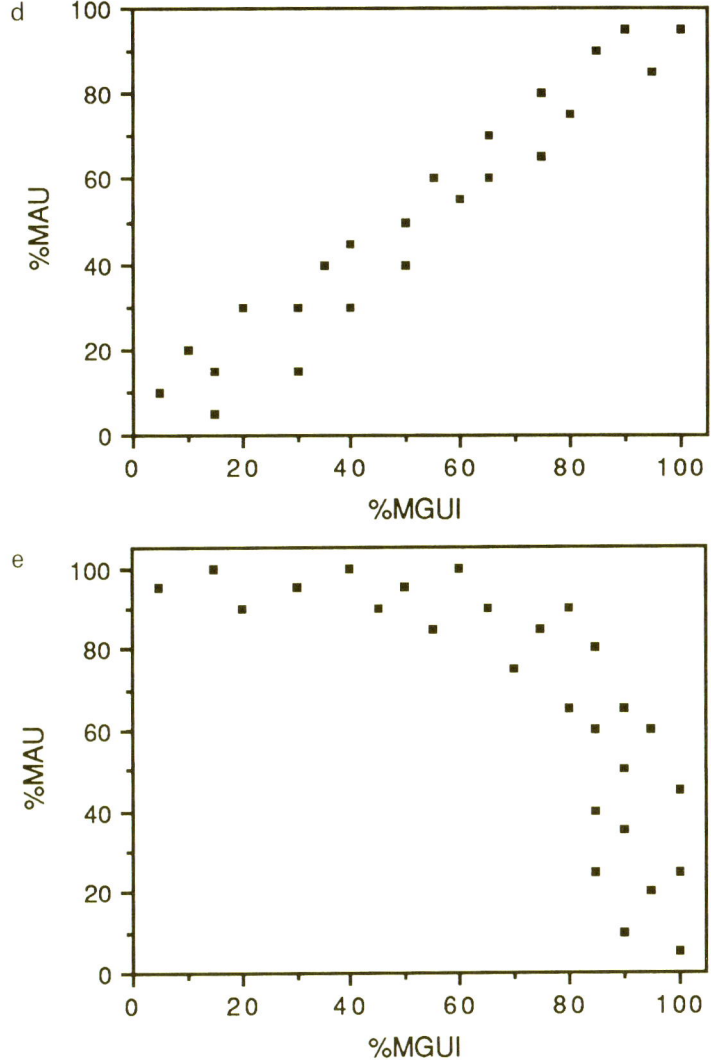

Figure 7.1. A family of strategies for utilizing and/or transporting animal carcass parts based on the %MGUI (after Binford 1978; see also Thomas and Mayer 1983). a, reverse (bulk) strategy; b, gourmet strategy; c, bulk strategy; d, unbiased strategy; e, reverse gourmet strategy.

values for caribou remains from Anavik against the caribou %MGUI (Table 7.1) shows what is readily inferred to be a reverse utility curve based on the model in Figure 7.1. That the statistical relation between the two sets of variables is significant is indicated by a correlation coefficient (r_s) of -0.85 ($P < 0.001$). Thus, the analytic technique is simply performed, and produces what appear to be readily interpretable results. We return to whether the latter is in fact the case later in this chapter.

Perhaps not surprisingly, given the logic behind Binford's (1978) model of utility and transport curves (Figure 7.1), the explicit knowledge of the food

Table 7.2 *MNE and MAU frequencies of caribou bones for two ethnoarchaeological sites created by the Nunamiut and reported by Binford (1978:78, 323)*

Site:	Anavik			Ingested		
Skeletal part	MNE	MAU	%MAU	MNE	MAU	%MAU
antler (2)[a]	106	53	100.0	1	0.5	7.69
skull (1)	44	44	83	1	1	15.38
mandible (2)	78	39	73.5	3	1.5	23.08
atlas (1)	40	40	75.4	1	1	15.38
axis (1)	44	44	83	0	0	0.0
cervical (5)	210	42	79.2	4	0.8	12.31
thoracic (13)	260	20	37.7	13	1	15.38
lumbar (6)	144	24	45.2	1	0.16	2.46
sacrum (1)	19	19	35.8	no data		
innominate (2)	45	22.5	42.4	1	0.5	7.69
rib (26)	364	14	26.4	26	1	15.38
sternum (7)	105	15	28.3	5	0.71	10.92
scapula (2)	36	18	33.9	13	6.50	100.0
P humerus (2)	30	15	28.3	0	0	0.0
D humerus (2)	33	16.5	31.1	7	3.5	53.84
P radius (2)	39	19.5	36.7	4	2	30.77
D radius (2)	49	24.5	46.2	3	1.5	23.08
carpals (12)	360	30	56.7	8	0.75	11.54
P metacarpal (2)	63	31.5	59.4	3	1.5	23.08
D metacarpal (2)	63	31.5	59.4	1	0.5	7.69
P femur (2)	18	9	16.9	5	2.5	38.46
D femur (2)	18	9	16.9	0	0	0.0
P tibia (2)	26	13	24.5	0	0	0
D tibia (2)	27	13.5	25.4	5	2.5	38.46
astragalus (2)	44	22	39.6	4	2	30.77
calcaneum (2)	46	23	38.6	3	1.5	23.08
P metatarsal (2)	73	36.5	68.9	4	2	30.77
D metatarsal (2)	46	23	43.3	2	1	15.38
first phalanges (8)	292	36.5	68.9	6	0.75	11.54
second phalanges(8)	288	36.0	67.9	11	1.37	21.08
third phalanges (8)	284	35.5	66.9	20	2.5	38.46

Note:

[a] number of times a skeletal part occurs in one individual

utility of carcass parts (Table 7.1), and the straightforward analytic technique of generating a scatterplot that could be interpreted in terms of the model, the technique was quickly put to use by archaeologists (e.g., Landals 1990; Speth 1983; Thomas and Mayer 1983). Here was an algorithm that apparently granted significant insights to the human behavioral part of the taphonomic history of a bone assemblage. The manner in which the utility indices were derived came under close scrutiny (e.g., Chase 1985; Jones and Metcalfe 1988;

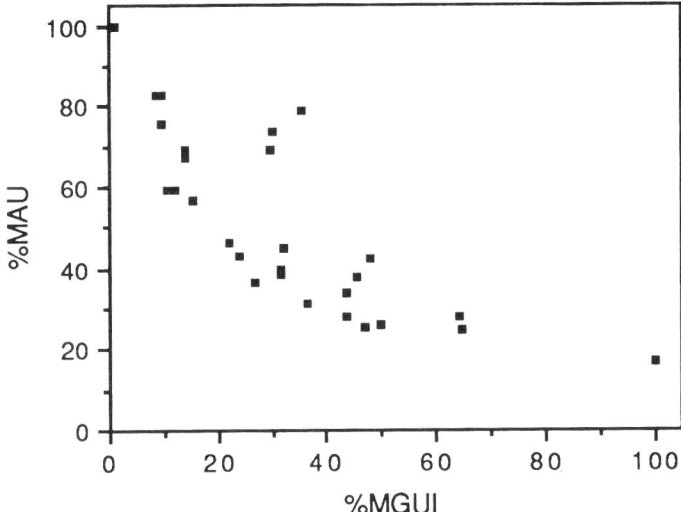

Figure 7.2. Scatterplot of caribou %MAU values from Anavik against caribou %MGUI values.

Metcalfe and Jones 1988), but no one could deny the utility of the model or invalidate the specific knowledge Binford had presented. Thus it should come as no surprise that researchers began to generate utility indices for other taxa (Blumenschine and Caro 1986; Borrero 1990; Brink and Dawe 1989; Emerson 1990; Kooyman 1984, 1990; Lyman *et al*. 1992b; Will 1985), many of which are summarized in Tables 7.3 and 7.4. Virtually all of these indices were derived in a manner similar to that used by Binford (1978). That is, the muscle and fat tissue associated with a skeletal part was weighed and, less often, the amounts of marrow and grease were estimated on the basis of the volume of each skeletal part. Those weight values were then mathematically converted to the index values. The indices have variously been called utility indices, economic utility indices, or food utility indices (FUI). Most of them have been derived by archaeologists who butchered carcasses and measured (typically weighing) the amount of food tissue (meat, marrow, fat, grease) associated with a skeletal part such as a proximal humerus, a complete skeletal element such as the scapula, and sections of the skeleton such as the ribs or thoracic vertebrae. Following Binford (1978) most analysts norm their indices to a scale of 1 to 100, and it is these normed values that are given in Tables 7.3 and 7.4.

Because most of these indices are founded on measurements from one or a few (always ≤4) individuals, I believe they are at best ordinal scale. The indices are based on average measurements of meat and fat amounts associated with, for example, the scapulae (left and right) of only one or a few individuals. They thus (a) mute individual variation such as that displayed by individuals of different age, sex, and nutritional status (typically correlated with season), not

Table 7.3 *Utility and transport indices for various taxa*

Skeletal element	Caribou FUI (Metcalfe and Jones 1988)	Phocid seal FUI (Lyman et al. 1992b)	Impala transport (O'Connell et al. 1990)	Alcelaphine transport (O'Connell et al. 1990)	Llama FUI (Mengoni-Gonalons 1991)
skull	9.1	27.4	13	3	14.75
mandible		with skull	13	3	
with tongue	31.1				9.95
without tongue	11.5				5.25
atlas and axis	10.2	with cervical	with cervical	with cervical	8.57
cervical	37.1	35.8	16	4	64.15
thoracic	47.3	24.9	16	4	61.75
lumbar	33.2	32.9	16	4	77.97
rib	51.6	100.0	9.5	3.5	100.00
sternum	66.6	2.7	with rib	with rib	99.35
scapula	44.7	19.8	12.5	3	41.66
humerus	36.8	10.7	11	2	36.68
radius-ulna	25.8	4.8	10.5	1	23.00
metacarpal	5.2		10	1	6.53
front flipper with carpals and phalanges		2.3			
innominate	49.3	44.5	15	4	40.18
femur	100.0	4.5	10	2	75.94
tibia/fibula	62.8 (w/tarsals)	16.5	10 (w/tarsals)	2 (w/tarsals)	43.04
metatarsal	37.0		9.5	2	11.46
rear flipper with tarsals and phalanges		7.7			
phalanges	19.4		15	2	4.78

Table 7.4 *Utility indices for bone parts of various mammalian taxa*

Skeletal part	Guanaco general utility (Borrero 1990)	Modified guanaco utility (Lyman 1992a)	Bison fat (Brink and Dawe 1989)	Bison modified total products (Emerson 1990)	Muskox modified general utility (Will 1985)
skull	10.0	10.0	no data	14.2	11.23
mandible					
with tongue	5.7	5.7		included with skull	36.55
without tongue	no data	no data			24.16
atlas	8.8	8.8	no data	6.4	20.47
axis	8.8	8.8	no data	7.8	20.47
cervical	51.3	51.3	no data	56.6	47.50
thoracic	22.1	22.1	no data	84.7	66.00
lumbar	44.9	44.9	no data	82.9	61.57
rib	100.0	100.0	no data	100.0	55.57
sternum	8.5	8.5	no data	52.9	83.54
scapula	38.4	38.4	no data	31.6	28.89
P humerus	23.8	38.4	40.5	31.6	55.58
D humerus	23.8	23.8	22.0	25.1	48.69
P radius	7.8	15.8	33.5	16.5	41.81
D radius	7.8	7.8	25.7	12.1	48.83
P ulna	15.8	19.8	no data	20.8	
D ulna	7.8	7.8	no data	12.1	
carpals	1.3	4.5	no data	6.6	42.54
P metacarpal	1.3	2.6	8.9	3.9	36.25
D metacarpal	1.3	2.4	15.2	2.6	49.82
innominate	40.2	40.2	no data	54.7	83.59
P femur	83.2	83.2	31.4	69.4	57.56
D femur	83.2	83.2	35.2	69.4	100.00
P tibia	21.3	52.2	33.5	40.8	80.62
D tibia	21.3	21.3	14.1	25.5	61.25
astragalus	1.7	11.5	no data	13.6	55.23
calcaneum	1.7	11.5	no data	13.6	55.23
naviculo-cuboid	1.7	11.5	no data	13.6	55.23
P metatarsal	1.7	6.6	12.4	7.5	49.22
D metatarsal	1.7	4.4	22.7	4.5	64.27
first phalanx	2.1	2.1	no data	2.4	31.61
second phalanx	2.1	2.1	no data	2.4	22.64
third phalanx	2.1	2.1	no data	2.4	7.45

to mention inter-population variation, and (b) are not average values for the complete range of variation that different individuals of a taxon may display because few individuals have been measured. Even if we eventually obtain the weight of muscle tissue for both sexes of each age class of individuals of a taxon and those individuals died during various seasons of the year, the fact that we cannot yet determine the sex, ontogenic age, and season of death of an animal represented by most isolated bones or teeth we find, forces us to treat the economic utility indices as ordinal scale. Thus while the scatterplots of Figures 7.1 and 7.2 are interval scale and are useful heuristics, ordinal scale statistics are called for when analyzing the data in the scatterplots.

Probably also as a direct result of Binford's (1978) efforts, other researchers began to monitor how modern hunter-gatherers differentially transport carcass parts of some taxa (Bunn *et al.* 1988; O'Connell *et al.* 1988; 1990; O'Connell and Marshall 1989) regardless of their food utility. Rather than, for example, defleshing a carcass and weighing the meat, marrow, and grease associated with particular bones to derive a utility index, these analysts recorded which bones were transported from each individual carcass of animals that were killed or scavenged by modern hunter-gatherers. They then ordered the skeletal parts from those most often transported to those least often transported. I (Lyman 1992a) term these "transport indices" and summarize them in Table 7.4, where the higher the number given for a skeletal part denotes a greater chance that a part will be transported; that is, these transport indices have not been normed. As with the utility indices, I believe the transport indices are best treated as ordinal scale.

The purpose of presenting all of the available indices here should be evident: they can be used to help explain varying frequencies of skeletal parts in an assemblage of bones. As Binford (1987:453) notes, these indices "can be considered quite literally as a frame of reference, that function much like a screen upon which slides are projected." For example, presuming the bones were culturally accumulated (Chapter 6), these indices can serve as one frame of reference or one part of an explanation for observed abundances of skeletal parts. Several such frames of reference (Voorhies Groups, Table 6.5; fluvial transport indices, Table 6.7) described in Chapter 6 pertain to fluvial transport processes. Below I consider other frames of reference and show how still other taphonomic variables must be considered to clarify whether or not differential utilization and transport of carcass parts by humans was responsible for any given set of bones, but the utility and transport indices are a good place to start the search for explanations of skeletal part frequencies.

Structural density of bones

> The investigator can begin with a concrete basis for comparison, since he knows the exact ratio of the different elements in the living animal . . . He has a precise ratio of

expectation with which to compare the observed, and, if the sample size is large, the differential representation of the elements can be studied in detail.
(R. D. Guthrie 1967:243)

Given an interest in explaining skeletal part frequencies by reference to transport or utility indices, the zooarchaeologist must be sure that those frequencies are indeed a reflection of economic decisions and not some other factor, such as differential preservation. As will become clear, the probability that a skeletal part will survive the rigors of various taphonomic processes is at least partially a function of that part's structural density (g/cm^3). It is to a consideration of that property that we now turn.

While the structural density of skeletal parts was referred to in the nineteenth century as potentially mediating various taphonomic processes, such as carnivore gnawing (see Chapter 2), that property did not receive serious attention again until Guthrie (1967) mentioned it as a possibly important factor affecting the frequencies of skeletal parts of Pleistocene mammals in Alaska. Guthrie (1967) did not, however, measure the structural density of bones. Better known to zooarchaeologists is the study by Brain (1967b, 1969), who measured the structural density of goat (*Capra hircus*) bones in an attempt to account for varying frequencies of skeletal parts in an assemblage known to have been ravaged by dogs (*Canis familiaris*) and people. At the same time, Voorhies (1969) noted that the structural density of bones seemed to exert strong influences on the transportability of skeletal elements by fluvial processes, but he did not measure the structural density of bones. Subsequent research by both those interested in zooarchaeological (Binford and Bertram 1977) and paleontological (Behrensmeyer 1975b; Boaz and Behrensmeyer 1976) problems expanded greatly on the results of the early studies, and several provided measures of the structural density of bones.

In measuring the structural density of bones, taphonomists have been intent on providing the explicit knowledge necessary to support statements such as those made in the nineteenth and twentieth centuries that "hard" skeletal parts tend to survive carnivore gnawing whereas "soft" parts do not (see Chapter 2). Thus Brain (1969) showed how his measurements of the structural density of goat bones correlated with the frequencies of skeletal parts remaining after the assemblage of bones had been ravaged by dogs and people. While Brain (1969) only measured the structural density of eight skeletal parts (Table 7.5), that was sufficient to show the strong mediating influence that structural density has on the survival of bones (Figure 7.3). The correlation between Brain's (1969) measurements of structural density and the frequencies of skeletal parts that survived the ravaging is strong and significant ($r_s = 0.80$, $P = 0.03$).

Some recent research has been founded in part on the argument that Brain's (1969) research was flawed because it "lacked systematic control [of] the exact frequencies of bone prior to ravaging by domestic dogs" (Marean *et al.* 1992:102). While Brain (1969) did not in fact report (and apparently did not

Table 7.5 *Frequencies and structural density of goat bones reported by Brain (1969), and Behrensmeyer's (1975a) measures of sheep bone density*

Skeletal part	Goat bone density	Frequency	Sheep bone density
P humerus	0.58	33	1.26
D humerus	0.97	336	1.75
P radius/ulna	1.10	279	1.64
D radius/ulna	0.97	114	1.59
P femur	0.75	28	1.47
D femur	0.72	56	1.42
P tibia	0.82	64	1.32
D tibia	1.17	119	1.64

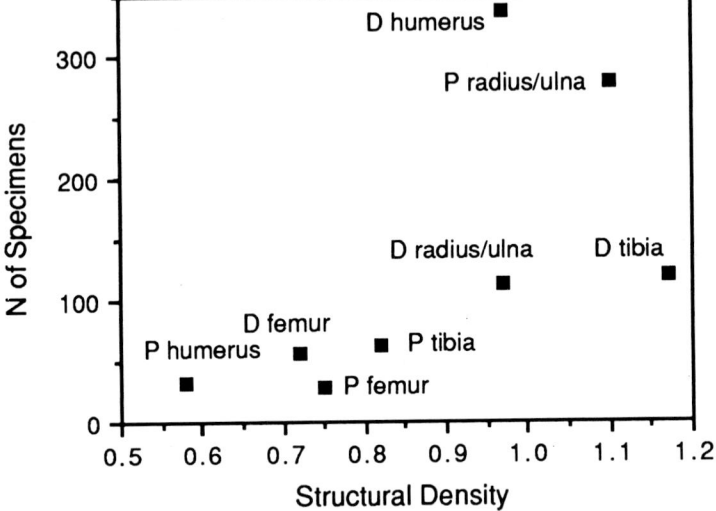

Figure 7.3. Scatterplot of Brain's (1969) goat bone structural density values against number of recovered goat bone specimens from a Hottentot village.

know) the pre-ravaging frequencies of bones, two important points must be made. First, the new research confirms Brain's conclusions that bones with low structural densities will more often be destroyed by carnivore ravaging than bones with high structural densities. Second, Brain's research began with the implicit assumption that the bones he studied represented once complete skeletons, the assumption all taphonomists *must* make in studies of skeletal part frequencies. That assumption is necessary because taphonomists studying prehistoric collections *never* know what was originally present "prior to ravaging." The *only* place to start in studying a skeletal part profile from a prehistoric context is with a model of a set of complete skeletons, the original number of complete skeletons typically being determined by the MNI for an assemblage. As we will see, the assumption may well be wrong, but it is a matter of starting with just this model, or none at all. Ultimately, then, Brain's (1969) pioneering research established an important analytic technique and a significant taphonomic relation: the structural density of skeletal parts can exert strong influences on bone frequencies.

The term structural density may seem cumbersome, but it is mandatory to discussions of taphonomic issues because the term *density* can denote a mass:volume ratio such as g/cm³, or, a frequency:unit area or volume ratio such as the number of bones per m² or m³. *Structural density* is meant to denote the ratio of the mass of a substance to its volume. If the material is homogeneous, structural density is a constant of the substance; if the material is heterogeneous, structural density is an average characteristic of the sample measured. Thus Shipman (1981b:23–25) suggests that many measures of bone structural density are in fact more properly labeled measurements of bone "composition" because bone as a material is more or less homogeneous, but its composition in the sense of the ratio of spongy to compact bone varies within and between skeletal elements. In other words, the porosity of bone varies, and hence Lyman (1984a:264), following materials science, distinguished *true density* from *bulk density* as follows:

$$D_t = M/V_s \qquad\qquad\qquad [7.1]$$
$$D_b = M/V_t \qquad\qquad\qquad [7.2]$$

Here, D_t = true density, D_b = bulk density, M = mass, V_s = volume of the substance exclusive of pore space volume in the measured specimen, and V_t = volume of the substance including volume of the pore space in the specimen measured. A measure of total porosity (f_p) can be derived from equations [7.1] and [7.2] with equation [7.3]:

$$f_p = (D_t - D_b)/D_t = 1 - (D_b/D_t) \qquad\qquad\qquad [7.3]$$

A synthesis and evaluation of much of the research on the structural density of bones was presented by Lyman (1984a), who showed that the measurements of density that had been taken to that time tended to be dissimilar due to differences in how density had been measured. Measures of structural density of bone were variously measures of bulk density, true density, or some hybrid of these because porosity and other important variables that influenced measurement results had been differentially controlled (Lyman 1984a). Thus it cannot be ascertained if, for example, Brain's measurements of domestic goat bone structural density differ from Behrensmeyer's (1975b) measurements of domestic sheep (*Ovis aries*) bone structural density (Table 7.5) because of differences between the taxa or because of differences between the measurement techniques used. This is a critically important distinction. The fact that Behrensmeyer's (1975b) density values correlate with both Brain's (1969) density measures ($r_s = 0.82$, $P < 0.03$) and with the frequency of goat bones in Table 7.5 ($r_s = 0.83$, $P < 0.03$) could result from either the structural density of bones, regardless of taxon, being a robust taphonomic factor, or dissimilar measurement techniques producing two sets of values (of different physical and structural properties) either or both of which are spuriously correlated with the frequencies of surviving bones.

Lyman (1984a) and other researchers (Chambers 1992; Elkin and Zanchetta 1991; Kreutzer 1992; Lyman *et al.* 1992a) have used a technique called photon absorptiometry or photon densitometry to derive what are called "bone mineral densities" for very specific parts of skeletal elements. The technique involves the measurement of how weak a photon beam of known strength becomes when it passes through the selected part of a bone. A *scan site* "is that area or part of the bone which was actually measured [or scanned] by the photon beam" (Lyman 1984a:272). The higher the mineral content of the measured part of the bone, the weaker the beam is that comes through the bone (the fewer the photons that pass through it). A photon detector counts the photons that pass through the bone, and this is converted to a measurement of bone mineral density by a computer module attached to the detector (see Kreutzer 1992 and Lyman 1984a for descriptions of the basic machine). Lyman (1984a:272–273) chose scan sites (1) that would allow assessment of known structural variation within each bone, (2) that were easy to locate and define on the basis of anatomical features in order to allow measurement of multiple specimens, and (3) that transected skeletal parts commonly found in archaeological sites. Subsequent researchers have tended to select the same scan sites (e.g., Kreutzer 1992). We may ultimately wish to measure other areas or parts of a skeletal element (see below), but for now the measured scan sites are sufficient for many analyses.

The measurements of bone mineral density are approximations of bulk density, and are given as g/cm^3 for the scan site. Unlike previous measures of structural density, the structural property being measured is identical between analysts because the measurement technique is identical. This enhances the validity of explanations for different patterns of bone survivorship across different taxa. If results like those in the example above using Brain's density measurements for goats and Behrensmeyer's measurements for sheep (Table 7.5) are found using bone mineral densities, the reason for the difference between the correlation coefficients calculated between bone survivorship and bone structural density and the less than perfect correlation between the two sets of density measures must reside in taxonomic variation in bone structural density (e.g., Kreutzer 1992).

The photon densitometry measurement technique is rather inexpensive and non-destructive; one has but to have bones to measure and a densitometer with which to measure them. Many such machines are housed in Schools of Medicine where they are used to study ontogenic changes in the mineral density of human bones; some may be found in hospitals where they are used as diagnostic tools. Thus far, bone mineral densities have been measured for deer

(*Odocoileus* spp.), pronghorn antelope (*Antilocapra americana*), domestic sheep (*Ovis aries*), North American bison (*Bison bison*), North American marmots (*Marmota* spp.), South American vicuna and guanaco (*Lama* spp.), and seals (*Phoca* spp.). Scan sites for all taxa that have been measured are shown in Figures 7.4, 7.5, and 7.6, and density values are given in Tables 7.6 and 7.7. Because other research indicates that the bulk density of bones varies between taxa due in part to variation in locomotor modes (e.g., Currey 1984; Stein 1989; Wall 1983), taphonomists should enlarge this list of measured taxa.

The structural density values in Tables 7.6 and 7.7 can serve much like the utility indices described in the preceding section: as a frame of reference for assessing the frequencies of skeletal parts in an assemblage of bones. That is so because experimental data (e.g., Behrensmeyer 1975b; Haynes 1980b, 1983a; Marean and Spencer 1991; Voorhies 1969) and ethnoarchaeological data (e.g., Binford and Bertram 1977; Brain 1969; Walters 1984, 1985) indicate that several taphonomic processes are mediated by the structural density of bones. For instance, those bones with the lowest bulk density also have the greatest porosity which in turn means they have the greatest surface area to volume ratios (the larger and/or more frequent the pores per unit volume, the greater the surface area per unit volume). Mechanical and chemical attrition should have greater effects on bones with low bulk densities (high porosity) simply because there is greater surface area to work on. And, in so far as greater porosity reduces weight per unit volume, whether or not a bone is transported by such processes as fluvial action will be influenced by the bulk density of that bone.

The analytic procedure for using the bone structural density values is a simple one and is similar to that used with the utility indices. The major difference is that, as originally constructed, one tallies the frequency of each skeletal part for which a structural density value is available. Thus for Brain's (1969) density values, one tallies the MNE of proximal humeri, distal humeri, etc. for the eight skeletal parts listed in Table 7.5. For the scan sites for deer listed in Table 7.6, one tallies how many of each scan site is represented in the collection. Those frequencies are then converted to %survivorship values based on the minimum number of individual animals (MNI) represented by the bone collection. If, for example, there is an MNI of 8, then if all bone parts survived there should be 16 of each paired element such as distal humeri, 8 skulls, 8 atlas vertebrae, 8 axis vertebrae, 40 cervical vertebrae (5 per individual), 208 ribs (26 per individual), 64 first phalanges, etc. The %survivorship values are derived by dividing the observed MNE frequency by the frequency expected given 100% survivorship (see the discussion of equation [7.4] below). Thus 4 observed distal humeri would represented 25% survivorship with an MNI of 8 (100% survivorship = 16). Note that such a norming procedure is not necessary for Brain's (1969) goat bone data because all skeletal parts are paired (Table 7.5); that is, all values plotted in Figure 7.3 would have the same relative

a

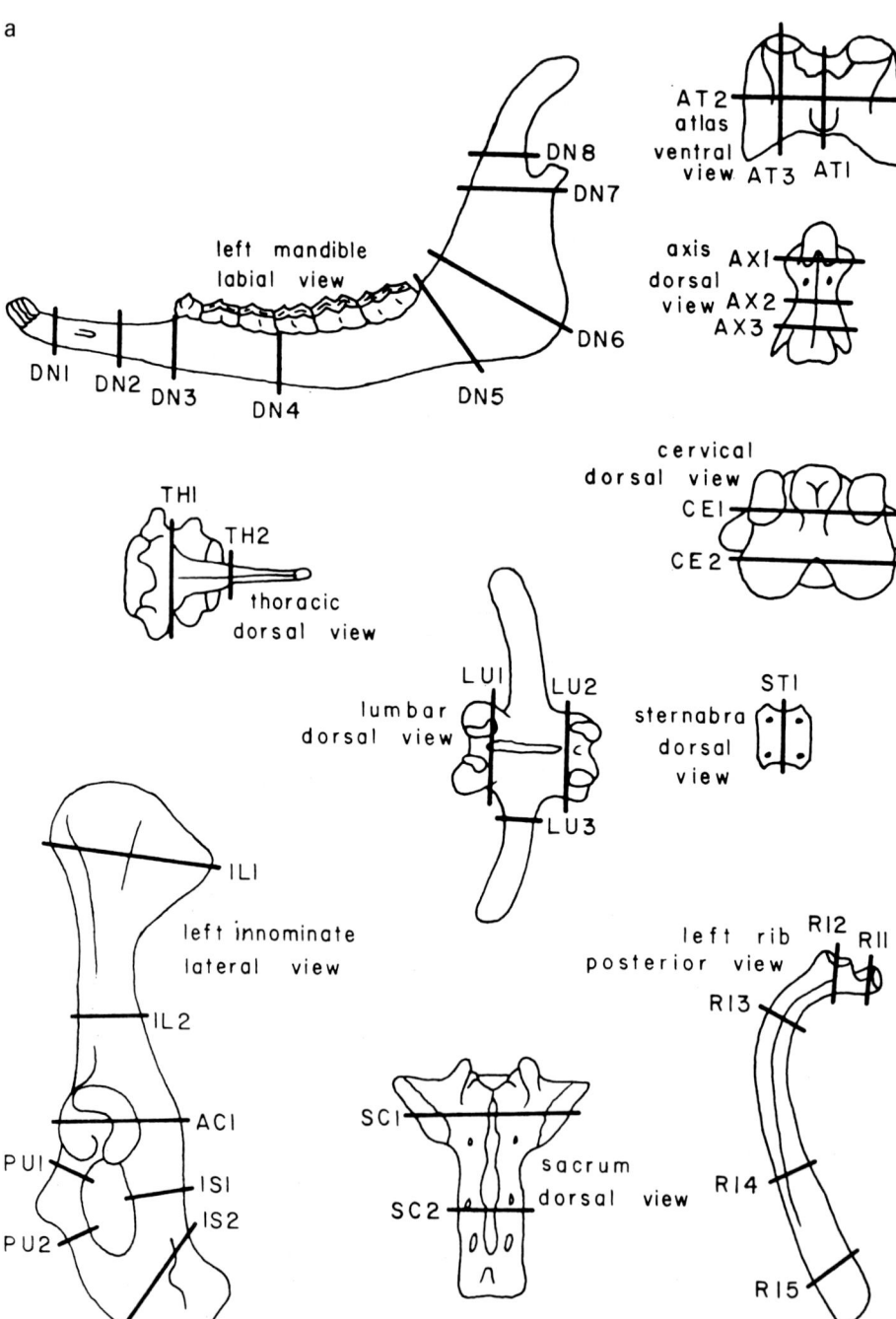

b

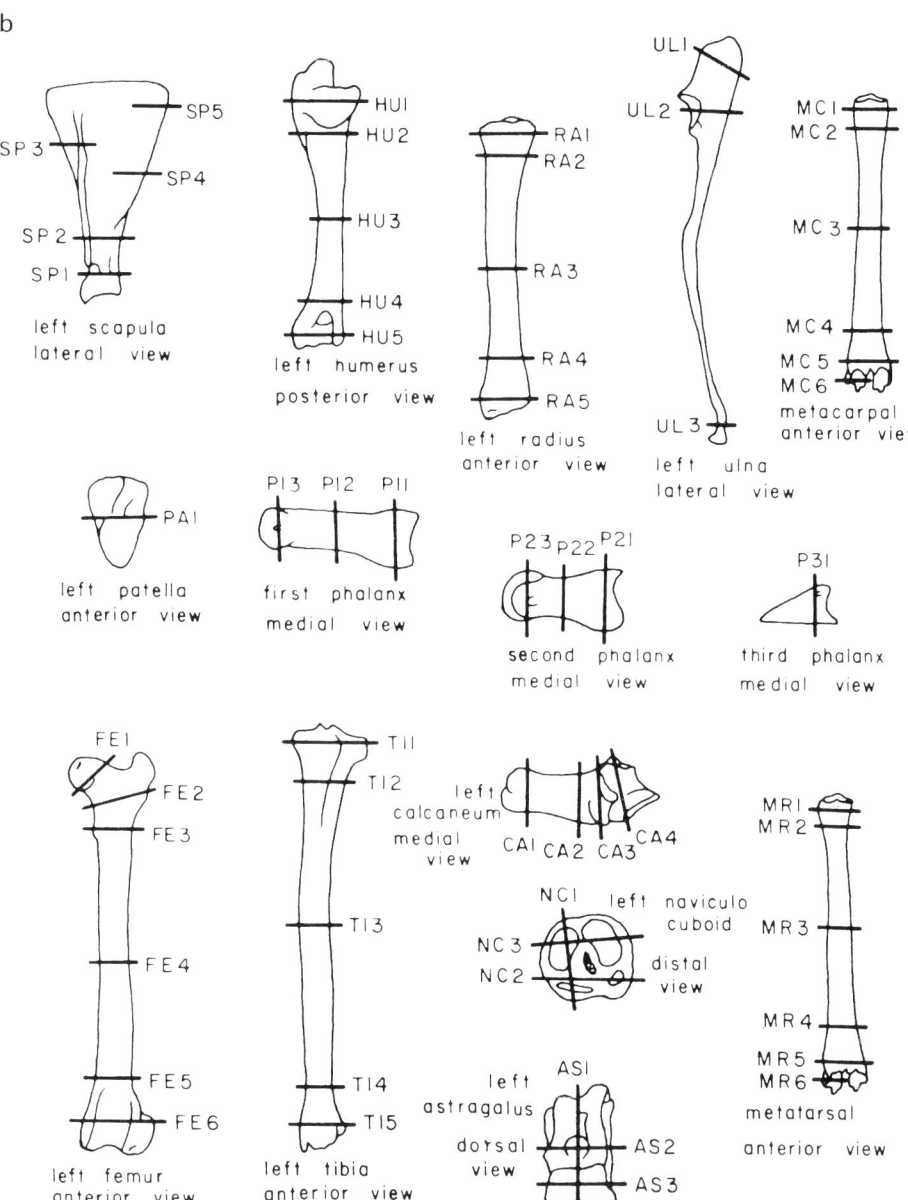

Figure 7.4. Anatomical locations of scan sites where photon absorptiometry measurements have been taken on ungulate bones (from Lyman 1984a:274–275, Figure 2; courtesy of Academic Press, Inc.).

a

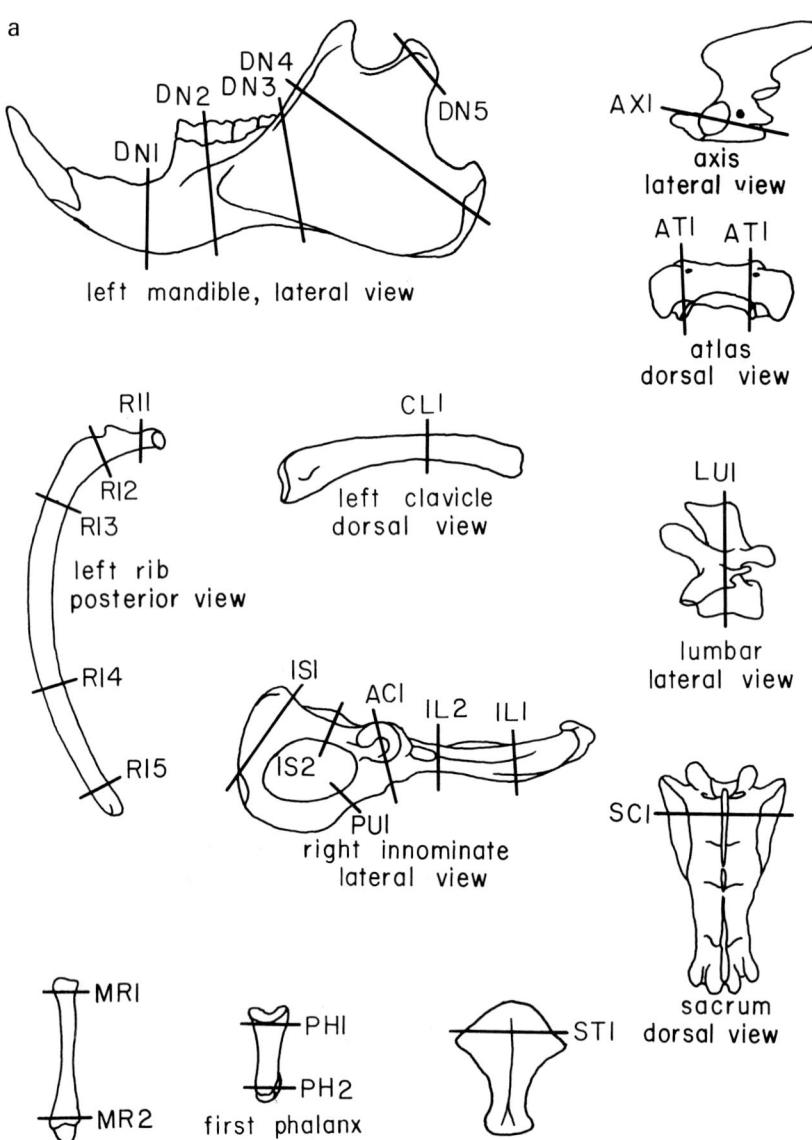

left mandible, lateral view

axis
lateral view

atlas
dorsal view

left clavicle
dorsal view

left rib
posterior view

lumbar
lateral view

right innominate
lateral view

sacrum
dorsal view

metatarsal
dorsal view

first phalanx
dorsal view

sternabra
ventral view

b

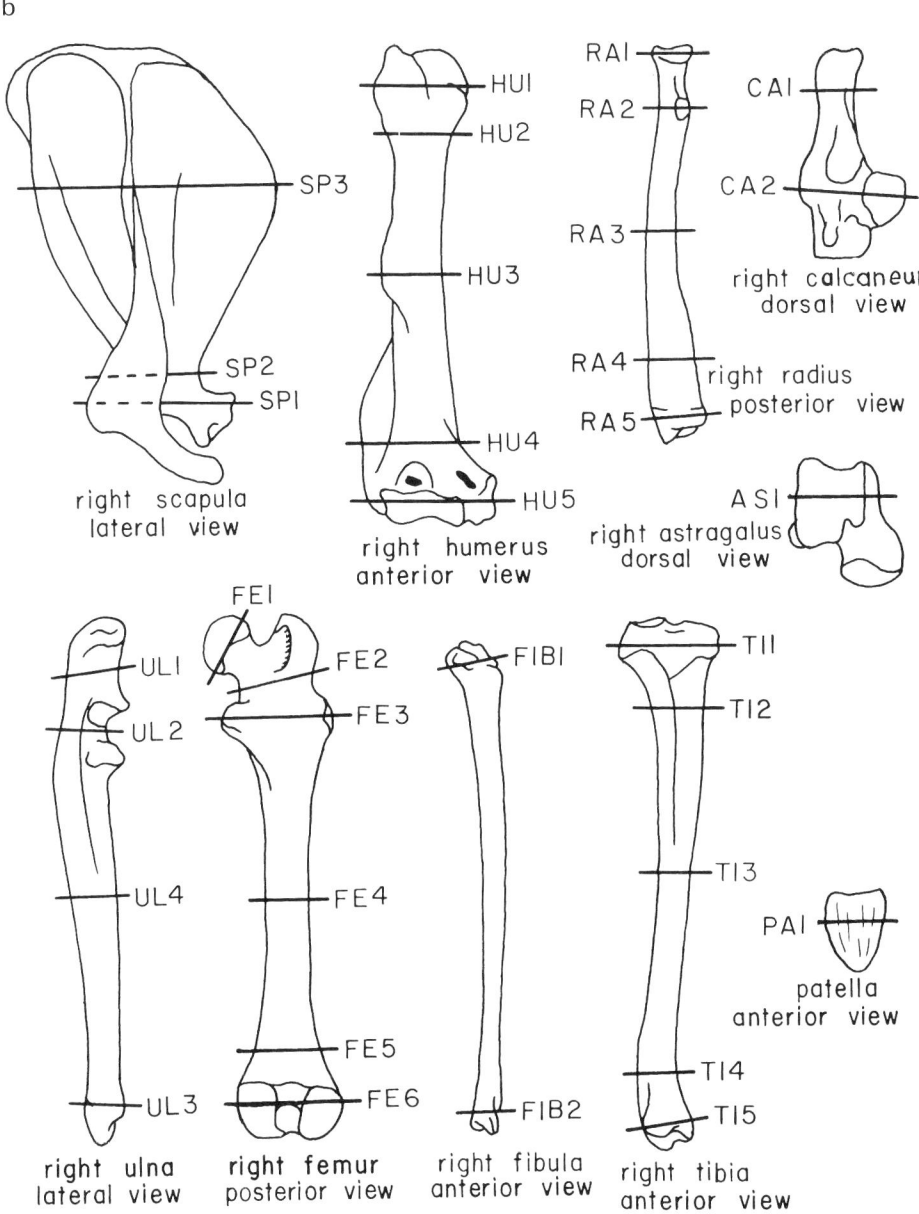

Figure 7.5. Anatomical locations of scan sites where photon absorptiometry measurements have been taken on marmot bones (from Lyman *et al.* 1992a:Figure 4; courtesy of Academic Press, Ltd.).

a

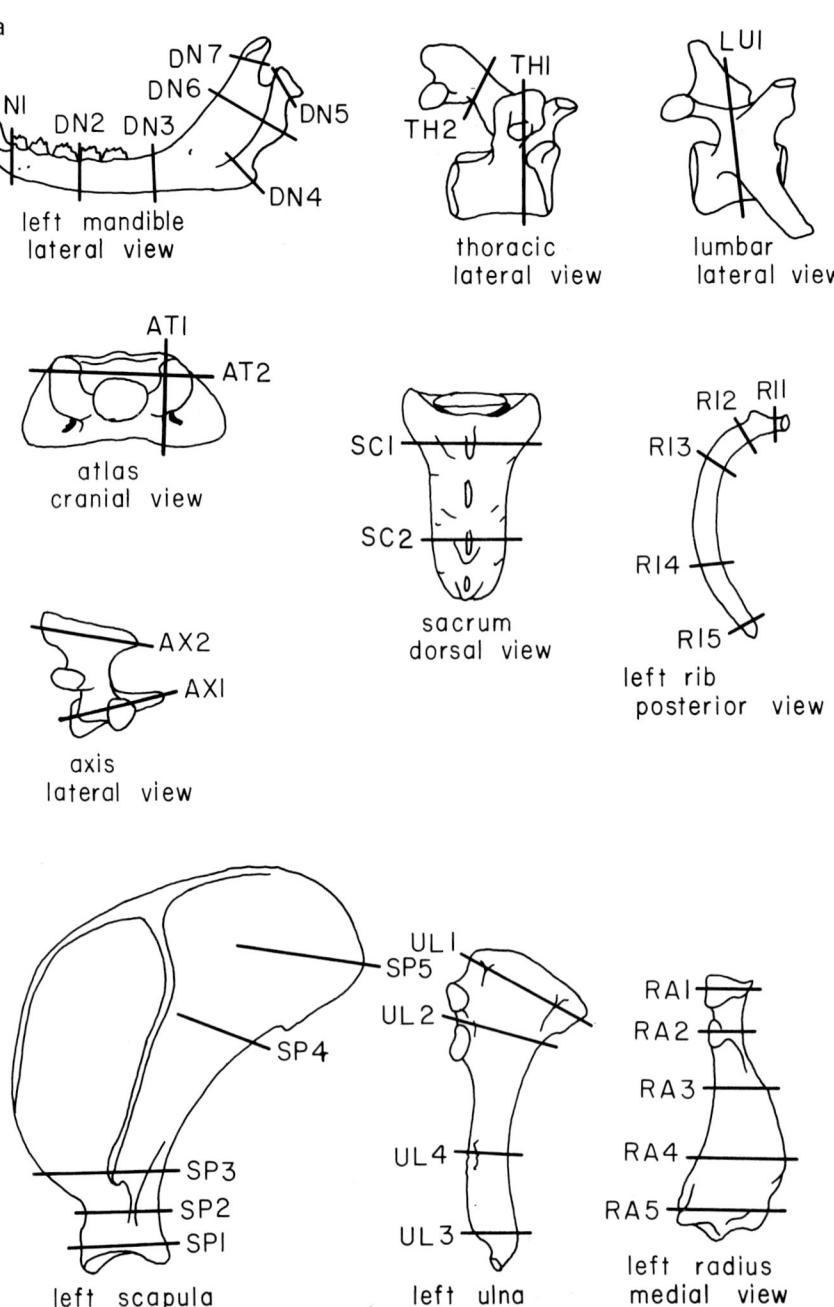

DN7
DN6
DN1
DN2 DN3
DN5
DN4
left mandible
lateral view

TH1
TH2
thoracic
lateral view

LU1
lumbar
lateral view

AT1
AT2
atlas
cranial view

SC1
SC2
sacrum
dorsal view

R12 R11
R13
R14
R15
left rib
posterior view

AX2
AX1
axis
lateral view

SP5
SP4
SP3
SP2
SP1
left scapula
lateral view

UL1
UL2
UL4
UL3
left ulna
lateral view

RA1
RA2
RA3
RA4
RA5
left radius
medial view

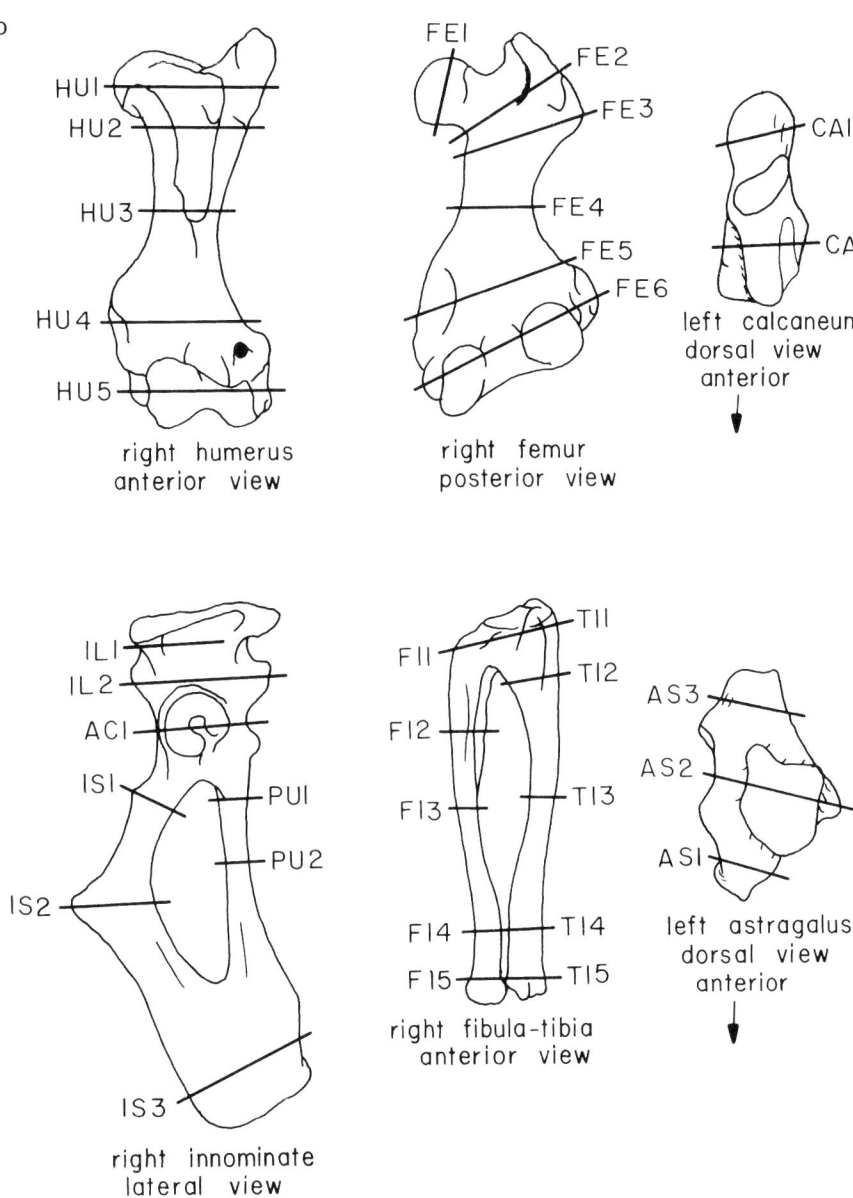

Figure 7.6. Anatomical locations of scan sites where photon absorptiometry measurements have been taken on seal bones (after Chambers 1992).

Table 7.6 *Average bone mineral densities for deer, pronghorn antelope, domestic sheep (Lyman 1982b, 1984a), bison (Kreutzer 1992), guanaco and vicuna (Elkin and Zanchetta 1991)*

scan site	bison	deer	pronghorn	sheep	guanaco	vicuna
2&3CP	0.50					
5MC	0.62					
AC1	0.53	0.27	0.14	0.26	0.22	0.18
AS1	0.72	0.47	0.39	0.54	0.65	0.55
AS2	0.62	0.59	0.48	0.63		
AS3	0.60	0.61	0.57	0.60		
AT1	0.52	0.13	0.12	0.07	0.17	0.18
AT2	0.91	0.15	0.13	0.11		
AT3	0.34	0.26	0.32			
AX1	0.65	0.16	0.13	0.13	0.17	0.16
AX2	0.38	0.10	0.11	0.14		
AX3	0.97	0.16	0.17			
CA1	0.46	0.41	0.29	0.43		
CA2	0.80	0.64	0.55	0.58	0.66	0.49
CA3	0.49	0.57	0.50	0.56		
CA4	0.66	0.33	0.20	0.43		
CE1	0.37	0.19	0.12	0.12	0.24	0.23
CE2	0.62	0.15	0.12	0.13		
CUNEIF	0.43	0.72	0.64			
DN1	0.53	0.55				
DN2	0.61	0.57				
DN3	0.62	0.55				
DN4	0.53	0.57			0.62	
DN5	0.53	0.57				
DN6	0.57	0.31				
DN7	0.49	0.36				
DN8	0.79	0.61				
FE1	0.31	0.41	0.16	0.28		
FE2	0.34	0.36	0.20	0.16	0.37	0.37
FE3	0.34	0.33	0.21	0.20		
FE4	0.45	0.57	0.33	0.36		
FE5	0.36	0.37	0.30	0.24		
FE6	0.26	0.28	0.27	0.22	0.29	0.23
FE7	0.22					
LATMAL	0.56	0.52	0.63			
HU1	0.24	0.24	0.06	0.13	0.28	0.23
HU2	0.25	0.25	0.12	0.22		
HU3	0.45	0.53	0.25	0.42		
HU4	0.48	0.63	0.44	0.37		
HU5	0.38	0.39	0.33	0.34	0.40	0.34
HYOID	0.36					
IL1	0.22	0.20	0.16	0.23		
IL2	0.52	0.49	0.33	0.47		
IS1	0.50	0.41	0.28	0.49		
IS2	0.19	0.16	0.32	0.11		
LU1	0.31	0.29	0.15	0.26	0.26	0.19
LU2	0.11	0.30	0.11	0.22		
LU3	0.39	0.29	0.10			
LUNAR	0.35	0.83	0.66			
MC1	0.59	0.56	0.33	0.40	0.60	0.54
MC2	0.63	0.69	0.41	0.55		

Table 7.6 (*cont.*)

scan site	bison	deer	pronghorn	sheep	guanaco	vicuna
MC3	0.69	0.72	0.57	0.67		
MC4	0.60	0.58	0.45	0.54		
MC5	0.46	0.49	0.40	0.38	0.45	0.39
MC6	0.53	0.51	0.44	0.50		
MR1	0.52	0.55	0.47	0.43	0.59	0.50
MR2	0.59	0.65	0.45	0.53		
MR3	0.67	0.74	0.57	0.68		
MR4	0.51	0.57	0.43	0.51		
MR5	0.40	0.46	0.39	0.31	0.43	0.38
MR6	0.48	0.50	0.44	0.39		
NC1	0.48	0.39	0.26		0.59	0.42
NC2	0.64	0.33	0.26			
NC3	0.77	0.62				
PA1		0.31	0.39	0.44		
P11	0.48	0.36	0.24	0.43		
P12	0.46	0.42	0.38	0.40	0.65	0.53
P13	0.48	0.57	0.45	0.55		
P21	0.41	0.28	0.23	0.34		
P22		0.25	0.24	0.39	0.55	0.40
P23	0.46	0.35	0.30	0.42		
P31	0.32	0.25	0.25	0.30	0.39	0.17
PU1	0.55	0.46	0.34	0.45		
PU2	0.39	0.24		0.25		
RA1	0.48	0.42	0.26	0.35	0.41	0.40
RA2	0.56	0.62	0.25	0.36		
RA3	0.62	0.68	0.57	0.52		
RA4	0.42	0.38	0.30	0.19		
RA5	0.35	0.43	0.34	0.21	0.37	0.38
RI1	0.27	0.26				
RI2	0.35	0.25				
RI3	0.57	0.40			0.37	0.31
RI4	0.55	0.24				
RI5	0.33	0.14				
SC1	0.27	0.19	0.11	0.20	0.20	0.20
SC2	0.26	0.16	0.25	0.16		
SCAPHOID	0.42	0.98	0.68			
SP1	0.50	0.36	0.27	0.25	0.38	0.30
SP2	0.48	0.49	0.10	0.33		
SP3	0.28	0.23	0.30	0.19		
SP4	0.43	0.34	0.15	0.32		
SP5	0.17	0.28	0.21			
ST1		0.22				
TH1	0.42	0.24		0.24	0.14	0.19
TH2	0.38	0.27		0.19		
TI1	0.41	0.30	0.18	0.16	0.33	0.26
TI2	0.58	0.32	0.26	0.20		
TI3	0.76	0.74	0.48	0.59		
TI4	0.44	0.51	0.40	0.36		
TI5	0.41	0.50	0.29	0.28	0.51	0.42
TRAPMAG	0.52	0.74	0.65			
UL1	0.34	0.30	0.28	0.18		
UL2	0.69	0.45	0.26	0.26		
UL3		0.44				
UNCIF	0.44	0.78	0.70			

Table 7.7 *Average bone mineral densities for marmots (Lyman* et al. *1992a) and phocid seals (Chambers 1992)*

MARMOT		SEAL					
scan site	density	scan site	density	scan site	density	scan site	density
AC1	0.44	AC1	0.47				
AS1	0.71	AS1	0.45	AS2	0.55	AS3	0.56
AT1	0.67	AT1	0.54	AT2	0.42		
AX1	0.45	AX1	0.56	AX2	0.49		
CA1	0.58	CA1	0.45				
CA2	0.84	CA2	0.45			CE1	0.35
CL1	1.09	CU1	0.56				
DN1	0.72	DN1	0.59				
DN2	0.58	DN2	0.84				
DN3	0.59	DN3	0.90				
DN4	0.59	DN4	0.84	DN6	0.89		
DN5	0.71	DN5	0.64	DN7	1.11		
FE1	0.56	FE1	0.50				
FE2	0.73	FE2	0.53				
FE3	0.66	FE3	0.52				
FE4	0.70	FE4	0.69				
FE5	0.39	FE5	0.45				
FE6	0.48	FE6	0.57				
FIB1	0.46	FI1	0.39	FI3	0.90	FI4	0.88
FIB2	0.55	FI2	0.78	FI5	0.76		
HU1	0.37	HU1	0.43				
HU2	0.44	HU2	0.39				
HU3	0.62	HU3	0.57				
HU4	0.77	HU4	0.67				
HU5	0.62	HU5	0.60				
IL1	0.46	IL1	0.60				
IL2	0.83	IL2	0.63				
IS1	0.58	IS1	0.67				
IS2	0.96	IS2	0.75	IS3	0.55		
LU1	0.34	LU1	0.38				
MR1	0.81						
MR2	0.85	NA1	0.57				
PAT1	0.83						
PI1	0.72						
PI2	0.79	PU1	0.70				
PU1	1.04	PU2	0.71				
RA1	0.79	RA1	0.63				
RA2	0.97	RA2	0.69				
RA3	0.95	RA3	0.71				
RA4	0.70	RA4	0.39				
RA5	0.51	RA5	0.45				
RI1	0.64	RI1	0.40				
RI2	0.73	RI2	0.50				
RI3	0.74	RI3	0.62				
RI4	0.79	RI4	0.63				
RI5	0.54	RI5	0.29				
SC1	0.33	SC1	0.43	SC2	0.34		
SP1	0.58	SP1	0.49				
SP2	0.51	SP2	0.48	SP4	0.63		
SP3	0.46	SP3	0.61	SP5	0.41		
ST1	0.26	TH1	0.34	TH2	0.37		
TI1	0.45	TI1	0.39				
TI2	0.53	TI2	0.47				
TI3	0.87	TI3	0.86				
TI4	0.74	TI4	0.56				
TI5	0.56	TI5	0.48				
UL1	0.66	UL1	0.44				
UL2	0.99	UL2	0.66				
UL3	0.40	UL3	0.35				
UL4	0.95	UL4	0.79				

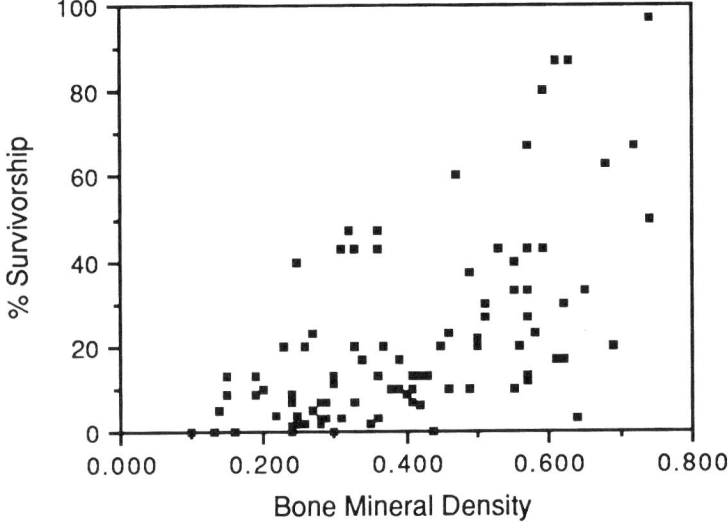

Figure 7.7. Scatterplot of %survivorship of deer skeletal parts from 45OK4 against bone mineral density values for deer.

values after derivation of %survivorship because all would have been divided by twice the MNI.

The %survivorship values can be plotted against the density values as in Figure 7.3, and a correlation coefficient between those frequencies and the bone density values can be calculated. As an example, the frequencies of each of 95 scan sites represented in an assemblage of deer (*Odocoileus* sp.) bones recovered from archaeological site 45OK4 in eastern Washington state are given in Table 7.8, along with their respective 100% survivorship values based on an MNI of 15, and their respective %survivorship values. The scatterplot of %survivorship frequencies for this assemblage against bone mineral density for deer scan sites (Table 7.6, exclusive of carpals and distal fibula) suggests that as bone mineral density increases, the frequency of skeletal parts increases (Figure 7.7). The correlation coefficient between the two variables is positive and significant ($r_s = 0.65$, $P < 0.001$), suggesting there is a strong relation between the two variables. A reasonable conclusion would be that some taphonomic process that is mediated by the structural density of the skeletal parts had a major influence on the frequencies of skeletal parts. Such processes include, but are not limited to, consumption by carnivores of low-density skeletal parts (typically called carnivore attrition or ravaging), human consumption of low density skeletal parts, differential fragmentation leading to less dense bone parts being crushed into unidentifiable powder in contrast to denser bone parts being simply broken into small but recognizable pieces during the extraction of grease and/or marrow, post-depositional crushing by sediment overburden, or some combination of these.

Table 7.8 *Frequencies of representation of scan sites of deer bones from archaeological site 45OK4 (from Lyman 1982b)*

Scan site	N obs.	N exp.	% survivorship	Scan site	N obs.	N exp.	% survivorship
DN1	10	30	33	DN2	10	30	33
DN3	12	30	40	DN4	13	30	43
DN5	10	30	33	DN6	13	30	43
DN7	14	30	47	DN8	5	30	17
AT1	0	15	0	AT2	2	15	13
AT3	3	15	20	AX1	0	15	0
AX2	0	15	0	AX3	0	15	0
CE1	7	75	9	CE2	7	75	9
TH1	2	195	1	TH2	9	195	5
LU1	7	105	7	LU2	12	105	11
LU3	3	105	3	SC1	2	15	13
SC2	0	15	0	SP1	13	30	43
SP2	11	30	37	SP3	6	30	20
SP4	5	30	17	SP5	2	30	7
HU1	2	30	7	HU2	12	30	40
HU3	13	30	43	HU4	26	30	87
HU5	5	30	17	UL1	4	30	13
UL2	6	30	20	UL3	0	30	0
RA1	4	30	13	RA2	9	30	30
RA3	19	30	63	RA4	3	30	10
RA5	4	30	13	MC1	6	30	20
MC2	6	30	20	MC3	20	30	67
MC4	7	30	23	MC5	3	30	10
MC6	16	60	27	RI1	6	390	2
RI2	17	390	4	RI3	36	390	9
RI4	35	390	9	RI5	21	390	5
ST1	4	105	4	IL1	3	30	10
IL2	13	30	43	AC1	7	30	23
IS1	3	30	10	IS2	0	30	0
PU1	7	30	23	PU2	0	30	0
FE1	4	30	13	FE2	1	30	3
FE3	13	30	43	FE4	20	30	67
FE5	6	30	20	FE6	1	30	3
PA1	1	30	3	TI1	0	30	0
TI2	14	30	47	TI3	15	30	50
TI4	9	30	30	TI5	6	30	20
AS1	18	30	60	AS2	24	30	80
AS3	26	30	87	NC1	3	30	10
NC2	2	30	7	NC3	5	30	17
CA1	2	30	7	CA2	1	30	3
CA3	4	30	13	CA4	6	30	20
MR1	3	30	10	MR2	10	30	33
MR3	29	30	97	MR4	8	30	27
MR5	3	30	10	MR6	13	60	22
P11	16	120	13	P12	7	120	6
P13	15	120	12	P21	3	120	2
P22	2	120	2	P23	3	120	2
P31	2	120	2				

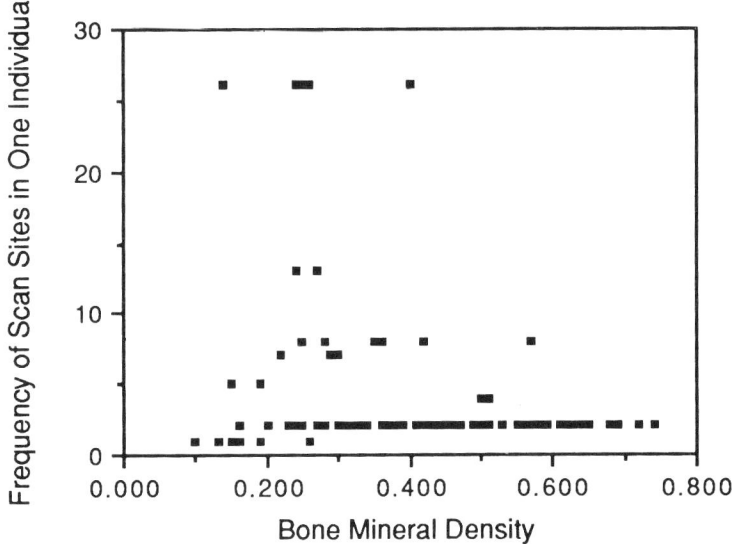

Figure 7.8. Scatterplot of frequency of individual scan sites in one skeleton against bone mineral density values for deer.

It is important to note that, while the MNE frequencies of scan sites in the 45OK4 assemblage correlate with the structural density of the scan sites ($r_s = 0.48$, $P < 0.001$), the conversion of those frequencies to %survivorship is a critical step in the analysis. It is critical because our interest lies not in how many bones there are, but in how many *survived* from the original carcasses, and a mammalian carcass has different MNE frequencies of different skeletal elements. An obvious example for artiodactyls such as deer is that they have 2 humeri (one left, one right, both in the front legs) but 8 first phalanges (2 per foot, with 2 hindfeet and 2 forefeet). Given an interest in the number of bone parts per carcass that survived, the observed frequencies must be converted to %survivorship values so that, for instance, a greater abundance of first phalanges than humeri reflects differential survivorship rather than the different numbers of skeletal parts per living individual. A scatterplot of bone mineral density values against the MNE frequency of scan sites in one complete skeleton (Figure 7.8) is suggestive of an inverse or negative relation between the two variables. There is no significant correlation between these two variables ($r_s = -0.06$, $P = 0.56$), but the scatterplot shows that the tendency would be for scatterplots with MNE values on the y-axis to appear as inverse relations if these values were not converted to %survivorship values. That the 45OK4 MNE values are imperfectly correlated with the %survivorship values ($r_s = 0.76$, $P < 0.001$) underscores the difference between MNE frequencies and %survivorship values. A significant correlation may be found between the MNE values for an assemblage and bone density, and not between the

%survivorship values and bone density, but the taphonomic significance of the former would be unclear.

I believe the structural density values listed in Tables 7.6 and 7.7 are at best ordinal scale. The reasoning for this belief is identical to that used with the transport and utility indices. The structural density values are averages of different individuals, essentially all skeletally mature (e.g., all epiphyses are fused) but of different ages, sexes, nutritional statuses, and genetic populations. It is well known that the structural density of bones varies with age, sex, nutritional status, and genetics, and different bones will vary differently. Thus the distal humerus may be 40% more dense than the proximal humerus in one individual but only 20% more dense in another individual. Again, because we cannot yet determine the sex, ontogenic age, and nutritional status of the individual represented by most of the bones we find, it is best to treat the structural density measures as ordinal scale and to use ordinal scale statistics when searching for evidence of density-mediated attrition.

Above it was suggested that, given the strong statistical relation between the %survivorship of scan sites in the 45OK4 assemblage and the structural density of the scan sites, it seemed reasonable to conclude that this assemblage had undergone one or more density-mediated attritional processes. By *density-mediated attrition* is meant the loss of skeletal parts due to their structural density; typically skeletal parts with low structural density are lost because they are more easily destroyed (but see the discussion of bone tools in Chapter 8). No specific attritional agent is identified, but rather only those taphonomic processes that are buffered or mediated by the structural density of skeletal parts are implicated. Thus, any one or more of a number of density-mediated taphonomic processes might be responsible for the frequencies of skeletal parts in the 45OK4 deer bone assemblage, although those that are not density mediated cannot be completely eliminated (see below). Several density-mediated processes were noted in the discussion of the 45OK4 assemblage. Others include fluvial transport, which tends to sort (remove) low-density bones (the winnowed or removed portion of the assemblage) from those of high density; the latter may remain (although their location may be changed) as a lag deposit (see Table 6.5 and Figure 6.5). Gnawing by rodents tends to focus on the densest skeletal parts (e.g., Brain 1980; Morlan 1980). Human selection of dense bones for tool manufacture (see Chapter 8), subaerial weathering, and diagenetic processes that more rapidly affect skeletal parts with low structural densities than skeletal parts with high structural densities (e.g., Lyman and Fox 1989) are other possibilities, although none of these has been studied actualistically. Obviously, determining whether a particular assemblage of bones underwent some density-mediated taphonomic process involves making scatterplots and calculating a correlation coefficient between values of skeletal part structural density and %survivorship. But it does not end there, as an example will make clear.

Lyman *et al.* (1992a) argue that neither the utility indices nor the structural

Table 7.9 *MAU values for the White Mountains marmots (from Grayson 1989) and the Salishan Mesa marmots (Lyman n.d.b), and corresponding scan sites for structural density values; skeletal part abbreviations are used in Figure 7.10*

Skeletal part	White Mountains	Salishan Mesa	Traditional scan site	Maximum scan site
Skull (SK)	55.50	13.00	none	none
Mandible (MAND)	84.00	23.50	DN3	DN1
Atlas (AT)	7.00	4.00	AT1	AT1
Axis (AX)	3.00	1.00	AX1	AX1
Innominate (PELV)	8.00	3.50	AC1	PU1
Scapula (SCAP)	24.00	5.50	SP1	SP1
P humerus (PH)	12.50	3.00	HU1	HU2
D humerus (DH)	33.50	7.50	HU5	HU4
P radius (PR)	52.00	9.00	RA1	RA2
D radius (DR)	28.50	1.00	RA5	RA4
P ulna (PU)	65.50	13.00	UL1	UL2
D ulna (DU)	12.00	2.50	UL3	UL3
Carpals (CARP)	7.57	n.d.	none	none
P metacarpal (PMC)	3.70	n.d.	none	none
D metacarpal (DMC)	0.50	n.d.	none	none
P femur (PF)	9.00	6.00	FE2	FE2
D femur (DF)	4.50	2.00	FE6	FE6
P tibia (PT)	3.50	1.00	TI1	TI2
D tibia (DT)	58.00	8.00	TI5	TI4
Astragalus (AST)	64.00	4.50	AS1	AS1
Calcaneum (CALC)	14.50	1.50	CA2	CA2
Other Tarsals (TARS)	3.86	n.d.	none	none
P metatarsal (PMT)	5.50	n.d.	MR1	MR1
D metatarsal (DMT)	1.40	n.d.	MR2	MR2
D metapodial (DMP)	1.25	n.d.	none	none
Phalanges (Ph)	1.98	n.d.	none	none

density values of skeletal parts should be used alone to interpret skeletal part frequencies. To do so would result in tacit acceptance of a simple correlation denoting a causal relation between two variables that might not in fact be causally related. That is, a correlation between bone frequencies and bone structural density is a necessary condition for inferring a causal relation between the two variables, but it is not a sufficient condition for such an inference. Lyman *et al.* (1992a) examine two assemblages of yellow-bellied marmot (*Marmota flaviventris*) bones (Table 7.9) and attempt to explain why one of them is and the other is not correlated with the structural density values for marmot bones (Table 7.7). They found that the MAU values for the marmot bone assemblage from the White Mountains of California (Grayson 1989) are not correlated with marmot bone density ($r_s = 0.3$, $P = 0.28$), as

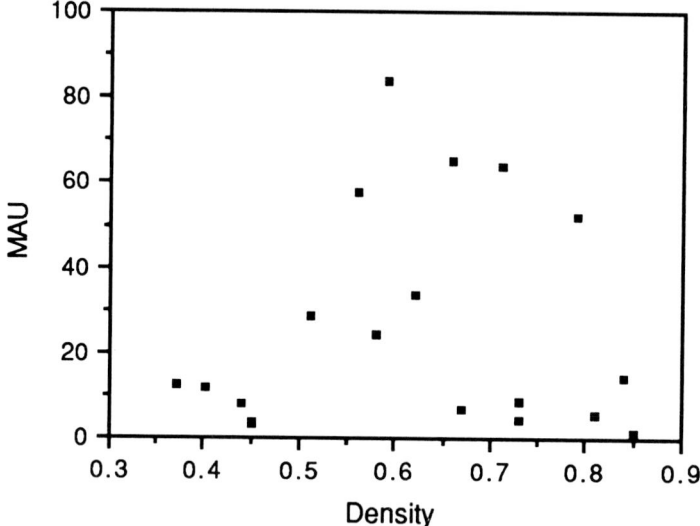

Figure 7.9. Scatterplot of MAU frequencies of marmot skeletal parts from the White Mountains against bone mineral density values for marmots.

implied in Figure 7.9, suggesting that density-mediated attrition was not responsible for the varying frequencies of skeletal parts (note the vertical scatter of points in Figure 7.9). The White Mountains marmot bones were, however, quite fragmented (Grayson 1989). Thus, Lyman *et al.* (1992a) suggest that differential fragmentation influenced the skeletal part profiles. They call upon the premise that smaller fragments will be less identifiable because they are less likely to have diagnostic landmarks (e.g., Lyman and O'Brien 1987). The major test implication of this hypothesis is that rare kinds of skeletal parts should have smaller average *proportional completeness values* (% of a complete specimen represented by a fragment) than the more common kinds of skeletal parts. While fragment size data are unavailable for the White Mountains materials, and thus the hypothesis cannot be tested, what is important here is that the initial projection of the marmot bone frequencies on the frame of reference of bone structural density directs us to other kinds of data that may help account for the varying frequencies of skeletal parts. Lyman *et al.* (1992a) found a significant correlation between structural density and the MAU values of marmot bones from the Salishan Mesa site in eastern Washington ($r_s = 0.46$, $P = 0.06$), as implied in Figure 7.10. Those marmot bones seem to be less fragmented than the White Mountains bones. Between these two assemblages, then, a question is raised about how structural density mediates fragmentation and the resultant identifiability of faunal remains, a topic touched on in Chapters 6 and 8.

It must be pointed out that the %survivorship value is the same variable as the %MAU value discussed in the preceding section of this chapter. This

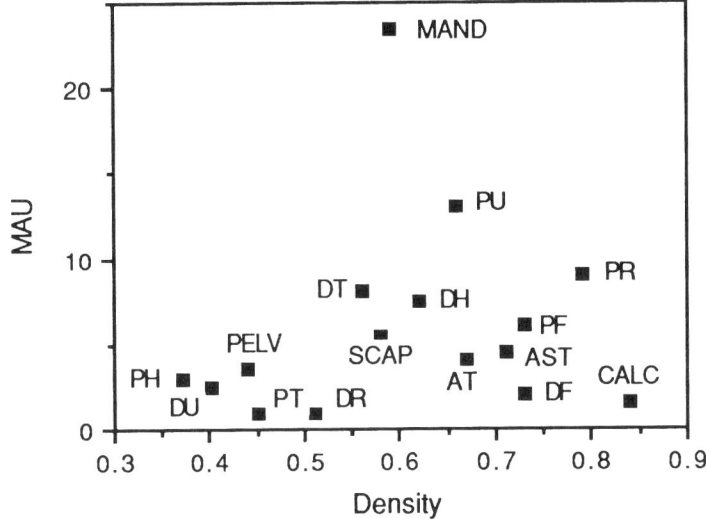

Figure 7.10. Scatterplot of MAU frequencies of marmot skeletal parts from the Salishan Mesa site against bone mineral density values for marmots.

equivalence can be shown as follows. Brain (1969, 1976) originally calculated %survivorship in an assemblage as

$$\frac{(MNE_i)\,100}{MNI\ (\text{number of times}\ _i\ \text{occurs in one skeleton})} \qquad [7.4]$$

The denominator in equation [7.4] tells us how many of skeletal portion $_i$ to expect given 100% survivorship, and it is equal to the maximum MNE value observed in an assemblage. It can thus be written as

$$\frac{(MNE_i)\,100}{\text{maximum MNE in the assemblage}} \qquad [7.5]$$

Equation [7.4] (and by implication equation [7.5]) is my reconstruction of Brain's (1967b, 1969, 1976) procedure, as he did not use the terms MNE or MNI. Brain (1967b:4), for example, describes the bone assemblage he studied as variously consisting of long bone "distal ends" and "pieces" of long bone ends and other skeletal elements and skeletal portions (in the senses defined in Chapter 4). He gives a good description of the derivational procedure he uses when he writes "in the case of ribs, 26 of which are found in a single goat skeleton, the original number contributed by 64 [= MNI] goats must have been 1664 [= 64 × 26]. Only 170 [= MNE] have been found, indicating a 10.2% survival" (Brain 1969:18). I presume Brain (1967b:3; 1969:17) used a quantitative unit equivalent to MNE in his procedure for measuring %survivorship because he gives the number of rib "parts" as 174. But, (170 ÷ 1664) 100 = 10.2%. Similarly, the NISP for mandibles is given by Brain as 188, what I believe to be the MNE is given by him as 117, and the %survivorship is given as 91.4% ($= [117 \div (2 \times 64)]\ 100$).

As defined by Binford (1978, 1981b, 1984b), %MAU is derived with the equation

$$\frac{[MAU_i]\ 100}{\text{maximum MAU in the assemblage}} \qquad [7.6]$$

Note that MAU_i is derived with the equation

$$\frac{MNE_i}{\text{number of times }_i\text{ occurs in one skeleton}} \qquad [7.7]$$

Substituting equation [7.7] into equation [7.6],

$$\frac{[MNE_i \div \text{number of times }_i\text{ occurs in one skeleton}]\ 100}{\text{maximum MNE in the assemblage} \div \text{number of times that skeletal portion occurs in one skeleton}} \qquad [7.8]$$

The denominator in equations [7.6] and [7.8] tells us how many to expect given 100% survival, and is dependent on the assemblage under study. The important point to note is that the expression "$\div$ number of times $_i$ occurs in one skeleton" in the numerator and denominator in equation [7.8] cancel each other out, and when they do, equations [7.5] and [7.8] are identical.

That equations [7.4] and [7.5], and equations [7.6] and [7.8], are measuring the same property can be illustrated by using Brain's data for the rib and for the mandible. For the rib, equation [7.8] is solved as

$$\frac{[170 \div 26]\ 100}{1664 \div 26} = \frac{[6.538]}{64} = 10.2\% \qquad [7.9]$$

For the mandible, equation [7.8] is solved as

$$\frac{[117 \div 2]\ 100}{128 \div 2} = \frac{[58.5]\ 100}{64} = 91.4\% \qquad [7.10]$$

These are precisely the results found when using Brain's procedure, or equations [7.4] and [7.5]. Thus it should be clear that %survivorship is measured precisely the same way as %MAU; both variables measure the same property, and are one and the same.

Recall that the transport and utility indices described in the preceding section are simply one "frame of reference" (Binford 1987) that can be used as a single step in constructing explanations for the frequencies of skeletal parts. The structural density values are another frame of reference and thus another step in such constructions. Thus, it is reasonable to ask if the %MAU values for caribou remains at the Anavik site (Table 7.2) correlate with the structural density of bone parts. We as yet lack structural density values for caribou, but we can use the density values for deer in Table 7.6 as that species is confamilial with caribou, and thus is the closest genetic relative for which we have density values. Then, we must decide which particular density values to use because the

Table 7.10 *Traditional density scan sites and maximum density scan sites typically correlated with MAU values*

MAU category	Traditional scan site	Maximum scan site	MAU category	Traditional scan site	Maximum scan site
mandible	DN4	DN8	innominate	AC1	IL2
atlas	AT1	AT3	P femur	FE2	FE1
axis	AX1	AX1	D femur	FE6	FE6
cervical	CE1	CE1	patella	PA1	PA1
thoracic	TH1	TH2	P tibia	TI1	TI1
lumbar	LU1	LU2	D tibia	TI5	TI5
sacrum	SC1	SC1	astragalus	AS1	AS3
rib	RI3	RI3	calcaneum	CA2	CA2
sternum	ST1	ST1	naviculo-cuboid	NC1	NC3
scapula	SP1	SP2	P metatarsal	MR1	MR1
P humerus	HU1	HU2	D metatarsal	MR6	MR6
D humerus	HU5	HU5	First phalanx	P12	P13
P radius	RA1	RA2	Second phalanx	P22	P23
D radius	RA5	RA5	Third phalanx	P31	P31
carpal	none	SCAPHOID			

%MAU values are for skeletal parts such as proximal humeri and there are two places on the proximal half of the humeri for which we have density measures, scan sites HU1 and HU2 (Figure 7.4). In my original work with this problem (Lyman 1984a), I selected the scan site that seemed best to typify the structural density of a skeletal part. Other researchers have suggested using the maximum density value recorded for a particular MAU skeletal category (Gifford-Gonzalez and Gargett n.d.; Rapson 1990). I (Lyman 1992a) subsequently referred to the former set of values as the *traditional density values*, and the latter as the *maximum density values*, both of which are listed in Table 7.10.

The Anavik caribou %MAU values are not correlated with the traditional deer bone density values ($r_s = -0.003$, $P = 0.93$) or the maximum density values ($r_s = 0.06$, $P = 0.75$). Using only these two frames of reference, utility indices and structural density values, then, might prompt us to conclude that the frequencies of caribou skeletal parts at Anavik are attributable to the operation of a particular transport strategy practiced by the human occupants of the site and not to density-mediated attrition. One might suspect such a case would be easily interpreted when found in an archaeological context. Similarly, the caribou remains Binford (1978:323) recovered from the ethnoarchaeological Ingsted site (Table 7.2) are correlated with the structural density of deer bones ($r_s = 0.38$, $P = 0.05$ for both traditional and maximum density values) but not with the caribou %MGUI ($r_s = 0.01$, $P = 0.9$). The frequencies of caribou bones at the Ingsted site seem to be attributable to density-mediated attritional factors rather than to a particular transport strategy followed by human hunters. Again, one might suspect similar statistically clear results found in archaeological settings would be readily interpreted. Unfortunately, as any serious student of taphonomy knows, not all zooarchaeological cases are so transparent.

Differential transport versus differential survivorship

The problem

Interpretations derived from the two variables measured by the transport or utility indices and the structural density values require data on bone frequencies, typically %MAU or %survivorship values, which we have seen are in fact the same. Given how interpretations are derived from, for example, scatterplots like those in Figures 7.2 and 7.3 and their associated (ordinal scale) correlation coefficients, the probability that a skeletal part will be transported and utilized by humans must be independent of (must not correlate with) the probability that this skeletal part will survive density-mediated attritional processes. If these two variables are somehow interdependent or are simply correlated (not necessarily causally related), then significant correlations between bone frequencies and both the transport or utility indices and structural density may result, and the analyst will be faced with a case of equifinality. That is, simply using the analytic procedures described thus far will result in the researcher not knowing whether to call upon differential transport and utility or differential destruction as the explanation for the frequencies of skeletal parts observed in a bone assemblage.

Several years ago I showed that there is a weak negative correlation between the bone mineral density values for deer and several of the utility indices listed in Table 7.1 (Lyman 1985a). I subsequently showed that many of the indices listed in Tables 7.3 and 7.4 are also negatively correlated with the bone mineral density values for deer (Lyman 1992a). These results prompt several conclusions. First, bones that have low structural densities tend to rank high in utility and bones that have high structural densities tend to rank low in utility. (Note that I use the term "utility" in the remainder of this discussion to denote both economic utility and the probability that a skeletal part will be transported; utility is thus a shorthand for both economic utility and transport indices.) Thus, both density-mediated destruction and a reverse utility strategy (Figure 7.1a) will produce a negative hyperbolic (L-shaped) curve when %MAU values are plotted against %MGUI values. Second, neither the %MGUI nor the structural density values, alone or in combination, will provide trustworthy bases for explanations of skeletal part frequencies. Other kinds of evidence are necessary to help sort out which process(es) (transport or destruction) was responsible for a particular bone assemblage (Lyman 1985a, 1991c, 1992a).

The problem is well illustrated by the guanaco general utility index and the modified guanaco utility index (Table 7.4). Both indices are negatively corre-

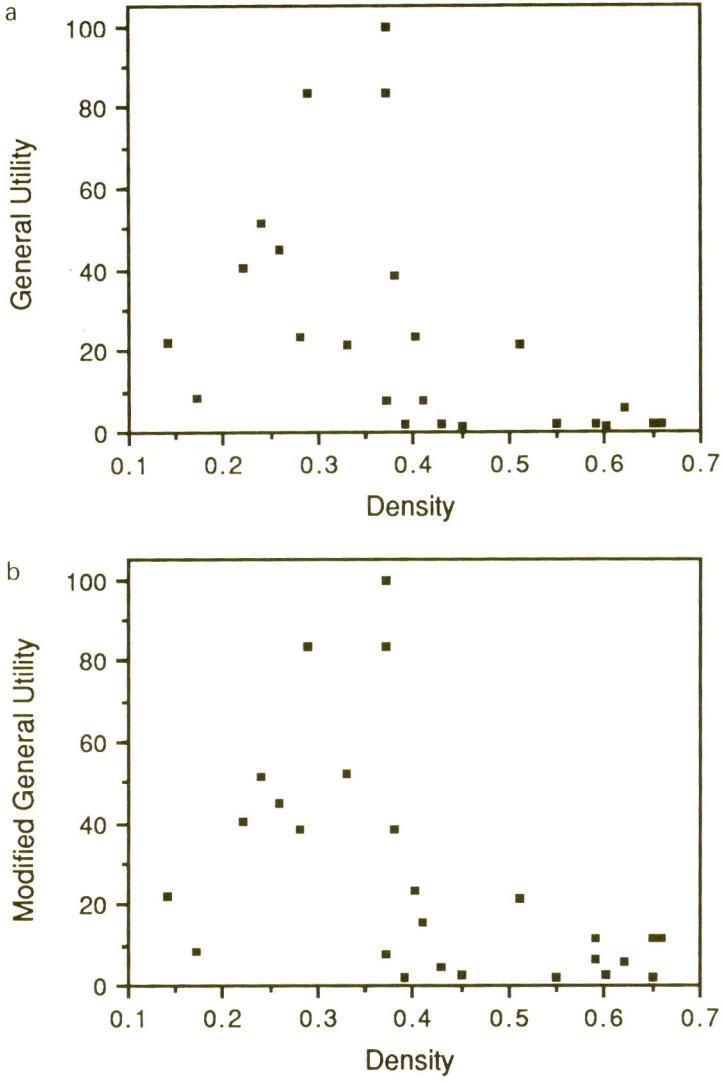

Figure 7.11. a, Scatterplot of the general guanaco utility index against guanaco bone density. b, Scatterplot of the modified guanaco utility index against guanaco bone density.

lated with the structural density of guanaco skeletal parts (Table 7.6). Scatterplots of the guanaco general utility index against bone density ($r_s = -0.71$, $P = 0.0005$; Figure 7.11a) and of the guanaco modified utility index against bone density ($r_s = -0.55$, $P = 0.004$; Figure 7.11b) clearly show the relation. The fact that guanaco bones of low density tend to rank high in utility while guanaco bones of high density tend to rank low in utility means, when a significant negative correlation between %MAU values for an archaeological collection of guanaco remains and the utility index for guanaco is found, the analyst may also find a significant positive correlation between those %MAU

Table 7.11 *%MAU frequencies for deer-sized animals for site 45CH302 (from Lyman n.d.a)*

Skeletal part	%MAU	Skeletal part	%MAU
mandible	100.0	P ulna	25.0
atlas	37.5	P metacarpal	43.8
axis	37.5	D metacarpal	68.8
cervical	13.75	P femur	12.5
thoracic	9.6	D femur	6.2
lumbar	44.4	P tibia	3.1
innominate	28.1	D tibia	78.1
sacrum	6.2	naviculo-cuboid	53.1
rib	27.5	astragalus	96.9
sternum	0.9	calcaneum	34.4
scapula	34.4	P metatarsal	59.4
P humerus	6.2	D metatarsal	56.2
D humerus	62.5	first phalanx	78.1
P radius	53.1	second phalanx	34.4
D radius	21.9	third phalanx	17.2

frequencies and the structural density of guanaco bones. Again, in such cases the interpretive problem is deciding whether the bone frequencies are the result of differential transport and utilization by people, differential, density-mediated destruction, or a combination of the two.

An example will make the significance of the preceding problem clear. The late Holocene-aged site of 45CH302 in eastern Washington produced a large sample of deer (*Odocoileus* sp.), bighorn sheep (*Ovis canadensis*), and pronghorn antelope (*Antilocapra americana*) remains (Lyman n.d.a). Because these three taxa overlap in size, some of their bones cannot be identified to taxon (such as ribs and most vertebrae), and many other bones cannot be identified to taxon if the specimens are incomplete or broken. I tallied the MNE of all remains for these three taxa, plus all remains that could not be identified to taxon but that could be identified as a member of this size class of mammals if the skeletal part represented by a specimen could be determined (Table 7.11). The %MAU frequencies for the 45CH302 assemblage of deer-sized bones not only correlate strongly with the structural density of deer bones ($r_s = 0.67$, $P < 0.001$), but also with both the caribou %MGUI ($r_s = -0.44$, $P < 0.05$) and the sheep %MGUI ($r_s = -0.55$, $P < 0.005$). These relations are shown in Figure 7.12, and the interpretive problem should be abundantly clear from these graphs: is the assemblage of deer-sized remains from 45CH302 the result of differential utilization and transport by humans, density-mediated attrition, or both? The correlation coefficients and the scatterplots do not help answer this question.

In attempting to produce an answer to the preceding question, Grayson (1988:70–71) suggested that when density-mediated destruction was the responsible process and differential transport was not, the statistical relation between %MAU and bone density will be positive and significant (typically, $P \leq 0.05$) while the statistical relation between %MAU and %MGUI will be insignificant (typically, $P > 0.05$). Conversely, Grayson argued, differential transport could be inferred if %MAU and %MGUI are significantly and positively correlated but %MAU and structural density are insignificantly correlated. This procedure of calculating and comparing correlation coefficients was used by Klein (1989) in an effort to show that remains of bovids of different size classes recovered from the African Klasies River Mouth site were variously attributable to density-mediated attrition and not differential transport. Grayson (1988:71) noted that other combinations of statistical correlations "will be more difficult to interpret."

All possible combinations of statistical correlations between %MAU values and bone density, and between %MAU values and %MGUI are shown in Figure 7.13. I have given each possibility a name: class 1, class 2, class 3, etc. (Lyman 1991c). Grayson's (1988) solutions to the equifinality problem noted above are indicated (Classes 2 and 4), and other possible interpretations of various of the possible classes are also noted in Figure 7.13. What is a taphonomist to do with this range of possibilities? To assess the significance and magnitude of this question, I calculated correlation coefficients between bone density values in Table 7.6 and the %MAU values, and utility indices in Tables 7.1 and 7.4 and %MAU values, for a sample of 184 ethnoarchaeological and archaeological assemblages of artiodactyl bones. Ninety-seven of the assemblages come from North America, one from South America, 44 from Africa, 30 from Europe, 10 from the Near East, and one each from the Caribbean and Asia. Forty-one are ethnoarchaeological, 90 are Holocene-aged, 52 date to the Pleistocene and one is pre-Pleistocene in age. When taxa were small or confamilial with deer, I used the deer (*Odocoileus* sp.) density values and the caribou and sheep %MGUI values; when taxa were large or were bison, I used the bison (*Bison bison*) density values and the bison total products index. I then identified each set of coefficients per assemblage as one of the classes, initially keeping ethnoarchaeological assemblages separate from archaeological ones (Figure 7.13).

Considering only those correlation coefficients for which $P \leq 0.05$ as significant, 84 of the 184 (45.7%) assemblages are positively correlated with bone density, and of those 84 only 9 (10.7%) are ethnoarchaeological assemblages. Only three assemblages, all ethnoarchaeological, are negatively correlated with bone density. These results suggest that post-depositional and post-burial destruction, which the archaeological assemblages have undergone but the ethnoarchaeological assemblages have not (or at least not to the same degree), are major taphonomic factors (see Chapter 11). Forty-seven (25.5%) of the 184 assemblages are negatively correlated and 14 (7.6%) are positively correlated with a utility index. About one third (67, or 36.4%) of the 184 assemblages

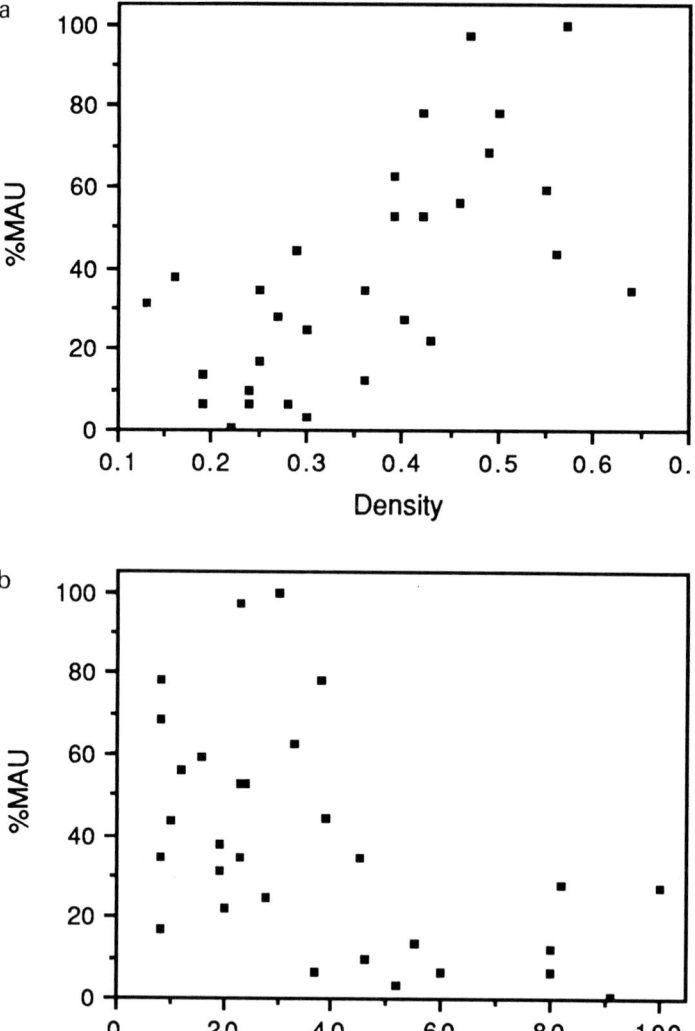

correlate with neither bone density nor the utility index (fall in Class 5). Thus, a third of the assemblages we might correlate with both bone density and utility may be inexplicable using only the variables of food utility and structural density. Further, only 74 (40.2%) of the 184 assemblages fall in Classes 2 and 4; this means that less than half of these assemblages might be explainable using only Grayson's (1988) statistical criteria. Finally, 16.8% (31 of 184) of the assemblages correlate significantly with both bone density and utility. That is, if this sample of 184 assemblages is representative of the total variability in faunal assemblages, then the equifinality problem will be encountered in one of every six assemblages we analyze. We have, then, reached a point in the search

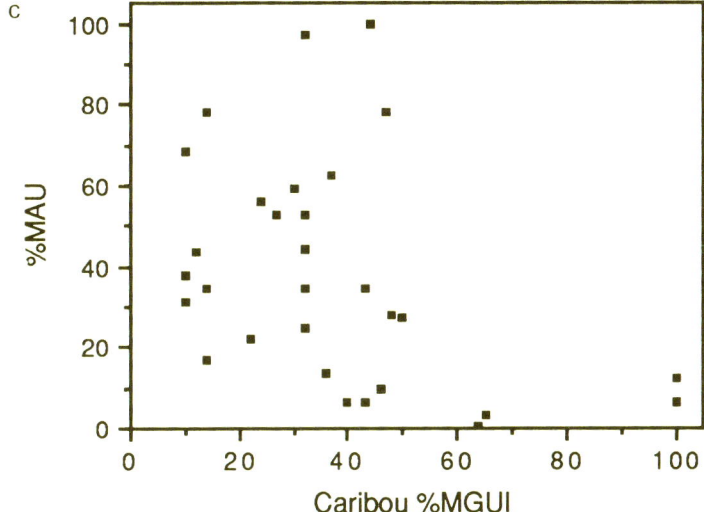

Figure 7.12. a, Scatterplot of %MAU frequencies for deer-size animal remains
from 45CH302 against the structural density of deer bones. b, Scatterplot of
%MAU frequencies for deer- size animal remains from 45CH302 against the
%MGUI for sheep. c, Scatterplot of %MAU frequencies for deer-size animal
remains from 45CH302 against the %MGUI for caribou.

for explanations of varying frequencies of skeletal parts that demands other
kinds of data and other kinds of analysis. But what kinds of data and what
kinds of analysis?

Toward a solution 1: counting units

We can begin to discern an answer to the preceding question if we first examine
the counting units used in studies of skeletal part frequencies. In most general
terms, the counting unit is some analytically specified portion or part of a
skeleton. How particular skeletal parts are defined is critical. The necessary and
sufficient conditions for identifying a specimen as representing a member of a
particular category of skeletal part or portion must be clear. The original utility
indices (Table 7.1) were determined for proximal and distal halves of long
bones, and portions of the axial skeleton consisting, with the exceptions of the
atlas and axis vertebrae, of more than one skeletal element. Examples of the
latter include the counting unit labeled "rib" and the counting unit labeled
"thoracic vertebra;" the former consists of 26 individual skeletal elements and
the latter consists of 13 individual skeletal elements in a typical ungulate (Table
4.1). Given the skeletal units (portions of the skeleton) for which utility index
values were derived, it was only logical to determine the minimum number of
each of those skeletal portions – each is an anatomically defined counting unit –

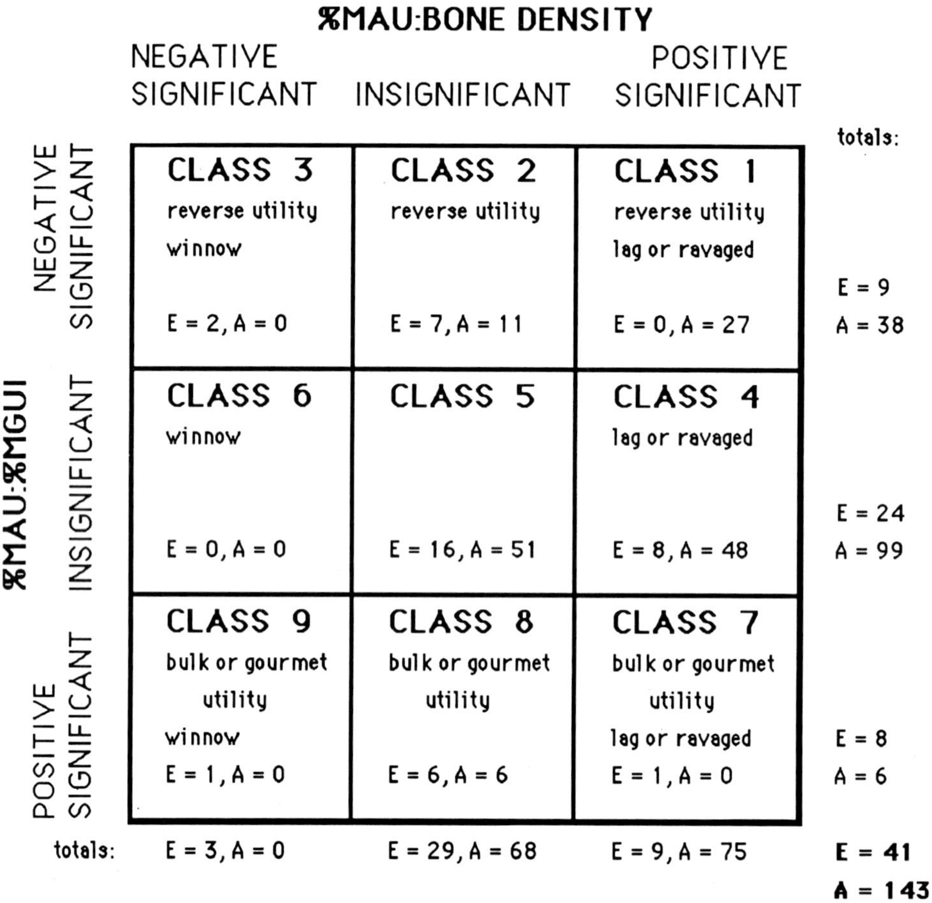

%MAU:BONE DENSITY

	NEGATIVE SIGNIFICANT	INSIGNIFICANT	POSITIVE SIGNIFICANT	totals:
NEGATIVE SIGNIFICANT	**CLASS 3** reverse utility winnow E = 2, A = 0	**CLASS 2** reverse utility E = 7, A = 11	**CLASS 1** reverse utility lag or ravaged E = 0, A = 27	E = 9 A = 38
INSIGNIFICANT (%MAU:%MGUI)	**CLASS 6** winnow E = 0, A = 0	**CLASS 5** E = 16, A = 51	**CLASS 4** lag or ravaged E = 8, A = 48	E = 24 A = 99
POSITIVE SIGNIFICANT	**CLASS 9** bulk or gourmet utility winnow E = 1, A = 0	**CLASS 8** bulk or gourmet utility E = 6, A = 6	**CLASS 7** bulk or gourmet utility lag or ravaged E = 1, A = 0	E = 8 A = 6
totals:	E = 3, A = 0	E = 29, A = 68	E = 9, A = 75	**E = 41** **A = 143**

Figure 7.13. All possible combinations (classes) of correlation coefficients between the %MAU of a bone assemblage, and both bone density and %MGUI. Classes that would represent a particular utility or transport strategy are indicated, as are those classes that represent ravaged assemblages, and fluvially sorted lag (not transported) and winnow (transported) assemblages. E = the number of ethnoarchaeologically documented assemblages that fall within a class; A = the number of archaeological assemblages that fall within a class.

in an archaeological assemblage (often referred to as the minimum number of elements, or MNE values), and it was only logical to use the structural density value which seemed to typify each particular skeletal unit (Lyman 1985a). Much archaeofaunal and paleontological faunal data were, at that time, being published in just such a form (Lyman 1984a). The suggestion to use the maximum structural density value for a skeletal unit rather than the traditional one in order to provide a more conservative test of the density-mediated destruction hypothesis is reasonable (Gifford-Gonzalez and Gargett n.d.), but these two sets of structural density values for deer are tightly correlated at an ordinal scale ($r_s = 0.89$, $P < 0.001$) as are the two sets of values for bison ($r_s = 0.75$, $P < 0.001$). The substitution of one for the other neither appreciably changes results like that in Figure 7.13 (Lyman 1992a, 1993b) nor does it help us determine whether skeletal part frequencies are the result of differential transport or differential destruction.

A taphonomist must ask if the skeletal units used to count skeletal parts are the appropriate ones. As we have seen, %MAU (or %survivorship) values are appropriate because they are typically interpreted in terms of the %MGUI and like measures which have as their counting unit proximal and distal halves of long bones. This does not mean, however, that proximal and distal halves of long bones are always or are the only appropriate measure. For example, Metcalfe and Jones (1988) and Lyman *et al.* (1992b) describe utility indices for complete long bones rather than proximal and distal halves. For these indices, complete long bone skeletal units are the appropriate counting unit (again, typically referred to as MNE values in archaeological cases). For the structural density values, the appropriate counting units are the scan sites (Figures 7.4, 7.5, 7.6); the fact that most bone frequency data had been (and still are) published as counts of proximal and distal halves or as complete skeletal elements prompted the selection of particular scan sites (Table 7.10) when correlating published bone frequencies with bone density (Lyman 1984a, 1985a). Thus, it seems we need to count our bones several different ways, or at several different levels of inclusiveness, in order to generate the frequency data appropriate to the frame of reference to which we wish to relate that data.

Because of the complexity and the significance of *counting* skeletal parts, it is critical at this juncture to review briefly the basic counting units presently in use. Recall from Chapter 4 that a *specimen* is a discrete, identifiable skeletal part such as a tooth, a complete phalange, or the proximal end of a tibia. A *skeletal element* is a complete, anatomically discrete unit, such as a first molar or a humerus. Thus, three specimens of the distal half of the humerus are three *specimens* that represent three *skeletal elements*, whereas two distal ends of humeri and a proximal end of a humeri are three specimens that minimally (assuming they are all from left elements) represent two elements. Complexity arises when tallying skeletal part frequencies to cast upon the utility or structural density frames of reference because we must convert specimen counts into %MAU values in such a manner that the quantitative units for both reference frames and the bone assemblage are the same (see the discussion of equations [7.4] and [7.6] above). That sameness comes from determining the minimum number of each skeletal unit represented by the bone collection. This demands that one determine which *specimens* belonging to a kind of skeletal unit are independent of one another; i.e., do not represent the same individual *skeletal element*. The skeletal units are of course determined and defined by the frame of reference (e.g., for some utility indices, the skeletal units for tallying long bones are the proximal and distal halves), and those units have tradition-

Table 7.12 *MNE and %MAU frequencies of hyena-ravaged domestic sheep bones (from Marean and Spencer 1991). Original MNE is 50*

Skeletal part	Surviving MNE	%Survivorship	Sheep bone density	Deer bone density	Sheep %MGUI
P femur end	9.95	19.9	0.28	0.41	80.58
P femur shaft	32.05	64.1	0.20	0.33	
Mid femur shaft	50.00	100.0	0.36	0.57	
D femur shaft	32.85	65.7	0.24	0.37	
D femur end	7.30	14.6	0.22	0.28	80.58
P tibia end	7.80	15.6	0.16	0.30	51.99
P tibia shaft	43.00	86.0	0.20	0.32	
Mid tibia shaft	50.00	100.0	0.59	0.74	
D tibia shaft	44.05	88.1	0.36	0.51	
D tibia end	25.20	50.4	0.28	0.50	37.70
P metatarsal end	26.48	53.0	0.43	0.55	15.77
P metatarsal shaft	30.65	61.3	0.53	0.65	
Mid metatarsal shaft	45.89	91.8	0.68	0.74	
D metatarsal shaft	31.90	63.8	0.51	0.57	
D metatarsal end	15.45	30.9	0.31	0.46	12.11

ally been called minimum numbers of elements (MNE). Obviously, the word "element" in the term MNE is not necessarily the same kind of skeletal unit as the term "element" as it is defined earlier in this paragraph and in Chapter 4.

Binford (1978:70–71) reasoned that because "our interest is in the actual use made of animals as food," MAU values will provide "undistorted conversions of the actual count of bones into animal units [that] accurately describe the relative proportions of anatomical parts" and avoid overestimating the amount of meat present at a site (see Lyman 1979b for a similar discussion). Recall that MNE is the minimum number of each skeletal part necessary to account for the specimens of each part, and that MAU is the MNE per skeletal part divided by the number of times that skeletal part occurs in a single skeleton. There are several ways to tally skeletal frequencies other than MAU and MNE, each with the express purpose of studying variation in skeletal part frequencies in the fossil record. For example, Lyman (1979b) suggests one can count the minimum number of butchering units; his suggestion is founded on an analysis of historic faunal remains that had been butchered with saws, and it has been followed by analysts studying historic faunal remains (e.g., Schulz and Gust 1983). Lyman (1979b) also suggests, following Read (1971), that one can count the minimum number of "skeletal portions" to obtain another measure of variation in how skeletons are represented. He distinguishs three skeletal portions: the forequarters (forelimbs), hindquarters (hindlimbs including pelvis), and rib–vertebrae skeletal portion. There have been other variants on this theme, such as Stiner's (1990a, 1991b) suggestion that skeletal parts can be tallied by what she calls "anatomical regions." She defines nine anatomical regions: horn/antler, head, neck (cervical vertebrae), axial column below the neck (thoracic vertebrae, ribs, sternum, lumbar vertebrae, pelvis, sacrum), upper front limbs (scapula, humerus), lower front limbs (radius-ulna, carpals, metacarpal), upper hind limbs (femur), lower hind limbs (tibia, tarsals, metatarsal), and feet (phalanges). All of these suggestions can be subsumed under the general rubric of a pattern recognition approach. That is, different scales of inclusiveness of portions of skeletons have been proposed in order to determine if detectable patterns in the frequencies of those portions occur across multiple assemblages.

Marean and Spencer (1991) provide data that are very conducive to illustrating the issue of appropriate counting units. In their study of the impact on bone assemblages by the spotted hyena (*Crocuta crocuta*), they found a strong correlation between the structural density of long bone parts and the frequencies of long bone parts of domestic sheep (*Ovis aries*) that survived gnawing. The correlation coefficient they describe was based only on the frequencies of proximal and distal ends of three long bones. They also report the frequencies of parts of shafts that survived hyena attrition for each of those long bones (Table 7.12). The frequencies of both shaft and end parts are almost significantly correlated with the sheep ($r_s = 0.48$, $P = 0.07$) bone density values and are correlated with the deer ($r_s = 0.60$, $P = 0.02$) bone density values (Figure 7.14). What is of greater interest to the question of how to disentangle those cases where %MAU values are positively correlated with bone density and negatively correlated with the %MGUI is illustrated in Figure 7.15. There, the difference between the scatterplots resides in the high survivorship of proximal and distal *shafts* of the three long bones and the low survivorship of their proximal and distal *ends*. The scatterplot of the proximal and distal *ends* against the sheep %MGUI (Figure 7.15a) has the appearance of a reverse utility curve (Figure 7.1a) whereas the scatterplot of the proximal and distal *shafts* against the %MGUI (Figure 7.15b) has the appearance of a bulk utility curve (Figure 7.1c).

Given that %survivorship and %MAU should be calculated on the basis of the *maximum* observed MNE for the skeletal part being tallied (recalling first how %MAU values are calculated [see equations 7.4–7.10], and second, that an MNE for a proximal or distal *half* of a long bone can be based on the minimum number of ends, the minimum number of near-end shaft fragments, or some combination of the two such that specimen independence is ensured), the difference in the two scatterplots in Figure 7.15 may seem irrelevant because the analyst should be plotting the values in Figure 7.15b in scatterplots such as those exemplified in Figure 7.2. That is so because the proximal and distal shaft parts outnumber the proximal and distal ends, so the former should be the basis for deriving the plotted MNE values rather than the latter. The difference between the scatterplots in Figure 7.15 is, however, quite relevant to our

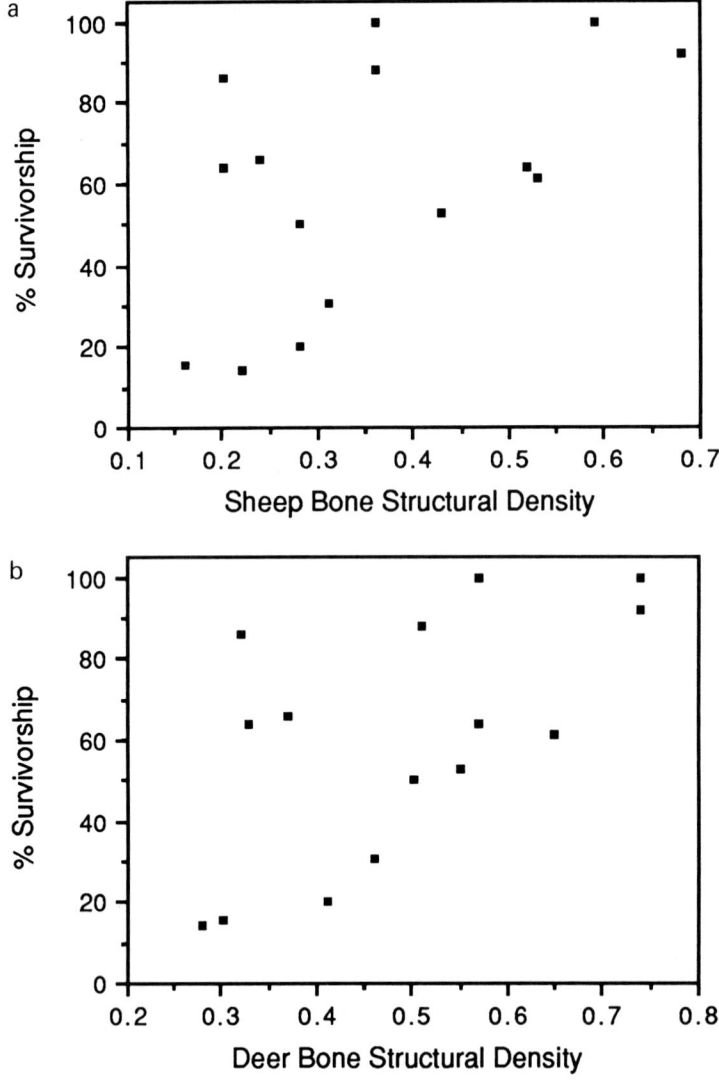

Figure 7.14. Scatterplots of %survivorship of skeletal parts after ravaging by hyenas (from Marean and Spencer 1991) against (a) sheep bone structural density and (b) deer bone structural density.

discussion for two reasons. First, that difference underscores the fact that correlations between bone frequencies and structural density should be calculated on the basis of the categories of skeletal parts that most closely correspond to the scan sites. Frequencies of proximal and distal *shaft* parts should not be compared to the structural densities of proximal and distal *ends*.

The second thing that is illustrated in Figure 7.15 is that density-mediated destruction has influenced the frequencies of some skeletal parts in this assemblage (long bone ends) but not others (long bone near-end shafts and

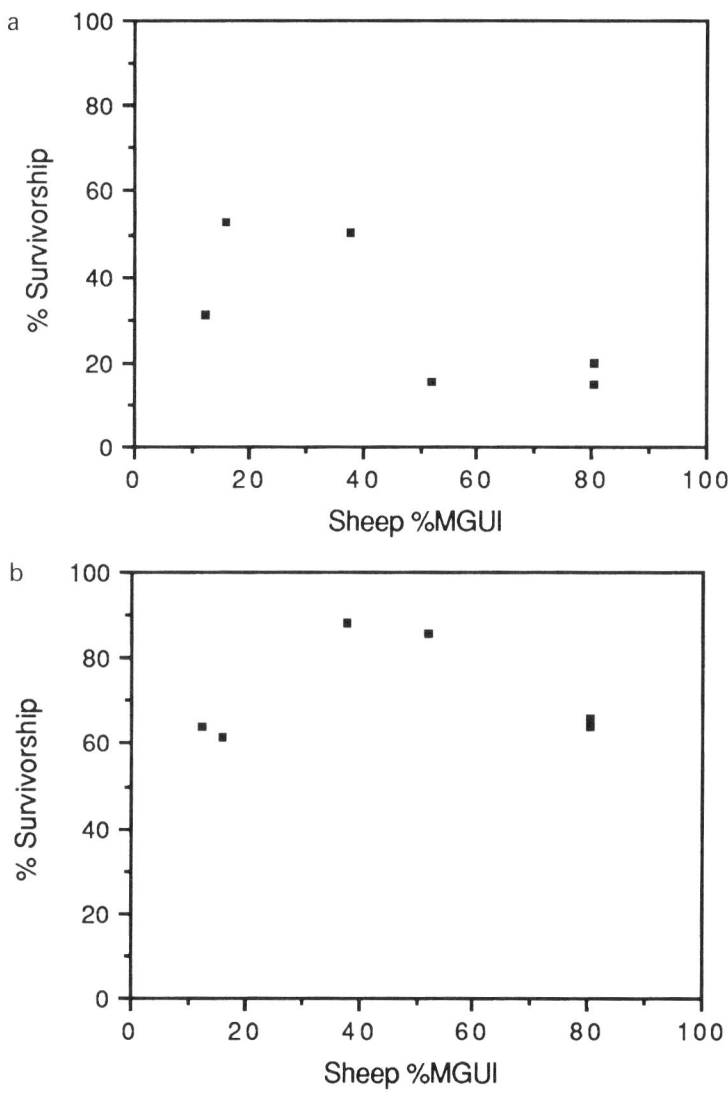

Figure 7.15. Variation in scatterplots of %survivorship of skeletal parts after ravaging by hyenas (from Marean and Spencer 1991; see Table 7.12) against sheep %MGUI for (a) long bone ends and (b) long bone shaft ends.

mid-shafts). This is important because it seems improbable that human butchers would transport, for instance, proximal femur shafts much more frequently than proximal femur ends, and the same goes for distal femur shafts and distal femur ends, proximal tibia shafts and proximal tibia ends, etc. If the kind of patterning in skeletal part survivorship shown in Figure 7.15 occurs across all major limb or long bones, I suspect the analyst would be safe in suggesting that the differential transport hypothesis may account for frequencies of *ends of shafts*, but density-mediated destruction accounts for the frequencies of long bone *ends*. If both the long bone ends (with low structural densities) and the corresponding shaft ends (typically having higher structural densities than adjacent ends) are both correlated with bone density, then it would be unwise to conclude that either measure of bone frequencies (ends or shaft ends) provide a valid measure of differential transport because both the bone ends and the shaft ends could have been subjected to density-mediated attrition.

Whether or not I am correct in these suggestions is largely irrelevant; their correctness can be assessed by ethnoarchaeological research, which shows us yet another direction for such research to proceed. What is important in the preceding example is that the scale of the counting unit was changed from the MNE of proximal and distal *halves* of long bones to the MNE of *scan sites* (Figures 7.4, 7.5, and 7.6). If archaeologically observed frequencies of the scan sites correlate with their structural density, then there is every reason to anticipate that the frequencies of the larger counting units of proximal and distal halves of bones will also, because the latter are (or should be) the maximum MNE of a scan site found on the proximal or distal half of a long bone. Similar arguments apply to all other skeletal parts when those parts are defined by the utility or transport indices in such a manner that they contain more than one scan site.

Rapson (1990) notes that long bone ends are rare but long bone shafts are abundant in the archaeological collection he studied, and he therefore suggests that density-mediated destruction of low-density long bone ends had occurred, but the high-density long bone shafts had not been destroyed. Fifteen years earlier Klein (1975:286) argued, following neotaphonomic observations reported by Sutcliffe (1970), that hyenas destroy long bone ends by gnawing and consumption, but leave long bone shafts, including the proximal and distal ends of the shafts. Klein suggested that hominid butchers seem not to destroy long bone ends because they break a bone open by impacting some part of the shaft rather than the end. Klein (1975:279) describes bovid remains from a late Pleistocene hyena den. That assemblage contained more long bone shaft ends (n = 290) than long bone ends (n = 170; both values are for the sum of the humerus, radius, femur, and tibia). This is a statistically significant difference (chi^2 = 31.3, $P < 0.001$).

Experimental work confirms the significance of Klein's (1975) and Rapson's (1990) suggestion. On the basis of their actualistic research, Marean and Spencer (1991:655) indicate that "if 100 percent of the [bone] sample has been ravaged by hyenas then large differentials will exist between MNEs calculated on limb-bone ends and MNEs calculated on middle-shaft pieces." They calculate a middle-shaft MAU to limb-bone end MAU ratio, and find that two to five times as many shafts will survive ravaging by carnivores as long-bone ends. (Note: in this case, for all skeletal parts considered, MAU = MNE ÷ 2, so end to shaft ratios are the same whether MAU or MNE values are used.) This in

Table 7.13 *Frequencies of skeletal parts at FLK* Zinjanthropus *(from Bunn and Kroll 1986, 1988) and complete bone utility index values (from Metcalfe and Jones 1988)*

Skeletal element	MNE based on ends only (proximal/distal)	MNE based on ends and shafts	Complete bone utility index
humerus	19 (5/19)	20	36.8
radius	14 (14/5)	22	25.8
metacarpal	15 (15/8)	16	5.2
femur	6 (6/6)	22	100.0
tibia	11 (10/11)	31	62.8
metatarsal	15 (15/10)	16	37.0

turn leads to the suggestion that such ratios allow the analyst to measure "the intensity of carnivore ravaging" (Marean and Spencer 1991:655), higher ratios (more shaft MAUs per end MAUs) indicating more intensive ravaging.

Setting aside for the moment the suggestion that shaft to end ratios might be used to measure how ravaged a bone assemblage is, Marean and Spencer's (1991) data indicate a possible way to escape the equifinality problem posed when skeletal part frequencies correlate positively with bone density and negatively with utility indices. Carnivore destruction should be evidenced by carnivore gnawing marks (see Chapter 6 and below), low long bone end to long bone shaft ratios, and a positive correlation of frequencies of long bone ends with bone structural density. When this combination of attributes is found, the analyst can determine the MNE of complete long bones on the basis of long bone shaft pieces (as well as overlapping long bone ends; see Bunn 1991; Bunn and Kroll 1986, 1988), and plot those values against a utility index such as that of Metcalfe and Jones (1988) for complete limb bones. That is, simply shift the inclusiveness of the counting unit, in this case from proximal and distal halves to complete long bones. Such a shift in scale for counting the bones requires, of course, a similar shift in the scale of the anatomical units in the utility index.

As an example of the preceding, we can use the bone assemblage recovered from the Plio-Pleistocene FLK *Zinjanthropus* site (Bunn 1986, 1991; Bunn and Kroll 1986, 1988). Relevant data are summarized in Table 7.13. Marean and Spencer (1991:655) interpret the ratios of hindlimb shafts to ends for this assemblage as "strongly suggest[ive of] intensive, perhaps 100 percent, carnivore ravaging." The MNE of long bone ends correlates strongly with the maximum structural density per skeletal part ($r_s = 0.81$, $P = 0.007$), also suggesting density-mediated destruction has influenced the bone frequencies in this assemblage. The MNE values as determined only from long bone ends are negatively (if insignificantly) correlated with the caribou MGUI ($r_s = -0.28$, $P = 0.35$). Thus it is not surprising that the frequencies of skeletal parts as determined from the long bone ends describes a negative relation when plotted

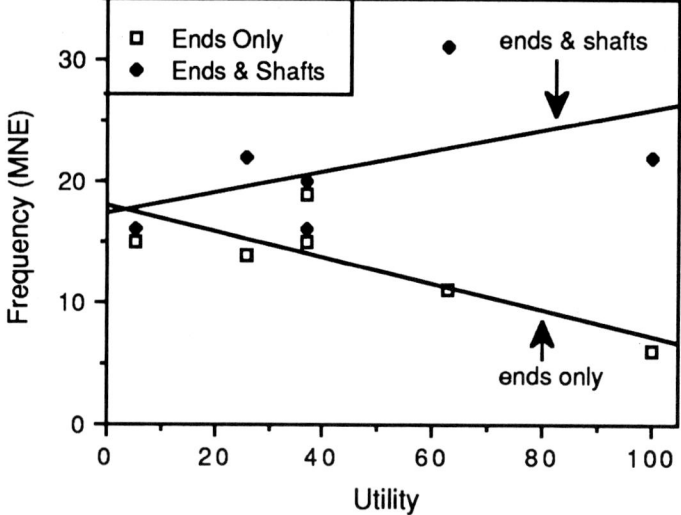

Figure 7.16. Scatterplots of MNE frequencies derived from long bone ends only, and from long bone ends and long bone shafts, for FLK *Zinjanthropus* assemblage (after Bunn 1986; Bunn and Kroll 1986, 1988), against food utility index (from Metcalfe and Jones 1988). Best fit simple regression lines are shown to emphasize the differences in the two sets of plotted points (compare with Figure 7.1).

against Metcalfe and Jones' (1988) complete bone utility index (Figure 7.16). But, including the shaft pieces along with the ends to derive MNE values results in skeletal part frequencies that describe a positive relation when plotted against that utility index (Figure 7.16). In this case, then, using complete bone MNE values rather than MNE values derived from only long bone ends seems to get us out of the interpretive dead end of equifinality. It appears that the faunal remains at the FLK *Zinjanthropus* site owe their presence there at least in part to their economic utility because bones of higher utility are more frequent than bones of lower utility.

Regardless of the correctness of the preceding interpretation, it is important to return to the notion that long bone shaft to long bone end ratios monitor the intensity of carnivore gnawing. Marean and Spencer's (1991) experimental data are compelling. For example, the maximum MNE for femurs based on ends is 9.95 but the femur MNE based on middle shafts is 50.47; the maximum MNE for tibiae based on ends is 25.2 but based on middle shafts it is 50.22; the maximum MNE for metatarsals based on ends is 26.48 but based on middle shafts it is 45.89. The ratios of end MAUs to shaft MAUs are 5.07, 1.99, and 1.73, respectively for femurs, tibiae, and metatarsals. These are impressive indications of the intensity of carnivore ravaging when it is realized that these ratios would be 1 if the entire carcass (total, if fragmented, bones) was present and unravaged. It is important to realize, however, that the significance of the ratios resides in the fact that entire bones (end to shaft ratios = 1) were given to the hyenas by the researchers. That is, the hyenas had the opportunity to eat *all* of *each* bone. The question then is, do the ratios reflect what was transported and given to the hyenas (what the hyenas had access to), or does it reflect what they did not consume? Would the ratios be as high if twice as many shafts as ends (end to shaft ratio = 0.5) had been given to the hyenas? This is *the* critical issue in the conflation of the effects of differential transport with differential destruction on skeletal part frequencies. Does the presence of a particular skeletal portion influence how carnivores exploit other kinds of skeletal portions in an assemblage? While Marean *et al.* (1992) apparently seem to think the answer to this question is negative (they must assume, if implicitly, that in prehistoric cases, as with studies of differential transport assuming complete skeletons were initially present, complete bones were initially present), there is no sound basis to think they are correct, and anecdotal data suggest they may be incorrect (e.g., Haynes 1980b, 1982). Additional experimental work is called for here. And, recall that while I suggested earlier that it seemed unlikely that human butchers would transport proximal ends, proximal shafts, mid-shafts, distal shafts, and distal ends as separate pieces and in different abundances, we do not know this for a fact.

Marean and Spencer (1991) found that of the 50 hindlimbs of domestic sheep they fed to spotted hyenas, an average of 70% (± 18%) of the ends but only 7% (± 4%) of the shafts were destroyed (averages of 12 measurements for bone ends, and 4 measurements for shafts). They are thus certainly correct that the frequencies of diaphysis fragments tallied as MNE values per complete limb bone will provide more accurate measures of the frequencies of those skeletal elements in an assemblage than MNE values based on frequencies of long bone ends. As noted in conjunction with earlier discussion of Table 7.12 and Figures 7.14 and 7.15, the %survivorship values for Marean and Spencer's experimentally generated assemblage are nearly significantly correlated with bone density when all skeletal parts (ends, proximal and distal shafts, mid shafts) are included ($r_s \geq 0.48$, $P \leq 0.07$). Omitting the epiphyseal ends and retaining only the proximal and distal shaft, and mid-shaft skeletal parts, the correlation disappears (for sheep bone density $r_s = 0.24$, $P = 0.5$; for deer bone density $r_s = 0.31$, $P = 0.38$). These results suggest the utility graph in Figure 7.15a is a function of density-mediated destruction whereas the utility graph in Figure 7.15b is not a function of density-mediated destruction (recall that in the previous section it was emphasized that such conclusions should be founded on more than just the correlation coefficient).

The example just reviewed is an important one. It indicates one way we might begin to disentangle the effects of density-mediated destruction from differential transport of skeletal parts. However, it is by no means going to work in all cases. Stiner (1991b, 1991e), for instance, claims that MNE counts based on long bone shafts may not always produce higher frequencies than MNE counts based on long bone ends. There are two ways to evaluate this claim indirectly, with an example, and with experimental data.

The skeletal parts in the 45CH302 bone assemblage (Figure 7.12, Table 7.11) were tallied by counting the MNE per long bone half; that is, the minimum number of proximal halves of humeri necessary to account for proximal ends, proximal shafts, and proximal-mid shafts of humeri were accounted for in deriving the MNE and thus the %survivorship of proximal humeri. Yet, that assemblage is significantly correlated with bone density ($r_s = 0.67$, $P < 0.001$). If only frequencies of proximal and distal halves of humeri, radii, metacarpals, femora, tibiae, and metatarsals are considered, and the maximum density values for the proximal and distal halves of these bones are used, the correlation is even stronger ($r_s = 0.81$, $P = 0.008$). Thus it is clear that the 45CH302 assemblage has been subjected to density-mediated attrition and a utility graph cannot be derived even from the frequencies of shafts of limb bones in that assemblage. This example does not directly substantiate Stiner's (1991b, 1991e) claim, but it does indicate MAU frequencies based on specimens of long bone shafts may sometimes correlate with bone density.

The experimental data with which to evaluate Stiner's (1991b) claim that MNE values based on long bone shafts and long bone ends will not always outnumber MNE values based only on long bone ends consist of shaft fragment to end fragment ratios. Bunn (1989) used a hammerstone and an anvil to break cow (*Bos* sp.) limb bones, and derived ratios of *identifiable* shaft fragments (specimens that could be identified to skeletal element represented) to identifiable end fragments. Bunn's (1989) ratios are NISP of shafts to MNE of each element, and are 2.6 for the humerus, 2.6 for the radius, 2.5 for the femur, and 4.7 for the tibia. Those ratios are higher if the shaft fragments that would be archaeologically unidentifiable are included: 5.1 for the humerus, 4.6 for the radius, 4.5 for the femur, and 6.8 for the tibia. Bunn's data can also be used to calculate NISP of shafts to NISP of ends ratios. For identifiable shafts those ratios are 1.3 for the humerus, 0.9 for the radius, 1.2 for the femur, and 2.2 for the tibia. For all shaft fragments, those ratios are 2.6 for the humerus, 1.6 for the radius, 2.2 for the femur, and 3.2 for the tibia. Binford (1978:155) recorded the average number of "splinters and chips" of long bone shafts per articular end produced by Nunamiut Eskimos fracturing caribou (*Rangifer tarandus*) long bones to extract marrow. The average NISP of shafts to MNE of ends ratios Binford (1978) reports are 17.5 for the humerus, 12.2 for the radius, 8.2 for the metacarpal, 16.6 for the femur, 19.7 for the tibia, and 6.6 for the metatarsal. Binford's (1978) ratios are about three times greater than Bunn's (1989) for the same variables (NISP of shafts to MNE of ends). That difference is perhaps attributable to taxonomic variation.

The point of Binford's (1978:156) ratios is that "if we observe more splinters and chips than we estimate [should be found given a certain number of ends], then we have reason to suspect that some articular ends were destroyed beyond recognition or some articular ends were removed from the site after the bones had been broken for marrow." That is, Binford (1978) suggests we can use those ratios to estimate the number of articular ends that should be present, a procedure he puts to use in a later paper (Binford 1988). The point of Bunn's (1989) ratios is to help substantiate his contention that determining the minimum number of bones represented by both long bone shafts and long bone ends will usually result in greater and thus more accurate MNE values than when just long bone end pieces are used (e.g., Bunn 1986, 1989; Bunn and Kroll 1986, 1988).

Both the ratios described by Bunn (1989) and those described by Binford (1978) indicate more shaft specimens (NISP) than end specimens (NISP) will result solely from hammerstone breakage of ungulate long bones. Removal of end pieces by carnivores will increase the ratio values. In the absence of carnivore ravaging or other taphonomic processes which selectively remove long bone ends, the ratios of pieces should be like those documented by Bunn and Binford, and, importantly, the ratios of MNE values based on long bone ends to MNE values based on long bone shafts should be approximately one. Thus it seems that each sample must be evaluated as to whether higher skeletal counts are in fact obtained using diaphysis-based MNEs or epiphysis-based MNEs.

I do not mean in the preceding to suggest that ratios of long bone shafts to long bone ends are not valuable analytical tools. Certainly they are. What I do mean to imply, however, is the fact that even these ratios (or, my frequencies of scan sites) may not be sufficient for analytically disentangling the impacts of differential transport and differential, density-mediated destruction on a set of prehistoric bones, and thus may be insufficient for gaining a taphonomic understanding of varying skeletal part frequencies. Other kinds of data are necessary to supplement the frequencies of scan sites represented, and shaft to end ratios.

Toward a solution 2: other attributes

Marean *et al.* (1992:117) suggest that sorting out the influences of density-mediated destruction from the influences of economically founded decisions of skeletal portion transport requires data on "the incidence of hammerstone percussion and tooth marking on bones in tandem with analyses of body part representation." That suggestion is founded on a number of actualistic experiments (Blumenschine 1988; Blumenschine and Selvaggio 1988, 1991; Marean 1991; Marean and Spencer 1991; Marean *et al.* 1992). While in most cases these experiments focus on the effects of spotted hyena (*Crocuta crocuta*) and concern the impacts of carnivore ravaging on bones of domestic sheep (*Ovis aries*), their results have broad implications for study of skeletal part frequencies.

In an innovative study of the effects of carnivores on skeletal part frequencies, Blumenschine (1988) created three kinds of bone assemblage. One, called "carnivore only" assemblages, consisted of bones of bovid carcasses killed, or scavenged, and consumed by large African carnivores. The second kind, called "hammerstone only" assemblages, involved fresh, defleshed bovid bones broken with a hammerstone while the bone lay on a stone anvil. Some of these assemblages were not further subjected to taphonomic processes, but others of them were placed on the landscape immediately following breakage and removal of marrow to produce the third kind of assemblage, called "simulated site" assemblages. The simulated site assemblages were scavenged by carnivores, mainly spotted hyenas, and the bones later collected for study.

Comparisons between the three kinds of assemblage reveal significant differences in attributes that might be used to help unravel the taphonomic processes we have been discussing. First, skeletal parts in the simulated site assemblages have significantly fewer tooth marks (8% to 45% of the specimens per assemblage display such marks) than the skeletal parts in the carnivore only assemblages (66% to 100%). Second, most epiphyseal ends of long bones in both the simulated site assemblages and in the carnivore only assemblages were consumed by carnivores, and thus the mere absence of epiphyses does not allow the analyst to distinguish bone assemblages ravaged only by carnivores from assemblages initially subjected to marrow extraction and subsequently to carnivore ravaging. Third, proportions of long bone end pieces with carnivore tooth marks are high, long-bone near-end pieces are mid-level in frequency, and proportions of long-bone mid-shaft pieces are low in frequency for both the simulated site and the carnivore only assemblages (Blumenschine 1988).

Blumenschine's (1988) experiments indicate the proportion of carnivore gnawed bone pieces is high in assemblages that did not have marrow removed and the bones were unbroken when carnivores encountered them (carnivore only), and low in assemblages in which the marrow had been removed and the bones had been broken prior to carnivore action. Thus, the proportional frequency of carnivore-gnawed or tooth-marked specimens in an assemblage may help indicate the pre-gnawing condition of bones processed by carnivores and, whether or not people had access to the bones prior to the carnivores. Further, Blumenschine and Selvaggio (1991) indicate that long bones in the simulated site assemblages tend to have many hammerstone percussion marks (see Chapter 8) located in the mid-shaft area, while tooth marks tend to occur on the epiphyses and near-epiphysis portions of the long bones.

While the presence of hammerstone or percussion marks on bones signifies a hominid taphonomic agent, the presence of a carnivore tooth mark similarly signifies a (quadrupedal) carnivorous taphonomic agent. Tooth marks on many of the epiphyseal or near-epiphyseal portions of long bones in conjunction with percussion marks on mid-shaft parts would suggest complete bones had been processed by humans and subsequently ravaged by carnivores. As with the ratio of long bone shafts to long bone ends, then, the frequency and distribution of tooth marks and percussion marks may help identify the taphonomic history of a bone assemblage, and thus help unravel the taphonomic meaning of varying skeletal part frequencies. Carnivore tooth marks, however, may not always be present on bones even if carnivores had unhindered access to them.

Kent (1981) shows that dogs (*Canis familiaris*), at least, can gnaw bones and yet leave no visible gnawing marks (importantly, she also shows that dogs redistribute bones; that is, dogs often move bones from their original depositional loci in sites). Haynes (1983a:171) shows that different carnivore taxa gnaw bones in distinct ways, but notes that whether an assemblage of bones displays gnawing damage or not depends on several variables that "affect an individual predator's behavior." Further, the presence of gnawing marks on bones does not necessarily mean the observed frequencies of bones are the result of carnivore-generated attrition. To illustrate this, it is necessary to assume that as a bone specimen becomes smaller, through fragmentation or attritional processes that remove or destroy bone tissue, that specimen becomes less identifiable as to the taxon represented and the skeletal element represented. Several studies indicate this is a reasonable assumption (Lyman and O'Brien 1987; Watson 1972).

As bone specimens become smaller and less identifiable, progressively more of them become "analytically absent" (Lyman and O'Brien 1987); that is, proximal humeri may be present in a collection but some of the pieces of proximal humeri are so small or modified by attrition that they are unidentifiable, and thus they are, for analytical (quantification) purposes, absent. The frequency of such analytically absent pieces should increase through time as taphonomic processes continue to affect a bone assemblage. This underscores the time-transgressive or cumulative nature of taphonomic processes; it also implies that there is a threshold at which a particular skeletal part will cease to be analytically present (identifiable). This is important in our consideration of the presence of carnivore gnawing marks on bones and the frequency of gnawed bones suggesting a bone assemblage has undergone density-mediated attrition because the extent of carnivore gnawing damage is cumulative.

Garvin (1987) fed domestic cow (*Bos taurus*) bones to dogs, and documented the amount of bone part weight loss over time due to gnawing. He fed the dogs defleshed, fresh proximal and distal halves of humeri, radii, femora, and tibiae (distal radii had carpals attached, and distal tibiae had tarsals attached). He recorded the percentage of total weight that was lost per specimen every two days over a 22 day period. I presume that any weight loss denotes the loss of bone material and thus the production of identifiable gnawing marks. Given that attrition of bones generated by carnivore gnawing seems to be mediated by the structural density of the bones, one can predict that skeletal parts with low structural densities will lose weight more rapidly than skeletal parts with high

Table 7.14 *Correlation coefficients (r_s) between percent weight loss of skeletal parts due to carnivore gnawing over time (data from Garvin 1987), and different measures of bone structural density. See also Figure 7.17*

Number of days gnawed		Bison density	Maximum bison density	Deer density	Maximum deer density
2	$r_s =$	0.02	0.06	−0.29	−0.46
	$P =$	0.91	0.85	0.45	0.23
4		−0.13	−0.17	−0.33	−0.48
		0.73	0.66	0.38	0.20
6		−0.20	−0.26	−0.46	−0.66
		0.59	0.49	0.23	0.08
8, 10, 12, 14		−0.54	−0.52	−0.60	−0.74
		0.15	0.16	0.11	0.05
16		−0.56	−0.57	−0.55	−0.81
		0.13	0.13	0.14	0.03
18, 20		−0.67	−0.62	−0.69	−0.88
		0.07	0.10	0.06	0.02
22		−0.78	−0.74	−0.71	−0.86
		0.04	0.05	0.06	0.02

structural densities. Thus, as more weight (more bone material) is lost through time (as dogs have progressively more days to gnaw on bone parts), the percent weight loss recorded by Garvin (1987) should be more strongly and negatively correlated with bone structural density. This is precisely what is seen in the correlation coefficients between percent weight loss and the bone density values (Table 7.14).

The percent weight loss over the first two weeks is not correlated with bone density (Table 7.14). This indicates that the mere presence of gnawing marks on bone parts does not necessarily mean carnivore-generated density-mediated attrition has produced the observed bone frequencies. The coefficients indicate that three weeks' worth of gnawing is necessary before such attrition will produce a statistically significant relation between bone frequencies and the structural density of bone parts. This time-transgressive or cumulative effect of carnivore gnawing is density mediated, as shown in Figure 7.17, where it is clear that bone material was lost more rapidly (and apparently was still being lost after three weeks) from the low-density proximal humerus than from the high-density distal humerus over the three-week gnawing period. Thus the mere frequency of gnawing marks may be misleading, by itself, as an indicator of carnivore-generated density-mediated attrition. The analyst may want to determine if the proportional (%) frequency of skeletal parts displaying gnawing marks varies directly with the structural density of the parts. That is, are more of the skeletal parts with low density (and high marrow and grease

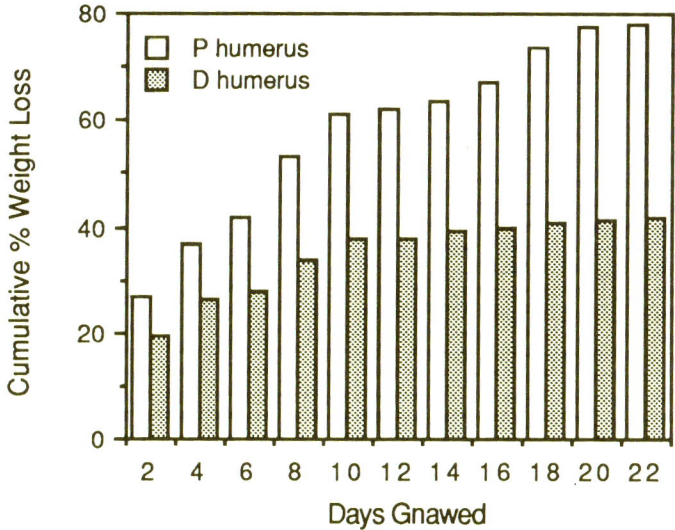

Figure 7.17. Bar graph of %weight loss of cow bones over time. Note that the low structural density proximal humerus lost more weight more quickly than the high structural density distal humerus over the first 10 days, and then continued to lose weight while the distal humerus lost almost no weight over the last 12 days.

content) gnawed than parts with high density (and low marrow and grease content)? If so, one would then have additional evidence suggesting that carnivores had in fact destroyed some long bone ends.

People break animal bones, sometimes intensively, in order to extract grease and marrow (see the discussion of "Within-bone nutrients" below). Thus, one might ask, if high frequencies of percussion marks indicate intensive fragmentation of bones by humans (e.g., Blumenschine 1988; Marean 1991), do they also indicate the analytical absence of certain skeletal parts due to the small size of the fragments? White (1956) suggested fracturing of bones during butchering may smash bones beyond recognition, and Binford (1981b) agreed that this could certainly happen (how Binford suggests we deal with an assemblage of bones that has undergone density-mediated attrition is described later). But neither of these authors nor any other taphonomist has attempted to find out how the frequency of percussion marks on bones might covary with density-mediated destruction. One might ask if there is a relation between how bones break and the structural density of bones. Again, while we seem to have learned a lot about how bones break, and why they break the way they do, in the past decade, no one has attempted to find a correlation between these two variables (bone fragmentation is dealt with in Chapter 8). Some have suggested selection of particular bones for tool material might result in the removal of bones with high structural densities (e.g., MacGregor 1985). But again, no detailed study of the covariation in utilized bone tool material and the structural density of utilized bones has been performed.

Stiner (1990a, 1991b) summarizes a set of attributes which she believes can be used to determine if the dominate biological cause of a bone assemblage was a non-human carnivore (especially a hyena), or a hominid. First, assemblages of faunal remains that are dominated by head parts (tallied with isolated teeth excluded) are produced by carnivores whereas hominid-created bone assemblages have approximately equal frequencies of all skeletal parts (are not head-dominated) and the animal carcasses are all more or less anatomically complete. Second, carnivores take more old adult prey as these tend to be scavenged as well as hunted; hominids tend to practice little selection with regard to age of prey (ambush hunting) or focus on prime-aged individuals (see Chapter 5). Third, bone assemblages resulting from carnivore activity display more evidence of gnawing than assemblages resulting from human activity. Fourth, "carnivore latrines" are associated with bone assemblages produced by carnivores but not with bone assemblages produced by hominids.

To the last of Stiner's (1991b) attributes we might add such things as carnivore coprolites in association with carnivore accumulations but not with bone accumulations created by hominids (e.g., Klein 1975), and more evidence of corrosive damage to bones that have passed through carnivore digestive tracts in the former assemblages than the latter (see Chapter 6). Many of the attributes summarized by Stiner (1991b) are precisely those that have long been in use by taphonomists to discern the taphonomic agent responsible for a bone accumulation (Chapter 6). What is important about these attributes in the context of explaining bone frequencies is that they bring to bear other lines of evidence to help explain why a particular skeletal part profile appears the way it does, lines of evidence that are to some extent independent of the structural density and utility of skeletal parts. Therefore, skeletal part frequencies that fall in Class 1 of Figure 7.13 may prove to be inexplicable given only the correlation coefficients between bone frequencies, and structural density and food utility, but with reference to other lines of evidence, such as demography of mortality, frequency of head parts, and frequencies of coprolites, gnawing damage, and digestive corrosion, those frequencies may in some instances be explicable.

Lyman *et al.* (1992a) perhaps said it well when they referred to the cliché that a statistical correlation does not necessarily denote a causal relation between two variables. After finding a significant correlation between the frequencies of marmot (*Marmota* sp.) skeletal parts and the structural density of those parts, Lyman *et al.* (1992a) went on to argue that differential recovery of small skeletal parts did not seem to be contributing to the correlation because small skeletal parts were excluded from the frequencies. Further, evidence of carnivore gnawing on some of the bones suggested that the potential existed for bone destruction due to such a taphonomic process. Finally, bone fragmentation did not seem to be contributing to the significance of the correlation because some of the marmot bones were complete. That is, they, like Stiner, used multiple lines of evidence to derive a conclusion. This returns us to Binford's (1987:453) statement quoted earlier: indices of utility, transport, and structural density of bone parts are "frames of reference," and as such their use as *explanatory algorithms* for skeletal part profiles is not advised; rather, they should be used as *one* of several steps in building a taphonomic explanation for those profiles.

Thus far in this chapter some of the simpler initial steps in explaining the frequencies of skeletal parts have been considered. There are, as the preceding two paragraphs imply, more complex issues involved as well. In the next section, several of these are reviewed.

Within-bone nutrients

Marshall and Pilgram (1991) suggest that human extraction of nutrients within bones, particularly marrow, may have a significant influence on skeletal part profiles. They propose that bones that rank high in within-bone nutrient utility will tend to be more fragmented than bones that rank low in within-bone nutrient utility. The initial effect of fragmentation during, say, marrow extraction, will thus be initially to raise the NISP per skeletal element. As fragmentation continues, however, some fragments will become so small as to become unidentifiable, and thus the later effect of fragmentation will be to reduce NISP values. Given that MNI is often a statistical function of NISP (Grayson 1984), the ultimate effect of fragmentation will be to reduce MNI values. Similarly, intensive fragmentation will reduce the observed MNE (and thus MAU) values for an assemblage (see Chapter 8).

Referred to as "analytical absence" by Lyman and O'Brien (1987), it is not at all clear how we might analytically overcome the effect of fragmentation and reduced identifiability of bone specimens (but see the discussion of refitting in Chapter 5). Probably the most important issue here, from a taphonomic perspective, is the attempt to identify such effects, and to measure their magnitude. This is precisely what Marshall and Pilgram (1991) do when they compare the degree of fragmentation of remains of two taxa. They measure fragmentation as the percentage of identified specimens that are complete for each skeletal element. Another way to measure fragmentation is to calculate the NISP:MNE ratio per skeletal element. Similarly, one could calculate an NISP:MNI ratio, to compare across taxa, if the same skeletal elements are used for all taxa (see Chapter 8 for more on fragmentation).

How would such ratios be used to assess the extent of or intertaxonomic variation in the exploitation of within-bone nutrients such as marrow and grease? One could plot the NISP:MNE ratios for marrow-containing bones against a marrow utility index, predicting that skeletal elements with high

Table 7.15 *NISP:MNE ratios for selected skeletal parts (from Marshall and Pilgram 1991)*

Skeletal part	Caprine ratio	Cattle ratio
P humerus	2.7	4.3
D humerus	2.9	3.5
P radius	3.5	3.3
D radius	2.2	4.8
P femur	2.2	2.2
D femur	3.9	6.3
P tibia	3.0	2.9
D tibia	2.7	6.7

marrow indices would be more broken and thus have higher NISP:MNE ratios than skeletal elements with low marrow indices. For example, the NISP:MNE ratios for caprine (sheep and goats) and cattle remains from an African Neolithic site are given in Table 7.15, and are plotted against the sheep marrow utility index (Table 7.1) and the fat utility index for bison (Table 7.4) in Figure 7.18, respectively. Neither pair of variables is significantly correlated (for caprines, $r_s = -0.13$, $P = 0.72$; for cattle, $r_s = -0.18$, $P = 0.64$), but both point scatters seem to define inverse relations, and that relation appears to be steeper for the cattle remains than for the caprine remains (recall that utility indices are ordinal scale, as are archaeological bone frequencies, so the appearance of steepness may be more apparent than real). But the cattle remains seem to be more fragmented than the caprine remains; the average NISP:MNE ratio for the former is 4.25 (SD = 1.6) and for the latter it is 2.9 (SD = 0.6). Recall that fragmentation has the initial effect of increasing NISP values, but as pieces are broken into progressively smaller pieces the second effect is to reduce NISP values. Figure 7.18 could reflect many of the cattle bones with high fat and marrow content being broken beyond recognition. Caprine bones may have been less intensively broken than cattle bones because the former have less marrow and fat content. This explains the steeper slope of the best-fit regression line for cattle and the nearly horizontal slope of the best-fit line for caprines (and thus the steepness may be real as well as apparent).

Regardless of whether or not the preceding is correct, it illustrates how the exploitation of within-bone nutrients might influence bone frequencies. It underscores the fact that fragmentation can produce an *analytical absence* of skeletal parts. Within-bone nutrients, such as marrow, grease, and the fat associated with neurological tissues of the axial skeleton (Stiner 1991b), represent critical variables that have not been intensively studied. If compared to such variables as NISP:MNE ratios, the proportional frequency of each skeletal unit displaying gnawing marks and/or percussion marks, and the like, within-bone nutrient indices may reveal details of the taphonomic history of a bone assemblage. Ultimately, like the frequencies of gnawing marks and long

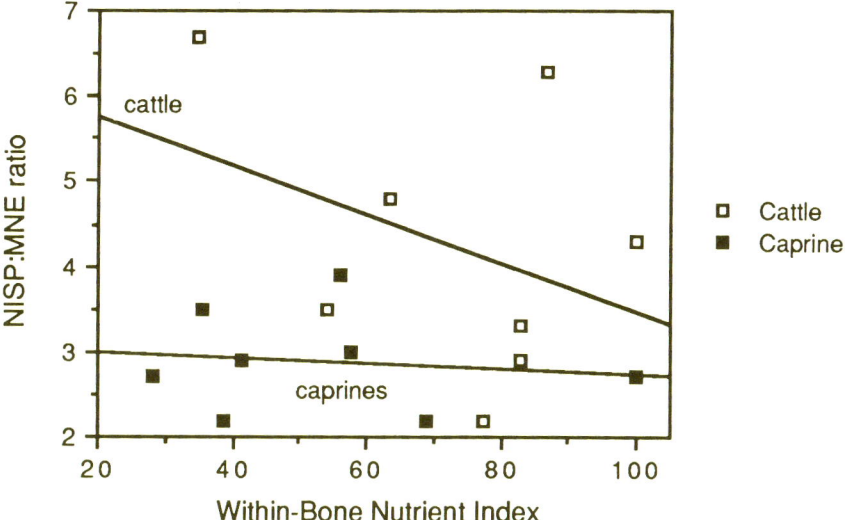

Figure 7.18. NISP:MNE ratios plotted against within-bone nutrient index for two taxa. Simple best-fit regression lines shown for reference. See text for discussion.

bone shaft MNE to long bone end MNE ratios, they may help us explain why some skeletal parts are relatively rare in some assemblages.

Reconstruction of ravaged assemblages

> We wish to recognize whether we are dealing with an assemblage that has been transported or with a residual population, that is, what remains after other parts are transported. Complicating the matter may be the presence and/or absence of destruction coupled with transport as well as different levels of destruction in different settings. How do we begin unraveling such a complicated set of possible conditions?
> (L. R. Binford 1981b:217)

Thus far, I have covered several analytic techniques and frames of reference developed to help explain varying skeletal part frequencies. Once we know that a particular bone assemblage has undergone carnivore ravaging or density-mediated destruction, can we still discuss differential transport of skeletal parts by prehistoric hominids? Can we perhaps reconstruct, for instance, the pre-ravaging assemblage? If we can answer this second question in the affirmative, then we can also answer the first question in the affirmative. It is with the second question that this section is concerned.

In a seminal contribution to the method and theory of modern zooarchaeology, Uerpmann (1973:319) urged zooarchaeologists to "develop models of bone disintegration so that the missing material can be determined and calculated scientifically." He suggested that the attritional processes acting on faunal remains "conform to certain laws which could be understood by the application of statistical methods [and eventually the analyst would] be able to calculate the missing quantity" of bones that had been subjected to those attritional processes (Uerpmann 1973:319). One of the principle criticisms of measuring abundances of faunal remains at the time Uerpmann wrote involved the effects of differential preservation and the fact that such preservation might be differentially distributed across taxa (see the review in Grayson 1984). If the bones of taxon "1" were well preserved but the bones of taxon "2" were poorly preserved, an inaccurate assessment of the relative abundances of the two would result from simple counts of bones and probably as well from minimum numbers of individuals. If a statistical model like that envisioned by Uerpmann could be developed, zooarchaeologists would have a powerful analytic tool allowing them to estimate more accurately relative taxonomic abundances.

Various researchers had been searching for regularities that might constitute such a model prior to the publication of Uerpmann's paper (e.g., Brain 1967b, 1969), and many continued to examine those regularities of bone attrition after 1973 (see references in Lyman 1984a). When, half a decade after Uerpmann's (1973) plea, a formal model of bone attrition allowing the "missing quantity" of bones to be estimated was constructed, that model was geared toward reconstructing the abundances of particular skeletal elements of a single taxon. The model is based on a polynomial equation derived by Binford and Bertram (1977:138) from an analysis of the statistical relation between the structural density of bone parts and the frequencies of bone parts in an assemblage known to have been ravaged by carnivores. The model has been used to estimate the "missing quantity" of bones by Binford (1978, 1981b, 1984b) several times. It has also been used by Blumenschine (1986a) to reconstruct what is thought to be a ravaged assemblage, and it has been mentioned with approval by other researchers (e.g., Marean 1991; Turner 1989).

Given the apparent logic of Uerpmann's original phrasing of the problem, and the availability of a model in the form of an easily solved equation, why is reconstruction of the missing quantity of bones not commonplace in taphonomic analysis? Perhaps, as Turner (1989:6) notes, the problem is that one must assume particular bone parts are rare due to destructive agents rather than faulty archaeological recovery. More generally, the reconstruction procedure requires the assumption that, when abundances of skeletal parts are positively correlated with the structural densities of those parts, rare skeletal parts are rare because they have low structural densities and they were thus originally present but were destroyed by density-mediated attritional agents. That is, one must assume that bones which are absent were in fact present prior to performing the reconstruction; more plainly, one is assuming precisely what one is trying to determine. In terms of ascertaining the economic utility or transport pattern (Figure 7.1), if we found a positive correlation between bone frequencies and structural density, we would be prompted to "reconstruct" the frequencies of the bones destroyed by density-mediated attrition. But to do so would *demand* that we assume those rare, low-density (high utility) bones were transported, utilized, and deposited on the site by people, only to be later destroyed. This is precisely what we want to determine; did the people actually transport and utilize those bones? I suggest this single requisite assumption is why the reconstruction technique developed by Binford and Bertram (1977) is not used by most taphonomists; no one is willing to grant the assumption.

But, for sake of discussion, let us grant the assumption. What happens next? Using structural density values they measured, Binford and Bertram (1977) derived a polynomial equation describing the statistical relation between the density values as the independent variable and %MAU frequencies in an assemblage known to have been ravaged by carnivores. They then solved the equation for each skeletal part using the structural density values to derive what they called "SP" values, or the proportional frequency of a particular bone part *expected* to survive density-mediated attrition. To reconstruct the assemblage of bone parts originally present, one divides the observed MAU per skeletal part (which, as we have seen, is the same as the survivorship value) by the SP value for that skeletal part to "yield an estimate of the MAUs originally present before the attritional agent destroyed the bones" (Binford 1978:210). That is, Binford is suggesting that when the analyst has an assemblage of prehistoric bone parts the frequencies of which are positively correlated with their structural densities, not only does it appear that the bone assemblage has undergone density-mediated attrition, but the analyst should reconstruct the assemblage to determine what was present prior to destruction. So, the question now is, what is the effect of such an analytic reconstruction on skeletal part frequencies? An example is the easiest way to answer this question.

Binford (1978:210–211) reconstructed bone assemblages created by Nuna-miut dogs using the bones recovered from "dog yards" and following the procedure outlined in the preceding paragraph. The observed frequencies of bone parts from two dog yards are given in Table 7.16, as are the reconstructed frequencies. It is important to note that the observed and the reconstructed frequencies are tightly correlated; for the Morry Dog Yard assemblage $r_s = 0.85$ ($P = 0.0001$) and for the Rulland Dog Yard assemblage $r_s = 0.94$ ($P = 0.0001$). These coefficients indicate the reconstruction has not significantly altered the relative abundances of skeletal parts in either assemblage. The final question to ask then, is, has the process of reconstruction resulted in frequencies of bone parts that would lead us to a conclusion different than that we might derive using the observed frequencies of skeletal parts? As Figure 7.19 makes clear, reconstruction would *not* lead us to a different conclusion. In that figure, both the scatterplot of observed frequencies and the scatterplot of reconstructed bone frequencies suggest a reverse utility strategy.

When Uerpmann (1973) called for the development of a statistical model

Table 7.16 *Reconstructing caribou bone assemblages from Nunamiut sites (from Binford 1978)*

Skeletal	Morry dog yard %MAU observed	Morry dog yard %MAU reconstructed	Rulland dog yard %MAU observed	Rulland dog yard %MAU reconstructed
skull	34.4	42.26	84.6	100.00
mandible	100.0	100.00	100.0	96.10
atlas	31.2	44.43	15.4	20.99
axis	31.2	56.88	15.4	26.92
cervical	14.7	35.32	15.4	36.34
thoracic	8.6	22.75	7.7	19.25
lumbar	15.5	30.88	20.0	38.09
pelvis	28.1	31.17	46.2	49.12
rib	5.2	18.94	1.6	5.65
sternum	0.0	0.00	0.0	0.00
scapula	0.0	0.00	38.4	62.31
P humerus	3.1	18.96	0.0	0.00
D humerus	12.5	20.30	23.1	35.93
P radius	18.7	39.64	23.1	46.83
D radius	25.0	48.40	30.8	57.20
carpal	7.5	22.75	7.7	22.34
P metacarpal	3.1	7.68	7.7	18.17
D metacarpal	3.1	8.36	15.4	39.57
P femur	3.1	7.51	0.0	0.00
D femur	6.3	21.05	0.0	0.00
P tibia	6.3	18.94	7.7	22.34
D tibia	9.4	12.91	15.4	20.45
tarsal	6.3	18.84	7.7	22.48
astragalus	9.4	23.04	15.4	36.34
calcaneum	6.3	15.35	15.4	36.34
P metatarsal	9.4	19.79	7.7	15.61
D metatarsal	6.3	18.94	7.7	22.48
first phalanx	2.5	12.63	1.5	7.53
second phalanx	2.5	17.52	1.5	10.36
third phalanx	2.5	25.26	1.5	14.94

that would allow the analytic reconstruction of frequencies of bones missing from an assemblage due to attritional processes, his plea made good sense. It still does. If we can build such a model, we can effectively erase much of the taphonomic overprint or bias that plagues the quantification of taxonomic abundances. Similarly, such a model, if properly constructed, would allow us to ascertain which skeletal parts were frequently transported and which parts were rarely transported in those cases when frequencies of skeletal parts seem to be at least in part the result of density-mediated attrition. The single model developed to date for the latter purpose, however, demands an assumption that seems unwarranted, the assumption that rare bones of low structural density

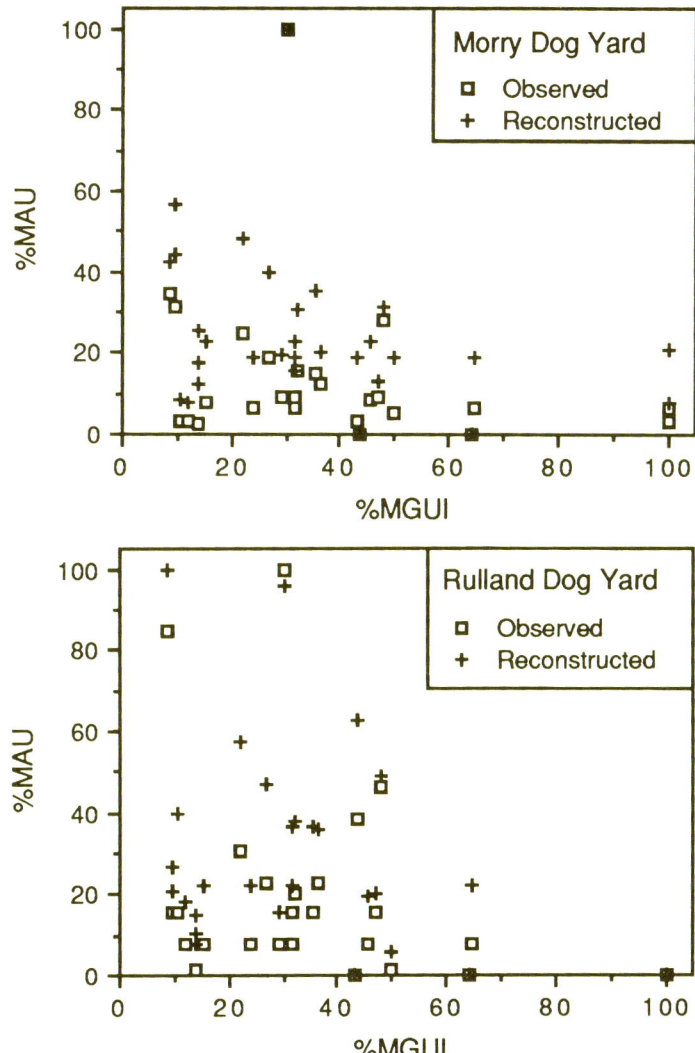

Figure 7.19. Scatterplots of caribou bone observed and reconstructed frequencies against the caribou %MGUI (data from Binford 1978).

were in fact once present in an assemblage. To grant the assumption results in presuming that what we are trying to show *might* have happened *did* in fact happen. Even granting the assumption in those cases where a strong positive correlation is found between bone frequencies and the structural density of bone parts, which are the logical cases to make such an assumption, the analytic process of reconstruction does not produce markedly different conclusions because the reconstructed frequencies are simply observed values increased proportionately to their structural density. Thus, to date, Uerpmann's (1973) plea has gone unfulfilled.

Other sources of variation in bone structural density

In his seminal study on bone density Brain (1969) notes that a bone's structural density correlates with the ontogenic age of the bone; in particular, he notes that skeletal parts with early fusing epiphyses tend to be denser than skeletal parts with late fusing epiphyses. Binford and Bertram (1977:105–109) also note a strong correlation between the ontogenic age of a bone and that bone's structural density. The significance of their observation is that they documented with empirical data a property recognized at least 15 years earlier by Guilday *et al.* (1962) that the bones of immature individuals preserve less well than the bones of mature individuals. It should also be clear that in at least some taxa, the structural density of bones not only increases as the animal matures, but it decreases as the animal approaches senility (e.g., Perzigian 1973). Thus, remains of ontogenically young and very old individuals may be rare because their bones did not preserve. As noted in Chapter 5, ontogenically young individuals tend to be abundant and ontogenically old individuals tend to be rare in living populations. The precise nature of the effects of ontogeny on the structural density of bones and the resultant effects on bone preservation have yet to be studied in the same detail as the within-skeleton variation in bone density described earlier in this chapter. Lyman (1984a), for example, selected skeletally mature (all epiphyses fused) skeletons to measure, as did Kreutzer (1992). We need, then, data on the structural density of bones from skeletally immature (e.g., epiphyses unfused) individuals.

   The data in Tables 7.6 and 7.7 make it clear that the structural density of skeletal elements varies within a single skeleton. But all scan sites thus far documented are for complete cross-sectional areas of a part of a skeletal element. Marean (1991:688) notes that, for example, the medial half of the proximal radius of small to medium-sized bovids is probably more dense than the lateral side, "presumably in response to greater body weight loading on the medial articular facet." Thus, different scan sites may eventually be measured to help better to account for fragments of the bone portions defined by the scan sites (Figures 7.4 and 7.5). Documenting the variation in density across scan site RA1 (Figure 7.4), for example, might help explain Marean's observation.

   Binford and Bertram (1977:117) also note that, given the mineral reservoir function of living bone, seasonal nutritional stress might result in the skeletal elements of animals that died when well nourished being of higher density than those same bones in an animal that died when under nutritional stress. This should be no surprise as it is clear that the nutritional quality of, say, a bovid in a temperate environment will vary with the seasons (e.g., Speth 1983), but to date few data on seasonal variation in the structural density of bones are available. Horwitz and Smith (1990) report that well-nourished domestic sheep (*Ovis aries*) ewes show no seasonal variation in the cortical thickness of the diaphysis walls of metapodials, but ewes under dietary stress display such variation. Rams show no such variation whether well nourished or seasonally under dietary stress, leading Horwitz and Smith (1990) to suggest that the added stress of gestation and lactation on ewes living under conditions of seasonally restricted food supplies can result in seasonal fluctuations in bone mass. That translates, of course, into seasonal fluctuations in the structural density of bones. This in turn suggests that a further step in research on the structural density of bones may require the generation of density data for individuals that died at different seasons of the year and under different nutritional regimes.

We have learned much in the last two decades about how the structural density of bones mediates and buffers the effects of taphonomic processes and agents. But we still have much more to learn. Why, for example, do bones of small mammals apparently not preserve well compared to the bones of large mammals (e.g., Behrensmeyer *et al.* 1979; see Chapter 9)? Is it because the surface to volume ratio of bones is greater in small mammals? It does not seem to be because of variation in the structural density of bones, as Lyman *et al.* (1992a) report that most scan sites for marmots are denser than their homologues in deer, and thus one might expect on this basis alone that bones of small mammals will withstand density-mediated taphonomic processes better than bones of large mammals. The structural density of skeletal parts is an important frame of reference that warrants further study, but it is not the only one. Covariation of the structural density of bones with other structural and morphometric properties should be studied simultaneously with our continued learning about structural density.

A final comment

Regardless of the quantitative unit used to tally frequencies of bone specimens, it is critical to remember that the *reason* for tallying bone frequencies and constructing skeletal part profiles is to study, and hopefully to explain, the differences and similarities between the archaeologically observed skeletal part frequencies and the frequencies of skeletal parts in a set of complete skeletons. The success of Marean *et al.*'s (1992) discussion of the impacts of carnivore ravaging on bone survivorship and thus skeletal part frequencies, for instance, resides in the fact that they *knew* what the original, pre-ravaging skeletal part profile looked like. Thus, they could measure exactly the proportion of bone parts remaining after ravaging. Archaeologically, we never know what was originally present, or what the pre-ravaging skeletal part profile looked like, and thus we cannot measure the proportion lost due to carnivore ravaging. We are forced to start with the model of a complete skeleton (or set of MNI complete skeletons).

The significance of the preceding can be made clear by further consideration of Marean *et al.*'s (1992) data. Analyzing those data (Table 7.17) just as one

Table 7.17 *Experimental data for bone transport and survivorship (from Marean et al. 1992), and how those data would be treated in an archaeological context. SFUI from Metcalfe and Jones (1988). See text for discussion*

Skeletal part	Marean et al.'s experimental data		Experimental data treated as if recovered from an archaeological context				Maximum Structural Density (deer)
	Original MNE	Surviving MNE	Original MAU	Surviving MAU	%MAU Surviving	SFUI	
mandible	0.0	0.0	0.0	0.0	0.0	12	0.61
atlas	6.0	2.0	6.0	2.0	8.0	10	0.26
axis	6.0	1.0	6.0	1.0	4.0	10	0.16
cervical	30.0	2.0	6.0	0.4	1.6	37	0.19
thoracic	18.0	0.0	6.0	0.0	0.0	47	0.27
lumbar	56.0	0.0	28.0	0.0	0.0	33	0.30
pelvis	49.0	25.25	25.5	12.6	50.0	49	0.49
rib	36.0	1.0	1.4	0.04	0.2	52	0.40
sternabra	0.0	0.0	0.0	0.0	0.0	67	0.22
sacrum	140.0	5.0	28.0	1.0	4.0	49	0.19
scapula	0.0	0.0	0.0	0.0	0.0	45	0.49
humerus	0.0	0.0	0.0	0.0	0.0	37	0.53
radius	0.0	0.0	0.0	0.0	0.0	26	0.68
metacarpal	0.0	0.0	0.0	0.0	0.0	5	0.72
femur	50.0	50.0	25.0	25.0	100.0	100	0.57
tibia	50.0	50.0	25.0	25.0	100.0	63	0.74
astragalus	50.0	18.0	25.0	9.0	36.0	63	0.61
metatarsal	50.0	45.9	25.0	22.9	91.6	37	0.74
first phalanx	100.0	4.0	25.0	0.5	2.0	19	0.57

would if they had been derived from an archaeological context underscores some of the problems discussed thus far. First, Marean *et al.* (1992) did *not* begin with a set of complete skeletons, as evidenced by the MAU values one can derive from their MNE values for the pre-ravaged assemblage. The pre-ravaging MAU values range from 0 to 28, and even disregarding the skeletal parts known to have pre-ravaging values of 0, the range is 1.4 to 28. In an archaeological case one might assume that all of the MAU values started as the same, maximum observed (MNI-based) value, or, one might assume that skeletal parts not present archaeologically (such as mandibles in this case) were not present to be ravaged and so should not be considered during analysis. Unfortunately, the literature is silent on which alternative is the correct one, and it is also rather quiet about which assumption is made by particular analysts in particular situations. (I have, in compiling the data summarized in Figure 7.13, omitted from analysis those skeletal parts not tallied or discussed by the original investigators, and used only those zero values suggested to be accurate by the original investigators.) Second, if the original, pre-ravaging MAU values are treated as representing what was transported to a site, those values do not correlate with the standardized food utility index for complete bones (SFUI; Metcalfe and Jones 1988) ($r_s = 0.28$, $P = 0.23$) nor do they correlate with the maximum structural density value recorded for each skeletal element ($r_s = -0.08$, $P = 0.72$). Even omitting the MAU values known to be initially zero does not result in significant statistical relations between the pre-ravaging MAU values and the SFUI or bone structural density ($r_s = 0.19$, $P = 0.52$; and, $r_s = 0.28$, $P = 0.34$, respectively). This is good because it indicates that there is little chance that the post-ravaging bone frequencies will be correlated with bone density or the SFUI simply because the pre-ravaging bone frequencies were correlated with those variables.

The third thing revealed by Marean *et al.*'s (1992) experimental data is that the post-ravaging %MAU values, when treated as an archaeological collection might be (zero values included in statistical analysis), do *not* correlate with bone structural density ($r_s = 0.16$, $P = 0.5$), nor do those values correlate with the SFUI ($r_s = 0.36$, $P = 0.13$). If we omit all skeletal parts for which the surviving (post-ravaging) %MAU is zero, then a significant correlation is found between %MAU and bone structural density ($r_s = 0.65$, $P = 0.04$), and a significant but slightly weaker correlation is found when only those MAU values *known* to be zero in the pre-ravaging assemblage are omitted ($r_s = 0.58$, $P = 0.04$). If all of the post-ravaging %MAU values equal to zero are omitted, as might happen during analysis of an archaeological collection, the remaining bone frequencies are not correlated with the SFUI ($r_s = 0.45$, $P = 0.15$), although the scatterplot of these values has the appearance of an unbiased utility graph (Figure 7.20). The point here simply is that if Marean *et al.*'s data had been derived from an archaeological context, they would not know which (post-ravaging) MAU values equal to zero should be omitted (those skeletal parts were never present

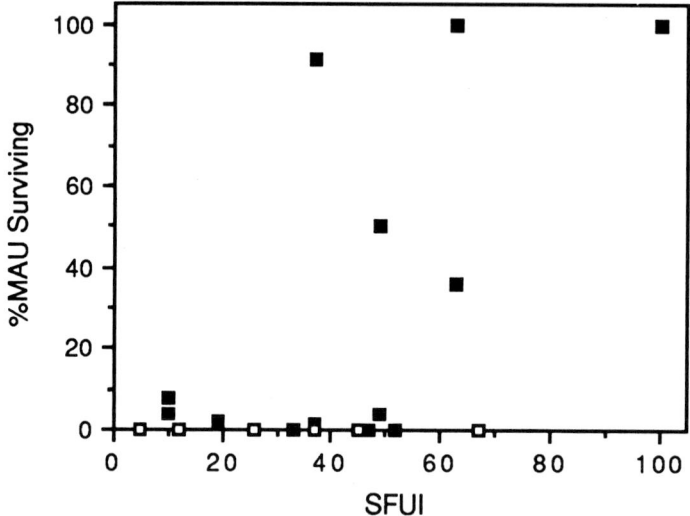

Figure 7.20. Standardized food utility index (SFUI) for complete bones (from Metcalfe and Jones 1988) plotted against the %MAU of surviving (post-ravaging) sheep bones (from Marean and Spencer 1991). Filled symbols are of elements originally introduced to carnivores; open symbols are of bones not introduced to carnivores.

to be ravaged) and which should be retained (those skeletal parts transported to the site but destroyed by carnivore ravaging) in the statistical analysis. This underscores the facts that (a) in an archaeological context one typically begins with a model of a complete (set of) skeleton(s) as a model, and (b) differential transport and differential destruction of skeletal parts can produce the same kind of skeletal part profile (Lyman 1985a). Clearly, more than just the economic utility, transport index, and structural density frames of reference are required to help us produce explanations of skeletal part frequencies.

Summary

> The major flaw in inferential arguments based on excavated data is the assumption, always implicit, that the absence of evidence is evidence for absence.
> (M. B. Schiffer 1987:356)

Since at least the middle of the twentieth century, archaeologists have been trying to perfect methods to explain why some portions of animal carcasses are abundant and other portions are rare in sites. On one hand, perhaps because of the focus on archaeological collections, Theodore White suggested that people might have transported carcass parts differentially based on the economic value of the parts. This was followed by Dexter Perkins and Patricia Daly's proposal that the distance carcass portions had to be transported may have influenced which carcass parts would be transported. Both of these suggestions subsequently found empirical support in the ethnoarchaeological record (e.g., Binford 1978; and O'Connell *et al.* 1988, 1990; respectively). On the other hand, perhaps because paleontologists are not concerned with the human behavioral meaning but rather the paleoecological meaning of skeletal part frequencies, members of this group of researchers such as R. Dale Guthrie and C. K. Brain sought explanations in another arena. Perhaps, they suggested, differential preservation of skeletal parts was in part a function of inherent structural properties of the bones. The structural density of bone parts was one such property, and it was easily measured.

To date, there seems to have been a great deal more research on the economic, differential transport type of explanation than on the differential preservation explanation. For example, economic utility indices applicable to historic archaeofaunas have been developed and applied to various collections (Huelsbeck 1989, 1991; Lyman 1987d; Reitz 1986; Schultz and Gust 1983; Singer 1985). However, no one to the best of my knowledge has yet measured the structural density of cattle or goat bones, although I have measured the structural density of a few domestic sheep bones (Lyman 1982b). If this chapter accomplishes anything, I hope it illustrates the necessity of exploring how we might measure the influences of natural, non-human taphonomic processes on bone frequencies, and that such exploration must keep pace with our ethnoarchaeological research on the influences of human taphonomic processes.

Numerous utility indices and measures of the structural density of bones have now been published. But these two taphonomic variables are not necessarily independent. Thus, additional attributes of the faunal remains are now regularly studied, attributes such as the ratio of long bone shafts to long bone ends, to provide other lines of evidence that tend to substantiate either the differential transport or the differential destruction interpretation. While some variables such as the human selection of particular bones for making tools have not been explored in detail (see Chapter 8 for a consideration of this problem), sufficient data exist to demonstrate that a mere correlation between one frame of reference and the frequencies of skeletal parts in an assemblage is not a sufficient warrant for arguing the particular frame of reference is the reason for those frequencies. I have, in this chapter, outlined the taphonomic variables that must be considered, and the major analytic techniques for building explanations of skeletal part frequencies.

8

BUTCHERING, BONE FRACTURING, AND BONE TOOLS

Introduction

The manner in which animal carcasses and skeletal elements come apart or are taken apart is an important taphonomic variable. Humans butcher animals and that behavior often, but not always, variously modifies bones. In fact, it might be argued that butchering animal carcasses is the single greatest taphonomic (and biostratinomic) factor in the formation of humanly created fossil assemblages. Humans exploit animals for a variety of reasons, but basically to extract resources, whether energy (food) or materials for tools or clothing. During that exploitation, skeletons are disarticulated and bones are broken and variously modified. But as we have seen in previous chapters (especially Chapter 6), non-human taphonomic processes can result in the disarticulation of skeletons and fragmentation of bones. In this chapter, I review these processes, focusing on the modification of skeletal elements for which hominids in particular are responsible.

Butchering

> The fragments of Aurochs exhibiting very deep incisions, apparently made by an instrument having a waved edge ... in which I thought I recognized significant marks of utilization and flaying of a recently slain animal, were obtained from the lowest layer in the cutting of the Canal de l'Ourcq, near Paris ... I have obtained analogous results by employing as a saw those flint knives found in the sands of Abbeville.
> (E. Lartet 1860 [1969:122])

The term *butchering* tends to hold different connotations for different analysts. Perhaps that is because it has seldom been explicitly defined. Lyman (1987a:252) defines *butchering* "as the human reduction and modification of an animal carcass into consumable parts." In this definition, "consumable" is "broadly construed to mean all forms of use of carcass products, including but not restricted to consumption of products as food" (Lyman 1987a:252). It is important to note the inclusion of the word "human" in the definition because Lyman (1987a:251–252) distinguishes butchering from what he calls *faunal processing*, defined as "the reduction and modification of an animal carcass into consumable parts." As we have seen in earlier chapters, many organisms

Table 8.1 *Carcass resources exploitable by a faunal processor or human butcher (from Lyman 1987a:252)*

hide	marrow	fat/blubber
hair	grease	meat
sinew (tendon, ligament)	blood	juice
bone	teeth	brains
horn/antler	viscera (and/or their contents)	hooves

Table 8.2 *Selected carcass-processing activities directed towards extracting consumable carcass resources (see Table 8.1) (from Lyman 1987a)*

evisceration	skinning, hide removal
disarticulation, dismemberment	defleshing, filleting, meat extraction
bone extraction	brain extraction
marrow extraction	blood extraction
bone grease extraction	bone juice production
periosteum removal	sinew or tendon removal

other than hominids process carcasses into consumable parts (e.g., carrion insects, bacteria, hyenas). Only hominids, however, *butcher* an animal carcass.

Russell (1987:386) states that the goal of butchery "is removal of meat," and thus implies that butchery is solely the removal of meat from carcasses. This perception is, I think, too narrow. Binford (1978:63) writes "butchering is in reality a task of dismemberment. Through it the anatomy of a large animal is partitioned into sets of bones that may be abandoned, transported, or allocated to different uses." This aligns well with my definition if the words "sets of bones" are taken to denote archaeologically visible results of a process involving not just bones but muscle, fat, and hide as well as bones. Binford (1978:48) also states that "butchering is not a single act but a series of acts beginning when the animal is killed and continuing at varying junctures until the animal is totally consumed or discarded." Thus while transport of skeletal parts (see Chapter 7) may occur between butchering acts, transport is not technically a part of butchering. Nor does butchering include the human behaviors of cooking and consumption. Of particular interest, given the definition, are (1) the carcass resources a butcher (or any faunal processor) seeks to exploit (Table 8.1), and (2) the processes or human activities necessary to render a carcass into those resources (Table 8.2). There are many kinds of potential resources to extract from, say, a 200 kg artiodactyl, and many processes that can be used to extract them. Ethnoarchaeological research indicates there are many factors which influence the particular resources exploited and the manner in which they are exploited (Table 8.3).

Butchering consists of a set or series of sets of human activities directed

Table 8.3 *Factors that influence utilized butchering techniques (modified and expanded from Lyman 1987a:253)*

Natural factors
   Prey animal: taxon, size of carcass, age and sex of animal, health status of animal
   Nature of procurement:
     Scavenged: condition of carcass (rancid?), completeness of carcass
     Hunted: number of animals killed, number of people present, type of kill site (location,
       accessibility, geological conditions, geographic conditions)
   Spatial relationships of kill site, habitation site, and processing areas
   Time of day: heat, amount of light remaining, weather
   Season of the year: heat, precipitation (type and amount)
   Dietary status of people: immediate versus long-term nutritional needs
Cultural factors
   Technology: available versus used, curated versus expedient tools
   Gustatory preferences
   Preparation and consumption: cooking vessel size, preservation technology (if any), storage
     capabilities and kinds
   Ethnic group involved: first animal rituals, kin present at kill site versus kin present at
     habitation site, selective hunting

towards the extraction of consumable resources from a carcass. It has a temporal duration, made up of the set and order of activities carried out to extract resources from a carcass. The actions and activities involved in butchering have been termed the *butchering process* or *butchering techniques*, and the results of the process have been termed the *butchering pattern* (Lyman 1987a:252). Zooarchaeological research between 1950 and 1980 was aimed at inferring the human behaviors making up a butchering process and evidenced by the butchering pattern as determined from an assemblage of faunal remains (e.g., Wheat 1972). A rule of thumb in such endeavors was to presume that prehistoric butchers utilized the most pragmatic butchering process, and evidence of butchering could be interpreted as reflecting efficient human activity (e.g., Spiess 1979; White 1956).

  Prior to the extended discussion of taphonomic issues in the late 1970s and early 1980s, analysts interested in studying prehistoric butchering techniques assumed many attributes of bone modification could be attributed to human activity. The assumption seems to have been founded largely on the basis of the association of the animal remains with artifacts (Binford 1981b; Lyman 1987a). With the increasing taphonomic awareness of the late 1970s, zooarchaeologists more frequently studied modification attributes displayed by bones in attempts to establish clearly those modifications that could be unambiguously ascribed to human activities. Ways to establish the identity of some marks as butchering marks or stone tool cut marks were extensively discussed in the early 1980s (Bunn 1981, 1982; Olsen 1988; Potts 1982; Potts and Shipman 1981; Shipman 1981a, 1981b; Walker and Long 1977). Discussions and experiments concerning marks made during the fracturing of bones by various agents and the kinds of fractures produced by them were presented (e.g., Bonnichsen 1979; Morlan 1980; Shipman *et al*. 1981). Some of the latter were concerned with identifying bone tools that had been only minimally modified. The remainder of this chapter presents an overview of research concerning these topics.

Butchering marks

While not without precedent (e.g., Guilday *et al*. 1962), it was largely as a result of the work of Pat Shipman (1981a, 1981b, 1983, 1986a, 1986b; Shipman and Rose 1983a, 1983b, 1984) that analysts began examining the microscopic morphology of various scratches on bones. Shipman's work indicated that marks made by stone tools are morphologically distinct from those made by, for example, carnivore and rodent teeth (Figure 6.1). Shipman argued that *cut marks* made by stone tools will (1) be V-shaped to U-shaped in cross section, but tend toward the former, (2) be elongate, (3) have multiple, fine parallel striae on the walls of the mark, and (4) sometimes display what she called "shoulder effects" (small striae parallel to the main striation) and/or "barbs" (would have a small barb or hook at one end). These attributes, particularly the third one, have become the major criteria for identifying cut marks, but are seldom used alone to identify scratches on bones as cut marks. That is because some of these attributes can be created by non-hominid taphonomic processes such as the gnawing action of carnivore teeth (Eickhoff and Hermann 1985) and trampling of bones on sandy substrates (Behrensmeyer *et al*. 1986, 1989; Fiorillo 1989; Haynes and Stanford 1984). Haynes (1991:163) argues that the butchery marks Shipman produced experimentally were *not* made during the *butchering* of an animal carcass, but rather were *deliberately made* to be visible, often on defleshed bones. Thus, the criteria Olsen and Shipman (1988) list, for example, as distinguishing trampling-generated from butchering marks may be invalid as the latter were not created during the process of extracting resources from a carcass, especially by butchers with a working knowledge of the anatomy of the animal they were butchering and with some concern of preserving a sharp tool edge. Nonetheless, the micro-morphological attributes Shipman proposed have come to be the ones analysts examine first. Analysts then typically examine additional attributes of the marks, particularly their anatomical location and orientation (Lyman 1987a; e.g., Gibert and Jimenez 1991; Noe-Nygaard 1989). (Some researchers suggest the direction in which the cut was made and the handedness of the butcher can be ascertained by microscopic examination of the cut mark [Bromage and Boyd 1984]. Other researchers suggest that computer-generated three-dimensional models of surface irregularities on bones such as butchering marks may aid in determining whether or not a bone was fresh when it was cut [During and Nilsson 1991].)

Guilday *et al.* (1962:63) suggest that a cut mark can be identified using two criteria. First, a mark must occur on "specimen after specimen at precisely the same location on the bone." Second, they suggest that each category of mark should have a detectable anatomical purpose or reason for occurring where it does. The second criterion is related to the notion that animals will be butchered efficiently, and as a result each mark has a function or purpose for existing. Thus we read statements such as "animals were butchered primarily at the joints since this is the easiest method" (Read 1971:53) and "the simplest way for men without power saws to divide a carcass is to follow the natural paths of muscles and cut them only where they need cutting, at their origins and insertions" (Gilbert 1979:152).

Thus the additional criteria used by zooarchaeologists to identify marks on bones as cut marks involve the anatomical placement and orientation of a mark, and the function of a particular category of mark suggested by its location and orientation. Binford (1981b:46–47) suggests cut marks differ from gnawing marks (see Chapter 6) in that the former "rarely follow the 'contour' of the bone [surface]" whereas the latter do (see Noe-Nygaard 1989 for a recent overview). Cut marks, Binford (1981b:47) argues, "generally result from three activities: (a) skinning, (b) disarticulation, and (c) filleting." Skinning cut marks are found around the shaft of lower legs and phalanges, and along the lower margins of the mandible or on the skull. Disarticulation cut marks occur on the "edges, or articular surfaces of the ends of long bones, and on the surfaces of vertebrae or pelvic parts" (Binford 1981b:47). Filleting cut marks generally parallel the long axis of the bone. Here, Binford is suggesting some general rules for identifying the function of a cut mark from attributes of its location and orientation. Most analysts today record the location and orientation of cut marks by illustrating each mark on line drawings of skeletal elements, much like the drawings in Figures 7.4 and 7.5. I have found that several views of each element, minimally an anterior and a posterior view of long bones, often are sufficient (e.g., Lyman 1991a).

That people use hammerstones to break bones must also be considered in any discussion of butchering marks (the kinds of fractures generated are discussed later in this chapter). Morlan (1980:50) suggests that "the upper size limit of carnivore tooth contact area is smaller than the upper limit of a hammerstone contact area." Bunn (1982:44) notes that "a pointed hammerstone can produce a small, tooth-sized indentation where the bone breaks, and a series of overlapping tooth-induced indentations can resemble the broad, arcuate indentations which rounded hammerstones more typically produce." Potts (1982:215) suggests that a hammerstone will produce bone flakes that are broader than they are long and have platforms that are broader than they are thick. He believes that the resulting flake scars will be broad, arcuate indentations, whereas carnivore tooth-created flake scars will appear as small notches. Based on additional experimental work, Potts (1988:101) suggests there is major overlap between the length–width attributes of hyena-produced bone flakes and human-wielded hammerstone-generated bone flakes. Given this case of equifinality (see also Figure 6.25 and associated discussion), it is fortunate indeed that Blumenschine and Selvaggio's (1988, 1991) experimental work with tooth-created pits and hammerstone-generated percussion marks (see Chapter 6) provides criteria that seem to be distinctive of both kinds of marks. The anatomical locations of hammerstone-generated impact scars may be recorded on line drawings of bones just like cut marks.

A conceptual framework for analyzing butchering

Within the context of analyzing butchering practices, *transport* of carcasses and carcass parts can be defined as the human movement of animal products from one position on the landscape to another position (see Chapter 7). Transport of animal body parts usually implies carrying such parts from the procurement location to a residential or consumption location. Butchery and transport are not independent processes, nor are they totally and always interdependent. Ethnoarchaeological data indicate that once an animal carcass is procured, humans are faced with a myriad of decisions. Transport logistics may or may not dictate if and how the carcass is reduced or butchered into transportable pieces (Binford 1978; Bunn *et al.* 1988; O'Connell and Marshall 1989; O'Connell *et al.* 1988, 1990). Small carcasses may be transported whole, without butchery (Hudson 1990; Yellen 1991a:5), and thus small carcasses may "be brought to camp skeletally complete or virtually so" (Bartram *et al.* 1991:102). Large carcasses are typically butchered into small, transportable pieces only some of which may in fact be transported (e.g., O'Connell *et al.* 1990, 1992). In some cases nearly all of the skeleton of proboscideans – the largest of terrestrial vertebrates – is transported; in others nearly the entire skeleton is not transported (Crader 1983; Fisher 1992).

The zooarchaeological literature is, on one hand, full of discussions about differential transport of carcass portions. These discussions largely concern why certain portions are transported and others not, and how to recognize such transport in archaeological contexts (see Chapter 7). On the other hand the literature is relatively silent on the interplay of transport logistics and butchery (see Bartram *et al.* 1991; Gifford-Gonzalez 1989a; Yellen 1991a, 1991b; for recent additions to the discussion). Thus it is important to integrate the butchering process with the processes of transport. Butchering typically involves a set of activities that, between the time of carcass procurement and final disposal of carcass portions, occur in varying orders and frequencies or intensities for different carcasses.

Binford's (1978, 1981b) categorization of the various butchering activities as skinning, dismemberment or disarticulation, and filleting or removing meat from bones is a useful one that captures the essence of the main activities of a hominid butcher. Extraction of viscera, blood, brains, marrow, grease, bone, and sinew, and periosteum removal are other activities that might be considered subsidiary to, in particular, dismemberment and filleting or food extraction (Lyman 1987a). As implied earlier, all of these activities have the potential to damage bones; that is, all can produce what are typically called butchery marks. Because butchery of an animal carcass is a process, it can be conceived of as a set of stages of greater or lesser discreteness. I have found the following stages to provide a useful conceptual framework: (1) *kill-butchery stage*, followed by transport of variously butchered carcass portions, (2) *secondary butchery stage*, perhaps followed by another transport episode (even if only redistribution within the site of consumption), and (3) *final butchery-consumption stage*. The importance of these stages is that they imply that butchery marks will be added to the bones of a carcass as it passes through the butchering stages. One might expect, then, fewer marks to occur on bones disposed of early in the butchery process (prior to transport) whereas bones retained (and transported) have the opportunity to undergo further butchery prior to consumption and may display additional marks that were created during final processing or consumption.

Thomas and Mayer (1983) inferred, in the case of the Gatecliff Shelter Horizon 2 bighorn sheep (*Ovis canadensis*), that they were dealing with what I term a secondary butchery site based on varying frequencies of skeletal parts. That inference in turn resulted in their conception of the butchering marks they observed as representing early stages of butchering rather than final or intermediate stages. Thomas and Mayer (1983) use the concept of *archaeological monitoring perspective* in their discussion. Simply put, the monitoring perspective is the archaeologist's observational position within a prehistoric human behavioral context. Thomas and Mayer were examining the skeletal part frequencies resulting from a particular transport strategy. They were in the geographic place away from which bones were transported, and thus particular bones were rare whereas others were frequent. If their monitoring perspective had been in the location to which bones had been transported, the kinds of bones they observed to be rare would have been abundant and the kinds of bones they observed to be abundant would have been rare. Within the context of the butchery stages, the analyst's monitoring perspective will perhaps influence the anatomical placement and frequency of butchering marks. Yellen (1977), Binford (1984a), and Gifford-Gonzalez (1989b, 1989c) report ethnoarchaeological data for what I term final butchery-consumption sites. Their data indicate the bones, and the butchery-related damage they have sustained, in such contexts represent the final stages of processing (see also Yellen 1991a). Thus it seems that not only may the analytical monitoring perspective influence the kinds of skeletal parts an archaeologist may find (see Chapter 7), it may also influence the kinds and frequencies of butchery marks on those skeletal parts.

How a carcass is butchered and transported are not only influenced by one's monitoring perspective or which stage of butchery is represented by the studied assemblage, they are influenced by a number of other factors as well. Gifford-Gonzalez (1989b) correctly observes that the anatomical distribution and frequency of cut marks result from costs and benefits of transport, the technology used to butcher animals, and cooking and consumption practices (see also Lyman 1987b). Her (Gifford-Gonzalez 1989a) ethnoarchaeological data suggest the size and anatomy of a carcass may constrain butchery practices, but for any carcass size and anatomical category there are a number of different dismemberment and defleshing strategies that may be used. Similarly, the transport logistics involved in moving carcasses and carcass parts from primary and secondary kill-butchery sites to the locus of final butchery and consumption will influence how an animal is butchered and thus the butchery-related damage inflicted on bones. Binford (1978, 1981b) labels those factors which influence how an animal is butchered and transported *contingency factors* (see Table 8.3 for a list of these).

Unless one is dealing with a kill-butchery site where relatively minimal consumption occurred, the archaeofaunal remains with which we typically deal are probably consumption waste, such as those remains found in village or camp sites. Much as the lithic analyst may focus on the debitage resulting from producing a tool, the zooarchaeologist focuses on the debitage resulting from producing, typically, a meal, particularly when the site is one at which consumption was relatively more frequent than initial processing. The effects of consumption on an archaeological bone assemblage can be significant if, for example, processing includes bone fragmentation, thus reducing the identifiability of specimens (Lyman and O'Brien 1987).

If the subject of study is frequencies of butchery marks, then a major assumption must be granted (Lyman 1992c, 1993a). The assumption consists of the premise that, given some set of bones X, some subset X′ of those bones will be butchered, and of those butchered bones some subset X″ will sustain damage in the form of butchery marks. The assumption is that some proportion of each skeletal element was butchered and some lesser proportion will display butchery marks, and those proportions will directly and positively covary at least at an ordinal (but perhaps not an interval) scale. Note that while I have said "bones are butchered," this is a shorthand form of saying "carcasses and/or carcass parts, including hide, viscera, muscle, and other soft tissues are butchered; strictly speaking, bones are only butchered when they are broken for marrow or grease extraction." I use the shorthand form because it is the bones, not the soft tissues, carcasses, or limbs that we study in archaeological contexts.

If butchery marks are epiphenomena, that is, if they are in some sense an unintended, accidental, fortuitous, or incidental result of butchery activities, then frequencies of butchered bones are potentially ambiguous indicators of the quantitative aspects of human behaviors, and thus terms such as "butchery pattern" would be inappropriate given its human behavioral implications (see Lyman 1993a for additional discussion). A fictitious example makes clear the significance of the assumption. Let us say that 10 femora and 10 humeri were available for butchery ($X = 20$). Of those, six femora and five humeri were butchered ($X' = 11$). Of those butchered elements, four femora and two humeri display butchery marks ($X'' = 6$). The critical statistical relation here is that more femora than humeri were butchered, and more femora than humeri actually display archaeologically visible butchery marks; that is, even though 60% of the femora and 50% of the humeri were butchered, 67% of the butchered femora and 40% of the butchered humeri display archaeologically visible butchery marks. If this relationship fails to hold, such as a case in which six of the ten femora and five of the ten humeri were butchered but one femur (17% of those actually butchered) and four humeri (60% of those actually butchered) display butchery marks, then analysts attempting to discern , butchery intensity and human behavioral patterns may be misled by the archaeologically visible record of butchery practices. I consider the issue of preservation of butchery marks below.

It is important to realize that some unknown proportion of butchered bones (X') may be archaeologically invisible; only that portion of X' which consists of butchery-marked bones (X'') is archaeologically visible. Therefore, the assumption that X' and X'' are positively and directly related is generally operationalized by analysts who present the proportion (percentage) of all bones (X) that display butchery marks (makes up X''). That is, for any one kind of skeletal element (i), the analyst normally presents the absolute frequency of butchery-marked bones (X''_i) and the proportional frequency of butchery-marked bones $= (X''_i/X_i)100$. Interpretation of those frequencies usually involves comparing those values for different skeletal parts (i), or taxa or bone assemblages (j); X''_{ij} compared to $X''_{i+1,j}$, or $(X''_{ij}/X_{ij})100$ compared to $(X''_{i,j+1}/X_{i,j+1})100$, respectively. However, what one actually is seeking to measure is the human behavior indicated by X'_{ij}, which may, of course, be archaeologically invisible. Thus the assumption is required that the frequency of butchery-marked bones in an assemblage (X_{ij}'') is directly and positively related to the number of bones in that assemblage that were in fact butchered (X_{ij}') in order to infer human behaviors.

If the frequencies of butchery-marked bones are not positively correlated with the frequencies of butchered bones then frequencies of butchery-marked bones, either absolute or proportional, are potentially ambiguous indicators of the quantitative aspects of human butchering behaviors. When, whether, and how often the assumption that the frequency of butchery-marked bones is positively correlated with the frequency of butchered bones is warranted is another matter. Because ethnoarchaeological research has not yet been published on this issue, analysts interested in analyzing and interpreting frequencies of butchery marks must make the assumption that the two variables are positively correlated. Making the assumption then leads to consideration of how to count butchery marks.

Quantification of butchering marks

That the frequency of butchery marks is in fact a significant variable in the analysis of prehistoric butchery is apparent in the literature. Kooyman (1984:54) reports on the prehistoric butchery of moa (a large flightless bird of the taxonomic order Dinornithiformes) carcasses at the site of Owens Ferry in New Zealand. He suggests cut marks on moa femora "are too few and scattered to indicate any general processing strategy; they probably indicate occasional alternative [butchery] patterns." That is, the infrequent occurrence of cuts on the femora relative to frequencies of cuts on other bones suggests to Kooyman that moa legs were not all butchered in like manners. Rapson (1990:287) calculates the number of "cut marks per unit area" of bone specimen surface to provide a "measure of cut mark intensity which controls for differences in [bone] specimen size" and size of the butchered taxon. He then identifies differences and similarities in butchery based on variation in the density (frequency per unit area) of cut marks on bones of animals of different sex and taxon.

Bunn and Kroll (1986:432) state that "frequencies of cut marks on different skeletal parts can be directly linked to the skinning, disarticulation, and defleshing of carcasses." They infer, for example, that because 20% of the bones making up the elbow joint in their sample display cut marks, "repeated dismemberment of the elbow joint is documented." Binford (1986:446) suggests that high frequencies of cut marks on long bone shafts indicate "extreme difficulty in processing already partially desiccated limb parts that had been previously ravaged by carnivores," thus there was little meat for the butchers to extract. In contrast, Bunn and Kroll (1986:450, 449–450) argue that "the presence of many slicing marks on once-meaty limb bones indicates that hominids removed substantial quantities of meat from the bones" because such high frequencies "are most likely to occur when it is difficult or impossible to see where the bone is, as when a complete, meaty limb bone is being defleshed." Binford (1988:127) suggests that "the number of cut marks, exclusive of dismemberment marks, is a function of differential investment in meat or tissue removal." The greater the effort invested, the more cut marks will result. He further suggests "the numbers of cut marks and their frequencies on different bones may reflect very different processing operations;" skinning may result in more marks on metapodials and fewer on femora, all else being equal (Binford 1988:128).

Regardless of who is correct in the above, it is clear that the frequency of cut marks is felt by these analysts to be an important variable that in some manner reflects human behaviors. In all of the preceding examples we find implicit adoption of the assumption that frequencies of butchery-marked bones (X'') will be directly and positively correlated with the frequencies of butchered bones, marked and unmarked (X'). Given the interpretive importance ascribed to frequencies of butchery-marked bones and frequencies of butchery marks on particular bones, it is mandatory that some explicit method be used to count them (Maltby 1985a). The number of bone specimens displaying butchery marks is a rather straightforward quantitative unit: simply tally the NISP (number of identified specimens) that have or display butchery marks. But is NISP an appropriate unit for such quantification?

The counting units for tallying butchery mark frequencies typically, although not always, define some limited anatomical space on a skeletal element. Typically the total frequency of that space (X, above) and the frequency of that space with butchery marks (X'', above) are tallied, regardless of how fragmented or complete the individual specimens representing that space are. That is so because the analyst is interested not just in the frequency of butchery marks, but in their anatomical distribution because that distribution relates to butchering functions (e.g., Lyman 1987a; Maltby 1985a, 1985b). Thus one tallies X and X'' for units such as lateral distal tibia or proximal greater trochanter of the femur. The terms "specimen" and "NISP" thus have different connotations in butchery analysis than in, say, the analysis of taxonomic abundances (see Chapter 4 for discussion of the latter). In the latter, a specimen or an NISP of one represents some discrete archaeological object, whether it is a complete femur or just the distal end of a femur. In butchery analysis, the portion of the specimen displaying butchery marks is tallied. For example, two of five recovered distal humeri might display butchery marks, resulting in the conclusion that 40% of the recovered distal humeri have butchery marks. The fact that three of those five humeri specimens are complete bones, one consists of the distal end and shaft, and the fifth one consists of just the distal condyle is irrelevant to tallying frequencies of butchery marks. If the shaft of the fourth specimen displays a cut mark, it is tallied in the "distal humerus shaft" category. Thus the meaning of the terms "specimen" and "NISP" in the analysis of butchery resides in the anatomical area sense rather than in the archaeologically discrete object sense.

The number of butchery marks is a potentially difficult counting unit to operationalize. I tally each discrete, nonadjacent (> 1 cm apart) and non-overlapping mark as an instance of force application (Lyman 1987a, 1992c, 1993a). While somewhat subjective (e.g., a cluster of striae is tallied as one instance even though multiple instances of force application are clearly represented), this seems to be the practice generally followed. I prefer this procedure because when striae in particular, and flake scars to a lesser degree, overlap, counting them individually may be impossible as one mark tends to destroy traces of a mark made previously (Figure 8.1a).

Use of the proportional frequency of anatomical areas which display

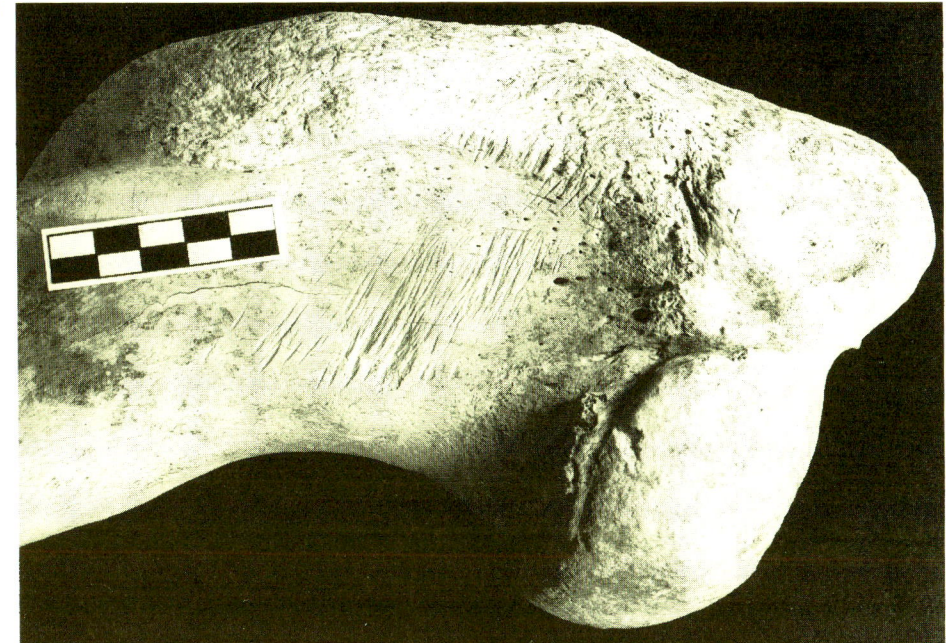

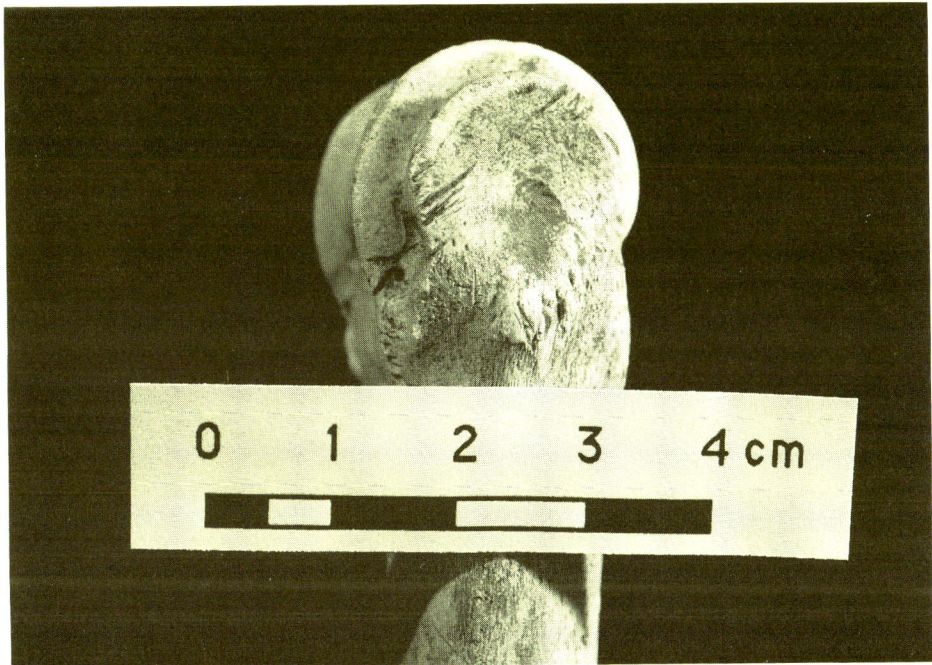

Figure 8.1. Examples of cut marks. (a) Steller's sea lion (*Eumetopias jubatus*) humerus with cut marks on proximo-lateral surface, scale bar is 5 cm (from Lyman 1992c:251, Figure 1; courtesy of the Society for American Archaeology); (b) deer (*Odocoileus* sp.) distal metapodial with cut marks on lateral surface.

butchery marks in analysis demands that those frequencies be tallied in such a manner as to take into account various taphonomic factors which destroy butchery marks. Weathering (Behrensmeyer 1978), gnawing by carnivores or rodents (Grayson 1988), root etching (Chapter 9), and other processes which alter the exterior surface of bones can obliterate butchery marks (Maltby 1985a, 1985b). Thus, the analyst tallies (a) the number of surfaces of each anatomical area that have the potential to display butchery marks, (b) the number of surfaces of each anatomical area that have been modified by non-butchery-related processes such that butchering marks that may have existed have been obliterated, and (c) the number of surfaces of each anatomical area that display butchery marks. Proportional frequencies of butchery-marked specimens are derived by dividing the third value by the first; the second value is ignored in butchery analysis, but when summed with the first value will provide a total count for the particular anatomical area under consideration (which may be important to derivation of MNE and MNI values). Thus, for example, 20 distal medial tibia fragments may be in a sample, but three are heavily weathered and two have been gnawed by rodents, leaving only 15 with the potential to display butchery marks. If five display such marks, then 33% [(5 ÷ 15) 100] is the proportional frequency of specimens with the potential to display butchery marks that in fact display them (the MNE of distal tibia would be 20). Unless there is evidence to suggest otherwise, it is probably reasonable to assume, and in fact must be assumed, that the destruction of bone surfaces which might show butchery marks is not related to the presence or absence of butchery marks on those destroyed surfaces.

Analyzing butchering practices

There have been several ways that zooarchaeologists have approached the analysis of butchering. One involves study of the fragments of bones, a topic I reserve for the next section. More recently, the focus has been on the butchering marks observed on bones. In the following, I review several approaches to analyzing and interpreting butchering by summarizing how various analysts have studied butchering marks.

Much of the most intensive study of butchery marks has been undertaken with bovid faunal remains recovered from the Plio-Pleistocene sites excavated at Olduvai Gorge, Tanzania. Here, the catalyst for such intensive study resides in the importance of understanding the early stage in hominid evolution and behavior represented by the collections. Study of butchering marks on bones recovered from sites at Olduvai began in earnest just over a decade ago (Bunn 1981, 1982; Potts and Shipman 1981; Shipman 1983), and continues today. Two researchers have been largely responsible for this research (others are, of course, also involved), and their approaches to the study of butchering marks differ somewhat. I review each in turn.

Shipman (1981a, 1983, 1986a, 1986b, 1987, 1988b; Potts and Shipman 1981) used SEM techniques to identify cut marks on some of the bones recovered from Olduvai Bed I sites. She then used chi^2 analysis to assess whether those marks were randomly distributed across "near joint" and "midshaft" locations on the bones, and whether those marks were randomly distributed across " meat-bearing" and "non-meat-bearing" bones, relative to (a) a sample of carnivore gnawing marks on 70 bones from Olduvai and (b) a sample of cut marks on bones recovered from the Neolithic pastoralist site of Prolonged Drift (Gifford *et al.* 1980). She found that "the Olduvai cut marks are distributed significantly differently from the Prolonged Drift cut marks and do not cluster near joints [and that] the Olduvai cut mark distribution is indistinguishable from that of the 70 carnivore tooth marks" (Shipman 1986a:30). Shipman (e.g., 1986b:698) concludes that the animals represented by the bones from the Olduvai sites were systematically butchered, and probably were scavenged rather than hunted.

Shipman's (1986a, 1986b, 1988b) assumptions include the following: (a) cut marks in near-joint locations signify disarticulation, (b) cut marks on meat-bearing bones (which are never defined or identified) signify defleshing or filleting, (c) cut marks on metapodials represent skinning, and (d) an appropriate analogue for a systematic butchering pattern can be found in the frequency distribution of cut marks across near-joint and midshaft loci on meat-bearing and non-meat-bearing bones in a sample of cut-marked bones from a Neolithic site. These assumptions have been criticized as poorly founded and in some cases refuted by ethnoarchaeological data (Bunn and Blumenschine 1987; Gifford-Gonzalez 1989c; Lyman 1987b). Shipman's analytical methods are, however, intriguing. They display a shift from the more common, to that point in time, procedure of illustrating a skeleton with the general anatomical locations of cut marks shown (e.g., Guilday *et al.* 1962), to use of the frequency distribution of cut marks across particular anatomical loci as an important source of information on hominid behavior. Bunn and Kroll's (1986) analysis is important because it illustrates the next step, beyond Shipman's work, towards increasing the intensity and extent to which data on butchery mark placement and frequency distributions are studied.

Bunn and Kroll (1986:436; see also Bunn 1981, 1982, 1983) begin their study of the bovid remains from the *Zinjanthropus* site at Olduvai with the statement "the location and frequency of cut marks on different skeletal parts can be used in conjunction with a knowledge of animal anatomy to identify patterning in the butchering techniques of present and past humans." On that same page, they go on to suggest that "cut marks on nonmeaty skin-covered bone surfaces, on or near epiphyses where connective tissues bind articulating joints, and on meaty bones at points of muscle attachment can provide unambiguous documentation of carcass skinning, joint disarticulation, and defleshing, respectively." They present data on the frequency distribution of cut marks

Table 8.4 *NISP and frequencies of cut-marked specimens in the FLK* Zinjanthropus *assemblage (modified from Bunn and Kroll 1986)*

Anatomical category	Small bovid			Large bovid		
	NISP	N cut	% cut	NISP	N cut	% cut
maxilla and cranium	49	0	0.0	46	2	4.3
mandible	20	2	10.0	160	11	6.9
vertebral centrum	29	1	3.4	57	1	1.8
vertebral process	28	0	0.0	59	3	5.1
rib	217	1	0.5	423	21	5.0
sternabra	1	0	0.0	1	0	0.0
scapula	7	0	0.0	22	2	9.1
P humerus	3	1	33.3	4	0	0.0
S humerus	13	3	23.1	45	8	17.8
D humerus	6	4	66.7	17	6	35.3
P radius	4	2	50.0	15	3	20.0
S radius	12	3	25.0	45	10	22.2
D radius	3	0	0.0	6	1	16.7
P ulna	6	0	0.0	10	2	20.0
S ulna	4	0	0.0	11	1	9.1
D ulna	0	0	0.0	4	0	0.0
carpals	13	0	0.0	17	0	0.0
P metacarpal	6	2	33.3	10	2	20.0
S metacarpal	12	1	8.3	20	1	5.0
D metacarpal	3	0	0.0	2	0	0.0
innominate	14	2	14.3	26	7	26.9
P femur	1	0	0.0	6	1	16.7
S femur	17	2	11.8	41	7	17.1
D femur	2	1	50.0	5	0	0.0
patella	1	0	0.0	1	0	0.0
P tibia	9	2	22.2	1	0	0.0
S tibia	36	11	30.6	92	7	7.6
D tibia	3	2	66.7	7	2	28.6
tarsals and D fibula	24	2	8.3	28	1	3.4
P metatarsal	12	2	16.7	6	0	0.0
S metatarsal	15	1	6.7	13	2	15.4
D metatarsal	2	0	0.0	4	1	25.0
phalange	27	0	0.0	13	0	0.0
sesamoid	16	0	0.0	21	0	0.0
S long bone	208	5	2.4	667	17	2.5
S metapodial	17	0	0.0	42	3	7.1
totals:	840	50	6.0	1947	122	6.3

across various skeletal parts of small bovids (10–110 kg live weight) and large bovids (> 110 kg live weight), summarized in slightly modified form in Table 8.4.

Bunn and Kroll (1986) present a number of inferences on the basis of their data (Table 8.4). Some of these can be summarized as follows: (1) Abundant cut marks on "meaty" limb bones (not defined by Bunn and Kroll) indicate large quantities of meat were cut from those bones. (2) Proportionately fewer meaty limb bones of large mammals have cut marks than meaty limb bones of small mammals, but fewer non-meaty nonlimb bones of large mammals have cut marks than non-meaty nonlimb bones of small mammals; Bunn and Kroll believe this is the case because there is more meat on the non-meaty, nonlimb bones in large mammals than in small mammals. (3) Cut marks on metapodials represent "skinning operations." (4) Cut marks in mid-shaft locations on meaty limb bones represent defleshing. Bunn and Kroll's (1986:436) assumption about the location of a cut mark being indicative of that mark's function (skinning, disarticulation, defleshing) may be correct in a general sense, given, for example, the ethnoarchaeological documentation provided by Binford (1981b) on the covariation of mark location and mark function. However, Lyman (1987a:263–265, 1987b:711) notes that while Binford (1981b) provided a functional typology for cut marks based on 108 anatomical locations *and orientations* of such marks, the general near-joint category of marks, for example, includes not just disarticulation marks but defleshing and skinning marks as well (Figure 8.2). Part of the reason for such variation may reside in whether the carcass being butchered is fresh and supple, frozen, or somewhat desiccated and stiff, such as a several-day-old carcass that is scavenged (Binford 1984b:110–112).

The frequency distribution of proportions of cut-marked specimens across different anatomical loci suggests to Bunn and Kroll (1986), among other things, a thorough and systematic butchering process was applied to both size classes of bovids (Bunn 1983), and much meat was removed from the bones by the butchers. Binford (1988:127) counters the latter by noting that "the number of cut marks, exclusive of dismemberment marks, is a function of differential investment in meat or tissue removal. When a butcher who is filleting meat seeks to get all the adhering tissue off the bones, there will be many cut marks; if little effort is made to clean the bones, relatively few cut marks result." Bunn and Kroll (1988:144) reply by suggesting that "a simple comparison of cut-mark frequencies among skeletal elements and animal size groups is ill-advised as the sole basis for establishing whether or not a substantial portion of the meat was present when the carcass was butchered," and indicate the frequencies of cut marks they recorded on the bones from Olduvai are similar to the frequencies documented at Neolithic sites where, it is presumed, cut marks were produced on bones with much meat adhering to them (e.g., Gifford *et al.* 1980; Marshall 1986).

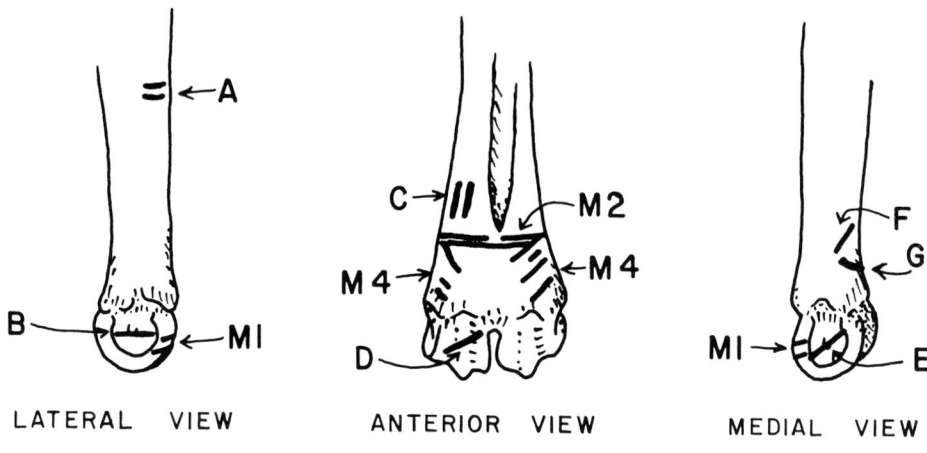

LATERAL VIEW   ANTERIOR VIEW   MEDIAL VIEW

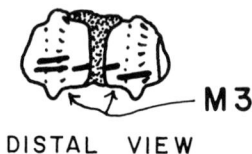

DISTAL VIEW

Figure 8.2. Distal metapodials showing locations of variously documented cut-marks (from Lyman 1987a:264, Figure 5.1; courtesy of Academic Press). M1–M4 from ethnoarchaeological contexts (Binford 1981b); A–G from prehistoric contexts.

    I present the preceding discussion without the intent of resolving the debate about the significance of the cut-marked bones from Olduvai, and without the intent of taking sides. What I hope to have done in these few paragraphs is offer the reader something of the flavor of the way analysts have approached butchery-mark data. There is, of course, much I have not said that can be found by reading the references cited. But there are also several other items here that can be touched on, items not to be found in the references cited. In particular, there are several ways to study the butchery data in Table 8.4 not explored by the authors who have previously discussed them. I review several of them here.

    First, the proportion of cut-marked bones of small bovids at Olduvai is not significantly different from the proportion of cut-marked bones of large bovids (arcsine $t_s = 0.3$, $P > 0.5$; see Sokal and Rohlf [1969:607–608] for discussion of the arcsine transformation statistic). This suggests that overall, the carcasses of small bovids were butchered just as intensively as carcasses of large bovids. Second, the proportion of cut-marked specimens per anatomical category for small bovids, as listed in Table 8.4, is not correlated with the total NISP per anatomical category for small bovids ($r = 0.19$, $P = 0.26$); the same is true for

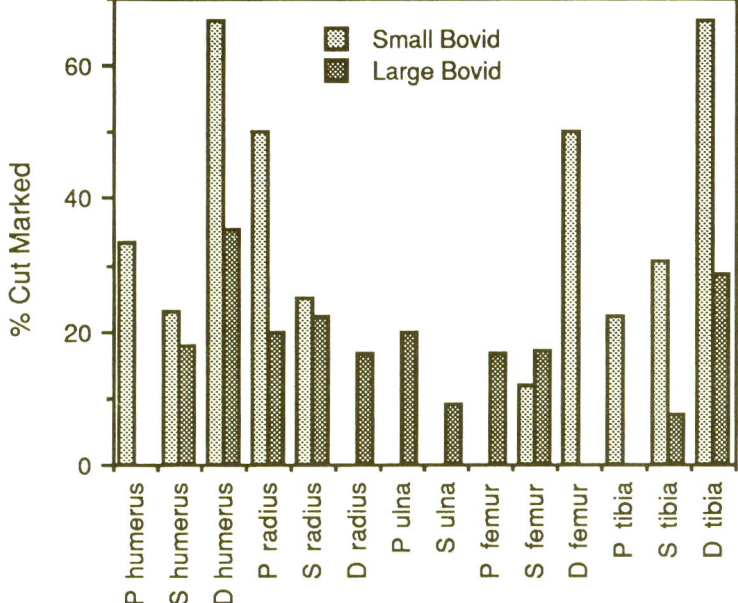

Figure 8.3. Proportional frequencies of cut-marked specimens in selected anatomical categories (from Table 8.4).

large bovids ($r = 0.15$, $P = 0.40$). Thus the proportions of cut-marked bones in the two animal size categories do not seem to be a function of sample size measured as NISP (note that the NISP per anatomical category for small bovids is strongly correlated with the NISP per anatomical category for large bovids; $r = 0.95$, $P < 0.001$). Third, the proportion of cut-marked specimens per anatomical category for small bovids is significantly but weakly correlated with the proportion of cut-marked specimens per anatomical category for large bovids ($r = 0.48$, $P = 0.003$). This suggests that there are, perhaps, differences between how large and small bovids were butchered, granting, of course, the assumption to any such analysis noted earlier that the frequency of butchery-marked bones is directly correlated with the frequency of butchered bones. That there may be differences in how these two size groups of bovids were butchered is suggested by Figure 8.3. This brings us to Bunn and Kroll's (1986) conclusion noted above that proportionately fewer meaty limb bones of large bovids than small bovids have cut marks, but fewer non-meaty nonlimb bones of large bovids have cut marks than non-meaty nonlimb bones of small bovids. How might this be studied analytically?

I presume meaty limb bones include the humerus, radius-ulna, femur, and tibia; the metapodials are non-meaty limb bones. There is no statistically significant difference in the proportional frequencies of butchery-marked metapodials for the two size categories of bovids (Table 8.5). There are,

Table 8.5 *Frequencies of cut-marked meaty limb specimens and metapodial specimens in the FLK* Zinjanthropus *collection (from Table 8.4)*

Anatomical category	Small bovid			Large bovid			Arcsine	
	NISP	N cut	% cut	NISP	N cut	% cut	t_s	P
Meaty limbs	119	31	26.0	309	48	15.6	2.398	<0.02
Meaty limbs + shafts	327	36	11.0	976	65	6.7	2.383	<0.02
Metapodials	65	6	9.2	97	9	9.3	0.022	>0.5

however, significant differences in the proportions of butchery-marked meaty limbs, whether or not Bunn and Kroll's unidentifiable long-bone shaft fragments are included. Proportionately more meaty limb specimens of small bovids display cut marks than meaty limb specimens of large bovids. To help determine why such a difference exists, I sorted the meaty limb specimens into categories that constitute the major limb joints: the shoulder includes all scapula specimens; the elbow includes the distal humerus, proximal radius, and proximal ulna; the wrist includes the distal radius, distal ulna, carpals, and proximal metacarpal; the hip includes the proximal femur; the knee includes the distal femur, patella, and proximal tibia; the ankle includes the distal tibia, tarsals and distal fibula, and proximal metatarsal (after Lyman 1991a, 1992c, 1993a). Relevant data are summarized in Table 8.6. Arcsine statistics indicate the only significant differences are in the hind leg; the knee joint was apparently more frequently butchered in small bovids than in large bovids, and the tibia shaft of small bovids was more frequently butchered than the tibia shaft in large bovids.

The perceptive reader will note two things about the preceding. First, I have not presumed or inferred anything about the function of particular categories of marks, such as near-joint marks representing disarticulation. Second, without more precise data on the location of the marks on scapula and innominate specimens, the statistical analyses in Table 8.6 may be faulty. That is so because in that table I included all scapula specimens even though some of the butchering marks may have been on the blade of the scapula in non-near-joint loci, and I excluded the innominate even though some of the cut marks may have been near the acetabulum in near-joint loci. Nonetheless, my purpose in presenting this analysis has not been to determine precisely what butchering activities were performed by Plio-Pleistocene hominids at Olduvai. Rather, I have simply illustrated how one might perform such an analysis were that one's goal. There are three other analytical avenues that can be described using the Olduvai data in Table 8.4.

First, are the proportional frequencies of cut-marked joints a function of joint tightness? One might expect the skeletal parts constituting a joint that is tightly bound by ligaments, tendons, and muscles to display more cut-marks

Table 8.6 *Frequencies of cut-marked specimens in joint, and meaty limb shaft locations (compiled from Table 8.4)*

Anatomical category	Small bovid			Large bovid			Arcsine	
	NISP	N cut	% cut	NISP	N cut	% cut	t_s	P
shoulder	3	1	33.3	4	0	0.0	1.509	>0.2
elbow	16	6	37.5	42	11	26.2	0.828	>0.4
wrist	22	2	9.10	37	3	8.1	0.132	>0.5
hip	1	0	0.00	6	1	16.7	0.779	>0.4
knee	12	3	25.0	7	0	0.0	2.203	<0.05
ankle	39	6	15.4	41	3	7.3	1.160	>0.2
S humerus	13	3	23.1	45	8	17.8	0.419	>0.5
S radius-ulna	16	3	18.8	56	11	19.6	0.071	>0.5
S femur	17	2	11.8	41	7	17.1	0.525	>0.5
S tibia	36	11	30.6	92	7	7.6	3.119	>0.01

than the bones of a less tightly bound joint. I calculated the average rank of disarticulation for the topi, wildebeest, and Grant's gazelle to provide a general measure of limb joint tightness from Table 5.5. That order is, from the first (least tight) to the last joint (tightest) to naturally disarticulate: shoulder, wrist, hip, elbow, and the knee and ankle are tied for last. That order of natural disarticulation does not correlate with either the proportion of cut-marked bones per joint of small bovids ($r_s = 0.06$, $P = 0.91$) or with the proportion of cut-marked bones per joint of large bovids ($r_s = -0.01$, $P = 0.98$), as listed in Table 8.6. If a significant correlation had been found, one might argue that (a) many of the marks are disarticulation marks, and (b) mark frequencies are related to the labor investment required to disarticulate joints.

Second, the proportions of individual cut-marked limb bones of both large and small bovids do not correlate with the food utility index (Table 7.3). That is so using two different data sets. In one, I calculated the proportions of cut-marked specimens for ends and shafts of the scapula, humerus, radius-ulna, metacarpal, innominate, femur, tibia, and metatarsal. In the other, I calculated the proportions of cut-marked specimens for the scapula and innominate, and the shafts only of the humerus, radius-ulna, metacarpal, femur, tibia, and metatarsal. One might be prompted, given such a result, to wonder if the cut marks are not related to meat extraction based on the expectation that bones with more meat on them would prompt more effort to remove that meat, resulting in more cut-marked specimens. This expectation is, in fact, one thing that Binford (1988) and Bunn and Kroll (1986, 1988) agree on in at least a general sense: more marks represent more investment to remove soft tissue, but whether that means there is much or little soft tissue present is unclear.

Finally, following the reasoning that the food utility of a skeletal part might influence the frequency of specimens of that part which are butchered and thus the frequency with which that part will display butchering marks, the frequency of flake-scarred bones in an assemblage should, perhaps, correlate with the marrow utility index, the grease utility index (e.g., Table 7.1), or some combination thereof. The frequency of flake-scarred specimens per anatomical category in the Olduvai samples has not been reported, precluding the analytical pursuit of this line of inquiry.

Summary

The available literature on butchering is extensive. Lyman (1987a:251) reported that over two dozen books and articles on this topic alone appeared in eight months during 1984 and 1985. A substantial number of articles appeared in 1986. The study of butchery has continued to be an important topic in the literature (e.g., Bartram *et al.* 1991; Yellen 1991a, 1991b). Analysis of what appear to be butchering marks has recently been used to identify ritualistic treatment of both animal remains (Noe-Nygaard and Richter 1990) and human remains (Torbenson *et al.* 1992), and human cannibalism (White 1992). The analysis and interpretation of butchery-marked bones is a research topic that is perceived by many zooarchaeologists as important. And, given the definition of taphonomy subscribed to in this volume and the definition of butchery used in this chapter, butchering should be considered an important research topic by taphonomists if there is evidence that hominids were involved in the taphonomic history of the bone assemblage under study.

In this section I have outlined and exemplified analytical procedures and assumptions that are necessary to the study of butchering. The conceptual framework for such study underscores the importance of, in particular, an assumption that is critical to the analysis of frequencies of butchery-marked bones: the frequencies of butchery-marked bones must be directly correlated with the frequencies of butchered bones, some of which may not be butchery-marked. If butchery marks are to some significant degree fortuitous epiphenomena, which they may well be in some cases (e.g., Lyman 1993a), then there is little point in doing more than simply recording that some bones in an assemblage are butchery marked. Here, I think, resides an area demanding many more ethnoarchaeological data than are presently available. We need to know not only how hominid butchering reduces animal carcasses into consumable parts, but how those behaviors are reflected archaeologically. We need details of how that reflection appears in terms of frequency distributions of marks across particular anatomical loci, and how and why variation between such frequencies can be created. We need to know more about how to count butchery marks (should they be tallied by skeletal element, by proximal end, distal end, and shaft). Without these and other kinds of information, our analyses of butchering marks will remain much like those described above: inductive pattern-recognition studies that identify patterns and variations in butchery mark data, the hominid behavioral significance of which is debatable.

In this first section, I have focused on butchery marks. But butchering often also results in broken bones (e.g., Binford 1978; Lyman 1978; Noe-Nygaard 1977, 1987; Yellen 1977, 1991a, 1991b). And while this section also tends to focus on butchering to extract food resources, it should be clear that other kinds of resources are also typically extracted from animal carcasses (Table 8.1). One of these resources is bone used to make tools. Making bone tools can result in broken bones. Thus bone fragmentation seems to be a reasonable bridge between discussion of butchering to extract food and discussion of bone tools. Thus, it is to the important topic of fractured bones that we now turn.

Fracturing of bone

> It cannot be assumed that all split and fractured bone on an archaeological site has been broken by man.
> (J. D. Clark 1972:149)

Perhaps it was Raymond Dart's (e.g., 1957) claim that Plio-Pleistocene hominids had broken some of the bones recovered from the Makapansgat limeworks in South Africa in distinctive ways that served as the catalyst for intensive and extensive investigation of how and why bones are broken the way they are. Certainly Dart's claims were then, and are now, controversial (e.g., Hill 1976; Maguire *et al.* 1980; Read 1971; Read-Martin and Read 1975; Shipman and Phillips 1976; Shipman and Phillips-Conroy 1977; Wolberg 1970; see Brain 1989 for a recent overview), as are similar claims for some broken bones found in North America (e.g., Bonnichsen 1973, 1979; Haynes 1983b; Irving and Harington 1973; Irving *et al.* 1989; Johnson 1983, 1985; Lyman 1984a; Miller 1969, 1975; Morlan 1980, 1983, 1984, 1988; Sadek-Kooros 1972, 1975). The controversy has led to a great deal of actualistic research on bone fragmentation. To understand fully the significance of bone fracture for taphonomic studies, I begin with a discussion of the fracture mechanics of bones before describing types of fractures and identifying agents of bone fracture.

Mechanics of bone fracture

Force must be applied to a bone to break it. *Strain* is the change in linear dimensions of a body resulting from the application of force. *Stress* is the ratio of the amount of force applied to the area over which the force is acting, and is often used synonymously with strength. *Elasticity* is the property that allows a body to return to its original shape and size after an applied force is removed. The *modulus of elasticity* is the ratio between unit stress and unit strain, and it measures the *stiffness* of a material, not its elasticity. Plotting the stress produced in a material against the strain, and increasing the stress, produces a curve indicating the stiffness of the material. The resulting curve is used to determine the breaking point and the amount of energy absorbed by the material prior to its fracture (Currey 1984; Davis 1985:55–56, 62–63; Johnson 1985).

> A *tensile* force is one which tends to pull a body apart, whereas a *compressive* force is one which tends to push a body together. A *shearing* force is one which causes one part of a body to slide in a direction opposite to that of an immediately adjacent part. *Torsion* or twisting combines tensile and shearing forces, while *bending* involves a combination of tensile and compressive forces.
> (Evans 1961:110–111)

Static loading involves the application of constant compressive pressure, generally with an even distribution of force (Johnson 1985:192). *Dynamic loading* involves focused sudden impact (Johnson 1985:170, 192). When either kind of loading exceeds a bone's tensile strength the bone fractures. The more rapid the loading rate, the less the maximum strain a bone can withstand and the less energy it can absorb before fracture (Davis 1985:63). Fracture begins in the outer layer of the bone and progresses inward. Stress waves created by the fracture front are variously absorbed, reflected and diffused by trabecular bone and thus fracture fronts tend not to pass through the epiphyseal ends of long bones; stress waves deform the dense cortical bone of long bone diaphyses and may fracture them. Dynamic loading imparts bending forces to a long bone, and "shearing is along a helical course that is inclined at a 45° angle to the longitudinal axis of the long bone" (Johnson 1985:171) to produce a spiral or helical fracture, which is a tensile failure. "Dry and mineralized bone exhibits horizontal tension failure in which the fracture front cuts across the diaphysis and produces perpendicular, parallel, or diagonal breaks" (Johnson 1985:172).

The composite nature of bone tissue – being part mineral and part organic – results in it being mechanically strong, stronger than either material alone (Chapter 4). A bone's microstructure governs bone fracture and the resultant kind of fracture, where *fracture* is "a localized mechanical failure" (Johnson 1985:160). Currey (1984:49) states that "bone deforms rather little before fracturing ... it does not fracture with a smooth surface, as a really brittle material does, and so cannot be truly brittle." Johnson (1985:160) elaborates that fresh, green bone contains moisture and marrow in the medullary cavity, and will "not behave in a brittle (or inflexible) manner, but rather, is a viscoelastic (i.e., flowable and deformable), ductile material capable of withstanding great amounts of pressure and deformation before failure [fracture]. Bone does behave in a brittle manner when it is well dried." A bone's stiffness, tensile strength, compressive strength, and hardness are all increased by drying (Davis 1985:66–67). Drying causes a bone to shrink, thereby increasing its bulk density and reducing its porosity. Thus dry bone behaves more like an inorganic material than wet bone, and "although dry bone might be stronger in static loading, it would be more likely to fail [fracture] at smaller forces during dynamic loading" (Davis 1985:67).

Johnson (1985:167) notes that osteons tend to decrease tensile strength (more osteons mean more cement, which is mechanically weak) whereas lamellae tend to increase tensile strength (see also Davis 1985:65). Fracture begins with what are called *microcracks*, which tend to follow cement lines around osteons (see Chapter 4). "As failure [fracture] occurs, the paths that the propagating fracture fronts take are the result of a dynamic interaction between the loading device, fracture dynamics, and bone structure" (Johnson 1985:170). Haversian and Volkman's canals reduce a bone's stiffness, and as the proportion of interstitial lamellae increases, the stress a bone can withstand without fracturing is reduced (Davis 1985:65–66).

This consideration of the biomechanics of fracture does not do justice to the extensive literature on the topic. But I have been brief because, as Davis (1985:70) notes, much of that literature concerns samples of bone tissue rather than bones or skeletal elements. Taphonomic agents fracture complete skeletal elements, and the literature on the biomechanics of whole bones is rather limited (see Davis 1985 and Morlan 1980 for introductions to that literature).

Davis' (1985) most interesting conclusions regarding the fracture of complete skeletal elements can be summarized as follows. (1) Bone size as sorted by the live weight of the animal contributing the bone does not influence bone fracture patterns. (2) Macrostructural differences between skeletal elements account for fracture location. "Differences in cross-sectional thickness of the cortical bone between elements and the presence of crests such as the anteriorly positioned tibial crest are the factors most likely to determine the location of a fracture" (Davis 1985:94). (3) Microstructural differences in skeletal elements exert some influence on fracture form. "For example, the collagen alignment in humeri is more likely to produce fractures with curved edges" (Davis 1985:97). Mid-shaft cross-sectional shape and diameter relative to cortical bone thickness also seem to influence fracture form. (4) Lightly weathered bone will fracture obliquely, but the proportion of such fractures decreases as the extent of weathering increases. There is a tendency for more intensively weathered bones to require less force for fracturing. (5) "In static loading the bone fails in tension usually at a point opposite to the point of loading. In dynamic loading, several fracture fronts form and radiate away from the point of impact" (Davis 1985:108). (6) "A high percentage of oblique fractures is expected in any assemblage due to properties inherent in the bones" (Davis 1985:129). (7) Statistically significant greater than expected frequencies of oblique fractures (spiral fractures, see below) occur under conditions of static and torsional loading, whereas dynamic loading produces fewer oblique fractures than expected. This is probably because under low strain rates such as static and torsional loading, fracture is between lamellae and obliquely oriented collagen fibers whereas the high strain rates induced by dynamic loading create fracture fronts that may cross lamellae (Davis 1985:133). (8) Dynamically loaded bones are more likely to display rounded fracture ends than statically loaded and fractured bones.

Davis' (1985) research results reflect the variables analysts study when examining broken bones. Taphonomists focus on the morphological attributes of fractures in attempts to identify the agent of bone fracture, the kind of force that resulted in a fracture, and the condition of a bone when it was fractured. We turn, then, first to types of fractures, and subsequently to identifying fracture agents.

Types of fracture

Johnson (1985:172) uses the following terms and definitions to discuss attributes of broken bones: *fracture location* – the area where failure occurred; *fracture front* – the leading edge of force and its direction (can be determined from features on the fracture surface); *fracture surface* – created by failure, a cross-section of compact bone exposed by the passage of force through it; *fracture shape* – the outline configuration of exposed compact bone that records the propagation path taken by the fracture front in planview. I utilize this terminology in the following discussion.

Shipman *et al.* (1981) describe and illustrate seven types of fracture. Their scheme was modified by Marshall (1989:14) to include an eighth type (Figure 8.4). Shipman *et al.* (1981:260) point out that "only rarely will a particular type of break identify the agent of breakage unambiguously." They note that the location and frequency of different types of breaks on different types of bones can be compared within and between assemblages. They suggest that comparison of such data for assemblages with known agents of bone fracture with prehistoric assemblages may reveal the agent of breakage in the latter. Most importantly, they argue, correctly I believe, that the analyst should "compare like with like" in such analyses (Shipman *et al.* 1981:259); that is, mammal bones of like-sized animals should be compared rather than, say, the broken bones of a rabbit with the broken bones of a cow or bird. This makes sense because different bones will break differently simply because of microstructural differences.

The types of fracture in Figure 8.4 are generalized. As well, the necessary and sufficient conditions for identifying a particular specimen as displaying one kind of fracture or another are not specified by Shipman *et al.* (1981) or by Marshall (1989). Gifford-Gonzalez (1989a:188) defines several of the fractures types: *perpendicular* or *transverse* fractures are at a right angle to the long axis of the bone; *longitudinal* fractures are parallel to the long axis of the bone; *spiral* fractures are curved in a helical, partially helical, or completely helical pattern around the circumference of the shaft. Shipman (1981a:371–372) and Johnson (1985:175), however, note that the type "spiral fracture" consists of at least two subtypes, one having a smooth fracture surface (fracture front apparently went between collagen bundles) and the other having a rough fracture surface (fracture front apparently went through or was perpendicular to collagen

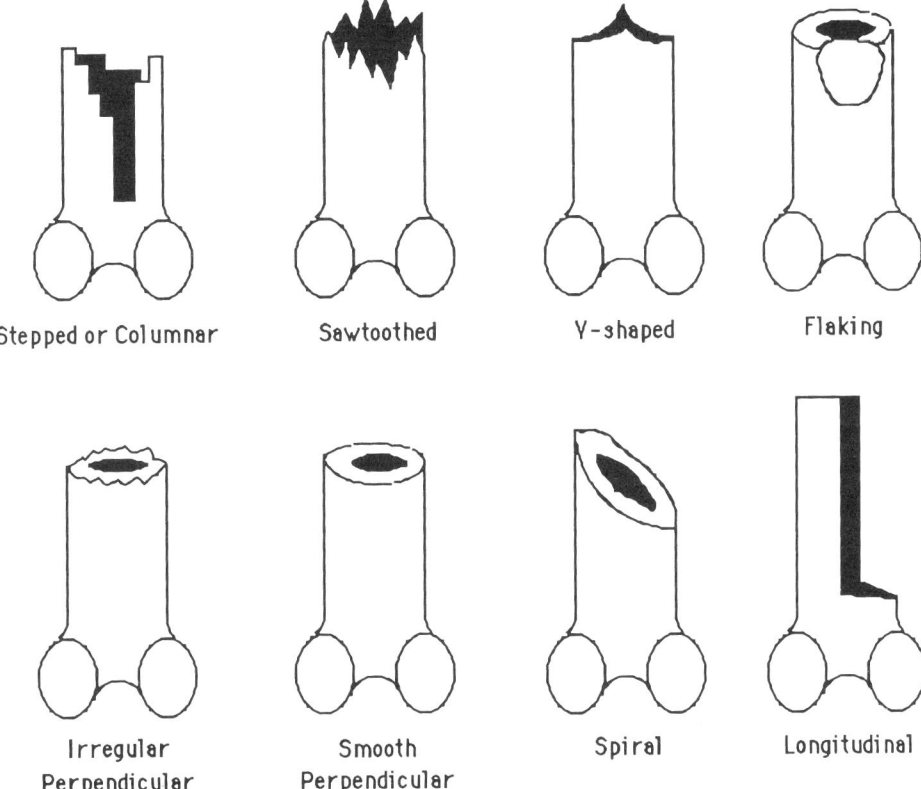

Figure 8.4. Fracture types (after Marshall 1989:14, Figure 1; courtesy of the author and The Center for the Study of the First Americans). All are schematic, with fractures occurring on the distal end of an artiodactyl femur shown in posterior view.

bundles). The former is called a Type I spiral fracture by Shipman (1981a) and is termed a *horizontal tension failure* or *fracture* by Johnson (1985); it is believed to result from the breakage of dry bone. The latter is called a Type II spiral fracture by Shipman (1981a) and a *true spiral* or *helical fracture* by Johnson (1985); it is believed to result from the breakage of fresh, green bone. The fracture planes of Type I/horizontal tension failures illustrated by Johnson (1985:174) are either perpendicular to the long axis of the bone or form an oblique angle relative to the long axis of the bone. Davis' (1985) term "oblique fracture" includes both Type I and Type II spiral fractures.

A more practical problem from an analytical standpoint concerns how one identifies the kind of fracture displayed by a specimen. Bunn (1982:43) points out that "on a specimen with many fractured edges, it is not obvious how to define the scale at which descriptive terms such as spiral, stepped, jagged, etc. should apply or how to define a boundary between the end of one such fracture shape and the beginning of another." Potts (1982:214), for example, reports that in one assemblage of bones he studied (NISP = 627) he found "possible segments of spiral fractures *and* longitudinal breaks on at least one side of over 90% of the specimens." These observations underscore the descriptive and rather general nature of the fracture types in Figure 8.4.

Another typology of fracture shape is described by Johnson (1985:172, 177). She illustrates six types of fracture shapes as seen in planview. These are: "transverse (straight edge), wide curved (rounded edge, sides far apart), narrow curved (rounded edge, sides close), pointed (converging apex edge), stepped (interrupted edge), and scalloped (series of semicircles or curves along an edge)." Respectively, Johnson's types of fracture shapes are most closely aligned with the smooth perpendicular, spiral, spiral (respectively), V-shaped, stepped or columnar, and irregular perpendicular types in Figure 8.4. However, there is minimal agreement in detail between the two typologies. This again underscores the subjective and intuitive nature of the typologies, and indirectly illustrates how difficult they may be to use.

Both typologies described above are meant to offer insight to the condition of the bone when it was broken. For example, based on experimental data, Johnson (1985:176) modifies a scheme originally proposed by Morlan (1980:48–49) and outlines a set of criteria many analysts use to discriminate between fractured bones that were broken when fresh or green and those that were broken when the bone was mineralized. She writes "fresh break fracture surfaces have the same color as the outer cortical surface, exhibit a smooth texture, and form acute and obtuse angles with the outer cortical surface. [Mineralized] break fracture surfaces have a contrasting color, exhibit a rough texture, and form right angles with the outer cortical surface" (Johnson 1985:176). The fracture surface of specimens broken when dry (not mineralized) may have a rough, bumpy texture and have angles that are a combination of acute, obtuse, and right relative to the outer cortical surface. Bones with spiral or helical (Type II) fractures that have a rough fracture surface seem to represent bones that were green or fresh when broken, as do flaked bones. Bones with spiral or helical fractures that have a smooth fracture surface were probably dry and/or slightly weathered; these fractures may be stepped in part due to the helical fracture front encountering a split line crack. Weathered and/ or fossilized bones tend to break in such manners as to fall in the stepped, longitudinal, and smooth perpendicular types. Fracture surfaces may pass through the epiphyseal ends of dry bones.

Davis (1985) broke tibiae, femora, and humeri of various African bovids. The broken bones were variously fresh and dry, and loading was variously static and dynamic but was consistently applied to mid-shaft locations. Some bones were also broken using torsional loading. Following recommendations by Hill (1980), Davis (1985:78–79) developed a system for classifying bone fractures. The classification uses a terminology that is intended to be purely

Table 8.7 *Fracture classification system of Davis (1985)*

Attribute
Attribute states (letter designations after Davis 1985:82)

Fracture orientation: relative to the long axis of the specimen; for long bones, the long axis is parallel to the long axis of the complete bone [NOTE: these attribute states are incorporated into the fracture morphology attribute below]

    X: mixed

    Y: parallel

    Z: oblique

Fracture surface location: based on the anatomical position of the area of maximum exposure of fracture surface with the specimen lying flat on a horizontal surface with the marrow cavity exposed (upward)

    A: anterior

    B: posterior

    C: medial

    D: lateral

Fracture morphology: the total form of the fracture by independent coding of the lateral, proximal, and distal edges of the fracture; the fracture surface closest to the epiphysis is the proximal end, the fracture surface farthest from the epiphysis is the distal end, the two remaining sides of the fracture surface are the lateral edges

    A: parallel and smooth

    B: oblique and smooth

    C: oblique and stepped

    H: curved in one or more planes

    I: irregular fractures

    J: V-shaped end morphologies

    K: horizontal to long axis

descriptive and not implicate a fracture agent (Table 8.7). Davis (1985:80–81) suggests that each fracture can be described with a four-letter code which describes the "form of the fracture. For example, a fracture with straight medial and lateral edges parallel to the long axis of the bone and proximal and distal edges straight and perpendicular [to the long axis of the bone], would be coded as an AAAA fracture." A helical spiral fracture illustrated by Davis (1985:82) is coded as a JJJJ and a saw-toothed specimen (Figure 8.4) as an IIII. Davis' (1985) treatment of fracture location circumvents the problem noted by Bunn (1982) and cited above about fractures that display varied morphologies. Davis' system allows the analyst to record four fracture morphologies in four adjoining locations along the fracture surface. Her system does not allow recording whether the proximal edge of the fracture surface is near an epiphysis, in a proximal shaft location, or in a mid shaft location. I have in Table 8.7 attempted to clarify some aspects of Davis' classification system, but several problems remain. For example, she mentions that several attribute states should be recorded relative to one or more of the following planes: the "z" axis, the "horizontal axis," and the "vertical axis." Davis' (1985) attempt

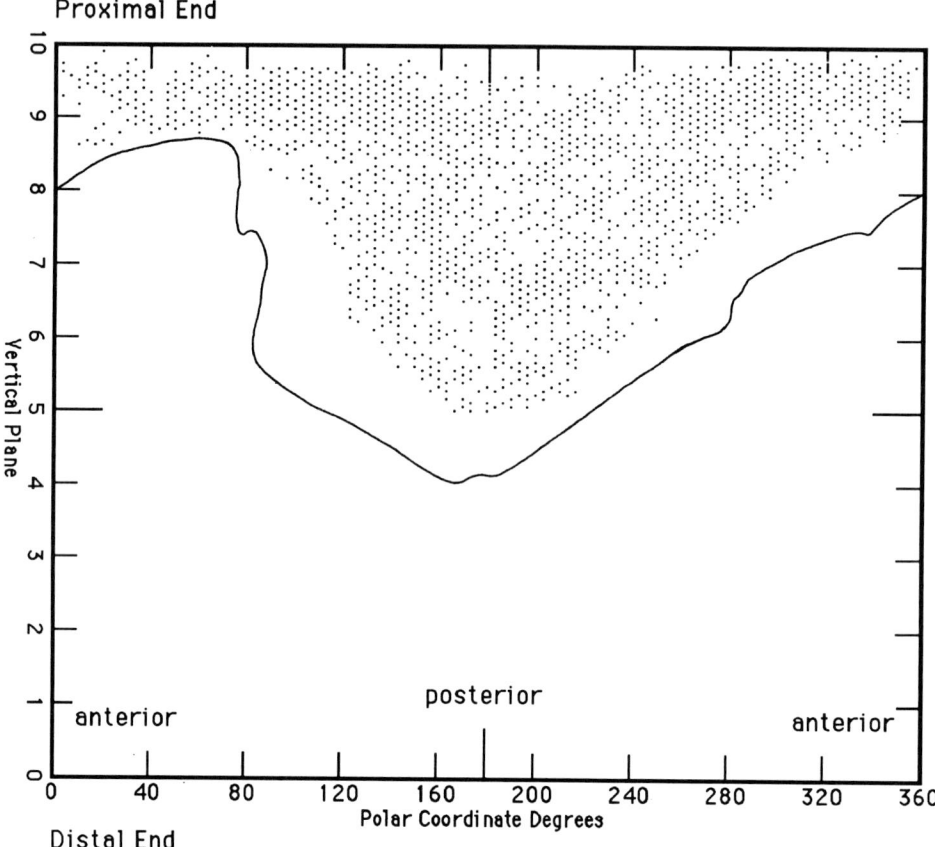

Figure 8.5. Fracture edge morphology of a broken metacarpal (shown in Figure 8.6) illustrated using Biddick and Tomenchuk's (1975) system of polar coordinates and vertical planes. Shaded area is the portion of the bone represented.

to circumvent some of the inter-analyst subjectivity she correctly perceives in typological schemes like those described in preceding paragraphs is effectively thwarted because she defines none of these planes. I believe, however, that the attributes and attribute states she uses are important ones, and with some clarification her classification could become a standard part of the recording and analysis of bone fractures.

Perhaps the most rigorous system for recording the morphology and location of a fracture is proposed by Biddick and Tomenchuk (1975). They describe a system of dividing a long bone into ten equal-sized length sections with what they term "vertical planes." This system utilizes polar coordinates to the nearest 10° at each vertical plane to record fracture shape. The polar coordinates are marked on elastic bands which are placed on a specimen at each vertical plane, with 0° aligning with the anterior sagittal plane and 180° aligning

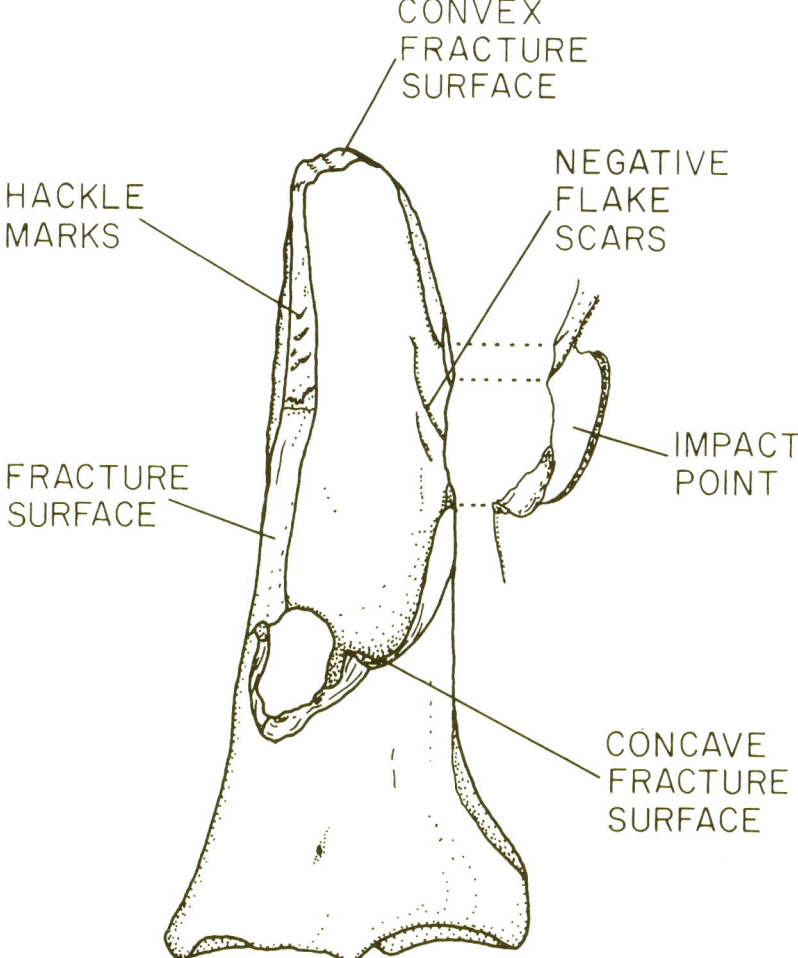

Figure 8.6. Features of fracture surfaces shown on a bovid proximal metacarpal (after Johnson 1985:177, Figure 5.5; courtesy of the author and Academic Press).

with the posterior sagittal plane. A reading is taken at each polar coordinate from the nearest vertical plane and recorded on a bivariate graph with vertical planes on the vertical axis and polar coordinates on the horizontal axis. Interpolation between plotted points produces a two-dimensional illustration of the location and shape of the fracture surface in the form illustrated in Figure 8.5, which shows the morphology of the exterior edge of the fracture surface of the bone illustrated in Figure 8.6.

While Biddick and Tomenchuk (1975:248) emphasize that their system provides data that are "precise, repeatable, computerizable, and statistically analyzable," no one to the best of my knowledge has employed it, although a decade later Münzel (1986) described a similar system for recording bone fragments using a three-dimensional coordinate system. Given the subjectivity of the typologies of fracture types described in preceding paragraphs, I think Biddick and Tomenchuk's (1975) system warrants careful consideration by taphonomists interested in studying details of the morphology of bone fractures.

A comment about spiral fractures

A majority of the literature on bone fracture concerns what have been generally labeled *spiral fractures*. That literature had its significant beginnings with claims by Dart (e.g., 1957) that early Pleistocene hominids utilized a "crack-and-twist" method of breaking bones to extract marrow and produce bone tools. The body of literature that resulted, both in Old World and New World contexts, is truly immense. A sampling of that literature is cited in the first paragraph of this section. What has become clear in the years since Dart's claims were made is that any number of taphonomic processes can create spiral fractures (Type II/true spiral fractures, above), including trampling (Haynes 1983b, 1991; Myers *et al.* 1980), carcasses falling some distance (Lyman 1984b), carnivore gnawing (Binford 1981b), and hominid activities (Bonnichsen and Will 1980; Zierhut 1967). It is also now clear that at least two kinds of fractures have been termed "spiral fractures," both horizontal tension failures and true spiral or helical fractures (Johnson 1985; see above). Most importantly, it is now known that because a bone displays a true spiral fracture (Type II) does *not* signify the agent of fracture. Johnson (1985:175) reports that a true spiral or helical fracture indicates only that the bone was broken when fresh; "it does not necessarily indicate the agency involved." She suggests that identifying the agent of bone fracture requires study of attributes displayed by the bone surface and features of the fracture. Determination of the agent of fracture thus now involves the study of multiple attributes, not just the kind of fracture displayed by a specimen. It is identifying agents of bone fracture to which we now turn.

Fracture agents

Identifying the taphonomic agent responsible for the fractured bones in an assemblage, when possible, can tell us much about the formational history of a bone assemblage. As listed by Davis (1985:7), major causes of bone fracture include (but are not limited to) hominids making tools, feeding hominids (e.g., Lyman 1978; Noe-Nygaard 1977), feeding carnivores (e.g., Binford 1981b), trampling (Chapter 9), factors related to climate such as subaerial weathering (Chapter 9), and post-burial factors such as compression forces induced by overburden weight (Chapter 11). Kurtén (1953:70) suggests that in bone assemblages with multiple individuals and a catastrophic mortality profile

(Chapter 5), broken bones may have "resulted from animals tumbling down a precipice when fleeing before a fire." Lyman (1984b) describes bones broken by processes related to volcanic activity. Certainly the list could be expanded to include pathological or antemortem fractures as well as other taphonomic processes that break bones.

The problem when studying an assemblage including broken skeletal elements is to determine how and why they are broken (and less frequently why some specimens are complete skeletal elements, see below). The analytical procedure generally involves the study of multiple attributes, especially those which appear on the basis of actualistic data to be diagnostic of taphonomic agents known at least occasionally to break bones. I review several of the better known and most intensively studied agents of bone fracture, and describe the diagnostic attributes they produce.

Feeding carnivores as fracture agents
The distinction between modification and destruction of bone on one hand, and simple fracture of bones on the other, is not always a clear one in the literature (e.g., Binford 1981b). I review the major kinds of modifications to bones (Chapter 6) and attributes of destruction of bones (Chapter 7) elsewhere. Those discussions supplement the following.

Bone-gnawing carnivores tend to break bones two ways, both of which involve static loading. One involves first removing, by chewing and gnawing, one or both epiphyseal ends of a long bone. This significantly weakens the structural strength of the bone. The diaphysis is then chewed, and it "collapses into long rectilinear splinters that generally follow the longitudinally aligned collagen bundles" (Johnson 1985:192). Gnawing marks in the form of furrows and punctures remain on the epiphyseal end, scoring and pitting occur on the diaphysis, and the end of the diaphysis may be scalloped (Figures 6.19, 6.20, and 6.23). The other way carnivores fracture bones involves simply chewing the complete bone, with force applied along the diaphysis until the bone fractures (Johnson 1985:192). Scoring and pitting of the diaphysis are common. The kinds of fractures that result from these two ways in which bones are gnawed depend on the skeletal element (microstructural and macrostructural features) and whether the bone is fresh and green or somewhat dry (for more details, see Binford 1981b; Bonnichsen 1973, 1979; Haynes 1980b, 1982, 1983b; Miller 1969, 1975).

In the only case reported in detail of which I am aware, Maguire *et al.* (1980) describe some sheep bones known to have been butchered and gnawed by humans. They indicate that in their limited sample, human gnawing only crushed and crunched fairly soft bone, such as along the edge of the ischium. Thus, the simple presence of gnawing damage on broken bones does not mean carnivores are the fracture agent. Channeled fractures and chipped-back edges on the diaphyses of long bones, for example (Chapter 6), do not seem to be created by hominid gnawing, although the latter may be occasionally produced by hominids making bone tools (see below). Thus the distributions as well as the kinds of gnawing damage must be considered.

Hominids as agents of bone fracture

Hominids can break bones several ways. Virtually all of them utilize dynamic loading. Direct percussion often creates "point loading" (e.g., Johnson 1985:192), percussion pits (Blumenschine and Selvaggio 1988, 1991), and/or flake scars (Lyman 1987a). Other features depend in part on such things as whether the bone, when impacted, is supported by a single anvil under the impact point, an anvil at each end of the bone, or an anvil at one end and the other end rests on the ground (Johnson 1985:209; Miller 1989:386).

Fresh long bones broken via dynamic loading tend to produce true spiral or helical (Type II) fractures. But such a fracture type does not, by itself, indicate a hominid fracture agent. The presence of a loading or impact point, in conjunction with the absence of gnawing marks and a loading point the diameter of which is greater than that produced by carnivore teeth (but see Figure 6.25 and associated discussion) seem to be the attributes used by most analysts to distinguish humanly broken bones from those broken by other processes and agents.

The *impact* or *loading point* is a circular or oval depressed area marked by incipient ring cracks or crushed bone (refer to Figure 8.6 in this and the following paragraph). The outer cortical edge, when viewed from the exterior, often will display a crescent-shaped notch at the loading point (Figure 8.7). Impact points are distinguished from *rebound points* which represent force redirected back into the bone from the anvil supporting it. In loading points, bone flakes analogous to lithic flakes may still be attached within the medullary cavity at their distal ends; alternatively, flake scars may be visible within the medullary cavity and beneath the loading point. A rebound point tends to be smaller than its related loading point, is located opposite its loading point, and displays little to no crushing (Johnson 1985:194, 210).

Fracture surfaces on bones broken while fresh tend to be variously weakly concave and convex, have a relatively fine texture macroscopically but a rough texture microscopically, and form acute and obtuse angles with the outer cortical surface of the bone. Other attributes of fracture surfaces are variously formed as a result of stress relief and impediments to the fracture front as it looses strength. *Hackle marks* are discontinuous curved grooves and ridges; *ribs* are semicircular or arcuate ridges that are usually continuous and concave relative to the origin of the fracture (Johnson 1985). Both hackle marks and ribs are stress relief features that spread outward from the point of impact and are diagnostic of dynamic loading. The presence of hackle marks and ribs indicates dynamic loading of green bone; their absence does not, however, indicate static loading. *Chattering* appears as highly accentuated, closely spaced, straight

a

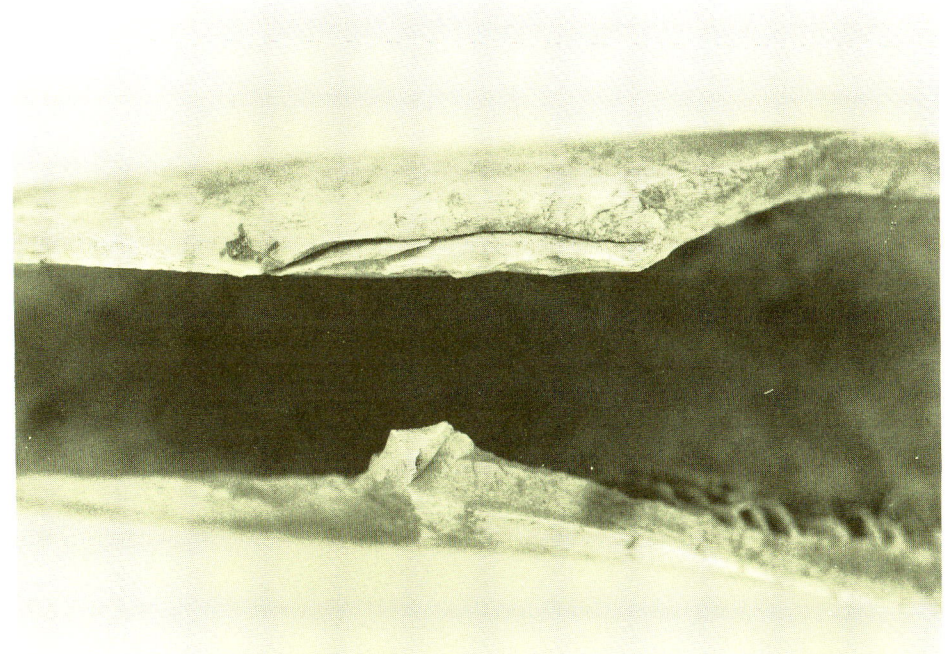

b

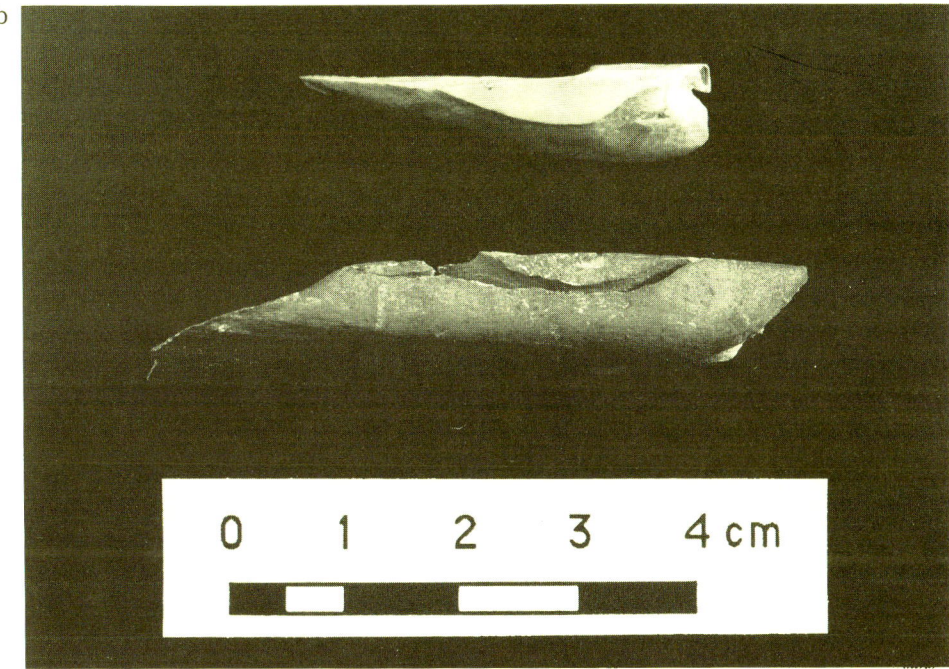

Figure 8.7. Loading points. a, deer (*Odocoileus* sp.) humerus shaft with flake still in place; b, lower, deer radius shaft with flake still in place; upper, flake removed from a loading point.

peaks and valleys; *stepping* is created by *split lines* causing an interruption to the flow of force and results in stepped or jagged fracture edges (Figure 8.4). Chattering and stepping are interference features. *Wedge flakes* are bone flakes that are removed from the exterior cortical surface of the bone created by bending failure when the bone flexes (Johnson 1985:194, 197).

Fracture due to other agents
Fracture due to trampling is discussed in Chapter 6. Crushing and fracturing of bones due to the weight of sediment overburden is mentioned in the following discussion and in Chapter 11. Subaerial weathering induced fracture is discussed in Chapter 9. The final topic of bone fracture to cover, then, is how one analyzes an assemblage containing some broken skeletal elements. In the following, several different kinds of analyses are summarized to illustrate various kinds of data and analytic techniques.

Analysis of fractured bones

Fracture types
In an intensive and extensive study of patterning in bone fractures, Villa and Mahieu (1991) describe a series of attributes of fractures, and compare the frequency distributions of those attributes across three prehistoric assemblages of human remains from France. The Late Neolithic Sarrians assemblage is from a collective burial and is made up of bones believed to have been broken *in situ* via overburden crushing as the bones dried. Conjoining fragments lay adjacent to one another, incomplete fractures or cracks were noted in some specimens, and breakage occurs in bones resting on concave or convex surfaces. The Early and Middle Neolithic Fontbrégoua assemblage is thought to represent an instance of cannibalism in which bones were broken while fresh by hominids. About 20% of the specimens have impact notches, half of which have microflakes adhering to the impact point, and 30% of the specimens have cut marks. Fracture surfaces have sharp edges. Conjoining specimens are separated by as much as 50 cm, and when adjacent they are not in correct anatomical position relative to one another. Stratigraphic boundaries closely define the extent of the bone cluster. Carnivore gnawing damage is not present. The Bezouce assemblage is from a collective burial and consists of bones broken during excavation. The bones were dry (not fresh) when broken.

   Villa and Mahieu (1991) compare the proportional frequencies of several attributes per assemblage. The *fracture angle* is the angle formed by the fracture surface and the cortical surface of the bone, and includes three attribute states: (1) oblique (obtuse or acute), (2) right, and (3) oblique and right (fractures with variable angles). The *fracture outline* or *shape* includes three attribute states: (1) transverse (fracture surface is straight and transverse to the bone long axis), (2) curved (spiral fracture combined with V-shaped or pointed fractures; complex,

Table 8.8 *Frequencies of fracture attributes in three assemblages of human bones. Values for attribute states are NISP (% of total NISP) (from Villa and Mahieu 1991)*

Fracture attribute	Sarrians	Bezouce	Fontbrégoua
Fracture angle, total NISP	269	253	174
oblique	22 (8.2)	27 (10.7)	114 (65.5)
right	176 (65.4)	174 (68.8)	47 (27.0)
oblique and right	71 (26.4)	52 (20.5)	13 (7.5)
Fracture outline, total NISP	358	287	261
transverse	193 (53.9)	144 (50.2)	92 (35.3)
curved/V-shaped	74/32 (29.6)	59/23 (28.6)	42/92 (51.3)
intermediate	59 (16.5)	61 (21.2)	35 (13.4)
Fracture edge, total NISP	358	287	261
smooth	111 (31.0)	158 (55.1)	163 (62.5)
jagged	247 (69.0)	129 (44.9)	98 (37.5)
Shaft circumference, total NISP	226	93	151
< 1/2	16 (7.1)	33 (35.5)	115 (76.2)
> 1/2, < complete	10 (4.4)	0 (0.0)	23 (15.2)
complete	200 (88.5)	60 (64.5)	13 (8.6)

multidirectional fracture surface), and (3) intermediate (straight morphology but diagonal to bone long axis, and stepped). *Fracture edge* refers to the texture of the fracture surface and includes two attribute states: (1) smooth and (2) jagged. *Shaft circumference* excludes specimens <4 cm long and has three attribute states: (1) maximum circumference is less than half of the original skeletal element, (2) maximum circumference is more than half in at least a portion of the length, and (3) complete circumference in at least a portion of the length. *Shaft length* concerns how much of the diaphysis is represented, excludes specimens <4 cm long, does not include consideration of the presence or absence of the epiphyseal ends, and consists of four attribute states: (1) shaft is < one fourth the length of the complete bone, (2) shaft is > one fourth but ≤ one half of complete length, (3) shaft is ≥ one half but ≤ three fourths of complete length, and (4) shaft is ≥ three fourths of complete length. (The shaft circumference and shaft length variables are similar to ones developed by Bunn [1983].) *Shaft fragmentation* is determined by plotting shaft circumference and shaft length in a bivariate graph (Villa and Mahieu 1991:42). Villa and Mahieu also examine the breadth/length ratios of shaft fragments, including only specimens with a breadth less than the original complete bone and ≥ 4 cm long because smaller specimens are difficult to identify to skeletal element.

Data for fracture angle, fracture outline, fracture edge, shaft circumference, and shaft length for the three assemblages are given in Table 8.8. Chi^2 analyses of the proportional frequencies of fracture angles and fracture outlines indicate the Sarrians and Bezouce assemblages are similar to one another and both

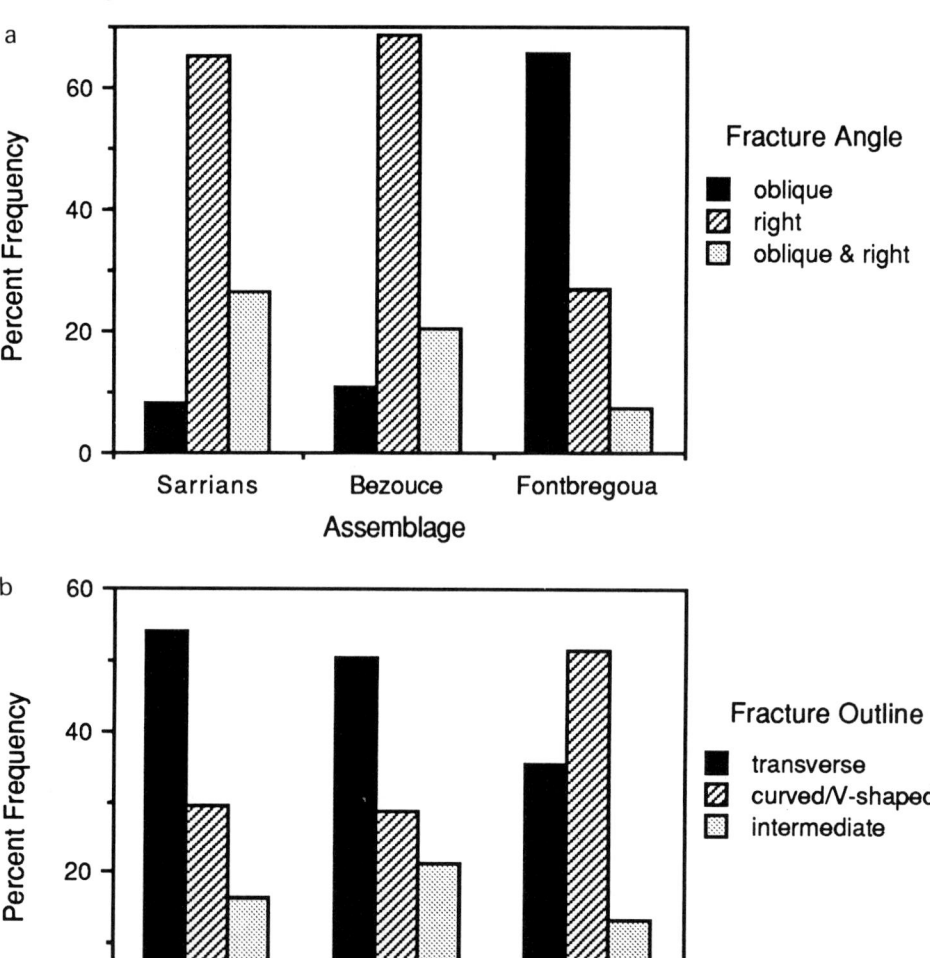

differ from the Fontbrégoua assemblage (Figure 8.8a and 8.8b). Thus Villa and Mahieu (1991) suggest that these two fracture attributes aid in distinguishing bones fractured when fresh from those fractured when dry. Chi² analysis of the fracture edge variable did not distinguish the assemblages in this manner, and Villa and Mahieu (1991:40) conclude that "this attribute will not discriminate between old and fresh bone breakage." They suggest that the kind of force applied – static or dynamic – may exert a stronger influence on this variable than whether a bone is old and desiccated or fresh when broken. The proportions of categories of shaft circumference completeness also indicate greater similarity between the Sarrians and Bezouce assemblages than between either of them and the Fontbrégoua assemblage (Figure 8.8c). Villa and

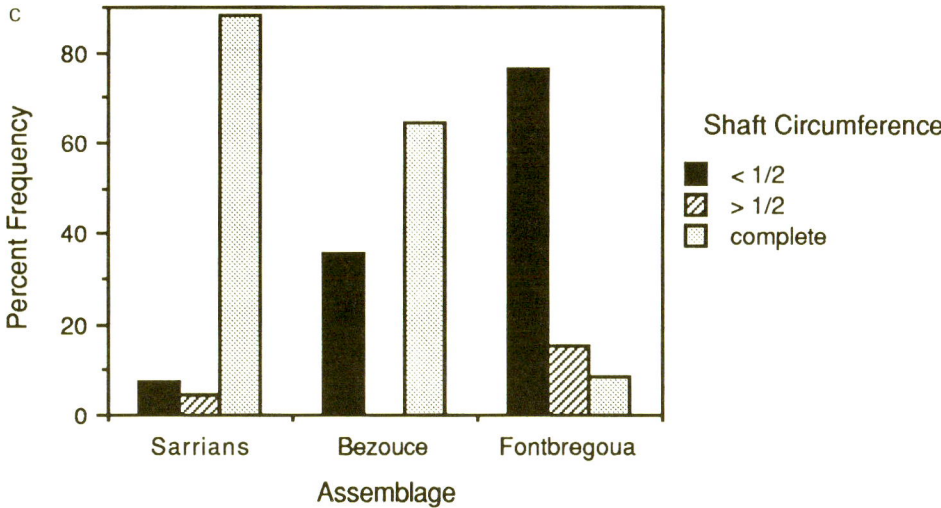

Figure 8.8. Bar graphs of three bone fragmentation attributes for three assemblages. a, percentage of total frequency per kind of fracture angle; b, percentage of total frequency per kind of fracture outline; c, percentage of total frequency per size of shaft circumference (from Table 8.8).

Mahieu (1991:41) conclude that "high frequencies of complete shaft diameters appear to characterize assemblages of post-depositionally broken bones" (see also Chapters 10 and 11). They conclude that post-depositional breakage tends to create short tubular (complete circumference) pieces of shafts whereas fracture of green bone creates splinters of various lengths with variously incomplete and complete circumferences. Thus the breadth to length ratio is lowest for the Fontbrégoua assemblage and is higher for the Sarrians and Bezouce assemblages (the latter two are not statistically significantly different).

Villa and Mahieu's (1991) analysis depends on the frequencies of various morphological attributes of the fractures. This raises the issue of how to quantify or tally such attributes. It seems that Villa and Mahieu (1991) tally the frequency of specimens or NISP that display each attribute. Thus a single specimen displaying a spiral (or transverse) fracture would be tallied as an NISP of 1. But two possibilities that might influence such tallies are evident. First, given that some specimens could be refit (e.g., Villa *et al.* 1986), the probability that a particular fracture type was tallied twice for the same skeletal element seems great. That is, a bone broken into two pieces and displaying an oblique fracture will be tallied twice, once for each piece. Is this the correct way to tally this and other fracture attributes? The specimens are anatomically interdependent, but should they be treated as independent units during quantification? The significance of this may be great if, say, oblique fractures are more easily refit than jagged or stepped fractures. More study of this aspect of the quantification of fracture attributes is required.

The second issue concerning the quantification of broken bones involves the

Table 8.9 Frequencies of skeletal parts in raptor pellets (from Hoffman 1988). N whole = number of whole specimens or complete skeletal elements. % whole = 100 (N whole/NISP). NISP:MNE calculated without whole specimens

Skeletal part	Great horned owl					Screech owl					Diurnal hawk				
	NISP	N whole	MNE	% whole	NISP:MNE	NISP	N whole	MNE	% whole	NISP:MNE	NISP	N whole	MNE	% whole	NISP:MNE
mandible	92	57	84	62.0	1.30	24	4	18	16.7	1.43	26	0	18	0.0	1.44
scapula	81	25	78	30.9	1.06	19	0	17	0.0	1.12	3	0	3	0.0	1.06
humerus	92	80	90	97.6	1.20	25	9	22	36.0	1.23	12	0	7	0.0	1.46
radius	69	53	66	76.8	1.23	17	11	16	64.7	1.20	4	0	4	0.0	1.10
ulna	81	64	80	79.0	1.06	10	0	6	0.0	1.67	5	0	4	0.0	1.46
innominate	73	30	73	41.1	1.00	18	3	18	16.7	1.00	5	0	5	0.0	1.00
femur	99	83	95	83.8	1.33	26	12	18	46.2	2.33	11	0	7	0.0	1.95
tibia	100	91	96	91.0	1.60	29	9	18	31.0	2.22	12	0	7	0.0	1.96
totals:	687	483	662	70.3		168	48	133	28.6		78	0	55	0.0	
MNI =	48					11					9				

fact that some broken bones may display one kind of attribute on one part of the fracture surface and another, distinct kind of attribute on a different portion of the fracture surface. The frequency with which this will occur depends, of course, on how attributes of fracture are defined. Thus, the attribute state "oblique fracture" is distinct from the attribute state "jagged fracture," but only by having an attribute state "oblique and jagged" would one be able to tally a specimen displaying both attribute states. The latter may occur when a diaphysis specimen is obliquely fractured on one end but has a jagged fracture at the other end. Again, the influence of this kind of possibility has not been considered in the literature. It should be in the future.

Proportions of broken bones

A potentially important taphonomic variable that should be considered in studies of bone fragmentation is the frequency of complete skeletal elements or unbroken whole bones. To address this topic, I distinguish the extent of fragmentation from the intensity of fragmentation. By *extent of fragmentation* I mean the proportion of the total NISP of an assemblage that consists of broken and incomplete skeletal elements and/or the proportion of the total NISP of an assemblage that consists of whole, unbroken, complete skeletal elements. By *intensity of fragmentation* I mean the size of the fragments. An assemblage of very intensively broken skeletal elements will have many small fragments whereas an assemblage of less intensively broken skeletal elements will have many large fragments. Of course, when measuring the intensity of fragmentation, one must be sure to compare fragments that come from bones of similar size. In the following I first consider measures of the extent of fragmentation. I then describe two ways to measure the intensity of fragmentation.

Extent of fragmentation

The proportion of the total NISP represented by whole skeletal elements is a rather quick way to measure the degree to which the bones in an assemblage are broken (e.g., Dodson and Wexlar 1979). For example, the data in Table 8.9 indicate that bones recovered from great horned owl (*Bubo virginianus*) pellets are more likely to be whole (70.3% of the total NISP are whole) than bones recovered from pellets deposited by screech owls (*Otus asio*) (28.6% of the total NISP whole); pellets cast by diurnal hawks contain no whole bones. The differences in proportions of whole bones in the total assemblages are statistically significant: great horned owl to screech owl arcsine $t_s = 9.94$, $P < 0.001$; great horned owl to diurnal hawk $t_s = 16.61$, $P < 0.001$; screech owl to diurnal hawk $t_s = 8.23$, $P < 0.001$.

The analyst could now inspect individual skeletal elements to determine if certain skeletal elements are more often complete when deposited by great horned owls than when deposited by screech owls. On one hand, the radius

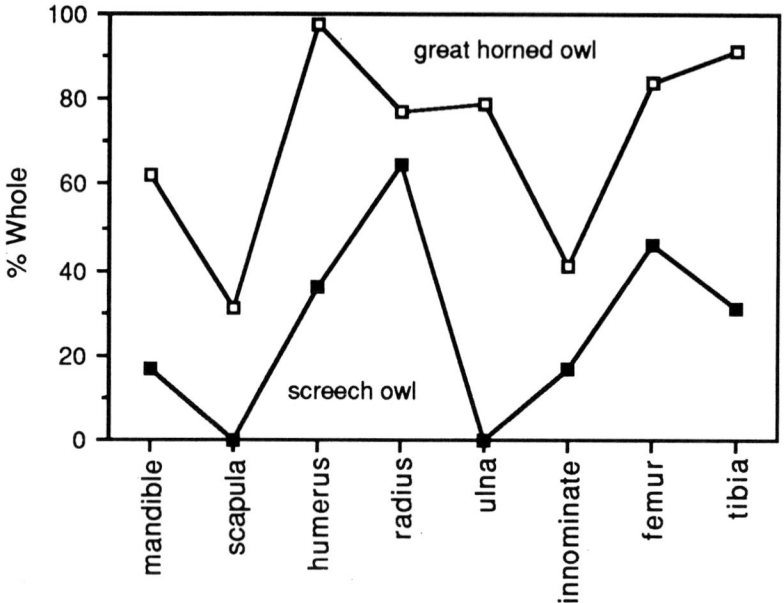

Figure 8.9. Variation in the proportion of complete skeletal elements (% Whole) deposited by two taxa of owls (from Table 8.9).

tends to have as much chance of being whole when deposited by great horned owls as when deposited by screech owls (76.8% whole versus 64.7% whole, respectively; arcsine $t_s = 0.99$, $P > 0.2$). On the other hand all other skeletal elements are much more likely to be broken by screech owls than they are by great horned owls (Figure 8.9). And, the proportions of whole specimens per skeletal element are not statistically similar between these two bone-accumulating and fracturing agents ($r = 0.55$, $P = 0.15$). While other data suggest these differences should probably not be used as diagnostic attributes of these bone-accumulating agents (e.g., Dodson and Wexlar 1979; Kusmer 1990), I have here shown how the proportion of whole bones can serve as an important measure of the extent of bone fragmentation (see also Todd and Rapson 1988).

Intensity of fragmentation
There are two obvious ways to measure the intensity of fragmentation. I consider each in turn.

Sizes of broken bones: As implied by Villa and Mahieu (1991), the size of diaphysis fragments is a good measure of the intensity of fragmentation. To illustrate this, I compare two samples of bone fragments recovered from late Holocene archaeological sites in North America. One sample comes from a site in eastern Washington (45OK11; Lyman, unpublished data); the other sample comes from a site in Missouri (23LN104; Lyman and O'Brien 1987). For each,

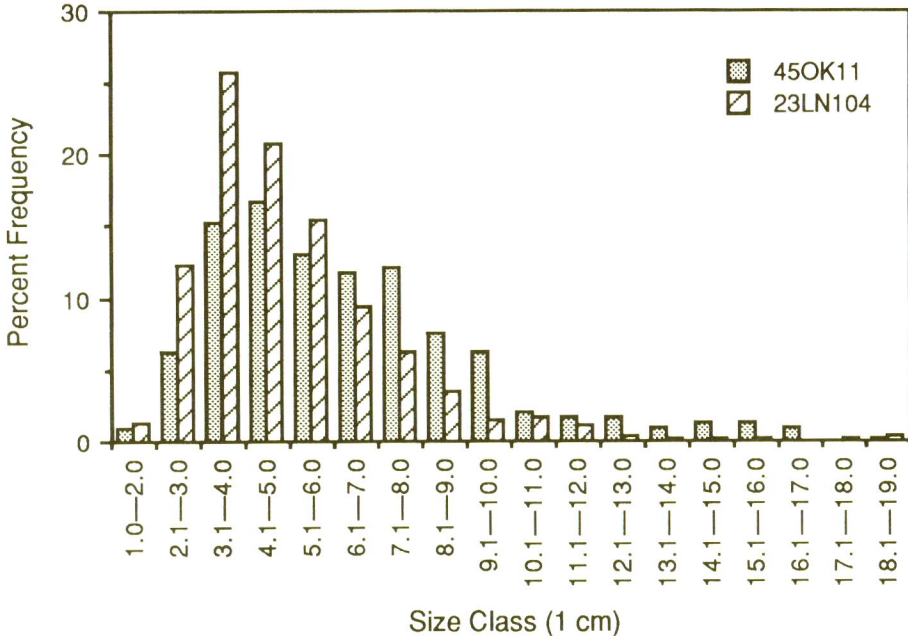

Figure 8.10. Proportional frequencies of 1 cm size classes of long bone diaphysis
fragments in two assemblages of deer bones.

the lengths of long bone diaphysis fragments of deer (*Odocoileus* sp.) and deer-
sized artiodactyls were measured to the nearest millimeter. The sample from
45OK11 consists of 306 specimens; the sample from 23LN104 consists of 972
specimens. Both samples appear to be largely attributable to the butchering
activities of human occupants of the sites. The frequency distributions of the
percentage of total specimens per 1 cm size class appear to be rather similar
between the two sites (Figure 8.10). However, the two distributions are
significantly different statistically (Kolmogorov–Smirnov two-sample
$D = 0.233$, $P < 0.01$). The greatest differences occur in the 2.1–3.0 to 4.1–5.0 cm
size classes; proportionately more fragments from 23LN104 occur in these size
classes than in the 45OK11 sample. As well, the 45OK11 sample has proportionately more specimens 6.1 to 10.0 cm long. The frequency distribution of
fragment size classes (Figure 8.10) suggests the 23LN104 assemblage is more
intensively fragmented than the 45OK11 assemblage because the former
assemblage has more smaller fragments than the latter assemblage. Why these
differences occur thus becomes the focus of further analysis. While I do not
have the data necessary to pursue such analyses, I would next inspect the
proportion of specimens that display carnivore gnawing marks, the proportion
of specimens displaying flake scars and percussion marks, the weathering
stages displayed by individual specimens, and other attributes known to covary
and be taphonomically interrelated with fragmentation processes. Another

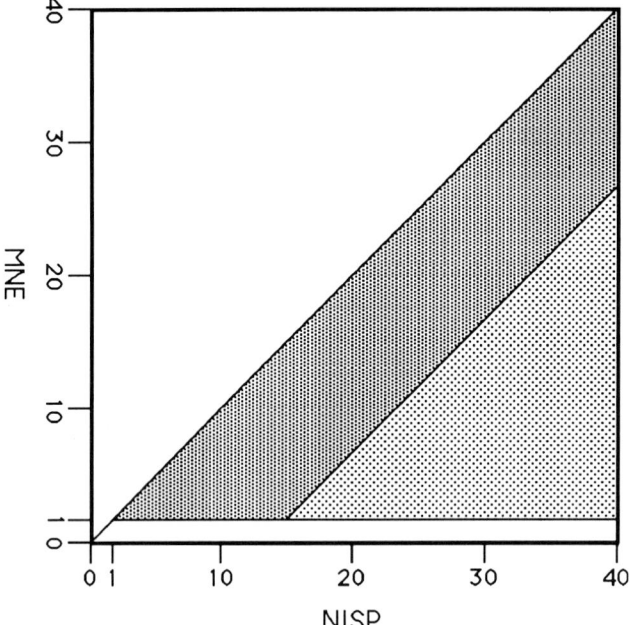

Figure 8.11. A model of the relation between NISP and MNE in an assemblage of bones. Plotted points will always fall on the diagonal (NISP = MNE) or in the shaded area (NISP > MNE). The dark shaded area represents the fact that a skeletal element can only be broken into a finite number of identifiable fragments, here modeled as 15 fragments per one MNE.

attribute of the assemblages that might help unravel the fragmentation history of these assemblages is NISP:MNE ratios.

NISP:MNE ratios: Richardson (1980:111) was one of the first taphonomists to calculate NISP:MNE ratios, or what he called a *fragmentation ratio*. Subsequent researchers have followed his lead (e.g., Schick *et al.* 1989). Recall that MNE is the minimum number of complete skeletal elements necessary to account for the specimens identified, and that those specimens may be variously complete and incomplete bones. If, for example, there are 2 proximal left humeri and 6 distal left humeri in an assemblage, the NISP for left humeri is 8 whereas the MNE for left humeri is 6 *if* none of the proximal specimens overlaps with a distal specimen. If one of the proximal specimens has a portion of diaphysis that is also represented on one of the distal specimens, the two specimens overlap and came from different skeletal elements. In this case the NISP would still be 8 but the MNE would be 7. Given how NISP and MNE are tallied, NISP will always be greater than or equal to the MNE for an assemblage (Figure 8.11). If NISP = MNE, then either all of the specimens are complete skeletal elements, or all of the specimens overlap or represent the same portion of a skeletal element, such as when all are the distal end of the tibia. If NISP > MNE, then some specimens are fragmentary and overlap of specimens is limited. Because any given skeletal element can be broken into some finite number of specimens that are identifiable to skeletal element (e.g., Lyman and O'Brien 1987), MNE will be < NISP to some limited degree (Figure 8.11). The intensity of fragmentation denotes the smallness of the fragments. Small fragments are less likely to overlap with one another than large fragments and be unlikely to be shown to be independent of one another. Thus, a large fragment will contribute one to NISP and probably will contribute one to MNE tallies whereas a small fragment will contribute one to tallies of NISP but will probably not contribute one to MNE tallies (the smallest fragments will be unidentifiable and contribute to neither).

NISP:MNE ratios should be calculated using tallies for NISP and MNE derived only from fragments; whole or complete skeletal elements should be excluded. Including whole bones in tallies of NISP and MNE reduces their proportional difference because both are increased the same absolute amount by inclusion of whole bones. For example, in an assemblage of bone fragments only that has an NISP of 10 and an MNE of 5, the NISP:MNE ratio is 2.0. Adding 2 complete skeletal elements to the assemblage produces an NISP of 12, an MNE of 7, and a NISP:MNE ratio of 1.71. Thus, the more complete bones in an assemblage (the greater the proportion of the total NISP that are complete bones), the less the proportional difference will be between NISP and MNE, and the lower the NISP:MNE ratio. In Table 8.9, the ratio for tibiae deposited by the great horned owl would change from 1.60 using only fragments to 1.04 if whole bones are included; the ratio for scapulae would change from 1.06 to 1.04 because there are fewer complete scapulae (30.9%) than there are complete tibiae (91.0%).

The NISP:MNE ratios in Table 8.9 were calculated without whole bones included in the tallies. Those ratios indicate that femora and tibiae deposited by screech owls and diurnal hawks (ratios all ≥ 1.95) are more intensively fragmented than those skeletal elements are when deposited by great horned owls (ratios ≤ 1.60). The average NISP:MNE ratio across the eight skeletal elements listed in Table 8.9 for great horned owl (avg. = 1.22) is more different from the average ratio for screech owls (avg. = 1.52; $t = 1.58$, $P = 0.14$) and diurnal hawks (avg. = 1.43; $t = 1.38$, $P = 0.19$) than the latter two are from each other ($t = 0.43$, $P = 0.67$), suggesting that screech owls and diurnal hawks more intensively fragment bones (break them into smaller pieces) than great horned owls.

Finally, it is important to note that most researchers working with bone assemblages deposited by raptors (Andrews 1990; Dodson and Wexlar 1979; Hoffman 1988; Kusmer 1990) report that these bone-accumulating and depositing agents tend most frequently to break the scapula and the innominate. The percent whole values for these two skeletal elements across all three taxa of raptors tend to be small (≤ 30%). NISP:MNE ratios for these skeletal elements, when calculated using only fragments, tend not to reflect this fact (all ratios ≤ 1.12). However, the percent whole values *in combination with* the NISP:MNE ratios for these two skeletal elements in turn suggest these raptors deposit large fragments of scapulae and innominates, and those fragments overlap one another.

Summary

Bones are made up of a composite, viscoelastic material that, when green and fresh, fractures according to biomechanical properties that have evolved through time. Studies of bone as a material grant insights to their fracture mechanics. Understanding those mechanics in turn allows us to discern the condition of the bone when it was broken. This is largely restricted to distinguishing fractures that occurred when the bone was fresh and green from fractures that took place when the bone was dry or mineralized. The important topic of how bones break when burned (see Chapter 9 for discussion of burning) has not been covered here because little research has been done on it. Typologies of fractures have been described by various researchers, but many of these suffer from having too generally or ambiguously defined attributes and no clear specification of the attributes that are necessary and sufficient to assign specimens to a type. This makes the typologies difficult to use and not readily replicable from analyst to analyst. Biddick and Tomenchuk's (1975) descriptive technique may provide detailed description and exact recording of fracture edges and surfaces, but is less readily converted to a typological or classification system. Clearly there is room for improvement here.

Study of how bones are broken is geared generally towards ascertaining the taphonomic agent responsible for the fragmentation. The examples of analyses outlined above focus on how fragmentation might be analytically manipulated during such ascertainment. And while several of those examples utilize bones deposited by raptors, there is no reason why those same analytical techniques cannot be applied to bones thought to have been broken and deposited by hominids.

Bone artifacts

> Artifacts are defined as objects that have been made or modified by human beings.
> (J. Bower 1986:14)

> Tool determination comes from recognition of a use-wear pattern after humans have been established as the agency of [bone] breakage.
> (E. Johnson 1985:175)

Artifacts and tools

The definition of *artifact* given above is similar to that found in virtually every introductory archaeology text. Its significance becomes clear when the implied definition of a *tool* given above is considered. *Tools* are artifacts that exhibit use-wear that was created because the artifact was used by humans to perform some function. Such a definition is no doubt far more narrow than most archaeologists will accept because artifacts may be used insufficiently to generate use-wear. And, of course, one could wonder if beads and ornaments made from bone and tooth are tools; they are used, but probably will not display use-wear. Similarly, because humans as taphonomic agents can *modify* bones during butchery by either generating cut marks on them or breaking them for marrow extraction or burning them as fuel in a fire, does that modification make those bones artifacts? By definition it does. However, most archaeologists tend to sort animal remains – humanly modified or not – from artifacts into the separate category *ecofacts*: "the material residue of the environment" (Bower 1986:14). Again, this is a typical definition. I believe that, given preceding sections of this chapter, humans do modify bones and thus by definition the modified bones would be artifacts. But I also believe that stone tool material and fine-grained sediment used to make projectile points and pottery, respectively, also are material residues of the environment. The distinction of artifacts, tools, and ecofacts, then, is not so clear-cut as it might first appear.

On one hand the preceding paragraph may simply capture the essence of a semantic quibble. We all rather intuitively know what artifacts are and how they differ from ecofacts, although the distinction between artifacts and tools may be less clear. On the other hand, the discussion hinges on the term "modification," and perhaps that is where we should start. For example, as noted earlier in this chapter, the creation of butchery marks – a form of modification – may well be an *incidental* byproduct of the human behavior of modifying animal carcasses into consumable resources. As such, butchery marks would be *unintentional epiphenomena* of the intended modification (see Lyman 1993a for additional discussion). Fractured bones resulting from marrow extraction would be an intended result of the activity, but the intent is to produce an edible resource, not a tool or artifact, although as we will see bones broken for the former may be used for the latter. Manufacturing an artifact or tool results from the *intentional* modification of a natural substance, whether that substance is clay for a ceramic, stone for a projectile point, or bone for an awl. Use-wear is an unintentional byproduct of use, regardless of the tool material.

Given that highly modified artifacts are easily recognized and identified as artifacts, the question of greatest moment for taphonomists is identifying those objects that *might* be artifacts given the limited modification they display. The remainder of this section, then, deals exclusively with this topic.

Identifying bone tools

The problem of distinguishing naturally modified objects from those modified by hominids and thus those representing artifacts, especially stone artifacts, has a deep history (Grayson 1986). The parallels between techniques developed for sorting stone artifacts from a collection of rocks, some naturally and some artificially modified, and bone artifacts from a pile of faunal remains, are remarkable. The basic technique involves study of two kinds of attributes, what I have termed attributes denoting *purposefulness* and attributes of *primitiveness* (Lyman 1984b). As Binford (1981b:4–8) documents, the fact that some modifications result in "redundant patterning producing a result to a design or plan" has long served archaeologists searching for purposeful modification of objects. Purposefulness reflects intentional modification in the sense noted earlier. Purposeful modifications are "repetitive, systematic, planned and controlled" (Ascher and Ascher 1965:244–245). The primitiveness of a modification "involves both the sophistication and degree of modification to the object. The less the degree of sophistication [and degree] of modification, the more tenuous the identification of an object as an artifact" (Lyman 1984b:325).

Purposeful shaping of an object by hominids to produce a desired morphometry can be termed *manufacturing modification* (Lyman 1984b:325). Manufacturing modification can range in sophistication and degree from the extensive modification necessary to produce a Solutrean biface or a Clovis point to the minimal or lack of manufacturing modification required of what have been called instant, impromptu, or expedient tools. Obviously, the more extensive the manufacturing modification to a bone specimen, the easier it will be to identify that specimen as an artifact simply because "no reasonable combination of conceivable agents other than people could have produced [the modification]" (Lyman 1984b:328). Semenov (1964:143–195) still provides one of the most extensive and detailed considerations of primitive bone tool manufacturing modification available, and MacGregor (1985) describes more modern techniques of make bone tools. Use-wear, because it is an incidental result of an object being used as a tool, may not be very extensive or obvious. In conjunction with minimal manufacturing modification, minimal use-wear can result in a bone object that has been only slightly modified, and that will thus be difficult to identify as an artifact.

Extensively modified bones, such as the scapulae shown in Figure 8.12, clearly could not have been produced by natural processes. There would be little argument that hominids were the major agent of modification, that the intended result was a tool (an awl) to be used, and that the same kind of tool was being produced by two distinct manufacturing procedures. The problem of identifying bone tools clearly resides with those bones that have been minimally modified. In these cases, the *distribution* of the attributes of modification becomes a critical variable (see below). While study of that variable may well increase the probability that minimally modified bones are identified as tools, Figure 8.13 illustrates specimens with minimal rounding and/or chipping of fracture points and edges that were not produced by hominids, but which might be misidentified as tools if found in association with indisputable artifacts.

Given that highly modified artifacts are easily recognized and identified as

a

b

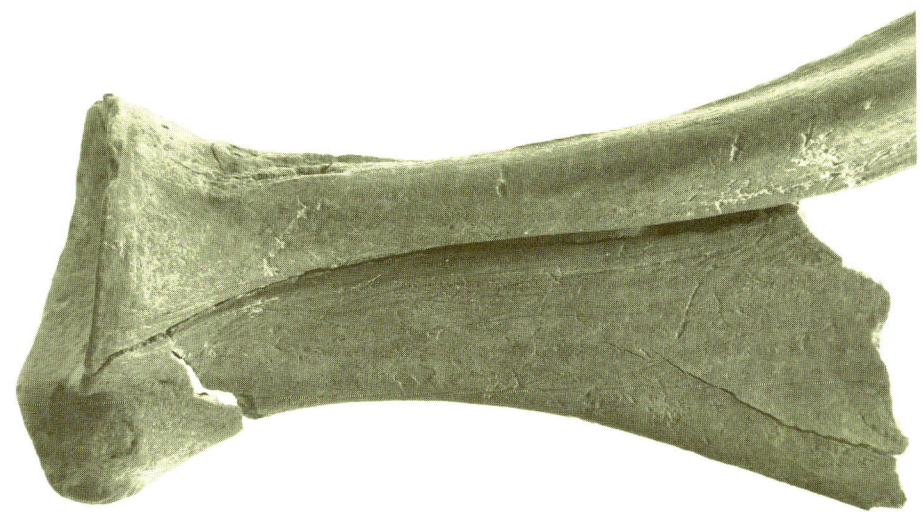

Figure 8.12. Prehistoric deer (*Odocoileus* sp.) scapula awls from eastern Washington state. a, upper three specimens are blanks formed by pecking and breaking posterior margin from the scapula blade, lower specimen is finished product; b, medial surface of a scapula showing the engraved groove used to remove the posterior margin from the blade.

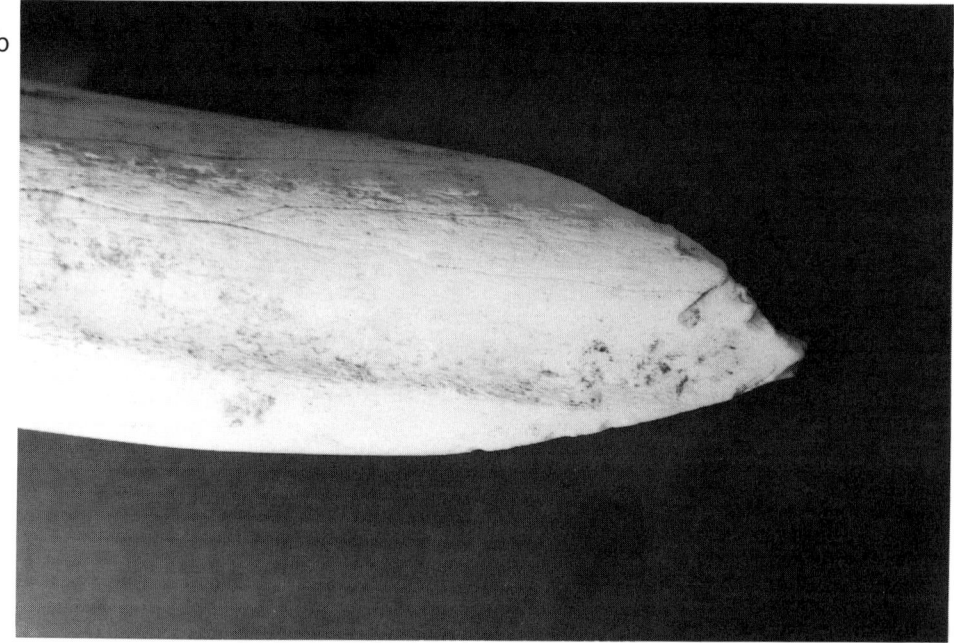

Figure 8.13. Pseudotools. a, naturally broken and flaked humerus shaft; b, naturally broken, flaked and pointed tibia shaft (from Lyman 1984b:321, Figure 4; and Lyman 1984b:320, Figure 3, respectively; courtesy of the Society for American Archaeology).

artifacts, how do we determine if objects that display limited modification are artifacts? In the following I review some of the criteria frequently used to sort skeletal remains into those modified by hominids for use as artifacts and tools and those used as tools, from all other specimens, regardless of whether the latter were modified (such as those that are butchery marked) by hominids or not.

Context

The *context* of a skeletal specimen denotes its spatial location and its spatial association with other objects, including cultural items such as undisputed artifacts and natural items such as sediments. An *archaeological context* is produced by a unique combination of natural and cultural processes operating on natural and cultural materials in a limited geographic area over some temporal span. An archaeological context is usually specified by labelling the spatial area an *archaeological site*. The most typical indication that one has found a site is the presence of *artifacts* (for discussion here, these include discrete portable objects as well as what are often termed features – sets of discrete objects which owe at least part of their artificiality to their spatial associations with one another), or objects that owe any of their attributes to human activity. When unquestionable artifacts are found in some frequency, usually in relatively dense (frequency per unit space) concentrations, the archaeologist declares that a site has been found.

Thus, a primary definitive criterion for an archaeological context is the presence of a site the definitive criterion in turn for which is the presence of artifacts, typically ceramic or stone artifacts. This results in a near-tautological line of reasoning when the archaeological context of bone items is used to assign them an artificial status. Binford (1981b:4–8), for one, has shown that the simple presence of bones in a site does not necessarily mean those bones were artificially deposited. This notion has a much deeper history when the search for ways to sort culturally deposited bone from naturally deposited bone is recalled (e.g., Thomas 1971). The archaeological context of bones, then, cannot be used alone to recognize either culturally deposited bones or humanly modified bones. Typically, context is only the first clue, often implicitly called upon, and suggests the *potential* that a particular bone was culturally deposited and modified by hominids.

Kind of bone

In North America in particular, analysts searching for what have been called *bone expediency tools* have used criteria regarding the kind of bone from which a purported tool is made to help identify these objects. Only bones of appropriate structure, weight, and strength were employed as expediency tools.

Yet no formal list of which skeletal elements are appropriate has been published. Review of the skeletal elements inferred to have been used as expediency tools indicates virtually any bone of the skeleton may be an expediency tool. That is so because, while only one part of a skeletal element might provide a working edge, weight and balance requirements can be met by using a joint consisting of several articulated bones. Therefore any bone may be a part of a bone expediency tool (see Lyman 1984b for additional discussion).

The bone expediency tool concept allows for the possibility that prehistoric hunters butchered an animal with its own bones. Thus, the butcher need only know how to produce a bone expediency tool rather than carry a butchering tool kit around the landscape. Experimental work suggests, however, that while this is certainly possible, a more efficient bone expediency tool would be made from a bone of an animal that died sometime previous to the butchering event (Frison 1982). Thus a bone expediency tool may be made from the bones of a taxon different from those being butchered. A single modified camel (*Camelops* sp.) bone in a bison (*Bison* sp.) kill site, for example (Frison 1982), is a likely candidate for an expediency tool. In such cases, the exotic bone may be more weathered than the common bones. Shipman and Rose (1988:308) report that bone flakes tend to absorb grease and moisture from the flesh of the butchered animal, effectively dulling the edge and "making it impossible to continue to use the same bone flake without considerable sacrifice in efficiency." This may indicate that when such tools are found they signify a scarcity of other, more efficient tool materials.

Modification attributes

The kind and distribution of modification on skeletal parts seem to be the major criteria analysts use to help identify bones made into and/or used as tools by hominids. Does abrasion occur only on the fracture surface, the distal end of the fracture, or the entire surface of a specimen? Is polish found only on the distal end of the fracture or is it found over the entire fracture surface plus the exterior cortical surface of the bone? The implications of answers to these and similar questions for identifying bone tools reside in the assumption that the modification will have a distribution restricted to the working edge if it is use-related wear and a less restricted distribution if it is the result of natural processes.

Kinds of modification

Archaeological evidence of manufacturing bone tools can take the form of chipped fracture edges, ground fracture edges (including striae), and the creation of detritus (bone flakes). Use-wear modification is restricted to the attritional loss of bone tissue and can consist of polish, rounding, smoothing, and microflaking of fracture edges and surfaces (Johnson 1985:213–217;

Lyman 1984b:318). Abrasion, polish, and rounding are discussed in detail in Chapter 9, and flaking is mentioned in earlier sections of this chapter. These modifications may be macroscopically visible (Figure 8.13), but some taphonomists have argued that microscopic examination of them is important (e.g., d'Errico *et al.* 1984; Shipman 1988a).

On the basis of their study of microscopic features of experimentally used bone tools and ethnographically documented bone tools, Shipman and Rose (1988:312) report that all utilized edges develop a macroscopically visible gloss or polish. Under microscopic examination, wear is "restricted to raised areas of the edge or rugosities that actually come into contact with the substance being cut; these initially raised areas become rounded, glassy [in texture], and smooth and lose surface detail progressively with continued utilization" (Shipman and Rose 1988:314). They suggest that the use-wear on a bone tool may show a preferred axis of motion based on microstriae, and that use-wear micropolish should not be unifacially distributed over the functional working edge or tip (Shipman and Rose 1988:329). Natural abrasion does not seem to produce the fine, glassy polish seen in use-worn specimens, but to help distinguish artificial wear from natural abrasion the sedimentary context of the specimen should be examined (Shipman 1988a:282).

Some natural processes can produce modifications to bones that mimic use-wear modification. Oliver (1989) documents what may be a common case when an animal suffers an antemortem bone fracture. Sixty percent (35 of 58) of a sample of bison (*Bison* sp.) bones displaying periosteal reactive bone growth or primary callus also display polish at the incompletely healed fracture edge. Noting that the periosteal sheath shields much of the cortical bone adjacent to the fracture from abrasive forces, Oliver (1989:87) suggests "the polish on these specimens was most certainly created by the movement of the fractured pieces across one another and the surrounding tissue in the live animal." Because primary callus is woven bone that is easily eroded, bones broken in live animals by natural process may leave broken bones with polished fracture edges in the fossil record.

Villa (1991:199–200) illustrates what she describes as an "elaborate biface made on elephant bone" from a middle Pleistocene site in Italy. This is an 18.7 cm long and 8.2 cm wide, bifacially flaked piece that in outline has the appearance of an Acheulian handaxe. It is more extensively flaked than a specimen illustrated by Agenbroad (1989:142) from the late Pleistocene Hot Springs Mammoth Site in South Dakota that was apparently created by trampling. Nor is the extent and frequency of flake scarring of the Italian specimen apparent in actualistically documented naturally broken African elephant (*Loxodonta africana*) bones illustrated by Haynes (1991:148–149). In the case of the Italian specimen, the frequency and patterned location – on more than one edge and more than one surface – of the flake scars seem to indicate hominid intent or manufacturing modification.

Olsen (1989a) discusses the differences between natural wear patterns on cervid antlers and cultural damage to these bony structures. Antler tine flakers used by hominids to work stone tend to have use-wear at the tip in the form of blunting. Microscopically the blunting results from heavy pitting due to crushing of the tissue; this sometimes produces a wear facet that is perpendicular to the long axis of the antler tine. Striae are visible microscopically on the antler tips, and occasionally a very small lithic flake is embedded in the antler tip. Sections of antler beam used as soft hammers for percussion reduction of stone tools display use-wear in the form of "heavy pitting, sets of fine parallel striae, and slanting V-shaped cuts compactly distributed over the location of impact" (Olsen 1989a:134). Natural surface modifications to antler resulting from cervid behaviors include polish and abrasion of sections of the antler beam and tines. Abrasion and polish on the tines tends to extend further down the tine toward the main beam than artificial wear. Randomly oriented shallow cuts widely dispersed over the tines also may be produced by natural behaviors of the cervids.

Distribution of modification attributes

Analysts typically examine the anatomical distribution of the various kinds of modification (e.g., Myers *et al.* 1980). But other attributes as well must be considered because it is becoming clear that certain processes can result in patterned and restricted distributions of what appear to be use-related wear modifications but which in fact are not use-related (e.g., Lyman 1984b; White 1992; see Chapter 9). For example, Johnson (1985:190) suggests that use-worn bone tools are distinguished by rounding and polishing on a localized or restricted area of the fracture edge and adjacent outer cortical and inner medullary surfaces, with no associated weathering or carnivore damage. Further, as implied by Shipman and Rose's (1988) experimental results cited above, Johnson (1985:191) notes that often the use-wear polish is restricted to the "convex fracture surface."

Lyman (1984b:328–329) suggests that the analyst examine the contextual distribution of various modification attributes of purported bone expediency tools. This effectively shifts the focus of analysis from one on the technological and use histories of the tools to one on their functional contexts. For example, are bone "choppers" used more frequently in sites formed during cold months and bone "fleshers" used more frequently in sites formed during warm months (e.g., Frison 1982)? Theoretically founded expectations, grounded in ethnoarchaeological observations, concerning the kinds of use-wear and manufacturing modifications that should be found given particular kinds of sites and particular circumstances of butchering (Table 8.3) could help zooarchaeologists know what kinds of modifications and modified bones to look for in particular cases.

Analysis of bone tools

Once bone tools have been sorted from an assemblage of faunal remains, their analysis can proceed much as with any category of artifact, whether the bone tools are extensively or only minimally modified by manufacture and/or use. For example, modification resulting from manufacture may reveal something about the technology used to produce the tools. Obviously technological analysis will be restricted to those specimens that have been made and not just used. The scapula specimens illustrated in Figure 8.12 are from a single site dating to about 1500 B.P. and located in eastern Washington state. They indicate that occupants of this site used two basic technological strategies to extract the posterior border of scapulae for production of what are termed L-shaped scapula awls: percussion or pecking through the thin blade of the scapula just anterior to the triangular-in-cross-section posterior border, and (probably burin cut) groove-and-splinter through the thin scapula blade. Functional analysis can examine, of course, the use-wear attributes.

In a rather innovative analysis, Shipman (1989:322) notes that, based on experimental data, utilized bone tools should display a clear distinction between utilized edges and unused edges, and that (microscopic) high points (convex surfaces) of utilized edges will be more highly polished than low points (concave areas). She examined 116 possible bone tools from Olduvai Gorge, and identified 41 of them as utilized tools based on microscopic use-wear attributes. She argues that four of the 41 tools were used as anvils and display "a series of peculiar punctures or depressed fractures on a single, broad, gently curving, natural surface" (Shipman 1989:325). All marks she recorded on the anvils are unlike carnivore tooth marks and are found on large specimens. For the remaining 37 tools, Shipman (1989:326) found no statistically significant tendency for particular skeletal elements to show a particular kind of use-wear. Specimens displaying use-wear have three times more "flaked fractures" than bone specimens from Olduvai that do not display use-wear, and use-worn bones appear to have been broken and flaked when fresh. The use-worn bone tools seem to have been regularly made from humeri, scapulae, and femora, and, most of the bone tools are from large rather than small mammals or non-mammalian taxa (Shipman 1989:328).

Bone tools and skeletal part frequencies

One of the taphonomic effects of making and using bone tools is the influence such a process has on frequencies of skeletal parts, a quantitative variable that is quite important in modern vertebrate taphonomy (Chapter 7). On one hand, if bones are sufficiently modified during the manufacturing of a tool, they will not be identifiable to skeletal element, and thus they will be analytically absent

Table 8.10 *MNE frequencies of bison bones from the Phillips Ranch site (from White 1952c)*

Skeletal part	MNE unmodified	MNE tools	Total MNE
scapula	12	56	68
P humerus	10	5	15
D humerus	20	0	20
P radius	15	0	15
D radius	11	0	11
P ulna	9	0	9
P femur	8	1	9
D femur	5	1	6
P tibia	7	1	8
D tibia	23	0	23

even though they may be lying on the laboratory table. There is little we can do about this type of bias in tallies of skeletal part frequencies. On the other hand, some bones made into tools are still identifiable to skeletal element. The difficulty in analysis here is that it must be determined whether or not these tools should be included in counts of skeletal parts. That is, the analyst must decide whether these modified skeletal parts owe their presence in an assemblage to their use as tools (which may have little to do with such things as the utility indices listed in Chapter 7), or if they owe their presence to having been part of an animal exploited for food. Without such determination, analysis of skeletal part frequencies can be seriously compromised, as the following example makes clear.

Several of the late prehistoric peoples who occupied the midwestern United States practiced horticulture. One of their commonly made and used tools was a hoe made from the scapula of a bison (*Bison bison*), but they utilized other bones of the bison as material for other tools as well. The MNE per skeletal part for 10 parts recovered from one site and reported by White (1952c) is given in Table 8.10. For this sample, conversion of the MNE values to MAU values is unnecessary because all included skeletal parts are paired bones, thus all MNE values would be divided by 2 and the results would thus not change. Note that for this particular site (Phillips Ranch), well over half of the bison scapulae recovered had been made into tools, and one third of the proximal humeri had also been made into tools. Figure 8.14 presents a pair of scatterplots of the MNE frequencies per skeletal part, with and without the tools included. The differences are remarkable, the most notable one being that without the MNE of tools scapulae rank fourth in abundance but with the MNE of tools scapulae rank first in abundance. If the tools are not included in the MNE tallies, the skeletal part frequencies are correlated with the bison food utility index ($r_s = -0.59$, $P = 0.07$) whereas if the tools are included in the MNE tallies the skeletal part frequencies are not correlated with that index ($r_s = 0.13$, $P = 0.73$).

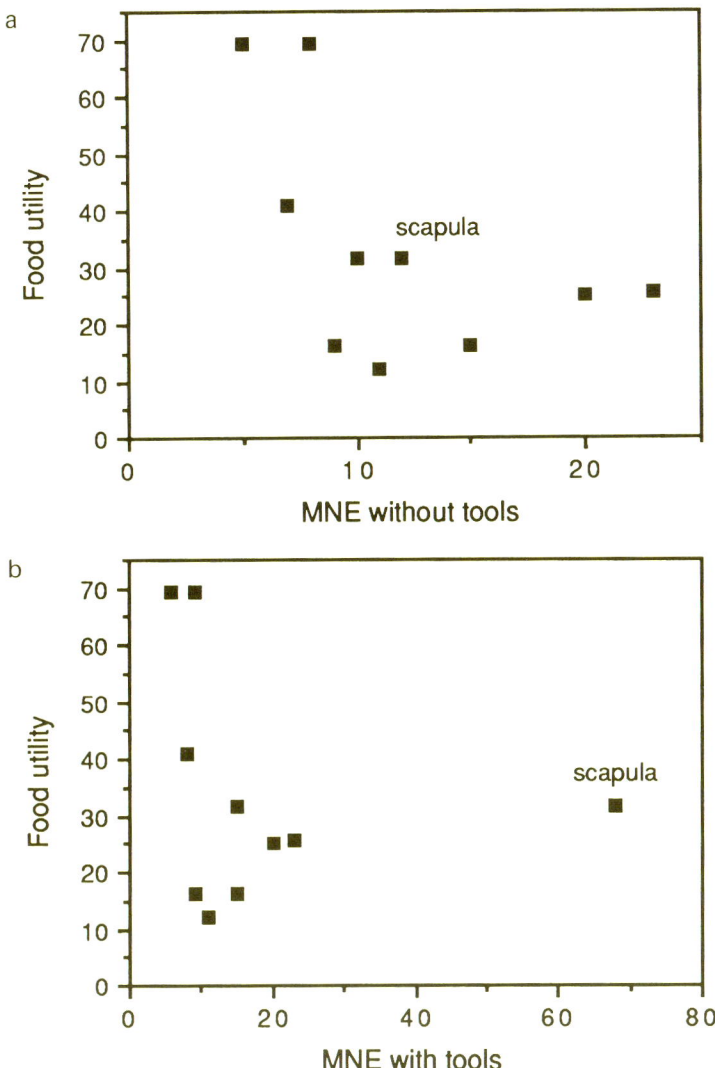

Figure 8.14. Scatterplot of MNE frequencies of selected bison bones against the bison food utility index (from Table 8.10). a, bones modified into tools excluded; b, bones modified into tools included. Note especially the relative position of the scapula in the two plots.

Including the bone tools weakens the correlation coefficient between skeletal part frequencies and the structural density of the skeletal parts (from $r_s = -0.36$, $P = 0.3$ to $r_s = 0.02$, $P = 0.95$).

Several zooarchaeologist colleagues have suggested that the utilization of bones as tool material may well be producing the L-shaped curves like that seen in Figure 7.2 and perhaps as well the positive statistical relationship between bone density and skeletal part frequencies such as that seen in Figure 7.12a. I

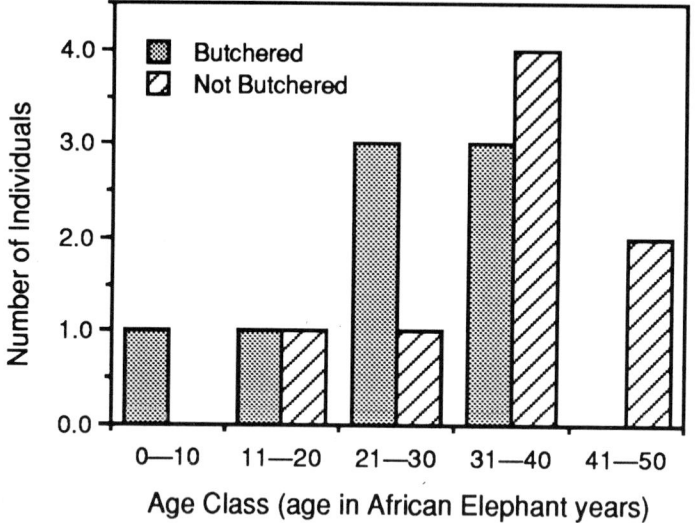

Figure 8.15. Demography of mortality of mastodon carcasses reported by Fisher (1987). Age is in African elephant years.

have not yet seen any data, either prehistoric or actualistic, to indicate that this is an accurate assessment of the situation, but the point is well taken. On this basis, and on the basis of the data in Figure 8.14, I suggest that if it can be shown that particular skeletal parts were being used as tools more often than other skeletal parts, then *all* of the former should be omitted from study of skeletal part frequencies. That way, the biasing effects of prehistoric hominids modifying some skeletal parts beyond recognition will be at least partially eliminated, and the additional biasing effects of prehistoric peoples selecting and curating particular skeletal elements for use as tools will also be at least partially eliminated from analyses of skeletal part frequencies.

Butchering, breakage, and bone tools

One of the finest examples of taphonomic analysis that considers evidence for butchering, fracturing of bones by hominids, and the manufacture of bone tools is found in Fisher's (1984a, 1984b, 1987; Shipman *et al.* 1984a) study of late Pleistocene mastodon (*Mammut americanum*) remains in North America. The 19 sites he examined each had the remains of one individual mastodon, but of those skeletons Fisher believes ten had been butchered by humans and the other nine were not butchered. He describes various lines of evidence to support his interpretations. The demography of the butchered mastodons is different from that of the non-butchered mastodons (Figure 8.15). The seasons of death for the two samples are completely different (Figure 8.16), suggesting to Fisher (1987:359) that the butchered mastodons were hunted because the

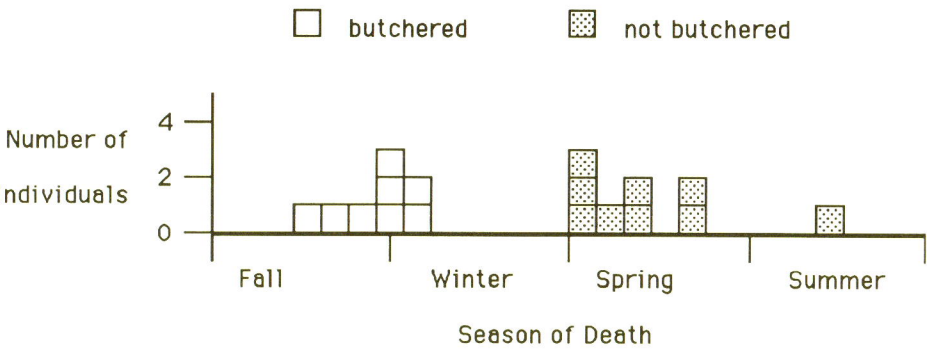

Figure 8.16. Season of mortality of mastodon carcasses reported by Fisher (1987).

seasons of death of the butchered individuals do not overlap with the late winter–early spring pattern of natural mortality.

The evidence for butchering consists of cut marks on some bones, and most impressively, there are several bones that were articulated in life which display marks on their conarticulating surfaces made by an intrusive implement. The bones of individuals that appear to have been butchered are more intensively and extensively broken than the bones of individuals that appear not to have been butchered. As well, some of the bones of the former group are burned whereas none of the bones of the latter group are burned. Some of the broken bones display fracture morphologies suggestive of breakage while the bones were fresh, and several also have percussion marks adjacent to the fracture edge. There are some carnivore gnawing marks on bones of the butchered group but not on bones of the non-butchered group. While there are no stone tools associated with the skeletons, there are several bone specimens associated with the butchery-marked carcasses which display polish suggestive of their use as expediency tools, and several specimens have had multiple flakes removed from their margins (Fisher 1984a, 1984b, 1987; Shipman *et al.* 1984a).

What is perhaps most impressive about Fisher's analysis is the detailed documentation of multiple attributes, all suggested on the basis of actualistic research to be exclusively or nearly exclusively produced by hominids. The presence of only butchering marks, or only green bone fractures, or only burned bones, or only a couple of specimens with polish restricted to one area, would not provide as compelling an argument for human intervention as the multiple lines of evidence marshalled by Fisher. In conjunction with the demographic data, the bone modification data allow him to build a convincing argument. The attributes reported by Fisher are not unambiguous signature criteria diagnostic of only hominid activity, for exceptions to each individual attribute can be found. Bone can be burned, for example, without human activity. What is convincing about the butchery marks is their purposefulness (especially those on tightly bound joints and found on conarticular surfaces) and their lack of primitiveness. The probability that the combination of attributes Fisher describes could have been produced by natural processes seems minute indeed.

Another reason Fisher's (1984a, 1984b, 1987) arguments are so compelling is perhaps less obvious. It concerns the fact that he is dealing with faunal collections that have relatively simple taphonomic histories, in part because they each represent a single mortality event and a single individual animal. Saunders' (1977) collection of the remains of some 30 mastodons from a single site had undergone various taphonomic processes (such as trampling) that Fisher's individual animals had not. Fisher's single-carcass assemblages made it much easier for him to establish the interdependence of skeletal parts (they all came from the same individual animal). In Binford's (1980, 1981b) terminology, Fisher is dealing with "fine-grained" assemblages whereas Saunders is dealing with a "palimpsest" or "coarse-grained" assemblage (see the Glossary). This simple fact underscores the notions that some taphonomic problems will be much easier to solve if the sample at hand has experienced a taphonomic history that does not obscure taphonomic traces critical to problem solution, and that taphonomic histories are cumulative. Also, sometimes taphonomic problems will only be solved with great difficulty, only be partially solved, or not be solved at all.

Summary

> Hunting man alone adapts some parts of the carcase which he is himself unable to eat to other ends, serviceable to his living, his comfort, his vanity or his whim. (I. W. Cornwall 1968:88)

In this chapter I have focused largely on hominid-related biostratinomic processes; Chapter 9 focuses on natural biostratinomic processes. The reason for this kind of separation in topics is that I suspect many who read this book will be archaeologists and zooarchaeologists, and they will have a strong interest in hominid modifications of faunal remains. Hominids variously butcher animals, and they may break bones during butchering or during the process of making bone tools. It is appropriate then, to conclude this chapter with a reiteration of some of the issues involved in establishing hominids as taphonomic agents that affected a particular bone assemblage.

Haynes and Stanford (1984:217) suggest there are three levels of "evidence" required for inferences of the hominid utilization of animal carcasses. First, the *contemporaneity* of hominids with the taxa must be established. Were the animals represented alive and in the geographic area when the humans were there? Stratigraphic mixing may result in an archaeological association of animal and human remains, and must be considered at this level. And, of course, the analyst should keep open the possibility that fossils may have been picked up and used by prehistoric peoples. The second level of evidence involves establishing the *association* of humans with the animal remains. Are there modifications to the remains that only hominids could have produced? These include such things as butchering marks. The third level of evidence involves ascertaining whether or not the hominids *utilized* the animal. Did they butcher and eat the meat, or did they simply make bone tools? Obviously, different kinds of evidence are required at each level, but as well the three levels are not mutually exclusive. Fisher's (1984a, 1984b, 1987; Shipman *et al.* 1984a) studies on the late Pleistocene utilization of mastodon (*Mammut americanum*) carcasses by North American Paleo-Indians warrant careful scrutiny and, in my opinion, emulation if the analyst is interested in establishing that a hominid taphonomic agent was part of the taphonomic history of a bone assemblage. But even then, and perhaps especially then, there are numerous natural biostratinomic processes that should be considered in analysis. It is the natural aspects of biostratinomy that we turn to in the next chapter.

9

OTHER BIOSTRATINOMIC FACTORS

Introduction

Many taphonomic processes may affect animal carcasses and bones between the time of animal death and burial of the carcass or bones. Several of the major human biostratinomic factors are discussed in Chapter 8. In this chapter other biostratinomic factors, several of which are natural processes, are reviewed. Discussion is limited to those factors that have been more or less extensively dealt with in the literature. As well, several basic comparative analytic techniques are described at the end of the chapter.

Weathering

> The degree of brittleness of the skeleton gives no information as to age; the nature of the place where it is found must be taken into account. The more the bones are exposed to air, the more quickly they disintegrate. The quantity of precipitation, the number of days below freezing, covering with clay, burial in sand or loam – all these factors play an important role in forensic medicine.
> (J. Weigelt 1927/1989:18)

Behrensmeyer (1978:153) defines the weathering of bone as "the process by which the original microscopic organic and inorganic components of bone are separated from each other and destroyed by physical and chemical agents operating on the bone *in situ*, either on the surface or within the soil zone." Weathering involves the decomposition and destruction of bones "as part of the normal process of nutrient recycling in and on soils" (Behrensmeyer 1978:150). Miller (1975:217) indicates weathering refers to "the effects on bone of saturation, desiccation, and temperature changes." Saunders (1977:104) records "weathering indices" that are "subjective interpretations of the conditions of at least three small anatomical crests or rugosities" per specimen. He lists four kinds of weathering ("unweathered, slight, moderate, and severe") but describes none of them (Saunders 1977:108). Behrensmeyer (1978:161) defines six stages of weathering of bones of mammals > 5 kg in body weight in subaerial/surface contexts, and uses readily observed, macroscopic criteria to distinguish those stages and to enhance their recognition in the field (Table 9.1, Figure 9.1). To record weathering data, the analyst records the maximum weathering displayed over patches larger than 1 cm^2 on each bone specimen, and "whenever possible shafts of limb bones, flat surfaces of jaws, pelves,

Table 9.1 *Weathering stages in large (after Behrensmeyer 1978) and small (after Andrews 1990) mammals. Additions to large mammal descriptions from Johnson (1985) in parentheses*

Weathering stage	LARGE MAMMALS		SMALL MAMMALS	
	Description	Range in years since death	Description	Range in years since death
0	greasy, no cracking or flaking, perhaps with skin or ligament/soft tissue attached (marrow edible, bone still moist)	0–1	no modification	0–2
1	cracking parallel to fiber structure (longitudinal); articular surfaces perhaps with mosaic cracking of covering tissue and bone (split lines begin to form, low moisture, marrow sours and is inedible)	0–3	slight splitting of bone parallel to fiber structure; chipping of teeth and splitting of dentine	1–5
2	flaking of outer surface (exfoliation), cracks are present, crack edge is angular (marrow decays, split lines well developed)	2–6	more extensive splitting but little flaking; chipping and splitting of teeth leading to loss of parts of crown	3–5+
3	rough homogeneously altered compact bone resulting in fibrous texture; weathering penetrates 1–1.5 mm maximum; crack edges are rounded	4–15	deep splitting and some loss of deep segments or "flakes" between splits; extensive splitting of teeth	4–5+
4	coarsely fibrous and rough surface; splinters of bone loose on surface, with weathering penetrating inner cavities; open cracks	6–15		
5	bone falling apart *in situ*, large splinters present, bone material very fragile	6–15		

a

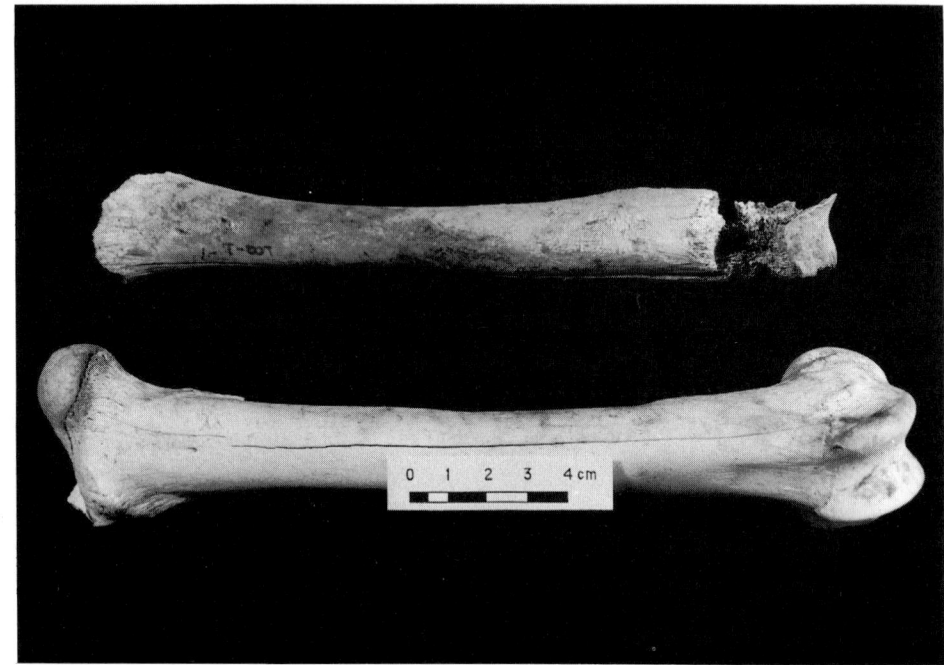

b

Figure 9.1. Bone weathering stages described by Behrensmeyer (1978). a, weathering stage 1 (lower) on two deer (*Odocoileus* sp.) femora showing split line cracks on both specimens; b, detail of weathering stage 1 (upper specimen in a) showing mosaic cracking; c, weathering stage 2 on a horse (*Equus caballus*)

c

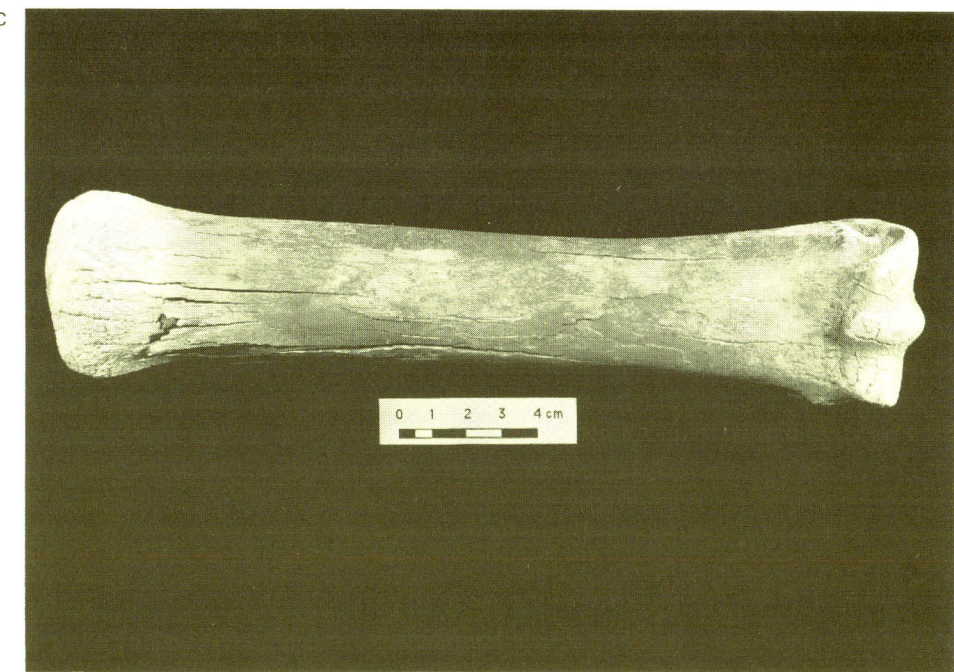

d

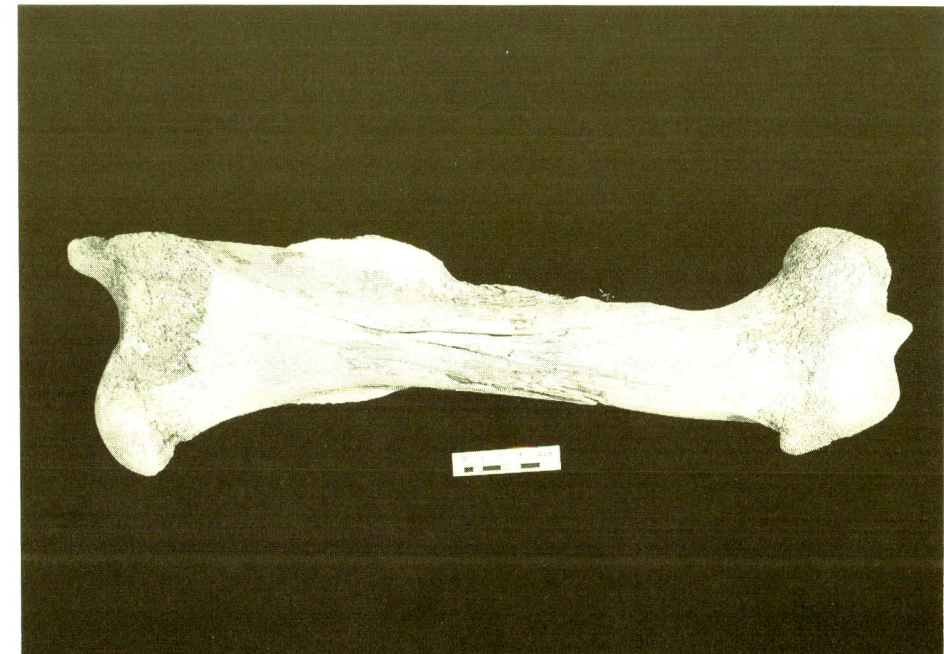

Caption for fig. 9.1. (*cont.*).
metacarpal showing flaking of outer surface and initial stages of exfoliation; d, weathering stages 3 and 4 on a horse femur, showing extensive exfoliation (near ends) and deep, multiple cracks (mid-shaft).

vertebrae, or ribs are used, not edges of bones or areas where there is evidence of physical damage" (Behrensmeyer 1978:152).

Andrews (1990:10–11) summarizes experimental results of weathering of small mammal bones, and compares them to Behrensmeyer's (1978) for large mammals (Table 9.1). He notes that bones embedded in an owl pellet are protected from weathering. Teeth show some splitting after two years of exposure, "probably due to the differential contraction of enamel and dentine" (Andrews 1990:11). Splits and cracks develop between collagen fibers, bones of the skull separate along sutures, and teeth fall out of their alveoli as weathering progresses. Chemical weathering begins on the bone surface and progresses into the bone tissue mass (Bromage 1984).

Behrensmeyer's (1978:161) hope was that once sufficient control studies were available, bone weathering features might "give specific information concerning surface exposure of bone prior to burial and the time periods over which bones accumulated." While she cautioned that her data were preliminary and their significance was conjectural, analysts have used them to infer the formational history of bone assemblages (e.g., Bunn and Kroll 1987; Potts 1986, 1988). Bone weathering is an important biostratinomic and taphonomic variable, and I devote some time to reviewing it.

Weathering and time

The weathering of bone is a historical process. The capitalization of "Time" signifies the passage of solar years. I distinguish this kind of temporal measurement from "time" or "taphonomic time." In the latter, time is measured on an ordinal scale; that is, we are able to say phenomenon A is older than phenomenon B but we are unable to say how many years older A is than B. The goal of analysis of bone weathering data is measurement of taphonomic time, or Time.

The stage of weathering displayed by a bone measures, in part, the rate and the duration of weathering. By *rate* I mean how quickly a bone passes through the weathering stages. At least three factors control the weathering rate. First, "small, compact bones such as podials and phalanges weather more slowly than other elements of the same skeleton" (Behrensmeyer 1978:152). Second, bones of different taxa, especially those of different body size, weather at different rates (Behrensmeyer 1978, 1982; Gifford 1981). Third, "the less equable the immediate environment (in terms of temperature and moisture fluctuations) of the bone, the faster it should weather" (Behrensmeyer 1978:156). The *duration* of weathering concerns the span of Time over which a bone is exposed to weathering agents. Exposure to those agents begins once soft tissues detach from the bones, and thus the manner in which hide, muscle masses, etc. are removed controls when exposure begins. Because bones may, by definition, weather in subsurface contexts, identifying the temporal end of exposure is difficult when it depends on the degree to which a bone is weathered.

One critical aspect of analyzing and interpreting bone weathering data involves the conversion of that data into at least taphonomic time if not Time. Behrensmeyer (1978:157) accomplished that conversion because she had knowledge of the years since death of the animal contributing the weathered bones. She found that the weathering stage displayed by the most weathered bone of a carcass' skeleton was related to the years since death. The correlation between each category of weathering stage and the years since death represented in her sample is strong and significant ($r_s = 0.802$, $P = 0.002$). That correlation prompted Behrensmeyer's suggestion that bone weathering data may reveal insights to the duration of prehistoric bone assemblage formation.

The correlation Behrensmeyer (1978) found can be expressed with the equation:

$$WS = f(YD) \tag{9.1}$$

where WS is the maximum weathering stage observed for an animal carcass, and YD is the years since that animal died. Lyman and Fox (1989:312) argue that when the analyst is dealing with an assemblage of bones consisting of multiple individual animals recovered from a single stratum, in order to interpret bone weathering data validly the equation that must be solved is:

$$YD_i, ED_{ij}, AH_{ij} = f(WS_{ij}, SE_{ij}, TX_{ij}, ME_{ij}) \tag{9.2}$$

in which YD and WS are defined as in equation [9.1], ED is the exposure duration, AH is the accumulation history, SE is the skeletal element, TX is the taxon represented, ME is the depositional microenvironment, i is the individual carcass that contributed the bone, and j is the particular bone of carcass i that is under study.

Equation [9.2] cannot in fact be solved for several reasons. It is, however, a useful heuristic that highlights the variables which influence the rate and duration of weathering. The variables on the right-hand side of equation [9.2] and the subscripts underscore the fact that prehistoric bone assemblages typically contain specimens of various skeletal elements from multiple carcasses representing multiple taxa, often surficially deposited across different microenvironments at different times. Variables on the left-hand side of equation [9.2] are related to the passage of time and underscore the fact that different carcasses will have died at different times relative to one another, that different bones (both within and between carcasses) will have different exposure durations, and that different bones (both within and between carcasses) may have been accumulated at different times. The analyst may record the weathering stage (WS) observed for, say, only left femora of one taxon, thereby analytically controlling SE and TX (Lyman and Fox 1989). The other four variables, however, cannot be analytically controlled in any strict sense. For example, if one desires to infer the Time over which a bone assemblage accumulated, one must assume that all the bones were accumulated when they were fresh or in weathering stage 0. While that may well have been the case, there are no actualistic data to substantiate that assumption and in fact limited data indicate it may not be warranted. To elaborate and clarify the interpretive issues raised by equation [9.2], I discuss each variable in that equation in some detail, beginning with those variables on the right side of the equals sign.

Weathering stage

Behrensmeyer's (1978) weathering stages (Table 9.1) each represent a point in Time along the continuous process of bone deterioration. Behrensmeyer (1978:152–153) states, for example, that "the six weathering stages impose arbitrary divisions upon what was observed to be a continuous spectrum." Johnson (1985:187), in fact, defines six weathering "phases" spanning Behrensmeyer's first three weathering stages. Gifford (1977:291) reports a sample of bones which progressed through weathering stages 0 and 1 to stage 2 within one to two years, but which then remained in stages 2 and 3 for several years, and suggests that we may be unable to assign "absolute Time values to different weathering stages" (Gifford 1981:418). That observation, plus Behrensmeyer's (1978:157) statement that "weathering stages are most useful in providing an estimate of the minimum number of years since death (or exposure)," indicates that the weathering stages are at best an ordinal scale measurement of time; they do not clearly measure Time.

Bones weather, by definition, in both surface and subsurface contexts (Behrensmeyer 1978; Frison and Todd 1986). Given an interest in only the weathering that takes place on the surface of the ground, it is surprising that so little effort has been made to distinguish subaerial (surface) from subsurface weathering (e.g., Frison and Todd 1986; Mehl 1966; Todd 1987b; Todd *et al.* 1987). While buried bones seem to weather much slower than exposed bones, the former do sometimes weather. Thus, until a reliable way to distinguish subaerial and subsurface weathering is developed, when weathering data are interpreted it must be with the assumption that subsurface weathering is insignificant.

Gifford-Gonzalez (1989a:192) reports damage to bone specimens that superficially resembles weathering cracks but which she believes was created by heating the bones, such as when they are exposed to a fire. Shipman (1981b:177) states that heating bone results in the "denaturing" of collagen fibers which are naturally "under tension in the bone;" the implication is that the tension in these fibers that is released by heating causes cracks to form, cracks which are "perpendicular to the direction of the collagen fibers and the long axis of the bone" (see the discussion of "Burning" below). This observation not only underscores the care that must accompany the identification of a weathering stage displayed by a bone, but the fact that weathering, as defined, is simply one form of bone deterioration.

Skeletal element

Behrensmeyer (1978) and Todd *et al.* (1987:68–70) suggest that different skeletal elements will weather at different rates, perhaps due to variation in their structural density (Lyman and Fox 1989:297; see Chapter 7 for discussion of structural density). We simply do not know, however, which skeletal elements weather fast and which ones weather more slowly. To control this variable, then, one could record the maximum weathering stage displayed by one kind of skeletal element, say, only the humeri or femora, to control for potential variation in the rate at which different skeletal elements pass through the weathering stages.

Taxon

Behrensmeyer's (1978:153) weathering stages were defined on the basis of mammals with > 5 kg body weight. Gifford (1981:417) observes "bones of roughly like-sized mammals of different taxa may weather at somewhat different rates due to constructional differences," and in her control samples "more heavily constructed equid bones weathered at a somewhat slower rate than homologous bovid bones." Schäfer (1962/1972:24) reports fences and tombstones constructed of whale bone last for centuries in Holland. Again, perhaps the structural density and/or the porosity of skeletal elements influences the weathering rate. To control for potential taxonomic variation in bone weathering, one should record the weathering stage displayed by the remains of each taxon separately (as well as each skeletal element separately). This would provide independent samples of data on the same phenomenon.

Depositional microenvironment

The microenvironment in which bones are deposited is discussed at two spatial scales by Behrensmeyer (1978). At a large scale one considers the general vegetational habitat of the depositional area, and at a fine scale one considers the specifics of the microenvironment at the location where a bone is deposited. The former concerns assemblages of bones as these occur over large spatial units whereas single bones each occupy a particular spatial point. It is relevant to consider each scale of depositional microenvironment.

Vegetation habitat in an area   Behrensmeyer (1978) presents data on 1,534 carcasses in six vegetational habitats, and reports the most advanced weathering stage per carcass. She found all six weathering stages represented in all six habitats, but the frequency distributions of representation of weathering stages differ between some of the habitats. Lyman and Fox (1989:298) calculated Kolmogorov–Smirnov two-sample D statistics between all possible pairs of habitat-specific weathering data (Table 9.2). While those statistics indicate some inter-habitat variation in the frequency distributions, Behrensmeyer

Table 9.2 *Kolmogorov–Smirnov* D *statistics between all possible pairs of carcass assemblages from major habitats, based on data in Behrensmeyer (1978). Number in parentheses next to* D *statistic is the weathering stage where* D *occurs (from Lyman and Fox 1989)*

	Swamp (N=322)	Dense woodland (N=312)	Open woodland (N=255)	Bush (N=184)	Plains (N=248)
Dense woodland	$D=0.14$ (1) $P<0.01$				
Open woodland	$D=0.16$ (1) $P<0.01$	$D=0.04$ (3) $P>0.1$			
Bush	$D=0.35$ (1) $P<0.01$	$D=0.21$ (1) $P<0.01$	$D=0.19$ (1) $P<0.01$		
Plains	$D=0.23$ (1) $P<0.01$	$D=0.09$ (1) $P>0.1$	$D=0.07$ (2) $P>0.1$	$D=0.12$ (1) $0.1>P>0.05$	
Lake bed (N=213)	$D=0.26$ (1) $P<0.01$	$D=0.12$ (1) $P=0.05$	$D=0.10$ (1) $P>0.1$	$D=0.09$ (1) $P>0.1$	$D=0.06$ (3) $P>0.1$

(1978:159) cautions that some of that variation might also be attributable to "changes in patterns of habitat utilization and/or mortality" of the animals in those habitats. This means that the accumulation history of bones may vary between habitats.

Microenvironment of a spatial point: Behrensmeyer (1978:158) argues that the "localized conditions (e.g., vegetation, shade, moisture) are more important to bone weathering than overall characteristics of the [different vegetation] habitats." The subaerial depositional microenvironment of a bone may inhibit or exacerbate the rate of weathering. Behrensmeyer's data indicate that the size and internal density of vegetation patches in which bones are deposited are important variables, and other data indicate that the magnitude of seasonal changes in weather and durations of the seasons are also important variables, as are the moisture content, temperature, and texture of the sediment on which a bone is deposited (Brain 1967a; Cook 1986; Miller 1975). Todd (1983a, 1983b) presents data indicating that the presence of multiple, closely spaced carcasses tends to create a different microenvironment than when a carcass is isolated from other carcasses. Further, in a late Pleistocene cluster of mastodon (*Mammut americanum*) bones Saunders (1977:104) found that "the bones occurring at the periphery and at the top of the bone bed were generally poorly preserved indicating severe weathering prior to final burial. [As well] the outer limit of the bone bed was defined by bones and tusks in generally poor condition." Lyman (1988a:103) and Lam (1992:401) demonstrate that bones deposited in a rockshelter or cave are less weathered than bones deposited outside of the rockshelter or cave; little else was different between the depositional environments of these specimens.

Granting that microenvironmental conditions may influence the rate of weathering, we still do not know the spatial scale at which that microenvironmental variation will create *significant* variation in weathering rates; that is, we do not know how far apart two bones must be (assuming they are of the same element and of the same taxon) in horizontal space to ensure that they occur in depositional microenvironments that are sufficiently different to cause one of them to weather significantly faster than the other. This problem seems to be dealt with analytically by assuming that all skeletal specimens within an assemblage were deposited in the same depositional microenvironment, or at least in one with variation that insignificantly influenced the rate of weathering. This may be reasonable as most assemblages are (analytically) specified as coming from some horizontally and vertically delimited spatial unit, typically a stratum or some other depositional unit. However, assuming that the microenvironment did not vary across that spatial unit probably becomes more tenuous as its size increases.

Years since death

A bone begins to weather after the animal contributing that bone dies and the bone is freed from the soft tissues in which it is embedded. Thus, bone 1 should be more weathered than bone 2 if bone 1 "died" before bone 2 and both were exposed within, say, a few days of death and all other variables in equation [9.2] are equal for the two bones. Intra-carcass variation in the weathering stages displayed by bones is attributable to variation in when those bones were exposed (freed from soft tissues). Thus the strongest correlation between the years since animal death and weathering stage should be found using the most advanced weathering stage displayed by the bones of a carcass, which is precisely the basis for the correlation between weathering stage and years since death described by Behrensmeyer (1978).

Years since animal death can only vary between individual animals; it cannot vary between the bones of an animal. That renders years since animal death a significantly different kind of taphonomic time than exposure duration and accumulation history (discussed below). To infer the years since animal death the analyst must distinguish which bones make up a particular carcass; that is, the interdependence of the bones must be determined and each set of bones from each carcass represented in an assemblage must be sorted from the complete collection. Hesse and Wapnish (1985:88) note that "since different bones are likely to weather at different rates, weathering stage data cannot be used for grouping specimens of the same skeleton." As well, such a procedure would certainly be tautological. Anatomical refitting (Chapter 5) may be the most promising technique for controlling the interdependence of skeletal parts, but as yet we lack sufficient control data to allow its application. The analyst might control specimen interdependence and thus measure variation in the years since death of the animals represented by recording the maximum weathering stage displayed by specimens of a single category of skeletal element, such as left femora, realizing that the left femur from a carcass may display a weathering stage different from that displayed by the right femur of that carcass (see below and Lyman and Fox 1989:299–300 for additional discussion).

Exposure duration

Because exposure begins when insulating soft tissues are removed from bones, two bones which die at the same time but which vary in when they are exposed may or may not experience similar exposure durations even if both are from the same carcass. Similarly, two bones which die at the same time and which are exposed simultaneously may not experience similar exposure durations if they are buried at different times. The influence of exposure duration as influenced by variation in when bones are first exposed is clear in a sample of 20 pairs of left and right domestic cow (*Bos taurus*) femora described by Todd (1983b). All pairs died at the same time, all were in the same depositional microenvironment, and all were in subaerial context when weathering data were recorded. Of the 20 pairs only eight displayed the same weathering stage; the members of eight other pairs differed by a weathering stage of one, the members of three pairs differed by two stages, and the members of one pair differed by three weathering stages. Further, all six weathering stages were represented by the 40 total femora even though all individual animals died at the same time.

Exposure duration can vary between the bones of one carcass as well as between the bones of multiple carcasses because exposure duration concerns individual bones. Exposure duration measured as Time (years) will always be less than or equal to the years since death, and thus it is a different kind of taphonomic time than years since death. This is especially so when factors influencing an organism's mortality (e.g., agent of death) are independent of factors influencing exposure of an organism's bones (e.g., scavengers). If an analyst wishes to infer years since death from an animal's weathered bones, they must assume that a carcass' bones were exposed immediately after death in order that the maximally weathered bone of a carcass be tightly correlated with years since death; in terms of equation [9.2], ED must equal YD.

Variation in the exposure of bones of a single carcass can be illustrated by examination of the weathering profile displayed by a carcass' bones through time. A *weathering profile* is "the percentage frequencies of bone specimens in an assemblage displaying each weathering stage" (Lyman and Fox 1989:300). Such profiles often serve as the focus of interpreting weathered bones (e.g., Boaz 1982; Potts 1986, 1988). When they are constructed for complete assemblages, they often consist of variously interdependent (bones from same carcass) and independent bones (those from different carcasses). There are no weathering data for individual carcasses that allow us to examine how the bones of a carcass pass through the weathering stages in the form of a

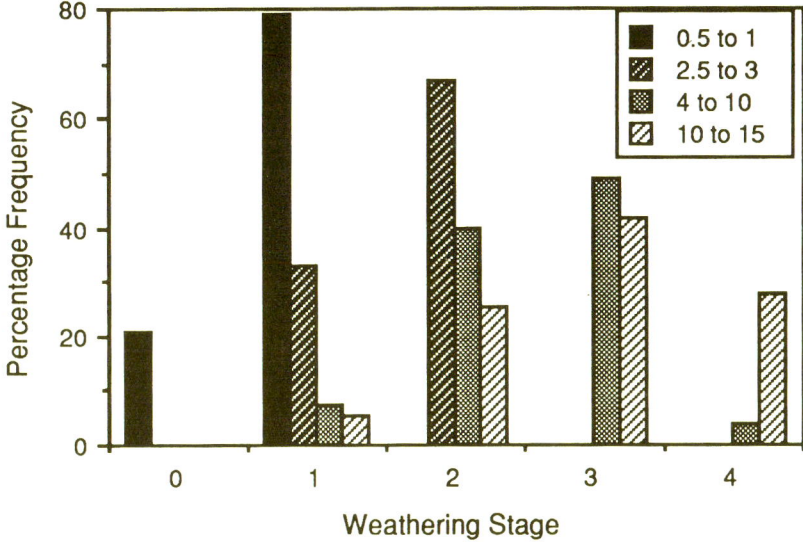

Figure 9.2. Weathering profiles for carcasses dead 0.5 to 1 yr, 2.5 to 3 yr, 4 to 10 yr, and 10 to 15 yr (after Gifford 1977, 1984).

weathering profile. Gifford's (1977, 1984) data for sets of multiple carcasses, each set having a known years since death, are plotted as weathering profiles in Figure 9.2. If all bones of a newly dead carcass display weathering stage 0, and all bones of a long dead carcass display weathering stage 5 (this kind of weathering profile will probably be rare as bones in this stage tend to disintegrate to dust), the weathering profiles in Figure 9.2 suggest a "wave model" (Lyman and Fox 1989:300) for the weathering profile of the bones of a carcass through time. The bones of the carcass weather progressively, but each bone weathers at a slightly different rate, and each experiences a slightly different exposure history. The ideal model thus takes the form of a unimodal wave which moves across the weathering stages from left (weathering stage 0) to right (weathering stage 5). Deviations from the ideal model are expected, of course, because bones of a carcass will experience varied exposure histories, will weather at different rates, and will probably be differentially dispersed and deposited in different microenvironments.

Accumulation history

Here we are particularly concerned with active accumulation of bones (see Chapter 6). The wave model of bone weathering for a single carcass (Figure 9.2) makes it clear that bones collected from a carcass dead only a few hours and that are deposited in a site and buried shortly thereafter will have a weathering profile different from that displayed by a carcass dead and exposed for a year-and-a-half. Thus, interpreting the duration over which a bone assemblage accumulated demands that the analyst assume all bones were accumulated when fresh (weathering stage 0). Is that a reasonable assumption?

I am unaware of any study focusing on the weathering stage displayed by bones collected by individual bone-accumulating agents. Anecdotal data indicate bones in weathering stage 0 are typically collected by carnivores and scavengers, but those data also indicate bones in weathering stage 1 are occasionally collected and bones displaying weathering stage 2 are sometimes, although rarely, collected by such bone accumulators (Lyman and Fox 1989:303). Brain (1980) indicates African porcupines prefer bones that are in weathering stage 1 and also apparently collect bones in weathering stage 2 (see Chapter 6). Frison (1982) suggests humans may have collected bones in weathering stage 1 rather than use fresh bones (in weathering stage 0) to manufacture bone tools. For the present, it may be reasonable to assume that the majority of the bones used to construct a weathering profile were accumulated when fresh (displayed weathering stage 0), but the analyst should be aware that some of the bones may have been somewhat weathered when accumulated. Note that Behrensmeyer (1990:233) reports that "weathered bones that have lost their organic material and become cracked are less buoyant" in water, and thus less susceptible to fluvial transport than fresh, green bone. This is probably because bone shrinks as it dries, thereby reducing its porosity and increasing its bulk density. Thus a weathering profile constructed from fluvially accumulated bones would have ambiguous significance for inferring the duration of accumulation.

Brain (1981:115–116) performed an experiment in which he placed the bones of a pig (*Sus* sp.) in a subaerial context and let them weather for seven years. Some of the bones were placed in the open and thus were exposed to direct sunlight, other bones were placed in the shade. Brain (1981:116) reports that even after seven years the difference in weathering stage is remarkable; the mandible placed in the shade "retained enough fat after seven years for dust to be adhering to it [was weathering stage 0]. The defatting process of the fully exposed mandible was complete within one year, and [perhaps] within three months. In a shaded situation, defatting may take decades." This observation prompted Yellen (1991b) to worry that such fat-rich bones on abandoned sites might be exploited by scavenging carnivores. These observations indicate how the weathering stage displayed by a bone may not be related to the years since animal death, but it also indicates that bones that have been dead for some time may be accumulated by carnivores.

Lyman and Fox (1989:304–307) explore some of the effects of an accumulation history involving the collection of bones in varied weathering stages and suggest that as the weathering stage increases (from 0 to 5) at the time of bone accumulation, the resulting weathering profile will suggest longer accumulation histories, longer exposure durations, and/or more years since animal death. Accumulation history is thus yet another kind of taphonomic time, distinct from exposure duration and years since animal death.

Weathering and spatial data

Behrensmeyer (1978) suggests that the analyst may be able to begin to control for variation in the accumulation history and depositional microenvironment of individual weathered bones by studying their spatial distributions and associations. She suggests that (1) if bones in an assemblage display all weathering stages and are "homogeneously mixed in a single deposit" the assemblage probably represents a long-term accumulation, and (2) if only one weathering stage is displayed by all of the bones and those bones are spatially clustered, then the assemblage may represent a short-term single accumulation event or, if the depositional microenvironment varied between the individual weathered bones, the cluster may represent a long-term series of accumulation events (Behrensmeyer 1978:161).

Bones are distributed within three-dimensional space (horizontal and vertical). This in turn demands that the analyst determine if the bones being studied come from a single, stratigraphically defined surface (there was minimal variation in vertical location), or if they come from a three-dimensional stratum (there was variation in both the horizontal and the vertical locations of bones). As well, the analyst should consider the spatial scale of the distribution. Were the bones scattered over an area 10×10 m, or 100×100 m horizontally? But as noted above, we simply do not yet know how much spatial variation in the locations of individual bones, either horizontal or vertical, might signify the potential for different depositional microenvironments. Thus not only are the two spatial areas just suggested somewhat arbitrary, they are, for interpretive purposes, probably ambiguous (see Lyman and Fox 1989:309–311 for additional discussion).

There seem to be two ways to remove some of the arbitrariness of the spatial units. The first should be obvious to an archaeologist. Because faunal assemblages tend to be spatially defined, the obvious spatial unit to start with is a depositional stratum. Stein (1987:340) notes that geologists define a *deposit* as representing "one depositional event [but] the duration of such a depositional event is not often known. A single deposit may represent either continuous or abrupt deposition over either long or short periods of time." I suggest, nonetheless, that because a depositional stratum (or perhaps an archaeological feature) represents a single *geological* event of deposition, it would seem to be a logical way to define an assemblage of fossils from which a weathering profile may be derived.

The meaning of a weathering profile derived from a single depositional unit may be determined by considering a second way to approach spatial data. The second approach is exemplified by Bower *et al.* (1985). They construct weathering profiles for two Middle to Late Stone Age bone assemblages from East Africa and examine the relation between bone size and severity of weathering. For the weathering profile labeled "E" in Figure 9.3 Bower *et al.* (1985) note that most bones displaying the most advanced weathering stages

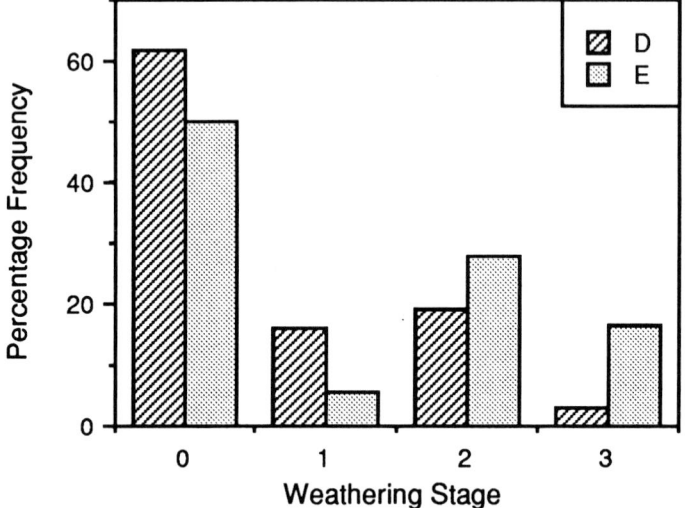

Figure 9.3. Weathering profiles for two assemblages of bones described by Bower *et al.* (1985).

are larger than less weathered bones, and those heavily weathered bones are all of the same taxon and probably from the same individual (an instance of taxonomic refitting). They also note that weathering profile "E" could be interpreted one of two ways: the heavily weathered large bones had been lying on the surface longer than the less weathered bones of smaller animals due either to (a) the greater sedimentation necessary to bury the large bones, or (b) a temporal disassociation between the accumulation of the large bones (accumulated first and thus exposed longer) and the small bones (accumulated sometime after the large bones). They conclude that "it is impossible to select the more likely of these two explanations" (Bower *et al.* 1985:51), but their use of bone size data lead them to a more detailed consideration of the accumulational history of the bone assemblage than is possible using only the weathering stages displayed by the bones.

Bower *et al.* (1985) also examine the relation of bone size to how weathered a bone is in their discussion of the weathering profile labeled "D" in Figure 9.3. They note that all sizes of bones are equally distributed across all weathering stages in profile "D" and interpret this profile as indicating more or less simultaneous deposition of all bones in that assemblage. Finally, they suggest that the bones in weathering profile "D" were exposed longer than those in weathering profile "E" even though the bones exhibiting weathering stages 2 and 3 in profile "E" make up only 22% of the total whereas bones exhibiting weathering stages 2 and 3 in profile "D" make up 44% of the total. The basis for this conclusion resides in the fact that some large bones and some small bones in weathering profile "D" display all weathering stages, so on average a bone in profile "D" must have been exposed to weathering longer than an average bone in profile "E."

Whether or not Bower *et al.* (1985) have offered correct interpretations is irrelevant (for example, their inferences demand the assumption that all bones were accumulated when they displayed weathering stage 0). What is of interest is their use of bone size as a way to consider the vertical dimension of deposition and exposure history. As well, they attempt to control specimen interdependence. It is perhaps indicative of the absence of sufficient interpretive analogs that Bower *et al.* (1985) cannot choose between alternative interpretations. Consideration of the variables influencing the duration and rate of bone weathering (defined in equation [9.2]) makes it clear that there are also various methodological weaknesses in how those variables might be analytically controlled when the analyst seeks to interpret weathering profiles. A more detailed example makes these points clear.

Analysis of weathering data

Potts (1986, 1988:48–56) provides the most detailed analysis and most extensive interpretation of bone weathering data yet available. Detailed review of his study provides important insights to the difficulties of analyzing and interpreting such data. My review is couched within the discussion presented above, and while Potts' interpretation may well be correct, there are reasons to suspect it may not be correct.

Potts (1986, 1988) presents data for weathered bone recovered from six Plio-Pleistocene sites at Olduvai Gorge. Data for the FLK *Zinjanthropus* site are summarized in Figure 9.4, along with data Potts presents for bones recovered from a hyena den and the sample of carcasses scattered across the landscape reported by Behrensmeyer (1978). Potts (1988:51) notes that the latter two samples represent "attritional, possible gradual, accumulations of bones rather than one major, short-term pulse of bone accumulation." Similarities between the two control samples and the FLK *Zinjanthropus* assemblage prompt Potts (1988:51) to suggest that "the time span represented by an accumulation of bones may be inferred from the distribution of weathering stages" and that the FLK *Zinjanthropus* assemblage probably represents "at least a 4-year period [and perhaps] at least a 5- to 10-year period" (Potts 1988:54). The frequency distributions shown by the three assemblages are not statistically significantly different (Kolmogorov–Smirnov two-sample D statistics are: Zinj to Landscape, $D = 0.33$, $P = 0.28$; Zinj to Hyena den, $D = 0.17$, $P = 0.39$; Landscape to Hyena den, $D = 0.33$, $P = 0.28$). Potts (1988:55) notes that the FLK *Zinjanthropus* bone assemblage occurred "as a layer within a thin, paleosol horizon about as thick as the bones themselves," and there was no patterned variation in weathering stages shown by bones in different horizontal contexts. He interprets this to indicate that differential burial (ED in equation [9.2]) has not influenced the bone weathering patterns.

Bunn and Kroll (1987:97) report that within the FLK *Zinjanthropus* assemblage they found "conjoining specimens of the same original bone that

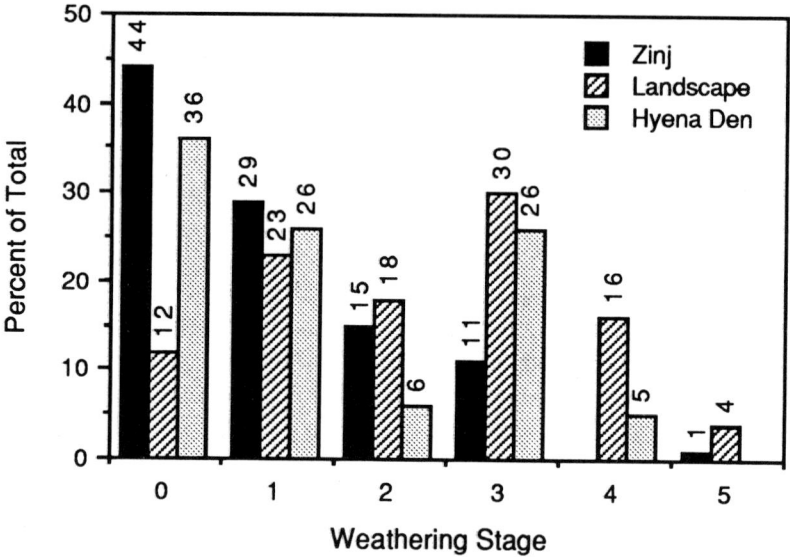

Figure 9.4. Frequency distribution of percentages of bones per weathering stage in three assemblages of bones (after Potts 1988:52–53, Figure 3.1; courtesy of the author and Aldine de Gruyter).

exhibit different weathering stages." Bunn and Kroll (1987:97) thus argue, contrary to Potts (1986, 1988), that the weathering data measure burial time, or ED in equation [9.2], rather than the duration of time over which bones accumulated. Given this debate, it is important to examine further the data Potts (1982, 1986, 1988) provides for the Olduvai bone assemblages. Potts (1986, 1988) attempts to ascertain the temporal duration over which six Olduvai bone assemblages accumulated (Table 9.3). He concludes that the bones making up the three assemblages found in thin stratigraphic layers accumulated over a 5 to 10 year interval whereas the bones making up the three assemblages found in relatively thick stratigraphic layers accumulated over a *minimum* of 5 to 10 years. Potts (1986:30) suggests that all of the Olduvai bone assemblages represent sites used over longer periods of time than sites used by modern hunter-gatherers who "tend to reoccupy campsites transiently for no longer than several months but only after several months of non-occupation." Given that all six of the assemblages consist of bones of multiple carcasses and numerous taxa, it is logical to suppose that more than one bone accumulation event is represented if it is assumed that early hominids would typically accumulate little more than one carcass or a portion of one carcass (and thus one taxon) per day. But this does not itself signify the temporal duration of the formation of the bone assemblages.

Potts (1986:29) reports that there is no correlation of the spatial distribution of bones and the weathering stages they display, and thus he suggests intra-assemblage variability in weathering "reflects the length of time bones were exposed on the land surface." To infer the temporal duration of accumulation

Table 9.3 *Frequencies of weathered bones in six assemblages from Olduvai Gorge (from Potts 1982:99–100). Frequencies are listed as NISP(% of total NISP)*

Site (deposit thickness, cm)	Weathering stage					
	0	1	2	3	4	5
FLK North-6 (50)						
all bones	403 (76)	77 (14)	20 (4)	32 (6)	0	0
long bones only	13 (27)	18 (38)	2 (4)	15 (31)	0	0
FLKNN L/2 (24)						
all bones	105 (46)	59 (26)	24 (10)	39 (17)	2 (1)	1 (0.5)
long bones only	9 (12)	27 (36)	12 (16)	23 (31)	2 (3)	1 (1)
DK L/2 (68)						
all bones	208 (52)	84 (21)	35 (9)	74 (18)	0	0
long bones only	17 (17)	20 (20)	16 (16)	45 (46)	0	0
FLK "Zinj" (9)						
all bones	771 (76)	147 (14)	63 (6)	36 (4)	0	1 (0)
long bones only	66 (44)	43 (29)	22 (15)	18 (12)	0	1 (1)
FLKNN L/3 (9)						
all bones	188 (64)	62 (21)	26 (9)	19 (6)	1 (0.5)	0
long bones only	11 (32)	9 (26)	6 (18)	7 (21)	1 (3)	0
DK L/3 (9)						
all bones	281 (60)	86 (18)	71 (15)	29 (6)	0	5 (1)
long bones only	29 (34)	24 (28)	17 (20)	15 (18)	0	0

of the bones, Potts notes the most advanced weathering stage in each assemblage, and then suggests that the minimum number of years necessary to attain that weathering stage based on Behrensmeyer's (1978) data (Table 9.1) provides an estimate of the minimal duration of bone accumulation. This of course presumes a more or less continuous and ongoing accumulation of carcasses and bones displaying weathering stage 0 over the time required for the first accumulated bones to reach weathering stage 3 or 4. A single or several temporally contiguous accumulation events over a single year could produce, for example, the FLK *Zinjanthropus* weathering profile (Figure 9.4), as Behrensmeyer (1978) noted, given sufficient variation in the exposure histories of the bones.

Gifford's (1977, 1984) data on the weathering stage exhibited by bones for which the years since death are known (Figure 9.2) are plotted on a three-pole graph, along with the six Olduvai assemblages, in Figure 9.5. Gifford's control assemblages for carcasses dead 4 to 10 years and those dead 10 to 15 years all appear to be much more weathered than the complete assemblages of Olduvai bones. Perhaps this is because Gifford's assemblages all derive from surface contexts whereas all of the Olduvai assemblages had been buried for nearly two million years. This leads us to a consideration of the geological context of the Olduvai specimens.

Potts (1982:102 103, 1986:30) suggests the bone assemblages found in thin

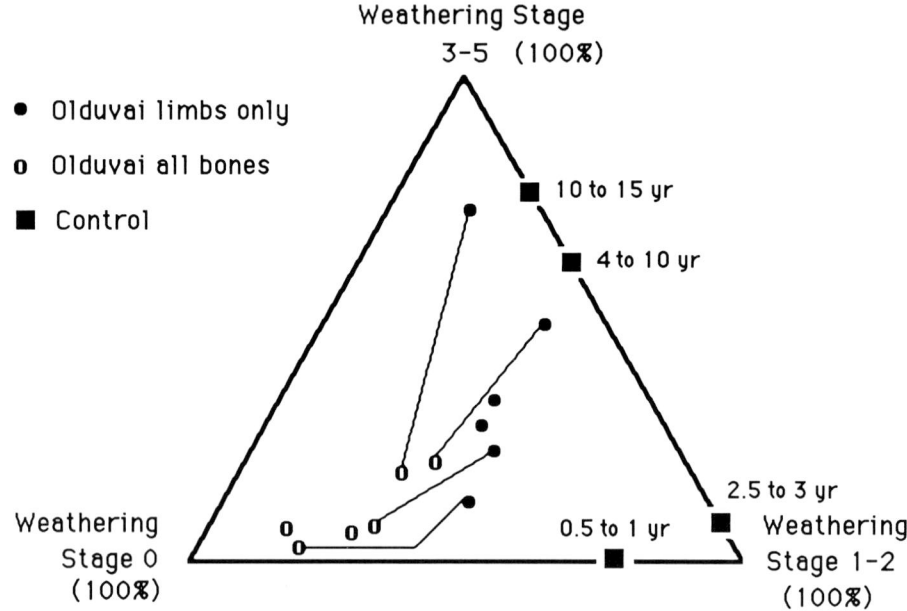

Figure 9.5. Three-pole graph of bone weathering data for six assemblages from Olduvai Gorge (after Potts 1982) and Gifford's (1977, 1984) control assemblages of carcasses dead for known numbers of years.

deposits accumulated over a shorter time period than the bone assemblages recovered from thick deposits. The basis for that suggestion resides in part in Potts' (1986:30) belief that "bones lying higher in the [thick] deposits may have started weathering years later than [bones] buried [more deeply]." While this presumes the same rate of deposition for thin and thick deposits (e.g., 1 cm per decade), it really only suggests that thicker deposits took longer to accumulate than thin deposits because the former are thicker than the latter; it is thus independent of the weathering data. It does *not* account for variation between the weathering profiles displayed by the assemblages of bones in thin and thick deposits. That variation is clear in Figure 9.6a where it is shown that, if all bones are included, there are proportionately more heavily weathered bones in the summed thick deposit assemblages than in the summed thin deposit assemblages (this pattern holds if each assemblage is considered individually). A Kolmogorov–Smirnov two-sample D statistic between the two is weak but significant ($D=0.078$, $P=0.05$), and indicates that the two are statistically different.

Differences between the summed thin deposit assemblages and the summed thick deposit assemblages are enhanced if we follow Potts (1986:23, 1988:53) and omit all but the limb bone diaphyses (Figure 9.6b; $D=0.217$, $P<0.01$). In eliminating all but the limb bones Potts is following Behrensmeyer's (1978:152) suggestion that small, non-limb bones weather more slowly than limbs. In so

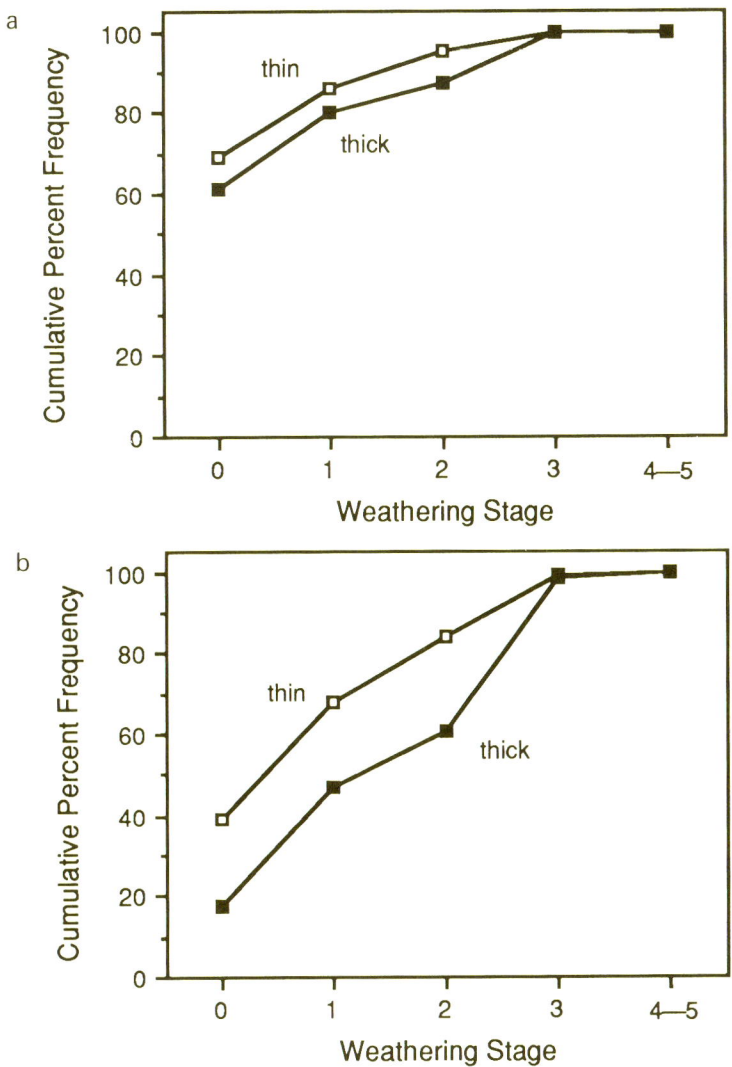

Figure 9.6. Cumulative percent frequency distributions for weathering stages of bones in summed assemblages of Olduvai Gorge thin deposit sites and summed assemblages of Olduvai Gorge thick deposit sites. (a) all bones included; (b) limb bones only included.

doing Potts (1986:29) hopes to "remove the bias toward weathering stage 0 by examining only the identified long bones with diaphyses." That this omission procedure in fact produces weathering profiles that appear to have been weathered for longer times is clear in Figure 9.5. There, the complete Olduvai assemblages of bones are to the lower right of the graph whereas the limb-bone only Olduvai assemblages are to the upper right. The question begged by omitting all but limb bones is whether the omission results in weathering profiles that appear to represent longer accumulation times simply because slower-weathering bones of the same carcass are omitted, or for some other reason.

Recalling that Gifford (1981, cited above) notes equid bones weather slower than bovid bones, I used Potts' (1982) data to determine that in the thin deposit assemblages, bovid specimens outnumber equid specimens 16.7 to 1 whereas in thick deposit assemblages bovid specimens outnumber equid specimens 28.3 to 1. Therefore the thin deposit assemblages should appear less weathered than the thick deposit assemblages simply because equid bones are relatively more abundant in the thin deposits than in the thick deposits. Further, the net change in weathering profiles for thick assemblages is greater than the change in weathering profiles for thick deposits when non-limb bones are omitted because, on average, more bones were omitted from the thin deposits than from the thick deposits, and this enhances the apparent difference between the two sets of assemblages (compare Figures 9.6a and 9.6b). These considerations suggest that variation between thick and thin deposit assemblages in terms of taxa represented in each (TX in equation [9.2]) and in terms of the bones included in each (SE in equation [9.2]) may account for at least some of the perceived variation in the two sets of weathering profiles, rather than solely variation in accumulation (or exposure) histories of the two sets.

A final comment concerns Potts' (1986:30) suggestion that his inferred "long-term accumulation of bones at each Olduvai site contrasts with" historically documented patterns of site use by hunter-gatherers. As yet, we do not have sufficient ethnoarchaeological data on (a) the duration of occupation of, (b) the history of reoccupation of, and (c) the weathering profiles of bones accumulated in camps and villages utilized by modern hunter-gatherers. While my discussion makes it abundantly clear that I am skeptical of Potts' (1982, 1986, 1988) conclusions regarding the bone-weathering data he presents, his analysis is extremely important because it helps underscore various ways to strengthen inferences built upon such data. To conclude our consideration of bone weathering, I now describe other kinds of data that should be recorded in conjunction with noting the maximum weathering stage displayed by a skeletal specimen.

Discussion

Behrensmeyer's (1978) seminal work on bone weathering reveals that weathering is a process, and thus it reflects the passage of Time. Her actualistic data indicate there is a strong correlation between the greatest weathering stage displayed by the bones of a carcass and the number of years since the animal died. However, as equation [9.2] makes clear, the relation of a weathering profile derived from a fossil assemblage and the passage of Time is obscure because of the several kinds of taphonomic time that are included. Actualistic research is necessary to allow us to disentangle the effects of these variables on our inferences.

Behrensmeyer (1978:152–153) suggests that analysts should record the most advanced weathering stage over areas greater than 1 cm² on each bone specimen. Lyman and Fox (1989:314) suggest such a procedure may mask precisely those data most relevant to assessing the various aspects of the formational history of a fossil assemblage. For example, Saunders (1977:104) reports that "in all instances where the effects of exposure were apparent, the exposed surfaces (skyward orientation) of the bone exhibited greater modification than the unexposed surfaces" in a collection of late Pleistocene mammalian remains he studied. Behrensmeyer (1978:153) notes that "bones are usually weathered more on the upper (exposed) than the lower (ground contact) surfaces." Thus, Frison and Todd (1986:39) suggest that bones subjected to minimal post-depositional and post-burial movement might not be uniformly weathered on all surfaces (see also Behrensmeyer 1981:611). If weathering data are to be analyzed, then, when bones are collected the weathering stage displayed by the uppermost surface of each bone should be recorded along with the weathering stage displayed by the lowermost surface. If a bone displays a broad range of weathering stages over various of its surfaces, it may have undergone slow burial, or possibly, multiple exposures and burials. One may eventually record a transect of weathering stages parallel to the long axis of a bone that deviates from the horizontal to help estimate sedimentation and burial rates. In conjunction with other evidence we now turn to, bone weathering may ultimately prove to be a highly significant source of taphonomic data (see also Chapter 11).

Root etching

The roots of many plants excrete humic acid, and often "dendritic patterns of shallow grooves" on bone surfaces "are interpreted as the result of dissolution by acids associated with the growth and decay of roots or fungus in direct contact with bone surfaces" (Behrensmeyer 1978:154). Morlan (1980:56–57) and Grayson (1988:30) indicate, however, that the etching may be caused by acids secreted by fungi associated with decomposing plants. "Fungi can not only decompose organic matter under relatively dry conditions, but can also produce a wide variety of organic acids during the process" (Grayson 1988:30). Perhaps it is a trivial distinction whether the roots themselves or the fungi associated with decomposing roots secrete the acids that etch bones, but for reasons noted below, this may not be a trivial distinction. In the following, I refer to the etching as "root etching" simply for convenience. Further, I consider root etching here under biostratinomy because while such etching can occur after bones have been buried, some mosses and lichens grow on bones prior to burial and can result in pre-burial root etching.

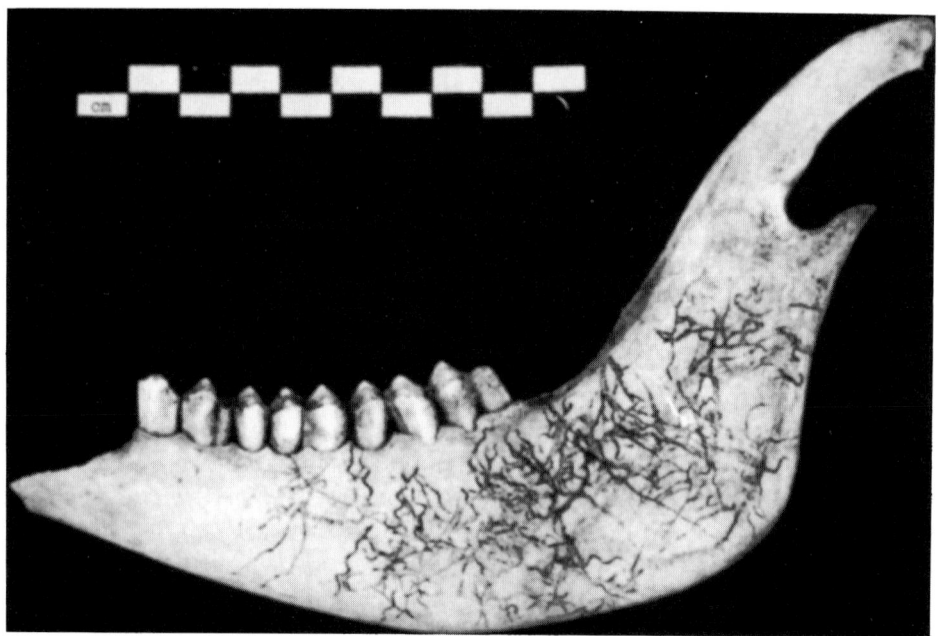

Figure 9.7. Root etching on a sheep mandible. Reproduced with permission from: Binford, L. R. *Bones: ancient men and modern myths*, Figure 3.07. New York: Academic Press, Inc. Copyright 1981 by Academic Press, Inc.

The wavy, "dendritic" (Morlan 1980:56), "sinuous" (Andrews and Cook 1985:685), "spaghetti-like" (Hesse and Wapnish 1985:85) patterns of the individual roots in contact with the bone are etched into the bone surface (Figure 9.7). Each rootlet lies in a groove "which presumably was formed by its exudates or by the microorganisms associated with rootlet metabolism" (Morlan 1980:57). Staining of the etched groove sometimes is no different in color from the unetched surface (Morlan 1980:57), sometimes the etched groove is lighter than the unetched surface (Morlan 1980:57), and sometimes the groove is darker than the etched surface (Binford 1981b:50). Microscopic inspection of root etching marks indicates they are "broad, smooth-bottomed, U-shaped [in cross-section] grooves" (Andrews and Cook 1985:685) that are internally etched (Cook 1986:157). Occasionally root etching has been interpreted as human-generated (Binford 1981b:49–51).

The presence of root etching indicates the bone existed in a plant-supporting sedimentary environment for at least part of its taphonomic history. Cook (1986:157) reports that root etching "affects bones as they are being buried and, in the case of lichen, indicates a period of at least partial exposure without much disturbance." Andrews (1990:19) implies that root etching occurs subsequent to the burial of a bone. However, we do not yet know precisely which kinds of plant roots create the etching, or even if it is the roots or associated fungi that create the etching, thus we do not know if bones must be buried, and if so how deeply they must be buried, to be susceptible to root etching. Further, we do not know the rate of root (or fungi) etching; that is, we do not know how long it takes for a root in contact with a bone to etch an obvious groove (and thus we do not know if different species of plants etch bones at different rates). Thus as I noted above, the distinction between whether the roots do the etching or fungi associated with decomposing roots do the etching may be a critical distinction if "root etching" data are to be used to infer exposure history. If we knew the kinds of things indicated about the agent responsible for the etching, this data may serve as a check on bone weathering data and aid in the determination of the exposure history of an assemblage.

Two other ways that root etching data may prove analytically useful concern the distribution of such etching in an assemblage. Grayson (1988) notes that the bone assemblages associated with five strata in a Utah cave display different proportions of etched bones. The deepest stratum contains no etched bones whereas from 1 to 17% (avg. $= 5.4 \pm 6.6$) of the specimens making up the assemblages from the four shallower strata are etched. Grayson (1988:30) reports that the deepest stratum is "the only stratum in the cave characterized by extremely low organic content," and, given his understanding of the mechanism which creates this kind of modification to bone surfaces he suggests "if this distribution is related to differential fungal activity through time, the distribution itself might have paleoenvironmental significance."

White (1992:119) suggests that "the presence of rootmarks on fracture surfaces or on the internal surface of limb-bone shafts can be essential clues about the relative timing of bone fracture." If root etching is present on a fracture surface, then the bone was broken prior to root etching, and thus perhaps prior to deposition. But, again, not knowing the (rate or) timing of root etching relative to time of deposition and timing relative to bone burial, precludes fine resolution regarding the timing of bone fracture. This is particularly the case when it is realized that root growth can fracture bones if the root(s) grow through the bone (Behrensmeyer 1978).

Distinguishing root etching from other kinds of acidic corrosion such as that created by acidic sediment matrices involves recognition of the individual channels or grooves formed by the roots (Andrews 1990:19). Digestive and sedimentary corrosion do not produce such grooves, although sometimes, apparently, root etching can be quite extensive and may resemble the former two kinds of corrosion (Andrews 1990:19). And, as noted above, root etching marks are sinuous and have smooth, U-shaped cross sections, allowing the easy distinction of them from human-created butchery marks (Chapter 8).

Trampling

Actualistic study of the effects of trampling by animals, including humans, on bones has tended to focus on three things: the creation of marks on bones, the fracturing of bones, and the spatial displacement of bones. Trampling also

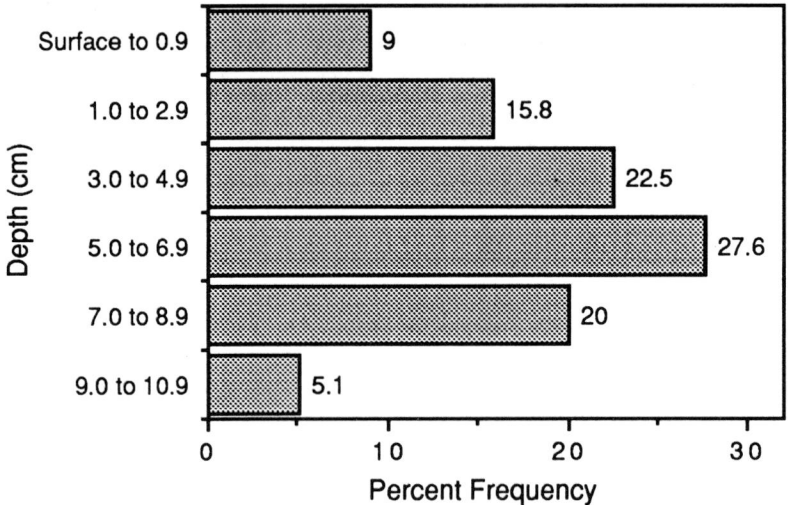

Figure 9.8. Vertical frequency distribution of trampled artifacts. Note the approximately normal distribution (data from Gifford-Gonzalez *et al.* 1985).

apparently abrades bones (e.g., Brain 1967a), but I reserve discussion of that kind of modification for the next section of this chapter because taphonomic processes other than trampling can also abrade bones.

Movement

Gifford-Gonzalez *et al.* (1985) summarize much of the literature on the effects of trampling on the vertical distribution of artifacts, and they performed experiments on those effects using stone artifacts and bones. While their results concern the former, those results are nonetheless intriguing. They report that many prehistoric assemblages thought to have been trampled and several experimentally trampled assemblages display an essentially normal frequency distribution of items against depth from surface (Figure 9.8). Such a distribution occurs when the items were first trampled while they lay on the surface, and when the substrate was not too fine, compact, or cemented. When the trampled items are located just beneath the ground surface prior to trampling, the frequency distribution of items after trampling tends to "more closely resemble a Poisson distribution in vertical configuration" (Gifford-Gonzalez *et al.* 1985:816). There is also some horizontal displacement of items, but Gifford-Gonzalez *et al.* present no data on that displacement, perhaps because it seems to be minimal. Olsen and Shipman (1988:536) report that "horizontal movement seems to be related to the compaction of the soil. A hard substrate enables bones to stay on the surface longer, which increases the probability of horizontal movement." Buried bones are less susceptible to horizontal movement resulting from kicking (see also Yellen 1991b).

Vertical movement of bones seems to be a typical, but variable result of trampling. It is not restricted to downward movement; it can also be upward as "when a foot placed immediately adjacent to the [bone] sinks deeply into the substrate and displaces the [bone] and adjacent sediments both laterally and upwardly" (Olsen and Shipman 1988:537). Vertical movement seems to depend on the intensity of trampling, the compactness of the sediments, the extent to which bones are already buried when stepped on, and the size (weight) and shape of the bone. Downward movement and resulting burial seems to occur rapidly in soft, sandy sediment but is slower and less extensive in silt mixed with gravel, the gravel apparently acting much like a "pavement" (Olsen and Shipman 1988:537). Trampled objects may sort by size and surface area, with small objects becoming more deeply buried than large objects or objects with large surface areas (Gifford 1977:183). Large bones such as many ungulate skulls may be simply stepped around. The orientation and plunge or dip of long bones may be altered by trampling, as when one end of a long bone is stepped on and forced beneath the surface while the other end raises up off of the ground surface (see Chapter 6 for discussion of bone orientation and plunge).

Fragmentation

Haynes (1991:253) implies that trampling may "destroy" some skeletal elements, especially those that are somewhat weathered and easily broken. This seems likely when it is realized that fragmentation, regardless of the taphonomic process or agent doing the fracturing, serves to reduce skeletal elements into pieces only some of which may be identifiable to skeletal element. Such "analytical absence" of skeletal parts (Lyman and O'Brien 1987) is equivalent to the actual destruction of a skeletal element. Yellen (1991b:165) suggests that "from an archaeological perspective destruction occurs either (1) when a specimen becomes so fragile that it cannot be exposed or removed from the ground for identification or (2) when it fragments into pieces too small to be recovered by normal archaeological means." Thus there are several ways to destroy bones but there are many ways to end up with skeletal parts being absent from one's analysis.

Andrews (1990:8–10) trampled single owl pellets in sealed plastic bags and found that excessive trampling resulted in the disintegration of the pellets, fragmentation of the mandibles and maxillae in the pellets (and thus trampling produced many isolated teeth), and smaller limb bones remained essentially intact while larger ones were somewhat fragmented. Many maxillae were destroyed. He notes that while encased in the pellet, rodent bones tend to be fairly well protected from the trampling forces. Thus, both large (elephant) bones (Haynes 1991) and small (rodent) bones (Andrews 1990) can be broken, and thus be effectively destroyed for analytical purposes, as a result of trampling. Yellen (1991b:165) suggests bone shape may influence the susceptibility of bones to fracture by trampling; "spherical" specimens should be less prone to breakage whereas "plate-shaped" and "cylinder-like" specimens should be more prone to breakage. While bone shape can be readily measured using techniques described in conjunction with Figure 6.6, I know of no data bearing on Yellen's suggestion.

Several authors suggest trampling can break bones, but only if certain conditions are met. Myers *et al.* (1980:487), on one hand, were able to break "slightly weathered" long bones of an adult domestic cow (*Bos taurus*), and note that fresh bone is a "tough, durable material and is difficult to break by trampling" (see also Olsen and Shipman 1988:537). Saunders (1977:105), on the other hand, argues that the fracture morphology of at least one mastodon (*Mammut americanum*) bone recovered from a late Pleistocene site in Missouri suggests the fracture occurred while the bone was fresh and was the result of trampling. All of these reports indicate (see also Chapter 8) that bones are more likely to be broken by trampling subsequent to some weathering (weathering stage 1 or greater) than when still fresh (weathering stage 0), although fresh bones can be broken by trampling. The critical aspect of analysis here, then, is to discern whether a bone was broken while fresh or after it was weathered (see Chapter 8 for additional discussion). To aid such analyses, one should also attempt to refit bone fragments mechanically (Chapter 5), and measure the distance between refit fragments. The interpretive assumption would be that closely contiguous refitting fragments were fractured after their deposition and thus by trampling or possibly sedimentary overburden pressures rather than their having been broken prior to final deposition (see Chapters 8 and 11 for additional discussion).

Olsen and Shipman (1988:537) suggest that "since breakage from trampling tends to occur in the weakest parts of the bone, as it would in most natural circumstances, there does not appear to be anything particularly diagnostic about the type or patterning of breaks created by this process." While they are perhaps correct, I suggest that because bones become structurally weaker as they become progressively more weathered, the temporal placement of when a bone is broken relative to its weathering stage may be an important bit of taphonomic information for unravelling the exposure duration (bones on the ground surface are more likely to be broken by trampling than buried bones) and the accumulation history of a bone (fresh bones are more likely to be accumulated by biological agents such as predators and scavengers than weathered bones).

Marks on bones created by trampling

Behrensmeyer *et al.* (1986, 1989) and Fiorillo (1989), while not without some non-experimental precedent (e.g., Andrews and Cook 1985), were among the first to demonstrate experimentally that trampling can produce scratch marks on bones. The problem being addressed was that the scratch marks created by trampling appear to be morphologically (on a microscopic scale) similar to stone tool-generated cut marks (morphological attributes of these marks are discussed in detail in Chapter 8). It suffices here to note that trampling marks tend to be more randomly oriented or multidirectional relative to butchering marks. Trampling marks tend to be located on the shafts of long bones rather than on the ends, and are shallow relative to butchering marks (Andrews and Cook 1985; Olsen and Shipman 1988).

Summary

Documentation of changes in bones in ethnoarchaeological and experimental contexts has indicated that the fossil record can be greatly modified from its depositional condition by trampling. Bones may be moved, broken, and scratched by trampling. Yet recognition of the effects of trampling in prehistoric contexts has not often been reported. Fiorillo (1989) suggests the shallow, sub-parallel scratch marks he observed on Miocene-aged equid bones from a paleontological site in Nebraska represent trampling marks because they more or less match experimentally generated trampling marks. The age of the bones Fiorillo studied precludes a hominid agent. Stahl and Ziedler (1990) infer that trampling by hominids resulted in the fragmentation and downward displacement of bones on the floor of a 4000 year old house in Ecuador. Bones on that floor which were not trampled were larger and had not been pushed downward through the floor surface. Stahl and Ziedler's inferences, like Fiorillo's, are founded on neotaphonomic and contextual data, and analogical reasoning. Their studies illustrate well the kinds of analyses necessary to inferring that trampling of a bone assemblage has taken place.

Abrasion

C. K. Brain (1967a) was one of the first modern taphonomists to describe the natural abrasion of bones. He reports that a collection of mammal bones scattered on sandy sediment around a waterhole had been trampled by goats (*Capra hircus*) and people coming for water. Such disturbance, Brain (1967a:98) believes, "serves to constantly abrade the weathered surface of the bone as it develops, producing a smoothness and polish of the sort that one would normally associate only with human agency." He also found naturally abraded bones around goat corrals and along paths, whereas abrasion was not evident on bones located in areas infrequently trod by animals or people. Brain (1967a) suggests that eolian-related abrasion does not seem to be the responsible taphonomic process because bones collected from sand dunes show "severe etching rather than smoothing or polish. The [bone] surface tends to be selectively [abraded] as a result of constant bombardment by the [wind-blown] sand grains" (Brain 1967a:99). Natural abrasion is "characteristically fairly general and not specifically restricted to any one part of the bone;" bone tools display abrasion (from manufacture and/or use) over restricted areas of the bone surface (Brain 1967a:99).

While a serious flaw can be found in Brain's (1967a) paper (he did not know whether the bones he suspected to have been abraded by trampling were *not* abraded prior to deposition, and he did not see the inspected bones be trampled), later experimental research confirms his observations. It has been found, for example, that fluvial transport of bones tends to abrade the entire surface of a bone specimen whereas abrasion by eolian activity seems to abrade only the exposed or top surface(s) of specimens (e.g., Shipman and Rose 1983a). Behrensmeyer (1990:234) reports that weathered bones are more vulnerable to abrasion and breakage in fluvial transport situations than fresh, green bone. Trampling creates deep scratches in bone surfaces (Olsen and Shipman 1988), something fluvial abrasion does not produce (see Chapter 6). Shipman and Rose (1983a:79) report that, based on their experiments of tumbling bones in a barrel with water and sediment, "(1) sedimentary abrasion of any significant duration will obliterate [stone-tool generated] slicing marks, (2) sedimentary abrasion will not produce marks that mimic slicing marks, and (3) sedimentary abrasion will occasionally produce marks that mimic carnivore tooth scratches." Shipman and Rose (1988:328) found that eolian abrasion of large mammal bone with loess does not create either a fine polish or scratches like those resulting from extensive utilization by hominids; for such abrasion to produce significant rounding of broken edges of bones, they suggest a great deal of time would be necessary (see Chapter 6 for additional details).

Given that manufacturing- and use-related damage to a bone should probably occur near the fracture, Sadek-Kooros (1975) lists the following possible categories for the distribution of damage displayed by a specimen, especially a broken long bone: (1) faint, irregular, randomly located over the surface of the specimen; (2) over the entire [exposed] surface of the specimen; (3) over the entire broken end of the specimen [presuming the other end is an articular end]; (4) over only the tip of the broken end; (5) over the entire fracture edge or surface; and (6) over the broken end and the shaft, but not the articular end. Few analysts have studied the distribution of abrasion modification across a sample of bones, but along with Brain and Sadek-Kooros, others (Lyman 1984b; Myers *et al.* 1980) suggest the distribution of abrasion and polish should be more spatially restricted on bone specimens used as tools than on naturally abraded bones.

White (1992) broke and then boiled several mule deer (*Odocoileus hemionus*) metapodials in a replica of a prehistoric ceramic vessel. While the water in the vessel never vigorously boiled, the bones were cooked for three hours and occasionally stirred. White (1992:124) inspected the bone fragments after cooking and found *pot polish* on 29 of 69 (42%) fragments. Longer specimens

(> 3 cm) were more likely to display pot polish. And, the polish was restricted to "projecting ends" of the fragments (White 1992:122), where beveling and rounding of the ends as well as microscopic striations were found. White (1992:122) suggests that this "abrasion" occurs due to "the contact between the broken edge of the bone and the pot side" which creates "facets" that are visible to the naked eye; the facets are "shiny relative to the remainder of the broken bone edge; they are placed at the most projecting points of the bone; and they are most easily recognized as modification to sharp broken edges." These abrasion-created facets more readily reflect light (and appear shiny) than unabraded bone surfaces. Importantly, White (1992:124) notes that the amount of pot polish in a collection "will exist along a continuum depending upon the length of cooking time, the amount of grit in the cooking vessel bed load, the roughness of the vessel's inner surface, and the amount of stirring."

Does White's (1992) pot polish weaken or compromise the validity of the distribution of abrasion damage over the surface of a specimen as an indicator of the use of broken bones as tools? White (1992:324) suggests it might. What seems to be called for, nonetheless, is more intensive and extensive study of the distribution of abrasion damage across the surfaces and edges of bone specimens with known taphonomic histories in order to establish the range of variation one might find in such distributions. This necessity is highlighted by Gilbert's (1979:185) observation that "handling of bone tools would likely raise a sheen in proportion to the length of time the tool was used." If Gilbert is correct, then use-wear related abrasion on bone tools may have a distribution on a specimen distinct from the distribution of abrasion or polish created by how and where the specimen was held.

All reference to abrasion to this point concerns macroscopic attributes. Bromage (1984) reports a series of experiments he performed to determine the microscopic effects of various abrasive forces on forming bone. Defining *abrasion* as "the result of any agent that erodes the bone surface through the application of physical force," Bromage (1984:173) notes that all of the abrasive forces he examined remove incompletely mineralized collagen fiber bundles. Such microscopic abrasion can occur in pre-depositional, depositional, and post-depositional contexts. Bromage (1984:175) found that sliding abrasion (with abrasive paper), brushing (with a toothbrush), rubbing (with the fingers; note Gilbert's [1979] comment in the preceding paragraph), and weight (such as sediment overburden) all produce smooth abraded surfaces with obliterated details of vascular canals. Rough abraded surfaces result from particles and water transmitted under pressure to bone surfaces; these surfaces appear shiny macroscopically as a result of "increased reflectance from many surfaces" (Bromage 1984:164). Bromage (1984:175) cautions that the precise taphonomic significance of his experiments requires further study and the establishment of ties with studies of macroscopic abrasion. One can wonder, for example, how much abrasion force (strength, duration) of each particular kind Bromage identifies is required to remove traces of other taphonomic processes such as butchering marks or gnawing marks.

A final observation by Martill (1990:282) is important; he suggests that abrasion may be mediated by "the slightly elastic surface of [fresh] bone which retains its organic matrix, thus absorbing some of the shock of impacting sand grains." The significance of this observation is crucial because it implies that bones displaying weathering stage 2 or 3 will suffer abrasive damage much more quickly than bones displaying weathering stage 0 simply because of differences in the degree to which the bones are weathered. If Martill's suggestion is correct, then analyses of abrasion damage should not be performed without consideration of the degree to which bones are weathered. In combination, perhaps these two variables will reveal details of the transport and exposure history of the bone assemblage. I am not aware, however, of any actualistic research on the covariation of weathering and abrasion.

Burning

It is important in any discussion of burned bone to clarify what is meant by "burned." Marshall (1989:17) notes that *cooking* involves the preparation of food for eating by heating that food by boiling, roasting, baking, or the like. *Heating* an object involves making that object warm or hot. *Burning* results from excessive heat and modifies or damages the heated object. Excessive heat can involve high temperatures, temporally long exposure to heat, or both. To suggest, then, that cooking will produce burned bone is, perhaps, too simplistic and may be unrealistic.

Interest in whether particular bones have been burned or not probably has roots in the typical use of such modification to infer that the bones were deposited by people and represent the remains of cooked meals (see Chapter 6), and in attempts to determine when early hominids first began to control fire. Regarding the latter, James (1989) reports that in 34 lower and middle Pleistocene sites in the Old World, 11 kinds of evidence of the use of fire are cited. One of those kinds of evidence is burned bone, which is reported at 10 of the 34 sites, and it is the exclusive kind of evidence at five sites. Both of these uses of burned bone are what I consider interpretive.

Before burning can be used in an interpretive sense one must know the attributes used to determine whether or not a bone specimen has been burned. I include discussion of burning in this chapter on biostratinomy because most bones are probably burned sometime between the death of an organism and burial of that organism's bones. However, it is important to note that while bones may typically be charred prior to deposition and burial or after deposition and prior to burial (e.g., Grayson 1988; Lyman 1988a), they may also be burned after burial if the matrix they are buried in is rich in organic material and dry, such as is found in many cave sites in the western United

States (see discussion in James 1989:9–10). Thus, there seem to be minimally two steps to interpreting burned bone: identifying bone as burned, and identifying when the bone was burned relative to its depositional and burial history. We consider each in turn.

Attributes of burning

On the basis of experiments "with bones in the ashes of campfires" Brain (1981:54) suggests "there are two distinct stages in the charring of bone." As collagen is carbonized, the bone turns black. With continued heating, the black carbon is oxidized and the bone becomes white and has a chalky consistency. Brain (1981:55) labels the black stage "carbonized" and the white stage "calcined." Johnson (1989:441) distinguishes four burning stages: unburned, "scorched (superficial burning), charred (blackened, towards charcoal), and calcined (blue-white, loss of all organic material, plastically deformed)." The differences between Brain's and Johnson's burning stages underscore that heating of a bone is a process because the bone's temperature (or perhaps the duration that it is exposed to heat) must increase for it to progress from "charred" to "calcined." It is my impression that typically it is the color of a specimen, and less frequently a specimen's lack of hardness when white and its brittleness when black, that analysts use to distinguish burned bone from unburned bone.

Kiszely (1973) suggests there are three basic stages of change in bone as it is heated. Water escapes from the bone at a peak rate when it attains a temperature of 137°C and slows to a minimum rate at 220°C. At the latter temperature the second stage of change takes over; it involves the liquification and decomposition of organic matter. The second stage peaks at about 330°C and is essentially completed when the bone attains a temperature of 380°C. Finally, virtually all organic matter is burned away at 600°C. Kiszely's experimental results were obtained with powdered cortical bone. Von Endt and Ortner (1984) report that the structural density of bone influences the accessibility of the molecular constituents of bone material. Thus, the temperature minima and maxima per stage may be lower for Kiszely's materials than for unpowdered bones and bone fragments.

On the basis of experiments in which bones were heated in a muffle furnace/ kiln Shipman *et al*. (1984b:314) conclude that color is a poor indicator of the *precise* temperature to which a bone was heated due to difficulties in recording color accurately and because bones may change color diagenetically and thus a specimen's color may have nothing to do with whether or not the bone was heated. They suggest, however, that a specimen's color can be used as an indication of the *range* of temperatures to which a bone was heated if diagenetic processes have not altered the specimen's color and it is clear that the bone was heated (Shipman *et al*. 1984b:314). Lightly heated bones (<400°C) tend

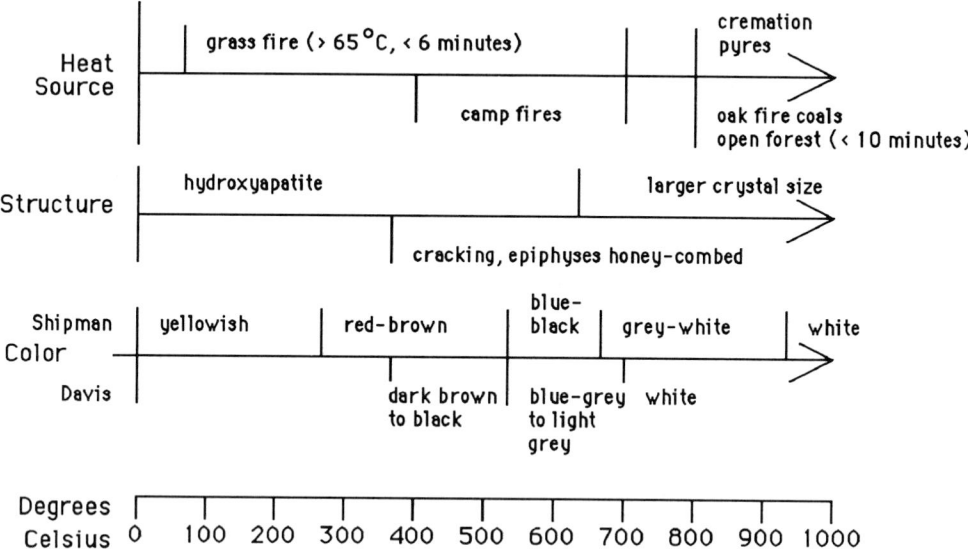

Figure 9.9. Summary of changes to bone subjected to heating. Redrawn after Shipman (1988a:279) with additions and modifications after David (1990).

toward neutral and yellow colors (Figure 9.9). Bones heated between about 300°C and 800°C tend to be yellow-red, and red to purple. Intensively heated bones (≥ 600°C) tend to be purplish-blue and blue. Bones that are "completely incinerated or calcined [can be] described as bluish-white or gray in color" (Shipman *et al.* 1984b:308).

Shipman *et al.* (1984b) report that the micromorphology of bone, enamel, and dentin changes with progressively greater temperatures, and these changes indicate recrystallization of bone mineral and possibly the melting of hydroxyapatite. Using scanning electron microscopy they found that a major change in hydroxyapatite crystal size occurs at about 645°C. Brain and Sillen (1988:464) report that in bovid bones "heated to 300–400°C the lamellar structure was greatly accentuated," but do not describe what this means in terms of appearance. Shipman *et al.* (1984b:321) suggest that the decomposition of the organic component of bone "probably occurs between 360°C and 525°C;" this range is higher than that reported by Kiszely (1973) and described above, although it is not incompatible with his results because Kiszely used powdered bone and thus the alteration he observed may have occurred at lower temperatures than that observed by Shipman *et al.* (1984b). Finally, Shipman *et al.* (1984b) report that bone shrinks more as it is heated to progressively higher temperatures. They also indicate that the amount of shrinkage may be dependent on the ratio of spongy to compact bone in the measured section, the amount of shrinkage increasing as the amount of spongy bone increases. Gilchrist and Mytum (1986) document an average shrinkage of 5 to 30% for cow (*Bos* sp.) and sheep (*Ovis aries*) bones, with the smaller (and apparently ontogenically younger, and thus less mineralized and more porous) bones of the latter taxon shrinking more than the bones of the former. The observations reported by Shipman *et al.* (1984b) are summarized in Figure 9.9, and are supplemented by observations reported by Brain and Sillen (1988), David (1990), and Gilchrist and Mytum (1986).

Buikstra and Swegle (1989) review data thought to indicate the condition of the bone at the time of incineration. They report that "most descriptive and experimental work has focused on distinguishing between bones burned in one of three conditions: fleshed, green (defleshed shortly before burning), and dry" (Buikstra and Swegle 1989:248). Experimental data indicate that bones burned when dry display surficial checking and cracking, a lack of longitudinal splitting, and no warping; bone that was fleshed and green when burned does not display these attributes. Fleshed (cremated) bone tends to have "serrated, transverse fractures totally through it, along with diagonal cracking accompanied by warping" (Thurman and Willmore 1981:281). Bone that was burned while green (moist, or weathering stage 0) displays "serrated fractures near epiphyses but otherwise parallel-sided fractures through it (along checking lines), and less pronounced warping" (Thurman and Willmore 1981:281). Buikstra and Swegle (1989:249) emphasize that the attributes used to determine if bone was fleshed, green, or dry when burned are tentative because they depend in part on the temperature of the fire, the length of time the bone is heated, and the amount of flesh covering the bone (see also David 1990 and below).

Buikstra and Swegle (1989) performed experiments to evaluate the criteria reviewed in the preceding paragraph. To approximate "open-air" heating, they heated some of each of fleshed, defleshed and green, and dry bones in a gas incinerator, and other variously fleshed, green, and dry bones in a wood fire. Their terminology for burning stages is: "unburned, smoked, and calcined" (Buikstra and Swegle 1989:250). I presume their "smoked" stage is equivalent to Brain's (1981) "carbonized" stage because they indicate that smoked bone is black and displays "incomplete combustion of organic materials" (Buikstra and Swegle 1989:252). The wood fire was built over bones laid on the ground to approximate human mortuary practices. Their conclusions and observations can be summarized as follows:

1. only defleshed bone is uniformly smoked (blackened);
   a. dry bone has insufficient organic substance to become uniformly smoked;
   b. flesh on bone insulates the covered areas which retain an unburned color whereas exposed surfaces become blackened;
2. fleshed bone that has been calcined cannot be distinguished on the basis of color from green bone that has been calcined;
   a. both fleshed and green bone that has been calcined exhibits deep longitudinal fissures (cracks), and transverse splitting is common on both; both are white, blue, and/or gray;
   b. dry bone that has been calcined exhibits shallow longitudinal fissures, and transverse splitting is present but rare; calcined dry bone is light brown or tan.

They conclude that taphonomists may distinguish bone that was burned when dry from bone that was either fleshed or green, but the latter two cannot be reliably distinguished. The ease with which the pre-burn condition may be diagnosed is dependent on the intensity of the burning: smoked bones are more difficult to diagnose than calcined bones.

Timing and agent of burning

Determining when a bone was burned relative to its accumulational and depositional history may be critical to analysis, and may help the analyst determine the mechanism of burning. It is important, then, first to outline how a bone might be burned. Following Buikstra and Swegle (1989), David (1990), and James (1989), there are several ways that bones might be burned. These can be summarized as follows:

      a. humanly burned bone (burning is intentional);
          1. cooking;
          2. disposal of food waste (perhaps to reduce attractiveness to scavengers; Gifford-Gonzalez 1989a:187);
          3. fuel for anthropogenic fires (for warmth and/or protection from predators);
          4. cremation (generally of human remains);
      b. naturally burned bone (burning is accidental or unintentional);
          1. nearness to one of the anthropogenic kinds of fires above;
          2. brush (grass, forest) fire;
          3. *in situ* burning of organic matrix.

The major distinction is that between naturally burnt bone and humanly burnt bone. Once this distinction is made, the analyst may want to determine if humanly or intentionally burnt bone was burnt during cremation, disposal of food waste, or cooking of food. I believe few would disagree with my suspicion that humanly intended burnings tend to occur more frequently when bone is fleshed or green than when bone is dry, although dry bone may occasionally have been used as fuel for a fire (thus the contexts of bone specimens may be important). Bones burned unintentionally or by natural fires may be fleshed, green, or dry. Thus, determining the condition of a burned bone prior to burning may provide a clue to the timing of burning.

    David (1990) reports three experiments: the burning of bones exposed to a brush fire, the burning of bones in an anthropogenic hearth for 25 minutes (maximum heat was 84°C for 15 minutes), and the burning of bones in an anthropogenic hearth in which the fire actively burned (flames visible) for 65 minutes and then smoldered (no flames visible) for an additional five hours (same fire as the second experiment). His results (David 1990:68, 71) indicate that "calcination (as evidenced by grey, white, blue, or bluish-green tints to a bone)" involves the oxidation of the carbon created during Brain's (1981) carbonization stage, and that calcination takes longer heating times, higher temperatures, or both, relative to carbonization. Calcination requires "temperatures of over 450°C to 500°C, or heating for over 3 to 4 minutes, or a combination of both" (David 1990:69). In the order listed, the three experiments produced the following proportions of the surface area of bones displaying attributes of being unburnt, carbonized, and calcined:

	brush fire	25 min. in hearth	6 hr. in hearth
unburnt	1.1	0.0	0.0
carbonized	98.9	75.5	5.0
calcined	0.0	24.5	95.0

David (1990:75) thus concludes that "natural conditions will regularly carbonise bones but will rarely calcine them. When large proportions of the surface area of a bone are calcined, one can safely infer (anthropogenic) prolonged fires under high temperatures." Thus the degree to which a bone has been burned may provide a clue to the taphonomic agent responsible for the burning.

Gifford-Gonzalez (1989a:193) suggests that "a bone burned all over was clearly subjected to burning after flesh had been removed, either by prior food consumption or by intensive incineration. Burning on articular surfaces only could have occurred when the rest of the bone was protected by soft tissues, as when a joint of meat is roasted." That is, the distribution of burning damage across skeletal elements may help one determine if the bone was burned during cooking. I suggest that such an inference requires some knowledge of the butchering pattern, especially how animal carcasses were disarticulated (see Chapter 8). Johnson (1989:441) suggests that the distribution of burning over a broken skeletal element may indicate if the bone was broken before, or after, burning. Evidence of burning on fracture surfaces, on the interior (medullary cavity) of specimens, or two conjoining pieces only one of which is burnt are good indications that a bone was broken (or disarticulated) before it was burned. Knight (1985:10) found, for example, that complete deer (*Odocoileus* sp.) bones burned in an open pit may be calcined on their exterior surfaces but only carbonized on their interior surfaces.

Other effects of burning on bone

Several of the authors cited in preceding sections suggest that burned bones tend to be more fragmented than unburned specimens (e.g., Johnson 1989). This is probably because burned bone is more brittle than unburned bone due to the removal of organic matter (collagen fibers) from the former; the effects of recrystallization (Shipman *et al.* 1984b) on the brittleness of a bone are unknown, but experiments reported by Knight (1985) suggest several things. He found that complete bones of deer (*Odocoileus* sp.) became more fractured and were more often fractured by heating than complete bones of beaver (*Castor canadensis*) and muskrat (*Ondatra zibethicus*) when all were heated under similar conditions. This may, however, be a function of the fact that the

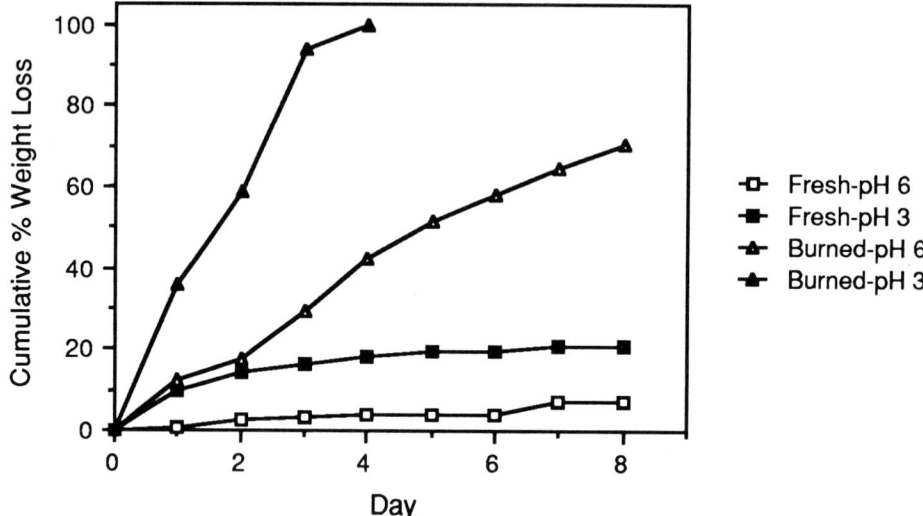

Figure 9.10. Cumulative percentage of weight loss of fresh and burned bone specimens placed in acid solutions (after Knight 1985).

deer bones were heated for six hours, the beaver skeletons for three hours, and the muskrat skeletons for one and a half hours, although Knight (1985:8) reports that all were "completely incinerated."

Knight (1985:22) measured the crushing load or compressive force (lb/in²) necessary to fracture completely calcined skeletal elements. He found that generally, burned bones from ontogenically old individuals require more compressive force to fracture than burned bones of young individuals. The compressive force required to fracture burned bones tends to be correlated with the structural density of a bone (see Chapter 7 for discussion of structural density). From this he concludes that "even after the destructive process of incineration each skeletal element has the same preservation potential relative to the other elements in the skeleton that it had in a fresh condition. A dense, fresh bone is still a dense bone after incineration" (Knight 1985:73).

Knight (1985) found that burned bone tends to dissolve more rapidly than fresh bone in acid solution. He placed completely calcined long bone diaphysis pieces of beaver (*Castor canadensis*) in acid solutions of pH 6, pH 5, pH 4, and pH 3. He placed similar pieces of fresh, unburned bone in solutions of the same pH, and recorded the daily weight loss of each specimen over eight days. Results for the specimens placed in the least acidic and the most acidic solutions are plotted in Figure 9.10 as cumulative percentage of weight loss over the eight days. Note that not only does the more acid solution result in more rapid weight loss for both fresh and burned bone, but that the burned bone in the pH 6 solution lost weight more rapidly than the fresh bone in the pH 3 solution.

Based on their experiments, Brain and Sillen (1988:464) report that burned bones contain more carbon (what they term "organic char"), which develops between 300°C and 400°C, than unburned bone. Further, the carbon-to-nitrogen ratio in fresh collagen ranges from 2.9 to 3.6 whereas in their experimentally burned bones that ratio ranges from 4.2 to 6.0. They used these criteria as well as histological ones to distinguish burned bones from those blackened by diagenetic incorporation of manganese dioxide. While they do not present the requisite data for statistical testing, the frequency distribution of burned bones across vertical space in an excavation they report seems to be a function of sample size or the frequency distribution of all bone specimens across vertical space (Brain and Sillen 1988:465). That is, the more bone specimens they found per vertical excavation unit, the more burned bones they found. I note that the significance of such a frequency distribution (or, the lack of a correlation between the frequency of bones and the frequency of burned bones) might reveal details regarding the burning history of the specimens.

Gilchrist and Mytum (1986) report that 10 to 50% of the bones they heated were "destroyed." Knight (1985) suggests bones are not literally destroyed by burning, but rather become extremely fragmented. Given that intensive fragmentation reduces the probability that a specimen can be identified (Lyman and O'Brien 1987), bones are probably not literally destroyed by burning but rather are analytically destroyed; that is, their pieces may be recovered but be unidentifiable due to their small size.

Analysis of burned bones

There seems to be a belief among at least some archaeologists that burned bones will preserve better than unburned bones. This belief apparently resides in indications that carbonized remains are chemically inert because they consist mostly of carbon (Gilchrist and Mytum 1986:30), carbonized remains are more resistant to biodegradation than uncarbonized organics (e.g., Bower 1986:94), or both. Knight's (1985) experimental results indicate the former is false (e.g., Figure 9.10). The latter may well be true considering that it is the organic fraction of bone tissue that is carbonized and at least some biological organisms, especially microscopic ones, feed on the organic fraction. The question remains, however, once an assemblage of faunal remains have been sorted into burned and unburned specimens, what is the analyst to do next?

Analysts often present counts of burned and unburned specimens, or proportions of the total NISP for each taxon that are burnt. These data are little more than descriptive, however, especially if bones burned intentionally have not been sorted from those burned accidentally. As emphasized by several authors (e.g., Grayson 1988; White 1992), determining the proportion of particular skeletal parts (e.g., distal humeri, first phalanges) that is burned may be more informative of the burning processes, and whether bones were fleshed or not when they were heated. Are only small specimens that cannot be identified as a part of a particular skeletal element burned? Do more burned bones display more butchery marks (Chapter 8) or evidence of carnivore gnawing than unburned specimens? Are skeletal elements that are only covered by thin layers of soft tissue in life, such as the metapodials of ungulates, more heavily and frequently burned than skeletal elements covered by thick layers of soft tissue, such as the humeri of ungulates? If multiple assemblages of burned bones are recovered from a site, are the same skeletal elements always burned in all of the assemblages?

If the stratum in which bones are found is organic rich and charred, and the bones in that stratum are also charred, then it is likely the bones were burned when the matrix in which they are embedded burned. If most of the burned bone specimens in an assemblage are spatially associated with fire hearths, fire pits, and/or concentrations of thermally fractured rock whereas unburned bone specimens are not associated with other evidence of fire or heat, then it is likely the burned bones were burned by humans, and probably intentionally. If the distribution of burned and unburned bones is identical, then one might argue that bones were burned prior to their final deposition (e.g., Stahl and Zeidler 1990).

There are few good examples of analyses of burned bones that are amenable to summary here (but see Gifford-Gonzalez 1989a; White 1992 for extended discussions). It should nonetheless be clear that the significance often ascribed to burned bones demands intensive and extensive analysis of faunal remains that have been burned. While actualistic research has gone far towards solidifying the modification attributes that signify bones were burned, that same research has only been minimally directed towards developing analytical techniques for unravelling the taphonomic histories of assemblages with varied frequencies of burned skeletal parts. Similarly, detailed analyses of burned bones which include the kind of data alluded to in preceding paragraphs are few in number at this time. Thus, it seems we need to explore the variability of burning modification along a number of dimensions (e.g., which skeletal parts are burned, where are burned bones found and what are they regularly associated with) to ascertain which of those dimensions are archaeologically visible, and which ones may help us identify human behaviors.

Other biological agents of bone modification

> A skull is found covered with mud firmly stuck on, and with the traces of the white ants' [termites] tunnels running through. If the mud is removed, large areas of the cranial walls may be found to be disappeared altogether. In less exaggerated cases, holes will be seen with white, gnawed edges, or perhaps only the surface of the bone has been attacked. The cranial sutures are a favourite site for the commencement of the termites' operations.
> (D. E. Derry 1911:245)

In earlier chapters we see that carnivores and rodents, in particular, variously consume and modify vertebrate faunal remains. There are several other

biological agents of bone modification, most operating during the biostratinomic phase of a taphonomic history, that should be mentioned. The major one
zooarchaeological taphonomists have been concerned with is herbivores
(mammalian carnivores are covered elsewhere in this volume), but there are
others as well. Martill (1990:279), for example, refers to studies on snails
known to eat whale bones. In the following I restrict discussion to three of the
better known kinds of bone consumers.

Insects

Derry (1911) was one of the first to report on the damage to bones that could be
created by the gnawing action of insects, especially termites. Behrensmeyer
(1978:154) brought the possibility of such damage to the attention of modern
taphonomists when she illustrated grooves gnawed in horn cores by moth
larvae which feed on the organic components of horns. The grooves she
illustrated are perpendicular to the grain of the horn core (Behrensmeyer
1978:156). In the only experimental study of which I am aware, Watson and
Abbey (1986) found that Australian termites gnaw bone and create "scratches"
in bone surfaces. Cancellous bone is more damaged than compact bone, and
"on compact bone, damage is concentrated on roughened surfaces or along
edges" (Watson and Abbey 1986:250). Cancellous bone is sometimes "tun-
neled into" (Watson and Abbey 1986:250). Overall, the chewing on bone by
termites is "superficial" and Watson and Abbey (1986:251) suggest bone is
simply "explored by chewing in the same way that termites explore a range of
hard plastic materials." Watson and Abbey (1986:253) found "no correlation
between the presence of visible organic material and the location or severity of
termite damage, but the high and uniform concentration of nitrogen in the
bones may have been attractive," although it appears that termites gnaw
"relatively fresh bone" more intensively than "old bones in [human] occupational deposits."

Over eighty years ago Smith (1908:524) described what he interpreted to be
gnawing damage to human bones created by beetles. He reports that the
gnawed bones typically show gnawing damage on the "under surfaces of the
bones as they happened to lie in the ground ... especially on those parts which
are pressed tightly against the soil," and on this basis he concludes that the
damage (originally thought to represent an antemortem pathological condition) had occurred postmortem. (Erzinçlioglu [1983:58] states that in forensic
examinations, "the soil beneath the corpse should be examined for larvae and
puparia.") Smith (1908:524) also reports that "a white powder, consisting of
pulverized bone, is often found sprinkled over the damaged part and the
adjoining soil; in many cases this is obviously fresh ... The burrows (usually
about 1 mm in diameter) of small animals can always be seen leading to the
[damage trace]." These burrows sometimes contained the remains of beetles.
Smith (1908:524) indicates that "the little grooves produced by the scraping of

the beetles are distinctly visible" on the edge of the damage traces with the aid of a magnifying lens, and sediment is often caked around the traces.

Many zoologists who study animal skeletons clean comparative skeletons by using dermestid (*Dermestes* sp.) beetles (Hildebrand 1968:21–23 with references). A caution often expressed when describing this technique of skeletal preparation is that as the bones are progressively defleshed, the skeleton should be closely checked to ensure that it is not damaged by the feeding beetles. Hefti *et al.* (1980:45) write that skeletons should be removed from the beetle colony when the skeleton is free of soft tissues "since longer exposure results in partial destruction of parts of the skeleton, the beetles attacking bone when they are deprived of other food." They do not describe the damage sustained by bones so attacked, but report that the damage "is easily seen by eye, since the process is such that pieces of conspicuous size become detached" (Hefti *et al.* 1980:47).

Kitching (1980) illustrates holes in fossil bovid bones he believes were gnawed by dermestid beetles. Following his lead, Jodry and Stanford (1992:111–113) illustrate and describe holes in bison bones recovered from a late Pleistocene-early Holocene site in Colorado. These holes are 9 to 14 mm in diameter, occur singly on a specimen, typically originate from a natural aperture such as a foramen, and appear to have been bored or chemically dissolved rather than punctured. Jodry and Stanford (1992:113) state that the holes they observed are larger than those gnawed by dermestids (which they indicate are ca. 6 mm diameter), and suggest, following Kitching (1980), that the holes in the fossils were gnawed by carrion beetle puparia. Holes bored or eaten through bone by insects "can be distinguished from punctures made by carnivore teeth by their larger size and the absence of crushed bone in the bottom" (Hesse and Wapnish 1985:85).

Several kinds of evidence must be recorded in order to infer the action of gnawing insects. Rogers (1992), for example, describes the size of bored holes associated with grooves in several dinosaur bones. He notes that the holes are filled with the same sediment as the "host matrix of the bone bed," that there are no scratches or grooves on the walls of the holes, and that fossil beetle puparial cases were stratigraphically associated (Rogers 1992:528–529). The location of the borings and/or grooves in the depositional context (e.g., on the upward or downward surface of the bone) and on the bone (e.g., in compact or trabecular bone) are also important considerations.

Herbivores

It was probably Sutcliffe (1973) who first brought to the attention of taphonomists that herbivores, particularly ungulates, gnaw bones and antlers, and create what might be mistaken for humanly-created tools. Sutcliffe's (1973) seminal paper has been followed by a number of other reports of various herbivorous taxa gnawing bones and antlers (Brothwell 1976; Gordon 1976; Krausman and Bissonette 1977; Sutcliffe 1977; Wika 1982; Bowyer 1983; Johnson and Haynes 1985; Warrick and Krausman 1986; Greenfield 1988). Some of these are by zoologists interested in the nutritional aspect of bone gnawing by herbivores. They believe such behavior, called *osteophagia*, is directed toward alleviating nutrient deficiencies, especially phosphorus and calcium. Regardless of the reason that herbivores chew bones, the effects of such chewing are taphonomically significant.

Sutcliffe (1973:428, 430) suggests that a bone is grasped in the cheek teeth "in a 'cigar-like' manner" and "chewing is with a sideways movement of the jaws." This chewing action results in "planing off the top and bottom [of a long bone shaft] until the marrow cavity or antler core is reached, leaving only the sides intact, [producing a] fork-like remnant" (Sutcliffe 1973:430). The prongs of the fork have a zigzag surface that "matches the alternating upper and lower cheek teeth of the chewer" (Sutcliffe 1973:430). Gordon (1976:121) describes the prongs of the fork as "undulating and [having progressively] thinned ends." He notes that within the undulations and on the crests between them there are "striae or wear patterns parallel to the valleys and crests; i.e., perpendicular to the longitudinal axis" of the specimen, and this in turn suggests the specimens were "chewed by a transverse grinding motion of the cheek teeth – a side to side movement normal to ungulates" (Gordon 1986:122). Brothwell (1976:182) suggests the chewing is a "grazing-sawing" motion that can produce "multiple and parallel grooved marks" on some bones.

Microscopic organisms

Using histological techniques, researchers have found various microscopic modifications to skeletal tissues apparently caused by the action of microorganisms. For example, it appears that some fungi penetrate bone within 25–30 days after exposure (Marchiafava *et al.* 1974). Fungi create tunnels in bone about 1 to 8 μm in diameter (one μm or micron = 1/1000 mm) as a result of attempts to gain access to collagen for consumption and metabolism (Hackett 1981; Marchiafava *et al.* 1974). Hydroxyapatite is redeposited in the tunnels as a new mineral – brushite – in the form of a "cuff, 3 to 6 μm thick, in the bone surrounding most tunnels, from 5 μm upwards" (Hackett 1981:247). Piepenbrink (1986:418) reports that "fungi can [also] decompose dead bone by more extensive destruction of hard tissues" (see also Garland 1987, 1988; Garland *et al.* 1987; Hanson and Buikstra 1987:554). In some experiments he found that "although in regions with extensive fungal growth the cortical bone surface is partially eroded, the fungi never actually penetrated or tunnelled through the bone tissue by means of distinct focal destruction" (Piepenbrink 1986:421). Like Hackett (1981) before him, he found "recrystallization in the tunnelled areas [of] previously dissolved calcium phosphate." Hackett (1981:247) notes such "mineral redeposition is not seen in pathological processes" and Piepen-brink (1986:424) suggests it "is never found in osteolytic processes that occur during life;" the mineral redeposition can thus be used to distinguish these postmortem changes from pathological ones.

Garland (1987:113) illustrates fungi within prehistoric bone tissue and implies their presence results in the "disintegration, disaggregation and dissociation of osteons. It is still possible to recognize the morphology of the lamellar bone adjacent to the Haversian canal but the bone towards the periphery of the osteon has an amorphous appearance." Bone mineral is lost, and a "granular" appearance to bone tissue is visible in thin sections as a result of fungal activity. All of these modifications "were found only in the outer cortical zone" of the specimens Garland (1987:113) examined. Fungi, rhizo-morphs, and bacteria lying within empty Haversian canals, medullary cavities, and trabeculae were found in some prehistoric bone specimens (Garland 1987:118). These as well as foreign mineral material in such spaces are termed *inclusions*, the "extraneous material within bone spaces" (Garland 1987:122).

Hackett (1981:264) suggests bacteria invading bone will create "tubules about 300 nm [nanometers, 1 nm = 1 billionth of a meter = 1/1000 of a micron] in diameter" (see also Garland 1988), and thus bacteria-formed tunnels can be distinguished from those formed by fungal activity on the basis of size differences. Piepenbrink (1986:424–426) suggests that fungal activity can also result in staining of skeletal material. Garland (1987:118–120) reports "*in terra* staining which had penetrated into the cortex [of prehistoric bone specimens] to various depths." Because the staining (and other modifications, see above) only occurred in the outer cortical area of the bone tissue or "the zone of bone lying in closest contact with the soil, there may be suggested a physico-chemical aetiology to the nature of the staining, with a necessary condition being the presence of ground water" (Garland 1987:121).

Hanson and Buikstra (1987:554–559) describe and illustrate various stages of bone destruction caused by micro-organisms. *Focal destruction* (the creation of tunnels) as described by Hackett (1981; see also Garland 1987:118) is the first stage, and the tunnels "gradually coalesce to form large patches of resorbed cortical bone" (Hanson and Buikstra 1987:555). The coalescence reported by Hanson and Buikstra may be the amorphous osteon borders reported by Garland (1987, see above). Focal destruction appears to begin intracortically in Haversian bone, and gradually spreads toward the periosteal and endosteal surfaces. "Secondary Haversian bone is mechanically weaker and less mineral-ized than interstitial and circumferential lamellar bone ... Micro-organisms appear to favor the less mineralized tissue for initial boring ... The abundance of vascular spaces in the mid-cortex facilitates the spread of organisms" (Hanson and Buikstra 1987:559).

The importance of these observations is that the small tunnels created by

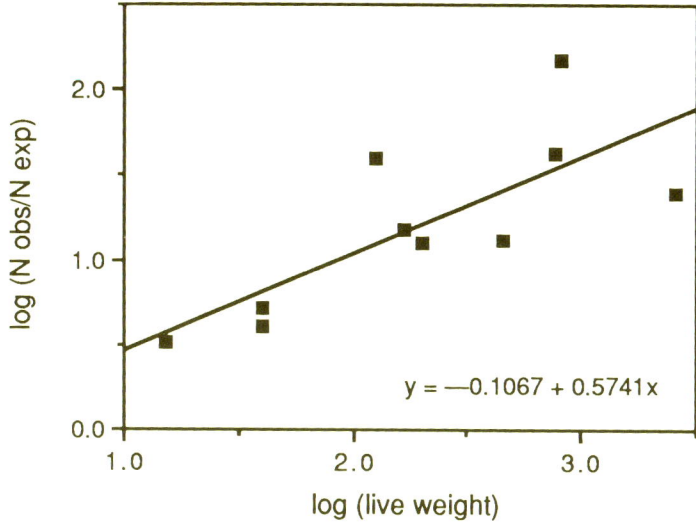

Figure 9.11. Regression of log 10 of live weight against log 10 of the ratio of number of individuals expected (N exp) to number of individuals observed (N obs) (after Behrensmeyer and Boaz 1980).

fungal and bacterial activity reduce the structural density and increase the porosity of the tissue (Bell 1990; Hanson and Buikstra 1987). This may exacerbate the effects of diagenetic processes such as crushing from sediment overburden weight and exchange of chemical ions (Garland 1987:122; see Chapter 11). Documentation of geological or biological staining underscores the fact that care should be taken when attempting to identify burned bone simply on the basis of color (Figure 9.9).

Preservation and size biasing

Comparison of surface-collected assemblages of naturally accumulated modern mammalian remains with the living community from which the assemblages were derived suggests that the remains of small animals preserve significantly less well than the remains of large animals (Behrensmeyer 1981; Behrensmeyer and Boaz 1980; Behrensmeyer *et al.* 1979). This is shown in Figure 9.11, where it is clear that as the log 10 of the average live weight increases for the 10 mammalian taxa plotted, the log 10 of the ratio of the observed to expected frequency of carcasses per animal size class also increases ($r = 0.77$, $P = 0.009$). This probably applies as well intra-taxonomically to small individuals, particularly juveniles of a taxon (Behrensmeyer 1981:600) due to the lower structural density of their bones. Behrensmeyer *et al.* (1979:17) suggest that this difference in preservation is largely due to greater destruction of bones of small mammalian taxa by carnivores and scavengers, and more rapid weathering and fragmentation due to trampling of small bones. In other words, they posit biostratinomic processes as the cause of the preservational bias against the remains of animals of small size, which is reasonable given that their study is founded on surface-occurring bones that have not yet undergone diagenetic processes. Retallack (1988:338), however, suggests a similar preservational bias against remains of animals of small body size may result from diagenetic effects because "bones of smaller mammals have a higher surface-to-volume ratio, and so are more prone to acidic dissolution than are those of large mammals."

Regardless of who is correct in the case above, and both may in fact be correct, Damuth (1982:441) argues that the slope of the regression line derived by Behrensmeyer *et al.* (1979), who omitted the two points farthest above the line in Figure 9.11 to derive a slope of 0.68, was sufficiently close to the surface-to-volume ratio of 0.67 to suggest to him the surface-to-volume ratio was controlling the relatively poor preservation of smaller bones. Study of fossil assemblages with this potential preservational bias in mind must, of course, presume bones of both small and large mammals were being accumulated and deposited in abundances proportional to their frequencies in the biotic community if the analyst wants to explain a relative paucity of small mammal remains as being the result of the surface-to-volume ratio bias, regardless of the mechanism of differential preservation. As noted above, similar preservational problems may accrue intrataxonomically; small skeletal elements of an individual animal may preserve less well than that individual's large skeletal elements, but this has not yet been studied in the same way that Behrensmeyer *et al.* (1979) studied intertaxonomic variation in preservation.

Comparative analytic techniques

Because we are still learning about which taphonomic processes and agents produce particular kinds of modifications to assemblages of faunal remains, some analysts have developed ways simply to compare various modification attributes between assemblages without specific reference to the taphonomic agents which may have created the modifications. I review several of these here to provide the reader with a feel for some of the attributes of bone assemblages that might be studied.

Percentage difference in long bone ends

In his study of the effects of carnivore destruction of bovid long bones, Richardson (1980) develops an equation for calculating the percentage difference between the frequency of proximal and distal ends of each individual long bone. That equation is:

$$\frac{|(\text{N complete}_i + \text{N proximal ends}_i) - (\text{N complete}_i + \text{N distal ends}_i)|}{(\text{N complete}_i + \text{N proximal ends}_i) + (\text{N complete}_i + \text{N distal ends}_i)} \qquad [9.3]$$

Table 9.4 *Frequencies of bone parts in selected sites (from Todd and Rapson 1988:310; Todd 1987a:235; Binford 1981b:174–175). Upper ratio of proximal:distal is raw values; lower ratio is standardized*

Site	%difference		MNE proximal:MNE distal	
	humerus	tibia	humerus	tibia
Jones-Miller	55.33	21.55	44:153	91:141
			28.8:100	59.5:92.2
Horner II	40.98	17.29	38.5:61.5	47.5:55.0
			62.6:100	77.2:89.4
Casper	40.74	12.50	20:47.5	31.5:24.5
			42.1:100	66.3:51.6
Olsen-Chubbuck	1.42	2.04	86.5:89.0	84.0:87.5
			97.2:100	94.4:98.3
Lamb Spring	54.55	17.24	2.0:7.5	5.5:8.5
			23.5:88.2	64.7:100
Wolf kills	60.00	37.50	2.5:9.5	5.0:11.0
			22.7:86.4	45.5:100

where i is the skeletal element of concern (generally the humerus, radius, metacarpal, femur, tibia, or metatarsal), N complete is the number of complete specimens of skeletal element i, N proximal ends is the number of specimens that are only the proximal end of skeletal element i, and N distal ends is the number of specimens that are only the distal end of skeletal element i.

Todd and Rapson (1988) use equation [9.3] to compare differences in proportional frequencies of long bone ends in several bone assemblages, and suggest when the differences are minimal, especially for the tibia and humerus, limited destruction of bones by carnivores is indicated. As an example, the data for percent difference values for humeri and tibiae from the six assemblages given in Table 9.4 (from Todd and Rapson 1988:310) are plotted in Figure 9.12. That graph indicates the "wolf kills" assemblage has greater differences in the abundances of both proximal and distal humeri and proximal and distal tibiae, unlike the Olsen-Chubbuck assemblage which has essentially equal abundances of all and which has apparently (on the basis of other evidence such as the paucity of gnawing marks) undergone minimal carnivore attrition.

Binford (1981b:219) uses a different technique for graphing data like those in Figure 9.12. He plots the standardized frequencies of proximal versus distal humeri, and proximal versus distal tibiae, on the graph in Figure 9.13a. As shown there, if the plotted points fall within the area labeled "Zone of No Destruction" then the conclusion is the assemblage has undergone very little density-mediated destruction (see Chapter 7) whereas if the plotted points fall in the area labeled "Zone of Destruction" then the analyst concludes the assemblage had in fact undergone some density-mediated attrition (in many artiodactyls the distal end of both the humerus and the tibia is denser than its proximal end). The frequencies of proximal and distal ends are standardized to

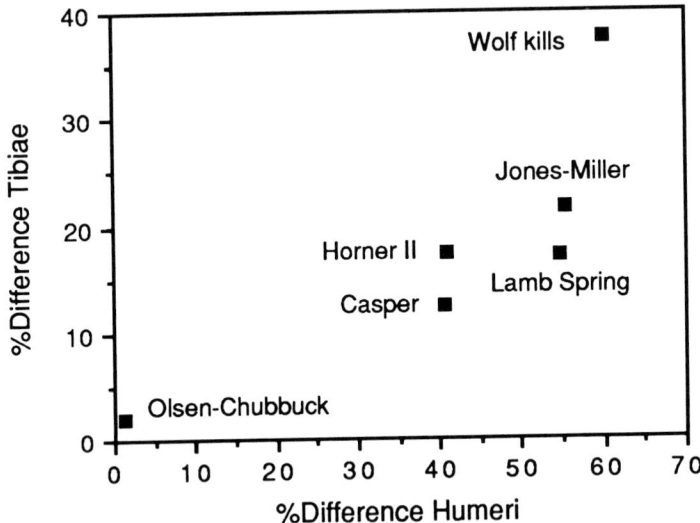

Figure 9.12. Scatterplot of % differences in frequencies of proximal and distal humeri against % differences in frequencies of proximal and distal tibiae (from Table 9.4).

allow plotting of both skeletal elements on the same graph. This is accomplished by determining the MNE (or MAU) for the proximal end and for the distal end of each bone, and then dividing all four values (proximal humerus, distal humerus, proximal tibia, distal tibia) by the largest of the four values and multiplying the results by 100. Binford (1981b:219) labels these "ratio values," examples from Table 9.4 (data from Binford 1981b:174–175 and Todd 1987a:235) of which are plotted in Figure 9.13b. This figure indicates that the Olsen-Chubbuck assemblage has undergone virtually no destruction whereas the wolf kills assemblage has undergone some destruction.

Long bone shafts

Todd and Rapson (1988:314–317) suggest using a standard measurement of the width of the articular end to calculate a shaft-to-end ratio as a measure of the amount of long bone shaft attached to the end. This procedure accounts for intrataxonomic sexual dimorphism and individual size variation. Blumenschine (1988:487) suggests recording the "limb segment" represented by a long bone fragment using the following categories:

1) epiphyseal fragment: has all or a portion of the proximal or distal articular surface;
2) near-epiphyseal fragment: lacks any articular surfaces, but has trabecular bone on the medullary surface indicating proximity to an epiphysis;
3) midshaft fragment: lacking articular surfaces and trabecular bone.

Todd and Rapson (1988:322) suggest differentiating between proximal and distal shaft portions (see also Hoffman 1988).

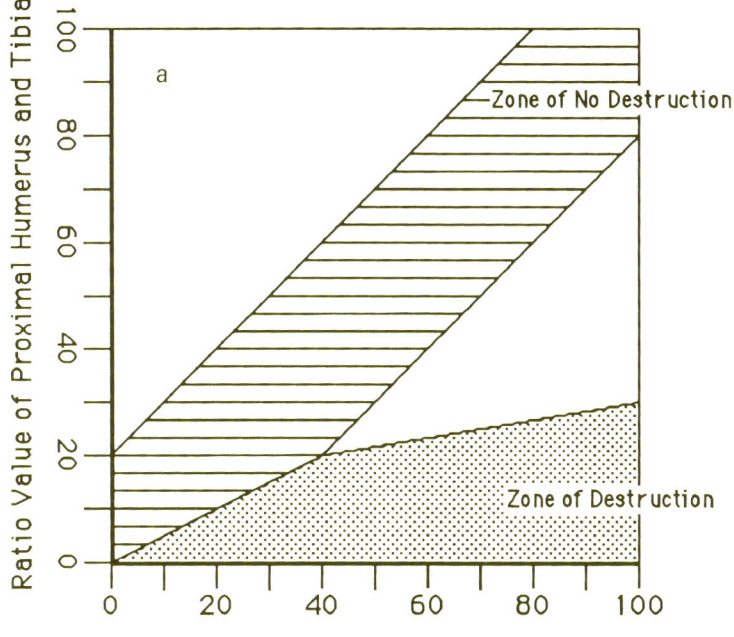

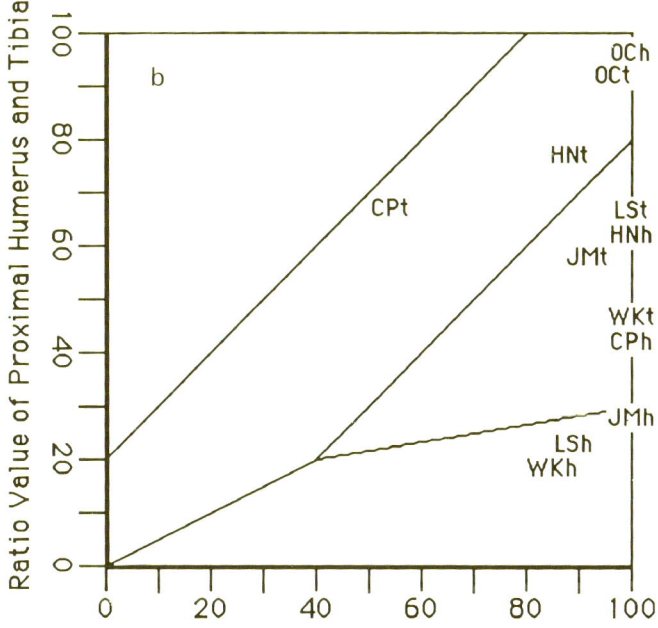

Figure 9.13. Bone destruction graphs. a, model graph of destruction of proximal and distal humeri and tibiae (after Binford 1981b:219, Figure 5.07); b, destruction graph for six assemblages in Table 9.4: JM, Jones-Miller; HN, Horner II; CP, Casper; OC, Olsen-Chubbuck; LS, Lamb Spring; WK, Wolf kills; h, humerus; t, tibia..

The frequency of shaft fragments generated by hominid agents of bone fracture has been discussed by Binford (1978) and Bunn (1989) (see Chapter 7). Todd and Rapson (1988) suggest the analyst may simply compare the NISP frequency of shaft fragments per skeletal element with the NISP frequency of articular ends for that skeletal element. In so doing, they identify what they take to be two different kinds of fragmentation, one in which the NISP of bone ends is inversely correlated with the NISP of shaft fragments and another in which the NISP of bone ends is positively correlated with the NISP of shaft fragments. While the precise taphonomic significance of such measures is unclear (although they are probably at least in part a function of the intensity of fragmentation, see Chapter 8), such values do provide a basis for comparing assemblages of broken bones.

Discussion

In this section I have mentioned some of the analytic techniques analysts use to record modifications to bone assemblages, and how the extent and frequency of those modifications might be compared between assemblages. I have, in fact, barely scratched the surface of this topic. It should be obvious, if the reader has read everything between Chapter 5 and this sentence, that there are numerous modification attributes that a taphonomist might choose to record and analyze. The number of such attributes is limited by only two things: (a) the actual modifications displayed by an assemblage of animal remains, and (b) the analyst's imagination. That is, there is no set or standardized procedure of taphonomic analysis; because taphonomic processes are historical and to some degree cumulative, each assemblage of bones is to a greater or lesser degree taphonomically unique and thus requires unique analyses. One works with what one has, and what one has is a model of a complete living skeleton, and a set of bones and teeth that are variously different from that living skeleton. Recording those differences is the first step in taphonomic analysis, and I have touched on what might be recorded in this section. I have also described various ways the recorded data might be manipulated during analysis. Assigning taphonomic meaning to the differences and analytical results is the second step. What created the modifications? Why were they created? How do these modifications add to or subtract from data that are relevant to the interpretive questions one wishes to answer? I have touched on how to answer these questions in other sections.

Summary

Biostratinomy is that portion of a taphonomic history which occurs between the death of an organism and the burial of that organism's remains. The taphonomic aspects of death are considered in Chapter 5, and burial and post-burial taphonomic processes are described in Chapters 10 and 11, respectively. Biostratinomy has occupied the discussion for three chapters because it tends to be the most studied aspect of taphonomic histories. This is especially so for zooarchaeologists with interests in human behaviors because those behaviors are generally biostratinomic. Thus we devoted some time to two of the most important hominid biostratinomic processes in preceding chapters, differential transport of skeletal parts (Chapter 7) and butchering (Chapter 8). In this chapter other significant biostratinomic processes have been reviewed.

Bone weathering is, generally, the mechanical and chemical deterioration of bone, mainly in subaerial or surface contexts. Bones tend to become more weathered through time, but the relation of how weathered the bones in an assemblage are to the formational history of that assemblage is neither simple nor straightforward. Root etching may occur as bones lie on the surface of the ground, and study of such modification may help us more fully understand the significance of bone weathering. Trampling tends to disperse bones horizontally and vertically, to scratch them, and to fracture them. Bones may be abraded by several different mechanisms, including trampling. Bones may be burned to various extents and by several different mechanisms, and microorganisms such as fungi, bacteria, and insects can all consume skeletal tissue.

While all of these biostratinomic processes variously modify bones, few analyses which integrate them all have been performed. As well, while many of these processes have been experimentally documented, their precise taphonomic significance is unclear because little has been written about how and why these variables should be recorded or quantified in prehistoric contexts. In fact, the development of analytic techniques for comparing attributes of bone modification seem to have outpaced our understanding of what those attributes signify regarding taphonomic histories. I return to this issue in Chapter 13. It is now time, however, to turn to the process of burial and its effects on vertebrate faunal remains.

10

BURIAL AS A TAPHONOMIC PROCESS

Asymmetrical bedding on either side of the carcass is of special significance, and we
must pay attention to it when we study fossil material.
(J. Weigelt 1927/1989:95)

Introduction

Most faunal remains paleobiologists and zooarchaeologists study are recovered from subsurface contexts. An important taphonomic problem then,
concerns gaining an understanding of burial processes. For example, humans
do *not* seem to be the only biological agent that buries animal parts. Sexton
beetles (*Microphorus* sp.) bury rodent carcasses (Milne and Milne 1976),
fossorial rodents often die in their burrows and thus are already buried, and
trampling (Chapter 9) by any number of biological agents can result in the
burial of animal remains. And there are numerous geological processes which
result in the burial of animal carcasses and remains (Behrensmeyer and Hook
1992). The burial process is important because not only are bones and teeth
placed in the sedimentary matrix in which they undergo diagenetic processes
(Chapter 11), but they may be variously moved, reoriented, broken, and/or
abraded during burial by the taphonomic agents of burial (Chapter 6).

Burial as a taphonomic process has undergone little intensive study relative
to biostratinomic processes that modify vertebrate remains. Straus (1990:261)
suggests that depositional and formational processes were initially ignored by
archaeologists in favor of the more immediate and interesting task of building
cultural chronologies once the deep antiquity of hominids was established in
the last half of the nineteenth century. Retallack (1988:342) indicates that in
paleontology the perception has, until recently, been that fossil concentrations
are the result of catastrophic burial events such as flash floods which simultaneously kill, deposit, and bury large numbers of carcasses and bones. Changes
in this perception during the early part of the twentieth century have resulted in
the necessity of altering our viewpoint from one of conceiving of fossiliferous
deposits as sedimentary units to conceiving of them as pedogenic units in order
to understand fossilization processes. For example, Retallack (1984, 1988:342)
notes that it is quite unlikely that fossils embedded in paleosols (fossil soils)
have been transported far "since a paleosol would be destroyed beyond
recognition if transported elsewhere," and the degree of development of
paleosols can be used as an indication of the time span of fossil assemblage

404

formation and depositional hiatuses. Animal remains must be deposited to become part of the fossil record; once buried, fossilization processes (chemical alteration, see the Glossary) affect them. Bones and teeth, then, begin as sedimentary clasts or particles. We are concerned here with their burial, having discussed their accumulation and deposition in earlier chapters. We return to diagenetic (or pedogenic) processes in Chapter 11.

Of major significance for taphonomists is the concept of a *taphonomically active zone*, or TAZ, proposed by taphonomists with interests in marine molluscs. Davies *et al.* (1989:208–209) define the TAZ as the "sediment–water interface and the bioturbate layer just beneath it," and suggest this is the depositional and stratigraphic zone where "most taphonomic loss is concentrated" (see Parsons and Brett 1990 for an overview of the TAZ and the invertebrate fossil record). A similar terrestrial TAZ can be designated for vertebrate remains. Taphonomic processes such as abrasion, weathering, and trampling occur at the sediment–air interface. Scavengers from bacteria to hyenas to humans variously move and modify animal carcasses that are deposited on the ground surface or at the sediment–air interface. As should be clear from preceding chapters, subsequent to the *initial* deposition of a bone (upon an animal's death), the sooner that a bone is buried beneath the sediment–air interface for good, the better preserved (more like it was in life) it will often be.

Many of the principles developed in studies of the burial of invertebrates also apply to vertebrates, whether those vertebrates live in aquatic or terrestrial environments. In this chapter some of the major aspects of burial are reviewed, with a focus on the geological attributes and context of the fossil record. Any serious student of taphonomy should have a basic understanding of geomorphic processes and ideally will have a geologist in the field to study the depositional contexts of the fossils being collected. As Wood and Johnson (1978:315) note, the context and spatial associations of archaeological (and zooarchaeological and paleontological) remains is the "foundation of our discipline. If we fail to record the context, or if we misread or misinterpret that context, proper archaeological [and taphonomic] interpretation is impossible."

Clark and Kietzke (1967:117) suggest there are at least "six major factors of burial that can, theoretically at least, result in a difference between the [deposited] assemblage and the [buried] fossil assemblage." These are: (1) the time interval between episodes of sedimentation; (2) the thickness of sedimentary increments; (3) the velocity of depositional forces in contact with bones or corpses; (4) the nature of the sediment, such as the amount of compaction and grain size; (5) the post-depositional action of roots and burrowing animals; and (6) the permeability of the sediment and chemical nature of the permeating solutions. In this chapter I consider basically the first four, reserving discussion of the last two for Chapter 11 as they largely concern diagenetic or post-burial processes.

Deposition and burial

If *burial* is to be considered a significant aspect of taphonomic histories, then it is important to distinguish two periods in the taphonomic history of faunal remains. The *deposition* of faunal remains refers to their dynamic placement either *on* a land surface or *in* an existing sedimentary unit. Deposition is simultaneous with burial only in the latter case; an example would be the catastrophic burial of an animal by volcanic ash or the death of a fossorial rodent in its burrow. *Burial* refers to the covering of faunal remains with sediments, which can be either mineralogical or biological, such as sand or vegetation, respectively. The distinction between deposition and burial may seem pedantic, but it is not. Too often one reads about "post-depositional" taphonomic processes that in fact refer to "post-burial" processes (e.g., Klein and Cruz-Uribe 1984:70–75). While it is true that post-burial processes are also post-depositional ones, the reverse is not necessarily true. A bone can be deposited yet remain on the ground surface and unburied for some time; it is this aspect of the timing of burial that allows us to talk about such things as bone weathering in subaerial contexts and, in part, trampling (see Chapter 9). The distinction of deposition and burial is what allows us to distinguish diagenetic (post-burial) processes from pre-burial processes, and within pre-burial processes we can distinguish biostratinomic processes that occur prior to deposition (such as butchering) and those that occur after deposition (such as trampling). The taphonomic process between deposition and diagenesis, then, is burial.

The burial process may influence how deeply faunal remains are buried as well as the types of sediments in which they are buried. Some diagenetic (post-burial taphonomic) processes are dependent on the depth of burial (e.g., sediment overburden weight). Thus not only are burial and rates of deposition important to a taphonomic history, so too are events of erosion or sediment removal (Henderson 1987). Sediment is not simply layered over bones. This may happen, but so might movement of sedimentary particles occur during sedimentation such that sediments lay over bones only briefly prior to their removal and replacement by other sediments. Bones may become sedimentary particles or clasts in some settings. Thus while I focus here on sediment deposition, the potential that erosion and (re)exposure of faunal remains occurs as part of the burial process should be kept in mind.

Sedimentation

The rate and mode of sedimentation seem to be the major variables that influence the content (the particular fossils represented) and structure (spatial arrangement of individual fossils) of the fossil record. Sedimentation can be slow or it can be rapid; sediments can be fluvially deposited or they can be

Table 10.1 *Standard sediment size classes*

Size class name	Particle diameter (mm)
Boulders	> 256
Cobbles	64 to 256
Pebbles	2 to 64
very coarse	32 to 64
coarse	16 to 32
medium	8 to 16
fine	4 to 8
very fine	2 to 4
Sand	1/16 to 2
very coarse	1 to 2
coarse	1/2 to 1
medium	1/4 to 1/2
fine	1/8 to 1/4
very fine	1/16 to 1/8
Silt	1/256 to 1/16
Clay	1/4096 to 1/256

deposited by wind. Each tends to produce particular taphonomic signatures. We therefore need to begin our discussion of sedimentation with several basic geological terms and concepts.

"*Sediments* are particulate matter that has been transported by some process from one location to another" (Stein 1987:339). A *geological deposit* or *stratum* is a three-dimensional unit distinguishable from other such units on the basis of unique physical properties, and is an aggregate of sedimentary particles (Stein 1987:339, 344). Grain size (Table 10.1) (grain size distribution, grain orientation or fabric, grain shape, and grain surface markings), composition (mineralogy), and structure (small-scale variations in grain size, grain shape, composition, or pore space) of sediments are influenced by the depositional environment (Stein 1987:357). Sediment grain size helps ascertain the magnitude of the energy of the sediment deposition mechanism, and is mentioned below. The deposition of sediments is "not an indiscriminate process that can happen anywhere" (Butzer 1982:44). Sediment supply, nature of ground cover, topography, and operative geomorphic processes influence if and where deposition occurs (Butzer 1982:44).

Butzer (1982:56–57) presents an outline of depositional settings that includes basic details regarding the sediments and sedimentary units (facies) that result in these settings. Some of his observations are summarized in Table 10.2. I hasten to note this table is no substitute for having a geologist on the site during recovery of fossils. What I intend by presenting it is to provide the reader with a general impression of some of the variation in sedimentary environments in which fossils might be found, and some idea of the kinds of sedimentary data

Table 10.2 *Depositional settings and attributes of sediments and sedimentary units (from Butzer 1982:56–57)*

	Depositional energy	Bedding; texture	Sorting
Spring	low to high	lenticular heterogeneous, may be contorted; organic muck, sand, precipitates	poor to good
Karst	low	primarily massive, often heterogeneous; humic loams or gravelly wash, precipitates	mainly poor
Cave	detritus from roof fall, fluvial, eolian	varied	generally poor
Seacoast	highly variable	thin to massive, complex facies; clays to cobbles	generally good at and below watermark, variable above
Lakeshore and marsh	low to moderate	thin to massive; clays to sands, organic matter	poor to good
Delta edge	variable	extensive and massive; clay to silt, some sand	moderate to good
Flood plain	variable	complex vertical and lateral sequences; clays to gravels	moderate to good
Eolian	low to moderate	well stratified, massive; silt to sand	usually excellent
Slope	variable	thin to massive, poorly stratified; silt to rubble	generally poor
Volcanic	variable	massive; mixed	moderate to excellent

that should be recorded (texture, size, and shape of sedimentary unit) if the taphonomist hopes to gain some understanding of burial processes that might have modified the recovered fossils. These attributes, and others, are important when it comes to unravelling the rate and mode of sediment deposition, which are, after all, the factors that result in the burial of faunal remains.

Rate of sedimentation

Kidwell (1985, 1986) discusses the taphonomic significance of the rate of sedimentation. Her discussion centers around the density (frequency per unit area) of fossils in a geological depositional unit. The density of fossils is a function of the rate of fossil input relative to the rate of sedimentation; fossils occur in dense concentrations if the rate of fossil input increases relative to the rate of sedimentation, and fossils display a scattered, non-dense distribution if the rate of fossil input decreases relative to the rate of sedimentation. Thus not only is the rate of sedimentation important, but so too is the rate at which individual fossils are input to the geological record. The rate of fossil input is a function of the rate at which fossils are supplied to the depositional locus and the rate at which those fossils are removed by various taphonomic processes.

The sedimentation rate in turn is a function of the volume of sediment deposited per unit of time and the rate of erosion or volume of sediment removed per unit of time.

Assuming a constant rate of fossil input, Kidwell (1985, 1986) distinguishes four types of fossil concentrations. First, fossils may increase in density with decreasing depth with the upper termination of the fossil concentration at a stratigraphic boundary created by discontinued deposition of sediment. This denotes a decreased rate of sedimentation until it stops. Second, fossils may increase in density with decreasing depth with the upper termination of the fossil concentration marked by an erosional surface. In this case, erosion has removed sediment and the concentrated fossils occur as a lag deposit left behind by erosional processes that removed sediments but not bones. Inverting the first two scenarios, the third case finds fossils increasing in concentration with increasing depth; the lower boundary of the concentration is marked by an abrupt initiation of sedimentation and increasingly higher rates of sedimentation with decreasing depth. The fourth scenario has the lower boundary of the fossil concentration marked by an erosional surface, densely packed fossils, and fossil density decreases with decreasing depth as sedimentation rates increase. Kidwell (1986:11) suggests the first two types of concentrations consist of fossils that are more abraded, fragmented, eroded, and weathered than the fossils in the other two types due to the greater exposure duration of fossils in the former two concentrations.

Taphonomists working with fossil collections from within the temporal range of radiocarbon dating can obtain multiple dates on a fossil-rich stratum in order to extrapolate a rate of sedimentation and a rate of fossil deposition. Retallack (1984:60) suggests that comparing the degree of development of fossil soils (paleosols) with the temporal duration of the formation of similar modern soils "may give an indication of the duration of breaks in deposition within a [stratigraphic] sequence of fossil soils" and thus provide estimates of sediment accumulation rates in sequences of multiple paleosols. In conjunction with Kidwell's (1985, 1986) models, the taphonomist may be able to use such depositional rates to help distinguish passive mass accumulations that formed over long time spans in an area experiencing low sedimentation rates from active mass accumulations that formed over short time periods such as are found in various mass kill sites (e.g., Olsen 1988; Todd 1987a; see Chapter 8 for discussion of bone accumulation). Knowing the rate of sedimentation may also help unravel the taphonomic significance of bone weathering profiles based on specimens from thick stratigraphic units (see Chapter 9).

Mode of sedimentation

Geological sediments can be deposited by a number of geological processes. The two most studied in the context of taphonomy are fluvial and eolian deposition (Saunders 1977:68). Most studies of depositional contexts in the service of paleontological taphonomy have involved floodplains, deltas, and river channels (e.g., Behrensmeyer 1975a, 1975b; Clark *et al.* 1967; Voorhies 1969); those concerned with archaeological taphonomy have been focused on differences between bone assemblages deposited in on-site and off-site locations (e.g., Bunn *et al.* 1991). There are, of course, other processes and settings of sedimentation. In the following I consider some of these, citing examples of many of them along with some of the key attributes taphonomists use to help understand burial processes. As will become clear, often if one can identify the agent of bone accumulation (see Chapter 6), clues to the nature of the burial process will be suggested.

Fluvial deposition

Hydrological processes are a major factor in the formation of many archaeological sites (Schiffer 1987:243–256). Fluvial deposition in channels has been studied by Behrensmeyer (1988b, 1990 and references therein). She notes that a rapid rate of fluvial sedimentation can be marked by sediments of mixed texture (fine to coarse sedimentary particles) whereas slow fluvial sedimentation rates will be marked by sediments of fine texture, although fine sediments may also be deposited relatively rapidly (Behrensmeyer 1988b). Faunal remains "may be overtaken by moving bedforms (ripples, sand waves), and scour on their downstream [or leeward] side also promotes burial" (Behrensmeyer 1990:235). This is similar to Weigelt's (1927/1989:95) observation over 60 years ago: a sizable carcass "restricts the flow of oncoming water, causing an increase in its velocity, which in turn means increased force; consequently, the water on the shoreward side of the carcass can dig out a deep hollow, while sediment accumulates on the other side. This often causes the carcass to lie crooked or to sink in."

Fluvial action can bury bones, and it can also move or transport them (Chapter 6). Bones not transported by fluvial action tend to be robust, heavy specimens that may be variously abraded and broken (see also Figure 6.5), particularly when sediments are coarse. High-energy environments of fluvial deposition such as channel fills and lag deposits tend to have high ratios of teeth-to-vertebrae whereas low-energy environments such as deltaic and lacustrine settings tend to have low ratios (Behrensmeyer 1975b). Abrasion and the fluvial transportability of bones are the major attributes taphonomists examine in fluvial depositional environments.

Boaz (1982) provides an extensive overview and results of experiments with fluvial processes. Calling on the work of Voorhies (1969), Behrensmeyer (1975b), Korth (1979), Hanson (1980), and others, she recorded the following variables in her study of fluvially transported and buried bones: sedimentary environment, taxonomic composition and relative abundances of taxa in the assemblage, demography of the fossil population, degree of articulation, degree of spatial association and dispersal of skeletal parts, weathering,

skeletal part frequencies and their relation to structural density and transportability, horizontal and vertical distributions of skeletal parts, orientation of long
axes of specimens, and extent and types of damage such as abrasion and
carnivore gnawing. She notes that fluvial transport and burial can be recognized by patterned orientations of bones (Figure 6.7), the spatially restricted
occurrence of lag deposits (Figure 6.5), and abrasion causing the exposure of
trabeculae with sand and gravel embedded in fossae and trabeculae. Boaz
(1982:219) notes the lower the structural density and the greater the transport
distance of a skeletal part, the lower the probability that the part will survive the
rigors of fluvial transport.

Eolian deposition

Eolian sediments tend to be fine, and most taphonomic study of them concerns
the abrasive effects of wind-borne sedimentary particles on bones (see Chapter
9). Eolian processes, like hydrological ones, are often important contributors
to the formational history of archaeological sites (Schiffer 1987:238–243).
Wind not only deposits sediments, it removes them and creates lag deposits of
fossils. It is in the creation of lag deposits that wind moves bones, and this
movement is largely downward. Some downslope movement might result from
wind activity and entail horizontal as well as vertical movement, but this is
surely gravity aided. I suspect that typically only very small, light bones are
moved significant distances by wind action, although Schäfer (1962/1972:37)
reports seeing "vertebrae of *Phoca* [a seal] being carried by the wind, racing and
jumping over the slightly salt-encrusted surface of the [flat sand beach]."

Hominid deposition

Hominids, including some early anatomical forms of modern *Homo sapiens*,
seem to be unique in the animal kingdom for ritualistic disposal of conspecifics.
Often, such disposal included burial. Purposeful burial is marked geologically
by the occurrence of the human remains in a stratigraphically distinct unit,
usually a pit feature. What appear to have been pets of prehistoric humans,
such as dogs (*Canis familiaris*), were also sometimes purposefully buried in pits.
Intentional burial, then, seems to be marked by intrusive pits containing animal
(including human) remains. The disposal of food remains by hominids seems to
be a typical mode of adding vertebrate skeletal parts to the future zooarchaeological record. Such additions are perhaps more often buried by natural
sedimentation processes such as eolian and fluvial deposition than by hominid
activities.

Hominids "determine *who* is buried and *when* but also *how* and *where*"
(Henderson 1987:49). As noted earlier in this chapter and in Chapter 11,
extrinsic factors such as the chemistry and porosity of sediments can influence
bone preservation. Did prehistoric peoples dispose of their dead in well-
drained sediments or in swamps? That simple difference can result in major

preservational variation. As well, cremation, manual defleshing, exposure of bodies to the elements for a period of time prior to burial, interment in a coffin or container of some sort, and other pre-burial factors all influence how well human remains are preserved (Henderson 1987). Such factors should, of course, be considered when dealing with the hominid deposition and burial of any faunal remains.

Fossorial animals

Burrowing or fossorial animals, such as many rodent taxa, sometimes die of natural causes in their burrows. The occurrence of an essentially complete skeleton of a fossorial rodent, that is to some degree articulated, and in a burrow (when filled with sediment, it is termed a *krotovina*) is typically inferred to represent such a death and natural burial. One must therefore be well aware of the geological context of the remains and the spatial relationships of individual bones.

Other burial processes

One of the more unique, if not rare, types of burial concerns the entrapment of animals in bogs, marshes, springs, or tar pits and their subsequent death and burial via the carcass sinking into the sediment. The La Brea tar pits are one famous example (e.g., Stock 1956). Various spring sites in North America have produced many well-preserved fossils of animals that became mired in the sediment and eventually buried. The 26,000 year old Hot Springs Mammoth site in South Dakota represents a fossil assemblage formed within "a springfed pond within a karst depression" (Agenbroad 1984:119). Most of the remains belong to the Columbian mammoth (*Mammuthus columbi*). Transportation of skeletal elements was minimal, and seems to have involved mostly gravity and perhaps bloated carcasses floating in the water within the karst (Agenbroad 1984, 1989, 1990). The majority of the remains are concentrated around the edges of the karst, suggesting the animals fell into the steep-sided depression, could not climb out, and eventually died at the base of the slope. Nearly complete skeletons are common in the deposits, but the extent of articulation varies. The Hot Springs site appears to represent attritional mortality and gradual accumulation of skeletons of mostly immature individuals, perhaps bachelor males (Agenbroad 1990). Carnivore damage to the bones is minimal, but some of the bones have been fractured by trampling from other entrapped mammoths, slope collapse and rock fall onto bones, and post-burial sediment compaction and slippage (Agenbroad 1989).

The ca. 16,000 year old Boney Springs site in Missouri contains the remains of 31 mastodonts (*Mammut americanum*) that apparently died of drought and nutritional deficiency around the outlet of an artesian spring (Saunders 1977). The demography of the dead animals seems to represent an instance of catastrophic mortality. The long axis of most bone specimens was parallel to the horizontal bedding plane (showed no plunge) and no preferred compass orientation was apparent. Bones rested directly on each other and occurred in "nearly circular plans of concentration, diminishing in diameter with each successively deeper level" (Saunders 1977:73). Only three instances of articulated bones were noted in the sample of 517 elements. Small elements of low structural density were less abundant than large elements of great structural density. *In situ* decomposition of soft tissues is suggested by the presence of remains of carrion-feeding coleopterous insects. Patterned fracturing of bones is attributed to "lithostatic pressure" (Saunders 1977:104), weathering, and trampling. Sedimentation was sometimes gradual, sometimes rapid, and was intermittent as indicated by bone weathering. Covering sediment originated from flood events of an adjacent river (Saunders 1977:109).

Burial processes

There are other aspects of burial that warrant comment. Was the burial of the fossil assemblage under study dynamic and short-term, or was it gradual and long-term? What kinds of specific depositional processes lead to burial? After all, a skeletal part must minimally be deposited on the ground surface in order for it to become buried. Some of these issues are discussed in this section. Trampling as a process of burial is discussed in Chapter 9.

Kranz (1974a, 1974b) follows Brongersma-Sanders (1957) and defines an *anastrophe* as a catastrophe of limited scope and area that produces mass mortality in the affected area. An anastrophic burial event is thus a catastrophic, short-term, burial event *potentially* producing what is typically termed a catastrophic mortality profile (Chapter 5), but this mortality pattern may not be realized. As Kranz (1974b) emphasizes, several factors other than extremely rapid deposition play a role. In particular, the organism being buried may have a chance to escape the entombing sediments. The life habits and size of the organisms, the type of sediment cover, and the depth of burial all influence whether an individual organism may escape (Kranz 1974b).

Most archaeologists are aware of the site of Pompeii that was anastrophically buried in A.D. 79 by the volcanic eruption of Mt. Vesuvius. Lyman (1989b) and Voorhies (1981) report on separate instances of anastrophic burial of large vertebrates resulting from volcanic eruptions. Such anastrophically buried assemblages of animal remains tend to display minimal carnivore gnawing when complete carcasses are buried, and the carcasses are relatively complete and articulated. These observations on rapid burial shortly after the original deposition of a carcass or bones tend to confirm most taphonomists' belief that such burial effectively buffers faunal remains from the myriad biostratinomic processes that might otherwise modify them. This does *not*, however, mean that taphonomic processes stop modifying bones upon their burial (e.g., see the discussion of bone weathering in Chapter 9). Rather, the diagenetic processes that modify bones after burial are simply different from, and tend to have less macroscopic affects on the bones than biostratinomic processes, as we see in Chapter 11.

Anastrophic burial by roof fall events in caves can result in the fracture of bones (e.g., Thomas and Mayer 1983) and perhaps their scarification (e.g., Dixon 1984). Caves and rockshelters are unusual depositional environments because they are spatially bounded, and as such Straus (1990:273) refers to them as "bone boxes." In such settings, deposition and preservation tend to outstrip erosion and lack of preservation due to the bounded spatial unit defined by a cave or rockshelter and the depositional environment being protected from climatic factors in particular. This does not, however, mean that caves are the best places to recover faunal remains that have undergone minimal taphonomic modification resulting from biostratinomic, burial, or diagenetic processes (see Straus 1990 for a more complete discussion).

Sites where animals fell or jumped to their deaths may have the first-deposited carcasses buried by carcasses deposited moments later (e.g., Wheat 1972). The distinction of a fall from a jump is an important one. "A *jump* is an intentional leaping action which culminates in an attempted innate and predictable landing position. A *fall* is a descent with a random impact position" (Hughes 1986:55). In her study of carcasses of wapiti (*Cervus elaphus*) that jumped off a 20 m high cliff, Hughes (1986) argues that jumps from high ledges will result in broken front legs, lumbar vertebrae, and ribs, and bruised faces; falls will produce less patterning in the location of fractures and injuries.

Behrensmeyer and Hook (1992) describe some very general attributes of fossil deposits in 34 different kinds of depositional environments. Many of these parallel those noted or implied in Table 10.2. For example, Behrensmeyer and Hook (1992:21) describe four kinds of "coastal" settings – offshore, beach, lagoon, estuary – and note that vertebrate fossils seem to be rare in the first two, common in the third, are "present" in the fourth, and all consist of rarely articulated remains of mainly aquatic taxa. At a very general level, their scheme is useful for comparing fossil assemblages from similar depositional contexts. However, they list "archaeological sites" as one of the 34 kinds of contexts, and note that vertebrate fossils in such contexts are "very common, [consist of] whole and broken parts, [and represent] allochthonous [assemblages]" (Behrensmeyer and Hook 1992:61). This characterization is far too simplistic to describe the range of variation in bone assemblages from different kinds of archaeological sites or from different depositional contexts (e.g., strata, features) within a site. Behrensmeyer and Hook's (1992) scheme is nonetheless suggestive. It indicates that zooarchaeologists should be cognizant of when they compare, say, an assemblage of bones from a kill site with one from a habitation site, or an assemblage from a cave with an assemblage from an open site, given differences in the natural and cultural depositional and burial histories of the two. Perhaps we should follow Behrensmeyer and Hook's lead and develop a list of kinds of archaeological depositional contexts in an attempt to note within-context taphonomic similarities and between-context taphonomic differences of faunal remains.

Spatial distribution of faunal remains

Study of the spatial distribution of vertebrate faunal remains traditionally was a matter of noting which taxa occur in particular strata, perhaps with their abundance relative to other taxa also being noted. This vertical perspective was supplemented in the 1960s and 1970s coincident with the shift in archaeological focus from chronological issues to synchronic variability within archaeological cultures. Zooarchaeologists began to study variation in the horizontal distributions of taxa and skeletal parts, positing such things as kill-redistribution or sharing (Lyman 1980), covariation in the kind of bone refuse deposited and the time of site abandonment (Pozorski 1979), and covariation of depositional contexts and differential preservation (Meadow 1978). Almost immediately ethnoarchaeological studies were published which suggested caution was due when interpreting the horizontal distributions of faunal remains (e.g., Binford 1978; Kent 1981), although some more recent neotaphonomic research indicates these early interpretations may well have been correct (e.g., Bartram *et al.* 1991; Marshall 1993).

Spatial distributions of faunal remains have not been studied with the same intensity that skeletal part frequencies and butchering marks have been. Archaeologists regularly record detailed data concerning the spatial locations and associations of the faunal remains they recover, and virtually every introductory textbook on archaeological methods and techniques devotes a chapter or two to basic field recovery and mapping techniques. Detailed spatial data are now regularly recorded in paleontology (e.g., Abler 1984). Given that burial processes determine the final spatial position of vertebrate remains it is rather surprising that these processes have seen so little study. Perhaps the dearth of such research is, as Yellen (1991b:154) notes, a function of the fact that the long-term study of post-depositional and post-burial processes is "rarely possible." Or perhaps the lack of research on the spatial distribution of bones and teeth is due to a generally shared perception that "it is likely that much of the [spatial] patterning evident in bone refuse can only be understood in terms of features that are unlikely to be archaeologically detectable" (Bartram *et al.* 1991:143). Whatever the case, as I have noted throughout this volume, spatial and contextual data are important to many analytical problems, and may become important to new ones as the volume of neotaphonomic data increases. In particular, study of the horizontal and vertical distributions, orientations, and associations of vertebrate faunal remains may tell us much about the processes responsible for their burial. As noted in other chapters, spatial data inform us about processes of dispersal and accumulation (e.g., Chapter 6), and often these processes (e.g., fluvial action) are also burial processes.

Summary

It is an inviolable rule that collections from different areas and levels must be kept separate.
(I. W. Cornwall 1956:241)

I have reviewed some of the major geological processes and effects of burial in this chapter. What I have found truly amazing is that despite the fact that while faunal remains typically come to us in buried form, there is very little written about the burial process itself. The brevity of this chapter is a good indication of that depauperate literature; there was simply little on the taphonomic effects of the burial of vertebrates for me to summarize here. Given the geological mode of occurrence of fossils, the mechanisms by which faunal remains attain a geological mode of occurrence would seem to warrant much more discussion than they have received by zooarchaeologists in particular.

Andrews (1992:34) argues that taphonomic problems can be divided "into two issues, the nature of the physical environment or sediments in which animal remains are found, and the nature of the fossil assemblage itself." The latter has received much attention. Perhaps the lack of literature on the former results from its less than readily visible nature; that is, while we have a model of a living skeleton to which we can compare the fossils we find to determine how they differ from living bones, no such model is available for the sedimentological record. The geological context of fossils is studied largely by sedimentological analyses focusing on processes of sediment accumulation, dispersal, and diagenesis (Andrews 1992:35–36). Thorough taphonomic analysis, then, demands consultation with a sedimentologist and/or geologist.

The burial of animal remains can be studied by examination of the geological context of the fossils. Rapid burial shortly after animal death may result in sorting (see the Glossary) of the remains, but it also removes the remains from biostratinomic processes many of which tend to modify bones at macroscopic scales. Thus, relatively complete, articulated, and largely unmodified skeletons will result. The longer the time span between deposition of the faunal remains and their burial, the greater the chance that those remains will be modified by biostratinomic processes. Once animal remains are buried, diagenetic processes take over. Sediment chemistry, porosity, and weight are important taphonomic factors, and are controlled in part by their original deposition rate and mode. It is to these topics that we turn in the next chapter.

DIAGENESIS

The nature of the bones, that of the soil, its dryness or humidity, its permeability by
air and water, the more or less ancient date of burial, the depth at which they lie,
have a considerable effect on the condition of the bones.
(N. Joly 1887:88)

Introduction

Sedimentary petrologists define diagenesis as the "alteration of sediments after
deposition" (Retallack 1990:129). It is, however, sometimes taken to mean
only "alteration after burial" (Retallack 1990:129). Throughout this volume I
take diagenesis to have the latter meaning for vertebrate faunal remains. The
importance of this distinction resides in the many post-depositional *and* pre-
burial taphonomic processes that can modify vertebrate remains, along with
modifications resulting from the burial process itself.

Once animal remains have been buried, a number of taphonomic processes
can act on them. Some of the more familiar ones are mineralization and
deformation. In this chapter, these and other diagenetic processes are discussed. Diagenesis of skeletal tissues is affected by *intrinsic factors* of the tissue
specimen, such as its size, porosity, chemical and molecular structure, and by
extrinsic factors such as sediment pH, water and temperature regimes, and
bacterial action (Von Endt and Ortner 1984). Intrinsic factors such as
hydrolysis of collagen by bone water may exacerbate or buffer extrinsic factors.
The post-burial history of animal remains involves their preservation as fossils
(e.g., Schopf 1975), and their chemical and mechanical alteration or destruction. Zooarchaeologists have explicitly recognized the potential effects of
diagenetic processes for at least two decades (e.g., Chaplin 1971), and most
zooarchaeological collections have undergone at least some diagenetic
modification.

Martill (1990:285–287) suggests that diagenesis as a process can be conve-
niently considered as taking place in three "environments:" the bone tissue
itself, the pore spaces and cavities within bones, and the sediment surrounding
the bone. In the first, ionic substitution of apatite minerals with other minerals
regularly occurs. In the second, pore spaces and trabeculae are filled with
diagenetic minerals; comparison of these with those of the surrounding
sedimentary matrix may indicate transport prior to final burial of the bone if
the two sets of minerals differ. If the two sets of minerals differ, the process is

417

sometimes referred to as "pre-fossilization" (Martill 1990:288). Finally, the sedimentary matrix in which vertebrate remains are buried affects (a) the outer surface of the bone, (b) the nature of ground water (both its chemical makeup and transportability), and (c) whether buried vertebrate remains are crushed or deformed based on the presence or absence of diagenetic cements in the sediment.

Rolfe and Brett (1969:233) arrange diagenesis into three phases. The first, *syndiagenesis*, involves bacteria in shallow sediments metabolizing organic matter in the skeletal tissue. *Anadiagenesis* "is the deep burial phase of compaction and cementation ... inorganic chemical reactions predominate" (Rolfe and Brett 1969:233). Finally, *epidiagenesis* can result in the disappearance of the fossil itself leaving only a mold; this phase seems to be of little concern to most zooarchaeologists and is not dealt with further here. Distinction of the first two phases may also be of little concern to zooarchaeologists, but I suspect would become more important as the age of the fossil assemblage increases.

One might think of diagenesis as being modeled by the equation

$$D = f(M, C, D, S, T) \tag{11.1}$$

in which D is the sum of the diagenetic processes influencing a particular fossil, M is the original physical and chemical composition of the *material* of concern (see Chapter 4), C is the *climate* of deposition, D is the *depositional* mode, S is the nature of the *sediment* in which the material is buried, and T is the *time* span over which a specimen is buried. I briefly consider each variable in turn.

The porosity of the skeletal tissue can greatly influence diagenetic chemical changes in those tissues. The "high density and low porosity of enamel makes it an attractive candidate for minimal [chemical diagenetic] alteration" (Sillen 1989:212). The porosity of dentine makes it fairly susceptible to such alteration (Carlson 1990:545). The greater the porosity of bone tissue the greater the susceptibility of that tissue to diagenetic change and the more rapid the chemical ion exchange between the tissue and the sedimentary matrix in which it is embedded (Hanson and Buikstra 1987:552).

Climate may have strong diagenetic influences, as in the contrast between arctic, temperate, tropical, and desert environments. Being frozen, alternately wet and dry, or constantly wet or dry, all have major effects on the rate and types of chemical reactions that can take place. For example, chemical reactions double in rate for every 10°C increase in temperature (e.g., Von Endt and Ortner 1984). Sillen (1989:220) suggests the action of micro-organisms such as fungi, known to attack collagen, is less pronounced in dry environments than in moist environments. Where a bone is deposited influences how quickly it will be buried after deposition, and bones that are weathered subaerially probably respond differently to diagenetic processes than unweathered bones. The acidity, chemistry, moisture content, permeability, and other properties of the embedding sedimentary matrix, as well as pedogenic (soil forming) processes if any, are also important (e.g., Retallack 1984; Tuross *et al.* 1989). Finally, as one might expect, a short temporal duration of burial may result in a bone not being exposed to various diagenetic processes that it would be exposed to were it buried for a long time. For example, Sillen (1989:212) writes, "the duration of interment is a critical variable in diagenesis, [but] diagenetic changes in a given environment may not bear a linear relationship to time." He goes on to make an observation which may relate to the distinction between syndiagenesis and anadiagenesis: "In the early post-depositional period, the mineral [of skeletal tissue] may be partially shielded from diagenesis by the organic phase ... since the apatite crystals become more exposed to water and soil ions" with the decomposition of the organic phase (Sillen 1989:219).

Equation [11.1] is a simplistic rendition of what is a complex suite of potential taphonomic processes. Carlson (1990:545) notes, for example, that the diagenesis of dentine in teeth may be linearly related to time, but given the variability in diagenetic processes that might affect dentine, the relation of time and diagenesis may not be statistically linear. At the outset I am compelled to note that my experience with diagenetically altered fossils is limited. But, many of the vertebrate remains zooarchaeologists deal with have undergone minimal (and probably insignificant) diagenetic modification. (The most extensive and intensive analysis of archaeofaunal remains recovered from Plio-Pleistocene contexts at Olduvai Gorge does not even list diagenesis or mineralization in the index [Potts 1988].) And, biostratinomy is certainly much easier (and more fun?) to study ethnoarchaeologically due to the often long time spans involved in and the subsurface (and thus largely invisible) nature of diagenesis. My treatment of diagenesis may thus seem less detailed than it might be. For example, I do not discuss the significant research on the effects of diagenetic processes on analyses of bone trace element content (e.g., Price *et al.* 1992; Whitmer *et al.* 1989), analyses which are becoming increasingly important in studies of human diet. The interested reader is encouraged to pursue the topics introduced here, and to read the references cited for additional information. A set of rather geochemically-detailed references on diagenesis and mineralization is found in Allison and Briggs (1991a), Lucas and Prévôt (1991), and Tucker (1990).

Mineralization, leaching, enrichment

> In the process of fossilization the bone's components are modified as follows: (a) gradual disappearance of all organic structures, i.e., the osteocytes and the [collagen]; (b) their replacement by material carried in the water of the ground; (c) substitution of the chemical elements constituting the crystalline lattice of the apatite.
> (A. Ascenzi 1969:527)

Basics and background

A *permineralized* or *petrified* fossil is formed by infiltration of mineral-bearing solutions into pores in skeletal tissue and deposition of minerals in those spaces. This process is also known as cellular permineralization or petrifaction. A *mineralized* fossil is formed by the replacement of the original hard parts of an organism with other minerals; the skeletal tissue's chemical constituents are dissolved and removed while minerals dissolved in ground water are simultaneously deposited in their place. This process is also known as authigenic preservation or replacement (Matthews 1962; Schopf 1975). Both mineralization and petrification result in a stony hard replica of the skeletal specimen. *Leaching* is the removal of soluble matter; *enrichment* is the addition of soluble matter. Thus mineralization and petrification, in a way, are enrichment processes. *Hydrolysis* is often a leaching process involving the combination of water with other molecules.

Bartsiokas and Middleton (1992:68) suggest the term "fossilization (of bone) may be defined as a naturally occurring physico-chemical process that *preserves* the gross morphology by ionic exchange and an increase in the size of the apatite crystallites; this process also involves the loss of organic material of the bone and the filling of voids with minerals" (emphasis in original). They indicate that fossilization refers to the "stoniness" of bone, and that the Greek word *apolelithomenon*, meaning turned to stone, was used to describe fossils over 2,000 years ago. Thus they argue that the term diagenesis is not a good descriptor of fossilization processes, and suggest instead the terms *apolithosis* to mean fossilization and the term *apolithotic effects* to refer to the changes brought about by fossilization processes. While it remains to be seen if their revisionist terminology is adopted, what is perhaps more important is that they document, as others have before (e.g., Tuross *et al.* 1989), that skeletal tissue increases its "crystallinity (which probably reflects an increase in [apatite] crystallite size)" during fossilization, and that may account for the "high compactness, hardness, stiffness and generally 'stony' appearance of heavily fossilized bones" (Bartsiokas and Middleton 1992:67). Preservation of the histological structure of skeletal tissue of reptiles and fish due to fossilization processes is reviewed by Ascenzi (1969).

A process we can term *encrustation* involves the precipitation of soluble salts on the surface of a bone or tooth. The soluble salts are derived from sediments and transported by ground water. Calcification – the precipitation of calcium-carbonate salts – is typical of some archaeological remains recovered from somewhat arid areas where moisture is insufficient to flush the salts from the sedimentary matrix. The coating of bones with such salts can obscure the details necessary to identify them and requires intensive cleaning (Stahl and Brinker 1991). Brain and Sillen (1988:464), for example, report both that fossils they examined were stained a dark color from "diagenetic incorporation of manganese dioxide" and that fossil surfaces were "obscured by heavy encrustation of calcium carbonate and manganese dioxide."

Diagenetic processes not only result in the creation of replicas, but they can significantly alter the chemical composition of buried skeletal tissues. Whitmer *et al.* (1989:244–245) outline a useful model of such chemical modifications by using the concept of *diffusion*–"a process that operates whenever mobile [chemical] elements occur in differing concentrations. In this event, a concentration gradient is established with the result that elements tend to migrate from areas of high to low concentration. This process continues between the areas until a concentration equilibrium is established." Thus, theoretically, if the skeletal tissue (M in equation [11.1]) differs in chemical composition from its surrounding sedimentary matrix (S in equation [11.1]) at the time of burial, the bone will tend to lose chemical elements in which it is rich (the bone will be *leached* of these elements), and gain elements in which it is poor relative to the sediment (be *enriched*). Pate and Hutton (1988) describe one method of studying variation in the chemistry of the fossil and the chemistry of the sedimentary matrix.

Whitmer *et al.* (1989) provide an excellent overview of the study of trace elements and skeletal tissue chemistry that goes far beyond the scope of coverage necessary in this volume; interested readers are encouraged to examine their account. It suffices here to summarize some general observations that are readily appreciated by those with minimal expertise in chemistry.

General observations

Chaplin (1971:16–18) suggests that at least four properties of the depositional matrix affect skeletal tissue preservation: pH, aeration, water regime, and bacterial action. He suggests (a) bacterial action will destroy the organic fraction of skeletal tissue (see Chapter 5), (b) bacterial action is inhibited in acidic sediment, (c) acidic sediment will dissolve the mineral fraction of skeletal tissue and the rate of dissolution depends on the degree of acidity and amount of percolating water, (d) skeletal tissue will preserve for only a few millennia in acidic sediment, and (e) preservation is better in basic sediments. Chaplin (1971) notes that the rate of decay and dissolution is dependent on the porosity of the skeletal tissue; more porous tissue decays more rapidly than less porous tissue. Thus the structural or bulk density of bone tissue (Chapter 7) is an important variable. The chemical breakdown and leaching processes progress from exposed surfaces into the bone tissue (Cussler and Featherstone 1981).

Rolfe and Brett (1969:226, 229) suggest (a) highly basic sediments inhibit bacterial activity, (b) the organic fraction of skeletal tissue decays more rapidly in an aerobic than anaerobic environment, and (c) phosphatic, keratinous, and organic skeletal tissues preserve in acidic sediments. Hare (1980:209) notes that "if a bone is left in a solution of a weak acid the mineral material will dissolve and there will be left a pseudomorphic model of the bone formed by its organic matrix." Bromage (1984:168, 175–176) reports that bones soaked in acetic acid have surficial mineral material dissolved and lacunae, canaliculi, and vascular and capillary spaces are enlarged. Eventually the bone takes on a pitted or etched appearance as a result of the enlargement of naturally present spaces. Knight (1985) reports that burned bone dissolves faster than unburned mammal bone when placed in an acidic solution (Figure 9.10).

Hare (1980:213–214) reports that water leaches soluble protein compounds from skeletal tissues by breaking collagen down into polypeptides ("subprotein units") which are subsequently leached. Ground water leaches the organic fraction of skeletal tissues at a rate that depends on how saturated the water is and the rate of ground water movement (flushing) through the sediment matrix (Hare 1980). Ground water may also leach the mineral component of skeletal tissues if that water is not saturated with the minerals making up bone tissue (Dodd and Stanton 1981:128). Such leaching tends to require permeable sediment and a sediment matrix that is mineralogically different from the bone tissue (Dodd and Stanton 1981:128). "If either the organic matrix or the mineralized components of bone are partially removed, the hardness and strength of bone will decrease" (Hare 1980:217).

White and Hannus (1983) suggest that hydroxyapatite has low solubility in alkaline (pH > 7.5) to slightly acidic (pH = 6.0) aqueous systems, while in highly acidic environments (pH ≤ 6.0) its solubility is very high. "Chemical weathering of bone is probably initiated by acids created as microorganisms decompose collagen" (White and Hannus 1983:321). They suggest that heating of bone apparently does not accelerate chemical weathering and appears to increase the proportion of phosphorus in bone (but see Figure 9.10 and associated discussion).

In a series of experiments, Von Endt and Ortner (1984) used fresh cow (*Bos* sp.) bone crushed to different sizes to assess the effect of temperature and bone size on the rate and nature of bone disintegration (see also Ortner *et al.* 1972). They found that small bone particles tend to preserve less well than large bone particles, and thus they suggest that the remains of small animals may be less well preserved than the remains of large animals (see also Chapter 9). They found that the protein in bone is lost more quickly at high temperatures. Von Endt and Ortner (1984:252) also note that the porosity of the skeletal tissue exerts a strong influence on its preservation; more porous tissue allows quicker diffusion of groundwater into the tissue and thus the quicker the protein-to-mineral bonds will be altered (see also Cussler and Featherstone 1981).

Analysis of chemically altered bone

> Change is the essence of the process of fossilization ... the cardinal prerequisite [is]
> that the object still be recognizable as the physical form of a plant or animal, or one
> of its parts.
> (S. F. Cook *et al.* 1961:356)

Gordon and Buikstra (1981) examine the correlation between the preservational condition of human bones recovered from sites in Illinois and the pH of the sediment from which the bones were recovered. They recorded the preservational condition of bone specimens as belonging to one of six categories. *Strong complete bone* displays no evidence of postmortem modification and specimens are whole. *Fragile bone* has external surfaces that may show some etching, could be fragmented, has some superficial destruction, and epiphyseal ossification centers of ontogenically young bone are eroded. *Fragmented bone* consists of cracked and broken skeletal elements, and surfaces are heavily etched and cracked. *Extremely fragmented bone* consists of specimens that are extensively and intensively broken (see Chapter 8) and quite eroded. The last two categories are similar; the *bone meal* and *ghost* categories include bone tissue that consists of a powdery substance that will not hold its anatomical shape without surrounding sediment.

Gordon and Buikstra (1981:569) found that ontogenically young bones were less well preserved than ontogenically old bones. They found a strong inverse correlation between sediment pH and the frequency of ontogenically mature bones preserved within particular preservational categories ($r = -0.92$, $P < 0.001$). That relation was also negative for the ontogenically young bones ($r = -0.48$, $P < 0.005$). They suggest the latter coefficient is less strong than the former due to the greater range of structural density (see Chapter 7) displayed by the ontogenically young bones. Together, sediment pH and ontogenetic age were strongly correlated with preservation ($r = -0.87$, $P < 0.001$).

Gordon and Buikstra's (1981) straightforward analysis is exemplary, with the exception that they did not publish the data from which their correlation coefficients were derived. They identified preservational variation, designed a way to categorize and quantify that variation, and sought and found a correlation between that preservational variation and a well-known diagenetic property. All such analyses of the diagenetic effects of taphonomic processes should be so clear (and include *all* relevant data).

Sediment overburden weight

As faunal remains become more deeply buried, the weight of the sediment overlying them becomes greater and underlying sediments become more compact and of greater bulk density because there are smaller and fewer pore spaces between sedimentary particles. Kidwell (1986:7) suggests differential compaction from sediment overburden weight can concentrate fossils, and this probably results from the greater bulk density (lower porosity) of the sedimentary matrix in general; more fossils per unit volume of sediment would result. This suggests that one may want to estimate the bulk density (porosity) of the sedimentary matrix along at least the vertical dimension of an excavation to ascertain if greater fossil density (frequency per unit volume) at greater depth is a function of decreased porosity with increasing depth (e.g., Retallack 1990:135).

Retallack (1990:132) suggests that the best evidence for compaction of sediments from overburden weight is "folding of vertical cracks [in paleosols] filled with distinctive soil material; comparison of the length of these [distorted and filled cracks] with their length in a straight line can yield a compaction ratio that is useful for reconstructing former soil thickness, density, and chemical composition." The following equation provides a way to determine the degree of compaction:

$$C_f = V_f/V_s = T_f/T_s \hspace{4cm} [11.2]$$

where C_f is the diagenetic compaction ratio or fraction of original thickness of the soil due to burial, V_f is the volume of the fossil soil or horizon in cm^3 including foreign material in the filled cracks, V_s is the volume of the fossil soil or horizon in cm^3 formed from parent material (excludes foreign material), T_f is the thickness of a fossil soil or horizon in cm derived from the soil or soil horizon, and T_s is the thickness of the soil or soil horizon in cm derived from parent material (Retallack 1990:69–70). Compaction and deformation may also be indicated by distorted root casts and crushed botanical materials (Retallack 1990:132–135).

Sediment overburden weight may also deform or fracture fossils. Shipman (1981b:172) suggests when the deposition of overburden sediment is rapid, the pressure from the weight may crush or break the buried faunal remains; "slower deposition may produce plastic deformation, in which the shape or dimensions of the original [fossil] are changed without breaking." Finks (1979:330) indicates that "mechanical distortion is accomplished by plastic flow of the [skeletal] material (mainly microfracturing and recrystallization). Simple compaction of the sediment after burial may distort fossils by shortening them in the direction perpendicular to the bedding plane." Because some degree of deformation precedes fracture (see Chapter 4), I consider deformation prior to crushing and fragmentation.

Deformation

The *deformation* of vertebrate remains refers to their distortion; that is, two or more anatomical points on a single skeletal element change their spatial locations relative to one another (see the Glossary). The degree and kind of deformation are determined by several factors, including both intrinsic and extrensic factors. *Intrinsic factors* include the original morphology, modulus of elasticity (see Figure 4.6), and orientation in both horizontal and vertical planes of the skeletal part being deformed. Henderson (1987:44), for instance, notes that the skull, when considered as a unit, is rather susceptible to deformation and crushing due to it being a large, hollow sphere. *Extrinsic factors* include the nature and timing of diagenetic factors such as mineralization and leaching (which alter the modulus of elasticity), and the grain size and composition of the sediment. "Coarser-grained sediments are more resistant to compaction resulting from the supporting effect of the grains and the lower pore-water volume" (Briggs 1990:244).

Deformation can take several forms. *Homogeneous deformation* involves changing the morphometry of a bone but maintaining the relative orientation of all planes; "all lines or areas of bone in one particular orientation are uniformly lengthened or shortened" (Shipman 1981b:179). Basically, the analyst superimposes an undeformed model of the fossil over a grid, and, determines the "strain ellipse" (Shipman 1981b:179) that would be produced if the undeformed fossil were a perfect circle and the deformed fossil were an ellipse. Determination of the forces necessary to create the ellipse from the circle indicates the forces necessary to deform the fossil into its present morphometry (see Doveton 1979; Sdzuy 1966; Wellman 1962 for more complete descriptions of the matrix algebra procedures). Computer-graphic programs can be used to help determine the original, pre-deformation shape of the fossil (Briggs 1990). Other forms of deformation include movement of anatomical points in different directions relative to one another, such as bending, which causes points on the concave side of the bend to become closer and points on the convex side to become farther apart.

For those less mathematically inclined, Briggs (1990:245) suggests one can take photographs of undistorted homologous skeletal parts from different angles to derive a model of the original morphometry of the distorted fossil; "these photographs are directly analogous to flattening the organism onto a bedding plane." Comparison of the distorted fossil with the photographs can help the analyst determine the direction(s) in which distortion forces were applied. Mirror imaging of paired, bilaterally symmetrical skeletal parts is also possible (Shipman 1981b:179).

Crushing and fragmentation

In Chapter 8 I describe analyses of fragmented bones performed by Villa and Mahieu (1991). One of the things they note in their analysis is the post-burial, diagenetic fracturing of some specimens. They argue that bones broken by sediment overburden weight are indicated by conjoining fragments lying adjacent to, and often in contact with, one another, and broken bones lying on convex or concave surfaces (thus the bones would be subjected to bending forces). As well, the basic morphology of fractures created after burial tends to be different than the morphology of fractures generated by biostratinomic processes. Post-depositional and post-burial movement of bone fragments due to diagenetic factors and turbation processes can be measured by refitting or conjoining pieces (see Chapter 5). For additional discussion of analytically detecting post-depositional fragmentation by study of fracture kinds see Chapter 8. The remainder of this section is devoted to other analytical techniques for detecting post-depositional destruction of bones.

Lyman and O'Brien (1987) suggest that fragmentation of bones can result in their "analytical absence." As a skeletal element is broken into smaller pieces, the probability it will be identifiable in the set of fragments decreases as fragment size decreases. This is so because the probability that any particular fragment will have an anatomical landmark that is diagnostic of the skeletal element decreases as the fragment becomes smaller. Thus, a collection of bone fragments may contain pieces of, say, femora, but if those pieces are sufficiently small and the anatomical landmarks are sufficiently incomplete, then the analyst will be unable to identify the femur. It is in this sense that fracturing of skeletal elements "destroys" them. The bone tissue is not destroyed, but the anatomical integrity of the skeletal element is sufficiently compromised that the effect is the same as if the bone tissue were destroyed when the analyst is attempting to identify skeletal parts. It is in this sense of "destruction" of bones – their analytical absence due to intensive fragmentation – that we turn to another method of detecting post-depositional destruction of skeletal parts.

Based on a study of several prehistoric assemblages with unknown taphonomic histories, Klein and Cruz-Uribe (1984:70–75) suggest that "post-depositional" destruction can be detected by study of a suite of attributes. I suspect that what they are attempting to measure is actually better labelled "post-burial" destruction, given their comment that the taphonomic processes involved are "the extent to which a sample has been leached and the extent to which it has been compacted in the ground [or] post-depositional leaching coupled with very slow sedimentation" (Klein and Cruz-Uribe 1984:70, 72). The attributes they study include the hardness and/or compactness of the stratum on which the bones lay because if they lay on hard substrates, pre-burial trampling will be more likely to crush the bones. As well, they suggest an abundance of isolated teeth and paucity of identifiable mandibles and maxillae, and a plethora of small, dense bones such as carpals, tarsals, sesamoids, and phalanges indicate the bone assemblage "has probably suffered greatly from post-depositional destruction" (Klein and Cruz-Uribe 1984:71). Harder substrates may produce (a) relatively high abundances of isolated teeth and tooth-bearing skeletal elements, and (b) high abundances of carpals and other small, dense bones, during early phases of an assemblage's post-depositional taphonomic history, assuming all skeletal elements were originally present to begin with (essentially

Table 11.1 *Ratios of NISP:MNI per skeletal part in two assemblages (from Klein and Cruz-Uribe 1984)*

Skeletal part	Level 4	Level 6	Skeletal part	Level 4	Level 6
maxilla	5.92	3.37	mandibular condyle	8.00	2.24
mandible	7.41	3.81	hyoid	3.50	3.14
atlas	5.00	5.00	axis	1.00	1.50
cervical	0.00	10.00	thoracic	6.00	11.00
lumbar	2.00	6.00	sacrum	3.00	1.00
rib	19.67	18.00	innominate	3.33	2.50
scapula	2.00	2.00	P humerus	1.00	5.00
D humerus	2.00	1.80	P radius	3.60	2.62
D radius	7.00	1.33	P ulna	2.14	1.43
D ulna	1.60	1.25	carpal	6.83	7.00
P metacarpal	6.00	5.45	D metacarpal	2.00	1.93
P femur	2.00	2.33	D femur	4.00	3.00
P tibia	1.50	1.00	D tibia	2.67	2.31
naviculo-cuboid	6.00	2.25	astragalus	1.60	2.10
calcaneum	2.78	3.00	small tarsals	2.92	3.14
P metatarsal	7.00	6.75	D metatarsal	1.33	2.00
1st phalanx	24.13	22.11	2nd phalanx	25.67	16.62
3rd phalanx	16.33	9.75	sesamoid	16.33	16.20

complete skeletons were deposited). But I wonder if during later phases of that history even small, dense bones and teeth will become destroyed. Data and arguments presented elsewhere (Lyman 1991c, 1993b) suggest that with sufficient time, many bones will be destroyed by density-mediated processes. For example, the data in Figure 7.13 suggest prehistoric, buried assemblages will be more prone to display such destruction, because more prehistoric assemblages (75 of 143; 52.4%) than ethnoarchaeological (9 of 41; 22.0%) assemblages are positively correlated with density.

We return to the issue of the frequency and preservation of small, dense skeletal elements below. First, however, it is important to review the other attribute Klein and Cruz-Uribe (1984:71) suggest will be displayed by post-depositionally destroyed assemblages: high NISP:MNI ratios per skeletal part. They indicate that two assemblages of bones from stratigraphically distinct contexts in the Lower Magdalenian El Juyo Cave have rather different NISP:MNI ratios per skeletal part. The ratios for the two total assemblages are: Level 4 – 97.76 (1662 NISP:17 MNI), and Level 6 – 38.47 (1462 NISP:38 MNI). Ratios for many of the individual skeletal parts are summarized in Table 11.1. It is important first to note that these are *not* NISP:MNE ratios as described in Chapter 8. Rather, they are NISP:MNI ratios per skeletal part, with MNI determined for each skeletal part (whether a specimen is from the left or right side is taken into account). The importance of this is that, as shown in Chapter 8, the NISP:MNI ratios may not measure fragmentation very well if

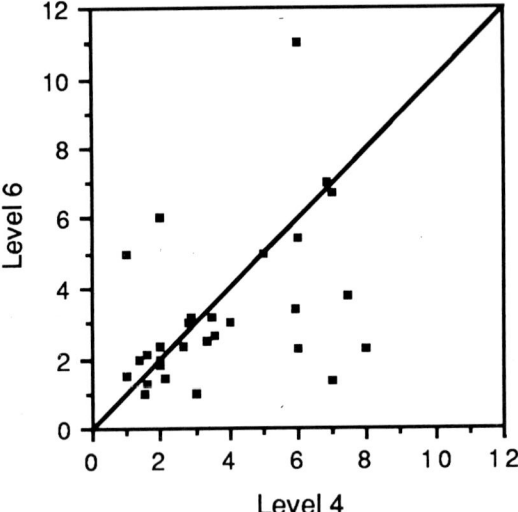

Figure 11.1. Bivariate scatterplot of NISP:MNI ratios per skeletal part for two
bone assemblages from the same site (from Table 11.1).

there are major differences in frequencies of left and right specimens of a
skeletal part and/or if there is a high proportion of complete skeletal elements
per anatomical part in the collections. Nonetheless, the ratios in Table 11.1 are
informative, and they are one of the criteria used by Klein and Cruz-Uribe
(1984) to argue that the Level 4 assemblage has undergone more "post-
depositional" destruction than the Level 6 assemblage.

Figure 11.1 shows that there are more high NISP:MNI ratios among the
Level 4 skeletal parts (points below the diagonal line) than among the Level 6
skeletal parts (points above the diagonal line). On the basis of such differences,
Klein and Cruz-Uribe (1984:75) conclude that NISP:MNI ratios are "a
succinct and reasonably objective measure of fragmentation, particularly when
they are presented for each skeletal part." Note here that these ratios *alone* do
not indicate whether fragmentation was biostratinomic, post-depositional, or
post-burial. But in combination with relatively high abundances of small,
dense bones and isolated teeth, Klein and Cruz-Uribe (1984:71) believe these
ratios are indicative of "post-depositional" destruction. The bones with high
NISP:MNI ratios were broken by post-depositional processes whereas the
small, dense bones and teeth "are particularly likely to survive mechanical
crunching against a hard substrate" (Klein and Cruz-Uribe 1984:71). The
NISP:MNI ratios for mandibles and maxillae are higher in the Level 4
collection (7.41 and 5.92, respectively) than in the Level 6 assemblage (3.81 and
3.37, respectively), and the former assemblage has virtually no mandibles or
maxillae but has many isolated teeth. Finally, the small, compact bones make
up more of the Level 4 total NISP (919, or 55.3%) than of the Level 6 total
NISP (544, or 37.2%; arcsine $t_s = 10.13$, $P < 0.001$).

In a later analysis addressing the same issue, Klein (1989:374–375) suggests skeletal elements of smaller taxa

> are more likely to retain their integrity during butchering and food preparation or during kicking and trampling across the surface of repeatedly occupied sites [than the skeletal elements of larger taxa]. Bones that are less damaged before burial are also more likely to survive leaching and profile compaction afterwards, which means that even a small pre-depositional difference in relative durability will be magnified post-depositionally.

That is, Klein (1989) is adding the variable of skeletal element *size* to the equation. He further suggests that skeletal parts with high structural density (Chapter 7) tend to withstand post-depositional destructive forces better than those with low structural density (Klein 1989:378). It is important to note, then, that the skeletal parts listed in Table 11.1 are all from red deer (*Cervus elaphus*), and thus skeletal element size is not influencing the results (e.g., Figure 11.1). And, while skeletal part abundances as measured by MNI in both assemblages are both correlated with bone structural density, the Level 4 assemblage is less strongly correlated with density ($r_s = 0.58$, $P = 0.001$) than the Level 6 assemblage ($r_s = 0.67$, $P < 0.001$) (frequencies of skull parts and mandibles are not included in the calculation of these statistics). Perhaps as much as 10% more of the variability in skeletal part frequencies in the Level 6 assemblage might be attributed to density-mediated destruction than might be so attributed in the Level 4 assemblage, which is the opposite of what might be expected on the basis of the NISP:MNI ratios.

Marean (1991) pursues the theme started by Klein and Cruz-Uribe (1984; Klein 1989). He focuses on the intensity of bone fragmentation (as defined in Chapter 8), but offers three important qualifications that must be met if this attribute is to be used to measure post-depositional and post-burial destruction. First, the bones used to measure destruction in the form of fragmentation should be those "that are never or very rarely fragmented by people or animals attempting to extract nutrition;" second, the bones "should be independent of the bone transport behavior of bone collectors;" and third, the method of calculating the degree of fragmentation "should be independent of the calculation procedure for archaeozoological measures of element abundance" (Marean 1991:680). Marean's (1991:681) experiments suggest that carpals and tarsals (excluding calcanea) would best meet the first requirement. Calcanea were sometimes fragmented by carnivores and display gnawing marks, and carpals and tarsals broken by humans with hammerstones and anvils display percussion marks (see Chapter 8); thus specimens displaying gnawing marks or percussion marks should be excluded from analyses of post-depositional destruction.

After feeding a set of tarsal-metatarsal and carpal-metacarpal limb sections of domestic sheep (*Ovis aries*) to spotted hyenas (*Crocuta crocuta*), Marean (1991:684–685) concludes that this carnivore swallows carpals and tarsals, and those bones may show digestive corrosion but only rarely display gnawing

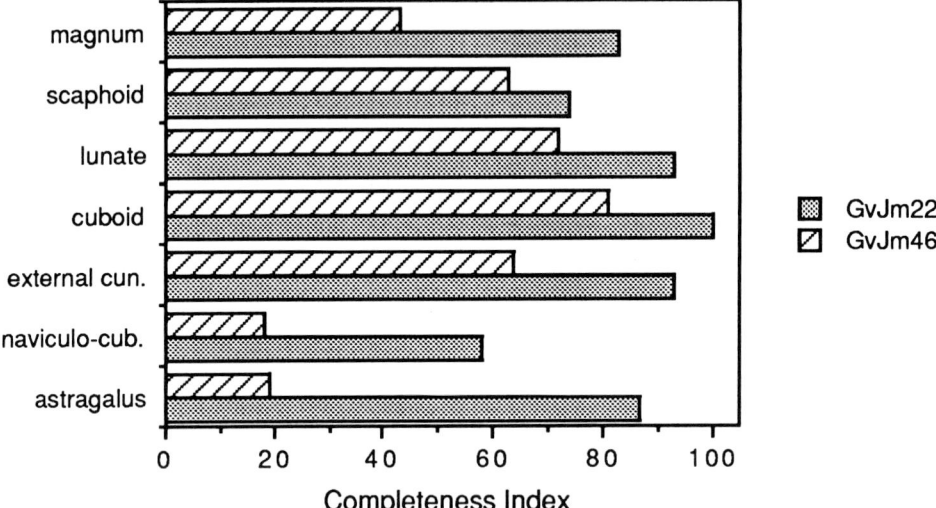

Figure 11.2. Bar graph showing variation in completeness index values across seven small, compact bones from two sites (modified from Marean 1991:688, Figure 2).

damage. As well, a statistic Marean calls the "completeness index" is typically high ($\geq 92\%$) for carnivore-deposited sets of carpals and tarsals. This statistic is calculated by "estimating for each specimen the fraction of the original compact bone that is present, summing the values, and dividing that by the total number of specimens ascribed to that bone and taxon. Multiplication by 100 provides a mean percentage completeness" (Marean 1991:685). Thus, if the analyst has a complete unciform, half of an unciform, and one third of an unciform, one calculates the completeness index for the unciform as: $100[(1 + 0.5 + 0.33)/3] = 100[1.83/3] = 100[0.61] = 61\%$.

Marean (1991) compares his experimentally derived completeness index data with completeness index data for two sites in Kenya. Site GvJm22 is a rockshelter; site GvJm46 is an open site. The small, compact, dense bones from both sites display no evidence of percussion marks, chopping marks, or gnawing damage. Marean (1991:687) concludes that "virtually all of the fragmentation is due to post-depositional destruction" and that the bone assemblage from "GvJm46 has undergone substantially more post-depositional destruction than [the bone assemblage from] GvJm22." As shown in Figure 11.2, the completeness index values indicate the GvJm46 fragments are, on average, smaller than the GvJm22 fragments.

Marean (1991: 687–690) follows Klein (1989) and suggests that bones of smaller taxa will be better preserved than bones of larger taxa. This is readily shown for Marean's two Kenyan sites in Figure 11.3. This bivariate scatterplot is between the completeness indices for six small, dense bones of two size classes

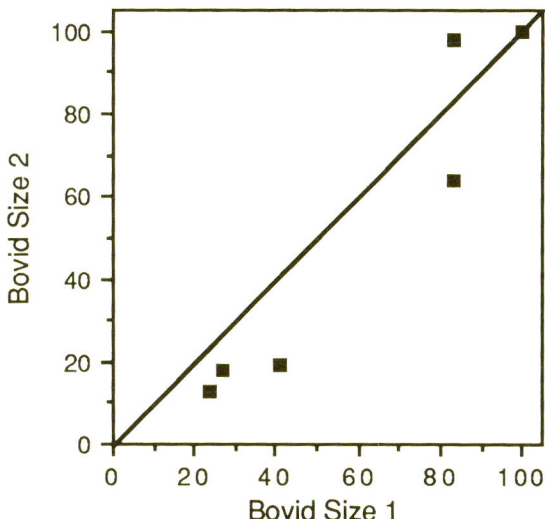

Figure 11.3. Bivariate scatterplot of completeness index values for six small, compact bones for two size classes of animals (from Marean 1991:689).

of bovids; size class 1 animals are 7 to 27 kg live weight and size class 2 bovids are 50 to 220 kg live weight. The plotted bones are the astragalus, calcaneum, naviculo-cuboid, external cuneiform, internal cuneiform, and distal fibula. If fragmentation is similar for the two size classes, then all plotted points should fall on or quite near the diagonal line. In fact, four of them fall below it and only one falls above it. With size 1 bovids plotted on the x-axis and size 2 bovids plotted on the y-axis, the positions of the plotted points indicate that four of the plotted skeletal elements have high completeness indices for the size 1 bovids relative to size 2 bovids, and only one skeletal element has a high completeness index for the size 2 bovids relative to the size 1 bovids. That is, overall, the completeness indices are higher for the size 1 (smaller) bovids than for the size 2 (larger) bovids, indicating bones of the former are better preserved than bones of the latter taxa, just as Marean and Klein suggest they should be.

Post-burial crushing and fragmentation have been little studied by vertebrate taphonomists and zooarchaeologists. Perhaps, as Marean (1991:678) suggests, the reason for this resides in any or all of the following: a lack of awareness of the problem, a belief that post-burial destruction is insignificant, and/or the lack of a readily applied analytical technique for detecting such destruction. The last is no longer applicable as Marean's (1991) technique and Klein and Cruz-Uribe's (1984; Klein 1989) criteria for recognizing such destruction are now available. More frequent application of these analytical tools may show precisely how widespread post-burial destruction is in various temporal, spatial, and geological contexts, and thus how significant a problem it is.

Table 11.2 *Turbation processes influencing burial, exposure, and movement of fossils (from Wood and Johnson 1978)*

Process class	Sediment mixing
Faunalturbation	animals, especially burrowing
Floralturbation	plants, root growth, and treefall
Cryoturbation	freezing and thawing
	frost heaving
	mass displacement
	frost cracking, ice wedges, sand wedges
	sorting
Graviturbation	mass wasting
	solifluction, creep, subsidence
	mudflows, earthflows, avalanche, landslide
Argilliturbation	swelling and shrinking of clays
Aeroturbation	gas, air, wind
Aquaturbation	water
Crystalturbation	growth and wasting of salts
Seismiturbation	earthquakes

Post-burial movement

Thus far discussion has focused on chemical and mechanical modifications of skeletal parts. Another category of modification does not influence the skeletal tissues but rather modifies the locations of skeletal parts. The detection of post-burial movement should be a major part of any analysis of the spatial distribution of faunal remains. This is so for any category of artifact or ecofact, as made clear by Wood and Johnson's (1978) extensive list of *turbation processes* (Table 11.2). Burial does not remove faunal remains or artifacts from processes that modify their locations. The discussion of burial processes in Chapter 10 makes it clear that burial can be a transitory phenomenon. Geological processes that bury bones may also re-expose them. But even when bones are buried and not re-exposed, various forces operating beneath the surface can move faunal remains.

Perhaps the easiest way to determine if there has been some post-burial movement of bones is the recognition of spatially disassociated refitting skeletal parts. But how does one determine if the spatial disassociation of refitting skeletal parts is the result of pre-burial or post-burial processes? As noted in Chapter 5, this can be accomplished with stone tools for which the order in which flakes have been removed from a core dictates the vertical ordering of flakes within a stratigraphic sequence: flakes removed first lying stratigraphically above flakes removed last must have been disturbed after burial. As also noted in Chapter 5, Villa *et al.* (1986) suggest that post-depositional disturbances result in variation in the vertical locations of refitting pieces of bone, but whether those disturbances were also post-burial is not clear. Bones may also move horizontally after burial, but that may be more difficult to detect. It may be that evidence of vertical movement can also be taken as evidence of horizontal movement.

One of the most intensive and well documented studies of refitting of bone fragments in an attempt to discern post-depositional (not necessarily post-burial) movement of mammalian remains is White's (1992:67–83) analysis of human remains recovered from a pueblo in the southwestern United States. He suggests "post-depositional disturbance" was minimal because most refits were between specimens within a single room and often within a single concentration of fragments; very few between-room refits were found (White 1992:83). White (1992:82) implies the numerous refits he identified indicate minimal post-depositional disturbance.

Another way to detect post-burial movement would be to note if the most weathered surfaces of bones were always the uppermost surfaces. In the absence of post-burial disturbances, this should be the case (Behrensmeyer 1981). As well, faunal remains that are differentially encrusted or coated with calcium carbonate may indicate post-burial movement. Calcium carbonate coatings often form on the lower or downward-oriented surface first. Thus, for both weathering and carbonate coating observations to inform one about post-burial movement, the fossils must be collected with appropriate orientational and provenience data.

Summary

> When a fresh bone becomes buried in the earth it undergoes chemical changes, differing in nature and degree with the chemistry of the surrounding matrix.
> (I. W. Cornwall 1956:205)

The post-burial modification of faunal remains can take several forms. Petrification, mineralization, corrosion, deformation, fracturing, and movement seem to be the major ones. Typically these are analytically detected by chemical and microscopic analyses. The influence of these processes on analysis depends, of course, on the analytical result desired. Mineralization can result in a perfect replica of the original whereas deformation can preclude the recording of accurate morphometric data. Fracturing can variously analytically remove specimens by making them unidentifiable, and it may obscure traces of human butchering activities and other processes.

Because most post-burial taphonomic processes are controlled in part by the context of the faunal remains, that context should be considered during analysis. This is true of virtually any of the taphonomic processes one wishes to study, but warrants emphasizing in the context of diagenesis because the geological context of faunal remains is the ultimate extrinsic factor of diagenesis, the more proximate extrinsic factors being sediment chemistry, texture, porosity, and the like.

12

TAPHONOMY OF FISH, BIRDS, REPTILES, AND AMPHIBIANS

Introduction

In preceding chapters I review various analytic techniques for assessing the taphonomic history of vertebrate faunal remains. That discussion focuses on mammalian remains because that taxonomic group has received the most attention in the literature. But mammals are not the only vertebrates with which zooarchaeologists deal. Birds, reptiles, amphibians, and some fish are also vertebrates. Many of the analytic techniques developed for mammalian remains are also applicable to these other vertebrate taxa. To illustrate this, in this chapter I review, with less attention to detail and fewer examples than in previous chapters, much of the literature on non-mammalian vertebrate taphonomy. The reader who has ingested and digested (keeping this somewhat taphonomic) the content of previous chapters will see many parallels among variables studied by taphonomists whatever the taxonomic subject. These include skeletal completeness, natural disarticulation sequences, inherent properties of skeletal elements such as size, shape, and structural density, and various kinds of modifications to bones.

This chapter is not meant to imply that non-mammalian vertebrates are less important taphonomically than mammalian remains. For example, it may prove very interesting to compare the taphonomic histories for each vertebrate category in assemblages rich in fish, birds, and mammals. I am unaware of any such comparative study, but this form of comparative analysis may prove enlightening beyond the details of, say, small mammal versus large mammal taphonomy. If this chapter prompts that kind of study, if it prompts others to perform more neotaphonomic research on non-mammalian vertebrate remains, or both, it will have fulfilled a greater purpose than my illustration of the intertaxonomic parallels in vertebrate taphonomy.

Fish taphonomy

> Fish carcasses are more vulnerable to decay than other vertebrates.
> (W. Schäfer 1962/1972:49)

Fish skeletons and natural processes

There have been probably fewer than two dozen articles explicitly concerned with the taphonomy of archaeological fish remains published in the last two decades (Colley 1990b). That is so, I suspect, because while fish remains are common in some archaeological contexts, they tend to be rare relative to mammal remains in most sites. Some argue that this is so because fish bones are more vulnerable to the effects of differential preservation than mammal bones due to the former's relative fragility (e.g., Wheeler and Jones 1989:63), or the generally small size of fish remains results in failure to recover them unless special techniques are employed. But it is also clear that some skeletal elements of fish preserve extremely well, and some fish elements are large (Colley 1990b), just as is found with mammal and bird skeletons. Whatever the case, Colley (1990b:215) is certainly at least generally correct when she states, "fish are subject to the same processes of taphonomy and site formation as other faunal remains." In part for that reason, my consideration of explicitly piscean taphonomy is relatively brief.

To begin, it is important to note that some (but not all) fish are not only vertebrates, and thus possess hard endoskeletons, they also have scales and otoliths as hard parts that might preserve and be recovered from archaeological contexts (Casteel 1976). Scales are formed in the dermal layers, and in teleost fishes are composed of a thin sheet of bonelike material that is somewhat soft and flexible (Hildebrand 1974:99; Romer and Parsons 1977:158). Otoliths are "ear stones" of the mineral aragonite that have an equilibrium-related function (Casteel 1976; Colley 1990b; Hildebrand 1974). The analytic and interpretive potential of fish remains is detailed in several publications (e.g., Casteel 1976; Colley 1990b; Wheeler and Jones 1989).

Wheeler and Jones (1989) present a useful synopsis of significant taphonomic factors that influence fish remains (see also Colley 1990b). The factors they consider can be summarized in a fashion similar to those that influence other vertebrate remains: those that alter the morphological or physical and chemical characteristics, or completeness of bones; and those that affect the spatial distribution of bones. These factors are mediated by such things as the "nature of the material forming the hard tissue" (Wheeler and Jones 1989:62). Wheeler and Jones note mineralized cartilage does not preserve well, within bony fishes some bones are more resistant to deterioration and modification than others, and the bones of some species are more resistant to modification processes than the bones of other species. Schäfer (1962/1972:56) reports that "carcasses of different species of fish do not [naturally] decay in the same way despite almost equal external conditions," and suggests "the ratio of the size of the body cavity to the mass of the body determines the mode of decay."

Wheeler and Jones (1989:63) suggest fish bone rarely preserves in acidic sediment, whereas sediments that are neutral or basic "are conducive to fish bone survival [but] otoliths rarely persist except in base-rich deposits." Coarse sediments can abrade fish remains (see also Smith *et al.* 1988). Schäfer (1962/1972:61) reports otoliths are unlikely to be transported far by fluvial processes due to "their compact shape" and Martill (1990:271) notes that "enhanced concentrations of otoliths relative to fish bone are thought to be attributable to the ability of otoliths to resist acid digestion in fish guts, where bone readily dissolves." Post-depositional and post-human behavioral factors include consumption of human waste by scavengers, trampling, and weathering, and these variously fragment, corrode, or remove bones from human consumption sites. These same sites can have fish remains added to them by various non-human processes, such as birds and otters depositing remains of their own meals of fish on archaeological sites, or in fish-consuming sea mammal stomachs if the human occupants exploited such mammals.

McGrew (1975) examines several attributes of a paleontological concentration of fish remains. He suggests the degree of disarticulation is an indication of how decomposed fish carcasses were prior to burial; the greater the disarticulation, the longer the carcasses were exposed to bacterial action while underwater but on the surface of the lake bottom. The extremely high concentration of fish carcasses in several stratigraphic layers indicates catastrophic mortality to McGrew (1975); he suggests that a period of climatic aridity increased the salinity of the lake beyond the point at which the fish could survive, prompting non-selective mass mortality (see Antia 1979 for other examples).

Antia (1979:140) implies that the weathering stages described by Behrensmeyer (1978; see Chapter 9) can be detected on otoliths. Antia (1979:143) argues that particular elements of fish skeletons will, like those of mammal skeletons, weather at different rates. Fish teeth and scales seem to weather more slowly than fin spines, headshields, and endoskeletal bones, and Antia (1979:143) hypothesizes that "the cyamine outer coatings of the teeth and scales may [help] to shield them from rapid weathering and disintegration, while the less well protected and more fibrous fin spines and bones [will] weather more rapidly." Antia (1979:144) proposes that constructing a bone weathering profile (e.g., Figure 9.2) may provide the analyst with "an estimate of the minimum residence time of [fish bones] on the substrate surface."

Elder and Smith (1988) suggest the inferred taphonomic history of a naturally deposited fish fauna can be tested with relevant paleoecological data when substantive uniformitarianism (Figure 3.1) is used to infer the ecology and behavior of long-dead fishes. As we have seen in previous chapters, the ecology and behavior of vertebrate taxa do, in fact, influence the taphonomic history of those taxa. After reviewing how one uses substantive uniformitarianism to reconstruct the paleoecology of fossil fish, Elder and Smith (1988) turn to "the use of taphonomy" which, given their discussion, adheres to the principle of methodological uniformitarianism.

Elder and Smith (1988) examine carcass (soft tissue) decay, transport, and burial, and focus on biological activity (e.g., bacterial decay) and hydrodynamic energy as entropic forces. They point out how fish faunas are influenced by near-shore versus off-shore ecological predilections of the taxa, how salinity, temperature, oxygenation, and overturn of water layers influence mortality, how scavenging of fish carcasses by bacteria (decay), crayfish, and insects can disarticulate carcasses, and how gas production within a dead fish results in floating, drifting carcasses; waves, currents, and winds move floating fish carcasses. They note that unburied fish carcasses will not remain on the bottom of a lake unless water temperature remains cold, and carcasses in near-shore locations where water tends to warm may float and disperse, dropping skeletal elements as bacterial activity removes soft tissues (see also Schäfer 1962/1972). They suggest that measuring the distances between paired skeletal elements allows determination of transport direction. When a fish carcass has been scavenged, bones are randomly distributed and have no preferred orientations. Fluvial currents tend to disperse all elements of paired bones in one direction, and many elements display a preferred orientation. Such observations can (and should) be supplemented with sedimentological, contextual, and associational data (Chapter 10).

In a series of experiments Jones (1984, 1986, 1990) found that fish bones fed to canids may or may not be preserved after passing through the digestive tract (see also Casteel 1971; Lyon 1970). "The most common signs that bones have been eaten are the presence of tooth marks and crushed vertebrae. Another feature is surface erosion caused by acid solution of otoliths, vertebrae and other elements . . . Otoliths show erosion clearly. They have a glossy surface and have lost most of their surface relief features" (Jones 1990:143). Fish bone is variously destroyed by trampling, some bones surviving better than others. Finally, "fish bone loses much of its mechanical strength when boiled" (Jones 1990:144).

Cultural taphonomic factors

Wheeler and Jones (1989) and Colley (1990b) list multiple human taphonomic factors that can influence the kinds of fish remains found and the species of fish represented in archaeological contexts. These include fishing techniques (where in a body of water, technology used), butchering techniques, cooking and consumption practices, and discard practices. Jones (1984, 1986) shows that consumption and digestion of fish bone by rats, a dog, a pig, and a human result in the loss of some bones, fragmentation of some bones (see also Casteel 1971), and only a few (about 13%; Wheeler and Jones 1989:72) of the originally ingested bones were recognizable upon recovery from feces. Mastication variously results in fragmentation and distortion, and digestion results in corrosion of bones. Richter (1986) shows that heating fish bone to 60°C with either soft tissue scraped off or soft tissue adhering to the bone results in melted ends of collagen fibers. "At 80°C the melted areas along the fibrils had increased and short sticks of collagen with swollen ends were seen. Very few undamaged fibrils were present" (Richter 1986:479). Boiling removed all evidence of native collagen from fish bone. Richter (1986:480) concludes by noting that diagenetic processes may also remove traces of heat-modified collagen (e.g., Marchiafava *et al.* 1974).

Retrieval of fish remains is notoriously difficult because many fish bones are relatively small. While some fish bones are large enough not to require screening, "experiments have shown that smaller fish remains can easily be missed unless [sediment] is passed through minimum mesh sizes ranging between 0.5 and 2 mm" (Colley 1990b:208). Smith *et al.* (1988) suggest that one reason fossil fish faunas are less taxonomically rich than modern fish faunas is because of the difficulty of recovering the (small) remains of small taxa and it is precisely the small fish taxa that often make up much of the taxonomic richness of a fish fauna. Several techniques have been described for contending with the differential recovery of fish remains (Colley 1990b; Wheeler and Jones 1989).

Fish seem to die attritionally in typical situations, but can die in large numbers or catastrophically under some conditions (Schäfer 1962/1972:51; Weigelt 1927/1989:163). Van Neer and Muñiz (1992) report an ethnoarchaeo-logical case in which fishermen cleaned their nets of fish too small to warrant processing, and in so doing a "fish midden," a dense concentration of articulated fish skeletons piled upon one another and about 3×3 m in horizontal extent, was created. They note that such anthropogenic accumu-lations, if excavated from a prehistoric context, might be confused with mass mortality of fish trapped in a seasonally desiccated pool as both kinds of accumulation tend to have few associated artifacts. Van Neer and Muñiz (1992) suggest that these two kinds of accumulation may be distinguished by the fact that an anthropogenic accumulation will have a greater number of species represented in it than a natural accumulation.

Wheeler and Jones (1989:78) report that "natural agencies depositing fish bones rarely produce substantial concentrations of the kinds of fish preferred as human food." Butler (1987, 1990) has, however, documented just such a case. The seasonal spawning habits of anadromous Pacific salmon (*Oncorhyn-chus* spp.) often result in the seasonal accumulation of numerous remains of these fish in quite small areas along river banks and islands. Butler (1987, 1990) was specifically concerned with developing analytical techniques and criteria that would allow her to distinguish such naturally created concentrations of salmonid remains from culturally created concentrations as salmon were (and are still) a major food source for many human groups along the Pacific rim of northwestern North America. Toward that end she collected all salmonid remains from a small point bar in the middle of a river in western Washington state. She then studied those remains, comparing various attributes of this natural assemblage with attributes of a standard salmonid skeleton and attributes of several archaeological assemblages.

Butler (1987, 1990, 1993) demonstrates that both the degree of carcass completeness and the frequency of particular skeletal elements allow distinc-

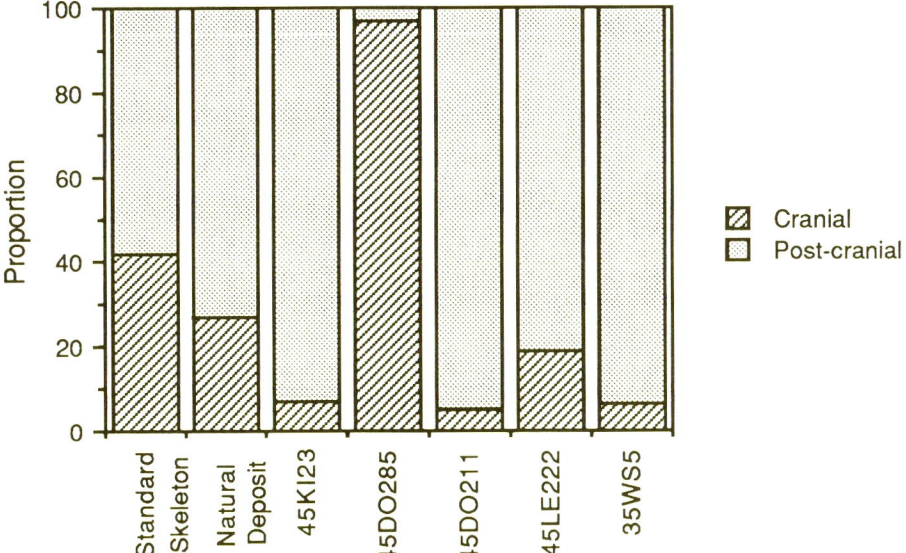

Figure 12.1. Proportional frequencies of salmonid cranial and post-cranial remains in a standard salmonid skeleton, a natural deposit of salmonid remains, and five archaeological sites (modified from Butler 1990:189, Figure 8.13).

tion between cultural and natural deposits of salmonid remains. Natural deposits contain relatively complete carcasses and thus about equal proportions of cranial and post-cranial elements. Cultural (archaeological) deposits contain relatively incomplete carcasses and post-cranial elements are much more abundant than cranial elements. She calls upon the structural density (see below) and size of bones as mediating natural taphonomic processes, and the distribution of edibile tissue, and butchery and disposal practices as influencing cultural taphonomic processes. Despite the fact that she examined thousands of fish bones from both natural and cultural deposits, she found no evidence of butchery marks, abrasion from fluvial processes, or gnawing marks of carnivorous scavengers, but she did find a nonrandom pattern of burning of skeletal elements suggesting cooking at one archaeological site. Bones less deeply buried in the soft tissues of the body tended to be more often burned than bones more deeply buried, suggesting cooking was of carcasses with the flesh attached to the skeleton.

Butler (1990, 1993) compared the frequencies of cranial and post-cranial elements between a standard skeleton, the natural deposit, and five archaeological deposits (Figure 12.1). The natural deposit varies in relative frequencies of cranial and post-cranial skeletal elements from a standard skeleton due, in part, to fluvial winnowing, weathering, and scavenging. Two of the archaeological deposits (45KI23, 45DO211) have relatively low frequencies of cranial parts due, Butler (1990:130) suggests, to "whole carcass deposition and *in situ*

Table 12.1 *Average skeletal completeness ratios for various sized horizontal units and sites (from Butler 1990)*

Assemblage	Spatial unit size			
	4 × 4 m	4 × 2 m	2 × 2 m	2 × 1 m
Natural Assemblage	0.65	0.68	0.71	0.63
45KI23	0.44	0.47	0.45	—
45DO211	—	—	—	0.64
45DO285	—	—	0.08	0.13
45LE222	0.85	0.66	0.72	—
35WS5	0.42	0.44	0.54	—

destruction." The archaeological deposit with high frequencies of cranial parts (45DO285) resulted from "almost exclusive deposition of salmon heads." The assemblage recovered from a rockshelter (45LE222), unlike the other archaeological assemblages, is in a non-riverine setting and appears to be similar to the natural deposit, leading Butler (1990:159) to suggest the ratio of cranial to post-cranial skeletal parts may be diagnostic of cultural and natural accumulations only in riverine settings, such as is the case with the other four archaeological collections. The final archaeological assemblage (35WS5) is the assemblage that prompted Butler's research, and she concludes that it appears to be a cultural deposit, although various problems with the sample available for her analysis preclude a firm conclusion.

Butler (1990:131, 1993) measured skeletal completeness by calculating the ratio of head to trunk MAUs with the equation:

MAU based on most frequent cranial element:MAU based on most frequent post-cranial element   [12.1]

with the largest of two values used as the denominator (Butler 1987, 1990:30). Ratios approaching one indicate essentially complete carcasses whereas ratios approaching zero suggest dissimilar proportions of front and rear ends. Butler (1990:131) predicted that fluvially (naturally) created deposits should have higher completeness ratios than cultural deposits, with the important qualification that "ratios would be larger at smaller spatial scales in natural sites than in cultural deposits." That is, natural deposits at virtually any spatial scale (in so far as the spatial unit is not significantly less than the average size of an individual fish) should have complete carcasses, while completeness ratios approaching one should be found more frequently at large than at small spatial scales in cultural deposits due to the butchery and bone-dispersal activities of humans. Given the limitations of the data available to her, Butler could not calculate completeness ratios at the same spatial scales for all assemblages. Basically, however, but not without exception, her prediction is born out (Table 12.1). Two archaeological assemblages have lower average completeness ratios than the natural assemblage, but two other archaeological assemblages have completeness ratios equal to or greater than the natural assemblage, prompting her to conclude that the completeness ratio is "not always a clear indicator of deposit origin" (Butler 1990:134).

Butler (1990:48–49, 1993) also assessed the potential taphonomic effects of skeletal element size, measured as the maximum length of a specimen. She distinguished five size classes by 1.5 cm intervals; size class 1 includes elements 0 to 1.5 cm in length, size class 2 includes elements 1.5 to 3 cm in length, and so forth. She found no correlation between the frequency of each size class as represented in her natural sample, and size class, concluding "small elements have about the same survivorship as large ones" (Butler 1990:69). Similarly, she found no statistically significant relation between these two variables in the archaeological assemblages, suggesting element size had not played a role in their taphonomic histories.

Butler (1990:46–47) measured the structural density for 51 elements of one skeleton, using the same photon densitometry techniques used by Lyman (1984a; Lyman *et al.* 1992a) and Kreutzer (1992) for mammals (Table 12.2; Stewart [1991] provides measures of bone density for three elements of each of several taxa, but how those measurements were derived is not clear). Butler's density values are, however, not the same as those available for ungulates as she did not divide the density value by bone thickness, thus her values are properly called "linear densities" (Lyman 1984a). She found a weak but statistically significant correlation between skeletal element frequencies in the natural deposit and bone density ($r_s = 0.308$; $0.05 > P > 0.02$ [Butler 1990:68]), suggesting to me the potential that some winnowing of less dense elements by fluvial processes had occurred. She also found statistically significant relations between bone density and skeletal part frequencies in two of four of the archaeological assemblages ($P < 0.001$), suggesting density-mediated survivorship influenced bone frequencies in those two assemblages.

Smith *et al.* (1988), like Butler (1987, 1990, 1993), call upon the reproductive behavior of fish as an important taphonomic factor, particularly the temporal correspondence of spawning migration and river level fluctuation. They found the probability that fish bones will be buried in floodplain sediments goes up as the abundance of fish increases (potential mortality is high, especially with taxa like anadromous salmonids that spawn in high frequencies and die afterward) and the rate of sedimentation increases. Predation of young might reduce the probability that their remains will survive to be deposited. Physiological variation between fish taxa may result in a high probability for some taxa to enter the fossil record and a low probability for other taxa. For example, during periods of drought, fish that need high levels of oxygen in the water are more apt to die than catfish which can survive water with little oxygen. Smith *et al.* (1988) suggest structurally denser bones are more prone to survive abrasion and acidic corrosion, and also tend to be the most taxonomically distinctive.

Table 12.2 *Structural density (g/cm²) of coho salmon (*Oncorhynchus kistuch*) skeletal elements (from Butler 1990)*

Element	Density	Element	Density
angular	0.063	basioccipital	0.023
basisphenoid	0.048	ceratohyal	0.040
dentary	0.042	ectopterygoid	0.027
epihyal	0.027	epiotic	0.067
exoccipital	0.142	frontal	0.044
hyomandibula	0.056	hypohyal, upper	0.049
hypohyal, lower	0.053	interopercle	0.024
lingual plate	0.056	maxilla	0.076
mesopterygoid	0.027	metapterygoid	0.016
opercle	0.036	opisthotic	0.017
otolith (sagitta)	0.072	palatine	0.040
parasphenoid	0.064	prefrontal	0.037
premaxilla	0.062	preopercle	0.038
prootic	0.077	pterosphenoid	0.057
pterotic	0.059	quadrate	0.084
sphenotic	0.054	subopercle	0.021
supraethmoid	0.023	supraoccipital	0.070
urohyal	0.062	vomer	0.053
cleithrum	0.037	coracoid	0.039
mesocoracoid	0.036	pectoral fin ray	0.036
postcleithrum[a]	0.020	posttemporal	0.048
scapula	0.060	supracleithrum	0.023
basipterygium	0.080	vertebra type 1	0.185
vertebra type 2	0.220	vertebra type 3	0.203
vertebra type 4[b]	0.128	caudal bony plate	0.033
hypural[c]	0.080		

Notes:

[a] middle element of series.

[b] average of vertebrae 4a and 4b.

[c] for H2.(Butler [1990:40; 1993:8] distinguished four basic types of vertebrae. From anterior to posterior these are: Type 1 is the first [anterior-most] vertebra in the column and has two prominent dorso-anterior facets for articulation with the exoccipitals. Type 2 lack parapophyses and fused neural and haemal spines, and have two distinct orifices on their dorsal and ventral surfaces. Type 3a has fused neural spines, type 3b has fused neural spines and abbreviated haemal spines, type 3c has fused neural and haemal spines. Type 4a has fused neural spines that are more robust than those of 3a, type 4b lacks neural and haemal spines and has a single orifice on both dorsal and ventral surfaces.)

Thus those taxa with dense bones will be more readily identified and more likely to be preserved in the fossil record.

Stewart (1989:79) was interested in developing criteria that would allow her to distinguish culturally from naturally deposited lacustrine fish-bone assemblages. She collected three assemblages from ground surfaces along the shoreline of Lake Turkana in Africa. Two were natural deposits given the designations PS1 (from an area 90 × 90 m) and PS2 (90 × 110 m), the third was a site where humans had roasted and eaten fish and was given the designation AS1 (11.4 × 9.5 m). The frequencies of collected fish bones per assemblage were 910, 360, and 651, respectively. Stewart examined four attributes of each assemblage: (1) frequency of bones per unit area; (2) skeletal part frequency; (3) taxonomic abundance; and (4) taxonomic diversity.

The frequency of bones per m^2 was higher in the cultural accumulation (6.01 bones per m^2) than in either of the two natural assemblages (PS1 = 0.11; PS2 = 0.03). The frequencies of skeletal parts were compared to their frequencies in a fish skeleton, and taxonomic abundances and diversity were compared to those values as reflected in the lake by living fishes. Stewart (1989:80–81) suggests that an average fish in Lake Turkana would consist of 65 (54.2%) cranial elements, 40 (33.3%) vertebral elements, and 15 (12.5%) non-vertebral post-cranial elements. The first value was reduced from the average of 102 cranial elements by omitting those elements Stewart deemed unlikely to survive because of their thin, delicate structure. As well, rays, ribs, and branchial elements were not included. Stewart's (1989:82) operating assumptions were that in natural assemblages, skeletal parts will occur in proportion to their natural occurrence in an average fish, and processing and consumption practices of humans will result in proportions of skeletal parts different from an average fish.

Stewart (1989) found that the ratio of cranial to vertebral elements was 1.6 for an average fish skeleton, 1.4 for the natural assemblages (PS1 and PS2), and 9.5 for the cultural assemblage (AS1). This results because the bone-accumulating habits of people lead to denser concentrations of fish bones and due to the multiple occupations of habitation sites (Stewart 1989:96). In the natural assemblages, non-vertebral post-cranial elements tend to be well represented relative to cranial elements and vertebral elements, and the latter tend to be rare. In the cultural assemblage cranial elements and non-vertebral post-cranial elements are relatively abundant and vertebrae are rare. On the basis of these observations, Stewart (1989:92) concludes that high proportions of cranial elements characterize cultural assemblages and distinguish them from natural ones. The ratio of cranial to vertebral elements is similar between the natural deposits and her average fish, and "is useful for characterizing the skeletal element composition of a naturally-deposited assemblage" because the ratio is lower for natural deposits than cultural ones (Stewart 1989:96).

Stewart (1989:97–98) concludes that natural deposits tend to reflect the natural fish population in terms of relative taxonomic abundances and taxonomic diversity, with the important caveat that fish taxa which seldom exceed 35 cm in length are rare and may be lost "in the fossilization process." Modern fishermen exploiting Lake Turkana catch fish "approximately in proportion to their present abundance." Using Stewart's (1989:94–95) data for the modern composition of Lake Turkana's fish fauna in terms of biomass, and her values for the biomass of fish represented by the cultural deposit (AS1) and the summed natural deposits (PS1 + PS2), if only taxa found in the AS1 assemblage are included, Spearman's rho between pairs of assemblages are:

modern catch	$r_s = 0.90$	
	$P = 0.07$	
cultural deposit	$r_s = 0.60$	0.30
	$P = 0.23$	0.56
	natural deposit	modern catch

These coefficients suggest the natural assemblage (PS1 + PS2) is similar to the modern fish fauna of the lake, and the cultural assemblage (AS1) is dissimilar to both the modern and natural assemblages. If only taxa found in the natural assemblage are used, coefficients are:

modern catch	$r_s = 0.86$	
	$P = 0.01$	
cultural deposit	$r_s = 0.68$	0.37
	$P = 0.01$	0.33
	natural deposit	modern catch

If all taxa for which Stewart calculated biomass are included, the correlation coefficients are:

modern catch	$r_s = 0.83$	
	$P = 0.008$	
cultural deposit	$r_s = 0.78$	0.50
	$P = 0.01$	0.11
	natural deposit	modern catch

These coefficients suggest that sample sizes are influencing the results in such a manner as to make the significance of these comparisons ambiguous. Species richness per assemblage, for example, is 15 for the modern sample of live fish from the lake, the summed natural assemblages have 9 taxa, and the cultural assemblage has 6 taxa. The coefficients increase due to ties in the rank of missing taxa, and the significance levels increase due to the increase in the number of categories being correlated. Thus the cultural assemblage is not correlated with the modern assemblage in any of the three comparisons, and is only significantly correlated ($P < 0.05$) with the natural assemblage when all taxa are included.

In a more extensive study that included additional natural deposits of fish remains, Stewart (1991) again focuses on establishing criteria distinctive of

Table 12.3 *Summary of criteria for distinguishing culturally from naturally deposited assemblages of fish remains around large lakes (from Stewart 1991)*

Naturally deposited assemblages	Culturally deposited assemblages

1. contain no small individuals
  a. small taxa and small individuals are absent (< 35 cm length of individual fish)
	1. contain "medium-sized" fish
	  a. size distributions of individual fish dependent on where fishing done and technology used to take fish

2. density-mediated attrition strongly reflected
  a. more strongly with increasing age
  b. bones of low structural density destroyed, taxa with only low density elements not represented
	2. density-mediated attrition weakly reflected, if at all

3. taxa living in near-shore zone most frequent	3. may be near-shore taxa, or others
4. many epaxial elements, but abundance relative to cranial and axial elements similar to relative abundances in an average fish	4. relatively few axial elements a. relatively more cranial and epaxial elements than axial elements due to differential butchery and transport

5. high taxonomic diversity
  a. slightly lower diversity than in living fish fauna
	5. low taxonomic diversity
	  a. notably lower than in living fish fauna

6. low density of fish remains per unit area[a]	6. high density of fish remains per unit area
7. more complete skulls	7. more fragmented skulls
8. no burning or butchering marks	8. some elements burned and some have butchering marks

Note:
[a] But see discussion here of Butler (1987, 1990), and McGrew (1975).

naturally and culturally deposited assemblages. Her results for fish faunas from large lake habitats are summarized in Table 12.3. I emphasize that the criteria deemed diagnostic by Stewart (1991) are founded on limited data; that is, the number of cases she (or anyone else, including Butler [1987, 1990]) has examined is not large, nor are these cases representative of the myriad possible variants. The data in Table 12.3 (along with those described by Butler [1987, 1990]) do, however, represent a major advancement in our knowledge about piscean taphonomy.

Both Butler and Stewart followed the lead of taphonomists focusing on mammals and explored attributes that would allow them to distinguish culturally from naturally deposited assemblages of fish bones. Their results are somewhat dissimilar as regards ratios of cranial to post-cranial elements, perhaps because Butler studied a riverine assemblage whereas Stewart examined a lake-shore assemblage or because the way the two of them defined "post-cranial" differs. They both have made significant contributions to our understanding of fish taphonomy that are in need of additional actualistic research for clarification and corroboration.

Avian taphonomy

> The limb bones are usually broken into either two or three pieces ... Many of the
> broken limb bones are charred at the broken end ... The broken state of the bones
> may indicate attempts at making whistles or other artifacts, or the manner of
> preparing the birds to eat.
> (H. Howard 1929:311, 384)

Interpretation of bird remains often mirrors that found in analyses of mammalian remains. There has been perhaps less taphonomic research on bird remains than on fish remains, although some important data generated by wildlife biologists are available. Perhaps there is a paucity of taphonomic research on bird remains because they tend to be rare relative to mammalian remains in most archaeological sites, but it is not clear what the overall typical relative abundances of fish and birds are in such sites. Bird bones are often found in sites, if in low abundance, and under exceptional preservational conditions such as dry cave deposits even feathers and quills have been found (Livingston 1988). As with most vertebrate groups, screening sediment results in the recovery of remains of small bird taxa that will not otherwise be found (Parker 1988).

Schäfer (1962/1972:41) reports that bird mortality can be catastrophic "at times of extreme weather conditions especially in the spring and autumn when many birds are migrating." When a bird dies and its carcass is on land, the soft tissues dry and shrink, thus the head and neck "bend backward and the tail with its feathers bend up" and the wings turn away from the body (Schäfer 1962/1972:46, 47), a phenomenon reported thirty-five years earlier by Weigelt (1927/1989:105–106). As bird carcasses deteriorate under natural conditions, the hindlimbs separate from the trunk, and then the pelvis from the lumbar vertebrae; bones of the forelimb and shoulder girdle along with the sternum "continue to hold together as a unit for a long time" (Schäfer 1962/1972:48). Hargrave (1970:59) suggests that during excavation of avian remains "care should be taken to collect some tracheal rings (windpipe), since [he] considers them to be proof that the bird was buried in the flesh. In like manner, the presence of calcified tendinal splints of the leg muscles of turkeys (*Meleagris gallopavo*) also indicate that the turkey was buried in the flesh." He is no doubt correct if these anatomical parts are in proper anatomical position. Hargrave (1970:59) also suggests that the presence of articulated wing bones, but "not the humerus ... could indicate that a wing was used as paraphernalia or for an ornament" by humans.

Rich (1980) suggests that variation in skeletal part frequencies she observed in a Tertiary deposit in South Africa was due to fluvial sorting and winnowing. She suggests coracoids, tarsometatarsi and tibiotarsi are abundant there due to their relatively high structural density, and small, elongate and slender elements, and those with high surface-to-volume ratios are rare due to their greater potential of fluvial transport. Radii, carpometacarpi, fibulae and furculae may be rare in Rich's sample due to their relatively greater fragility and the difficulty of identifying them if they are fragmentary (Livingston 1989:539).

Rich (1980) had virtually no actualistic data to inform her interpretations. Relevant actualistic research was performed by Bickart (1984), who documented the decay, disarticulation, damage, and fluvial transport of twenty-eight bird carcasses of three species. He found that unless protected in cages or rapidly buried, bird carcasses are quickly removed by scavengers. Undisturbed carcasses "became firmly stuck to the substrate within several days, possibly by a combination of body fluids and ground moisture" (Bickart 1984:527). "Stuck" carcasses were not subsequently moved by fluvial action except when "the most severe storm of the season" struck, and even then the bird bones were "little moved from their original positions" (Bickart 1984:528). Within a year about half of the bones of four, relatively unmoved carcasses were buried. Decay and disarticulation varied across individuals, taking about two weeks for some individuals and six months for others. Bickart (1984:528) reports that disarticulation took the following order: (1) ribs from sternum; (2) hindlimb joints; (3) vertebrae; (4) pectoral girdle (sternum-coracoid?); (5a) proximal wing joints (coracoid and scapula-humerus first; then humerus and radius-ulna, then coracoid-scapula); and (5b) distal wing joints (manus). Bird bones damaged by scavenging carnivores display "green fractures typical of fresh bone," and 84% of the recovered long bones had one or both articular ends removed (Bickart 1984:531). After one year of exposure, weathering damage was virtually non-existant, although the bones were in relatively moist, shaded settings so we should not expect them to be weathered (see Chapter 9).

Rosene and Lay (1963) report that of sixty bobwhite quail (*Colinus virginianus*) carcasses they placed in agricultural fields in Alabama and Texas, 25–50% disappeared completely within four days due to the activities of scavengers. Some individual quail disappeared completely within 24 hours, and after 30 days there were virtually no traces of any of the 60 quail. Balcomb (1986) placed carcasses of 78 songbirds (mainly passerines) in corn fields a few days after planting. After five days, 72 carcasses had been removed by scavengers, and from this and other data Balcomb (1986:819) concludes the average time a carcass survived was about 1.2 days, the rate of carcass disappearance was greatest during the first 24 hours, and over half the carcasses were totally removed without a trace (i.e., not even a feather remained). Tobin and Dolbeer (1990) found that bird carcasses disappear from cherry and apple orchards within 8–10.5 days, on average, due to scavenging activity. Ground cover did not affect the duration of carcass survival. Working with bird carcasses placed in cattail marshes, Linz *et al*. (1991) found that more carcasses were removed by scavengers as carcass density increased. As well, a greater proportion of carcasses was removed from shallow water (< 15 cm) depositional loci than from deep water (> 30 cm) loci.

Graham and Oliver (1986) studied a set of over 200 American coots (*Fulica*

americana) that froze into a lake. This produced a classic instance of catastrophic mortality. Coot carcasses were first scavenged by gulls (*Larus* sp.), but
while coot bones were exposed by this scavenging few were broken or
disarticulated. Heads did fall off, and many carcasses apparently disarticulated
into anterior (cervical vertebrae, wings, coracoid/sternum complex) and posterior (synsacrum, hindlimbs) halves. Later scavenging by mammals produced
more bone breakage and scattering, with "points of attack usually at the breast
or hindlimbs."

Ericson (1987) compared the ratios of anterior limb elements to posterior
limb elements in 54 assemblages, including two modern beach collections, one
paleontological collection and 51 archaeological collections. He calculated the
ratio as the number of wing elements (humerus, ulna, carpometacarpus)
divided by the total of wing and hindlimb elements (femur, tibiotarsus,
tarsometatarsus), times 100 to derive a ratio called the percentage of forelimb
elements statistic. Calculating the ratio for each of several taxa of birds,
Ericson (1987) found that in naturally deposited assemblages forelimb elements tend to be more abundant than hindlimb elements but in archaeological
assemblages hindlimbs tend to be more abundant than forelimb elements. He
interprets this to indicate humans were focusing their exploitation of avian
carcasses on the more meaty hindlimbs. This is different from a suggestion
made by another zooarchaeologist (cited in Bramwell *et al.* 1987), who suggests
that raptors would deposit many carpometacarpals, tarsometatarsals, and
coracoids (distal limb elements) relative to proximal limb elements, whereas
hominids would deposit many humeri, femora, and coracoids (proximal limb
elements) relative to distal limb elements of their avian prey. Bramwell *et al.*
(1987) present data that suggest the latter pattern of bone frequencies may
result from bone-depositing activities of the golden eagle (*Aquila chrysaetos*).
They also report this raptor damages the posterior portion of the sternum and
breaks the keel of the sternum of avian prey.

Livingston (1988, 1989) uses several lines of evidence and analytic techniques
to assess whether large assemblages of avian remains recovered from archaeological sites in Nevada represented culturally or naturally deposited materials.
One of those sites is a cave, in which the horizontal distribution and density of
avian remains suggest non-human accumulators of avian remains *may* deposit
more of those remains near the cave entrance and just outside the entrance and
deposit fewer remains inside the cave (Livingston 1988). She also notes that
many avian remains, including feather bundles, skins, and bones, have been
made into artifacts, suggesting people could have accumulated and deposited
some of the unmodified avian remains. Third, Livingston (1988) compares the
relative abundance of waterbirds and mammals from her cave site with the
relative abundances of those taxa in other cave and open sites in the region,
noting that her cave contains more remains of large waterbirds than other caves
but about the same relative frequencies as observed at several open sites in

marsh-side locations. The presence of feathers and avian bones and skin in human coprolites recovered from the cave, and the fact that local coyotes (*Canis latrans*, one potential non-human accumulator of the avian remains in the area) tend to focus their subsistence pursuits on lagomorphs, lead her to conclude at least some of the bird remains in the cave were accumulated and deposited by humans.

The avian remains from the cave site and a nearby open, marsh-side site exhibit minimal weathering and root etching (Livingston 1988). Livingston (1988, 1989) followed Ericson's (1987) lead and examined the relative frequencies of sets of skeletal elements. She found that for both her cave site and open marsh-side site avian forelimb and hindlimb elements occurred, on average across all taxa, in virtually identical frequencies. One individual bird taxon that displayed significantly low abundances of forelimb elements appeared, on the basis of other evidence, to have been culturally deposited, as Ericson's (1987) model predicts. But another bird taxon had relatively high but statistically insignificant abundances of forelimb elements; over one third of the remains of that taxon seem to have been deposited by natural mechanisms (raptors) on the basis of other evidence. That plus the fact that Rich's (1980) paleontological assemblage had a low abundance of forelimb elements lead Livingston (1989:542) to conclude that Ericson's (1987) percent of forelimb elements ratio "may not be a reliable indicator of human activity in the depositional process."

Taking a different analytic approach, Livingston (1988, 1989) compares the avian assemblages from her cave site and open marsh-side site and finds that relative taxonomic abundances are similar, suggesting to her that similar accumulational and depositional processes may be at work at both sites. She argues that the percent of forelimb elements statistic tends to be high for avian taxa in her samples that are strong fliers with more robust wing elements than avian taxa that wade, have trouble taking off from water, and have lightly built wing elements. Taxa which do not meet these expectations have either small and thus unreliable samples, or had wings that were used to manufacture decoys by the prehistoric occupants of the sites. She concludes that "if, in fact, there is an underlying property controlling element survivorship in avian remains, derived measures of survivorship based on element frequencies will also reflect that property" (Livingston 1989:545), and suggests structural density may be that underlying property and thus may vary with the functional anatomy of the taxon.

Unpublished research cited by Parker (1988:201) indicates that "microscopic examination" of bird skeletal parts can reveal whether or not the remains "were in fact eaten by humans," but he does not indicate the nature of the microscopic attributes. Many specimens in a large sample of bird bones from a site on the coast of Oregon that I examined display cut or butchering marks, indicating that birds in some prehistoric contexts were subjected to the same intensive level of butchery as the mammal remains with which they were associated (Lyman 1991a, 1992c, 1993a). Stallibrass (1990:158) reports unpublished experiments indicating domestic cats (*Felis cattus*) damage bird bones; two-month-old kittens can "consume whole long bones of juvenile chickens (*Gallus gallus*) and complete chicken carcasses can be consumed by adult cats."

Reptilian and amphibian taphonomy

If little is known about piscean and avian taphonomy, still less is known about reptilian (with the possible exception of dinosaurs; e.g., Dodson 1971; Sander 1992) and amphibian taphonomy. That may be because remains of these taxa are seldom found in abundance in archaeological sites, and/or because so few of these taxa have been studied actualistically. For example, Weigelt's (1927/ 1989) volume is still *the* classic one for learning about the more immediately postmortem effects of crocodilian taphonomic histories.

Meyer (1991) buried two marine (hawksbill) turtle (*Eretmochelys imbricata*) carapaces in intertidal sands of a lagoon. Decomposition of ligaments holding individual elements of the carapaces together allowed disarticulation and wave action dispersed individual elements within three weeks. Meyer (1991:92) found that orientations of the long axis of individual skeletal parts from other turtles primarily align perpendicular to the coast and secondarily parallel to it due to shore-parallel currents. Meyer (1991:92) reports that "direct predation by large vertebrates can be deduced from scratch marks on the shell." Meyer concludes that rapid burial is required if turtle carcasses are to be found in articulated condition.

Schneider and Everson (1989) summarize the biology of, and archaeological and ethnographic information concerning the desert tortoise (*Xerobates agassizii*) of southern California and adjacent Nevada and Arizona. They list behavioral traits and ecological predilections of this taxon, as well as the manners in which people utilized it, including for subsistence, ceremony, ritual, medicine, and technological and symbolic purposes. Schneider and Everson (1989:189) point out that distinguishing culturally from naturally deposited desert tortoise remains is difficult; they conclude burning is not a reliable indicator but "there is no question that ground, drilled, decorated, or otherwise modified specimens are an indication of cultural use." Their study is valuable not so much for the taphonomic information they provide, which is actually minimal (e.g., they do not describe any of the many ways people modify turtle shells), but rather for the fact that it illustrates that turtles at least, if not all reptiles, are subject to the same kinds of human taphonomic factors as mammals, birds, and fish.

Summary

Perhaps the most striking thing in non-mammalian vertebrate taphonomy is the fact that we know so little about it, although serious efforts are being made to learn about the taphonomy of fishes. Another very striking thing is that virtually all of the research being undertaken on the taphonomy of fish and birds by zooarchaeologists is following in the footsteps of taphonomists working with mammals. The attributes and variables the latter have studied, such as skeletal part frequencies and extent of disarticulation, are precisely the variables chosen for study by taphonomists working with non-mammalian vertebrates. While that is perhaps to be expected, and certainly is warranted, I wonder if such mimicking is limiting what we can learn about the taphonomy of these other kinds of organisms. The logical starting point for all taphonomic analysis (regardless of taxon) is the model of a complete, living organism. What I am thus asking is, should the different life habits of fish, which are aquatic, and birds, which are aerial and often arboreal, and mammals, which are variously terrestrial, arboreal, and aquatic, be taken into account in our taphonomic analyses? Certainly since the seminal work of Schäfer, Weigelt, and others, we should be at least thinking in these terms. Fish are more likely to be subjected to fluvial taphonomic processes than terrestrial animals. Similarly, alpine animals are likely to be subjected to fluvial transport and abrasion from sediment during fluvial transport whereas desert-dwelling animals are less likely to be subjected to such action. Knowing whether the faunal remains one is studying derive from arboreal, terrestrial, or aquatic taxa can thus be of major importance during taphonomic analysis. And, often these distinctions parallel taxonomic lines.

13

DISCUSSION AND CONCLUSIONS

> Bones are documents as are potsherds and demand the same scrupulous attention both on the site and in the laboratory.
> (M. Wheeler 1954:192)

Introduction

Modern taphonomic analysis is (sometimes excrutiatingly) detailed, it is extensive, and it is intensive. The number of variables the analyst should and perhaps must consider is large, and tends to increase as the complexity of an assemblage's taphonomic history increases. I noted earlier, for example, that assemblages representing one or a few individual organisms signifying one accumulational and depositional event often tend to be easier to interpret than long-term accumulations consisting of multiple taxa and multiple individuals. It seems that the latter kind of assemblage is more common than the former, therefore the taphonomist may typically be faced with a collection of vertebrate remains that had a complex taphonomic history that may not be analytically discernable.

"Taphonomic change is sequential" (Andres 1992:39). Because taphonomic processes are historical, they are cumulative (effects of some processes obscure or destroy effects of earlier processes) and in some cases, one process is dependent on a preceding one. For example, carnivore gnawing may obliterate butchering marks on bones, or a highly weathered bone is unlikely to be transported to the den of a scavenging carnivore. What a taphonomist studies is how a collection of fossils differs from the living skeletons of animals represented by those fossils, and if and how the population of skeletons differs from the natural biotic population of skeletons. For example, Andrews (1992:34) suggests "all taphonomic problems can be subsumed under the general question: what is the origin of the fossil assemblage, what are the [accumulational, depositional, preservational] processes involved, and how may these processes have changed the original composition of the [fossil] assemblage?"

To answer the question just posed, Andrews (1992:33) echoes earlier authors (e.g., Gifford 1981; Shipman 1981b) and suggests four criteria are necessary. First, taphonomic analysis should be problem oriented in order that, for example, we know which taphonomic characteristics of a fossil assemblage signify bias and which ones signify information relevant to the inferences we

452

wish to make. Second, relevant neotaphonomic or actualistic data (determined, in part, by the problem) must be available to help interpret taphonomic attributes of the fossils. Third, taphonomic attributes must be well defined so that they may be recorded and tallied. And fourth, taphonomic histories of fossil assemblages, with taphonomic events in proper sequence, must be written. The last – writing taphonomic histories – is, of course, what taphonomy is all about. What one does with that history, once it is written, depends on the kinds of inferences one wishes to make about the bone accumulators, the bone modifiers, or the paleoecological conditions indicated by the fossils.

Are the bones disarticulated? What is the demography of the represented organisms? Are complete skeletons present, or only certain parts of the skeleton? Do some bones have carnivore gnawing marks or butchering marks? Have some bones been weathered, burned, broken, or abraded? What are the orientations of the long bones? Have some bones been crushed by sediment overburden? These questions underscore Andrews' (1992) third criterion and summarize the gist of much of what has been said in preceding chapters. The question that remains concerns how an extensive and intensive taphonomic analysis might be performed. In previous chapters I present examples of analysis that have as their focus one, two, or (less often) three variables or modification attributes. How do we integrate these single variables into a multi-variate analysis?

Multi-variate taphonomic analysis

> Relatively unambitious taphonomic studies must assemble extensive multivariable descriptions, often in the form of dishearteningly large tables ... If archaeologists are to learn from taphonomy it is clear that they are going to have to handle descriptions in the hyperspace of many variables. [We need to determine] whether some observed variables consistently co-associate and others consistently avoid each other.
> (R. Wright 1990:260, 262)

> Taphonomy as a whole becomes more useful the more we think about processes beyond descriptive details. Narrative anecdotes and numerical or memorial data sheets are the necessary ingredients, but they make a meal only with conceptual recipes.
> (A. Seilacher 1992:123)

Gifford-Gonzalez (1991:217) underscores the significance of multi-variate taphonomic analysis with her distinction between the 1970s and 1980s research focus on identifying taphonomic agents and the recent emerging focus on "more complex analyses and more ampliative inferences about the life relations of prehistoric hominids." The former phase of taphonomic research has been quite successful in several respects, as evidenced by earlier chapters of this volume. The second and newer phase will be more difficult, Gifford-Gonzalez (1991) contends, for two reasons. One involves the search for appropriate levels and scales of analysis, and clarification of data categories. I suggest, as Gifford-Gonzalez (1991) does, that this problem is easily remedied by more interplay between ethnoarchaeological or neotaphonomic research and the analysis of prehistoric vertebrate faunas. Too often I read ethnoarchaeological research results that seem to have been written by someone who has never dealt with a collection of prehistoric faunal remains. Many of the published ethnoarchaeological data cannot be used as frames of reference for comparison with prehistoric materials because they are presented in a way that cannot be replicated in an archaeological context. For instance, Yellen (1991b:172) reports that "even a few days of continued site occupation after discard [of bones and bone fragments] will serve to protect the fragments from predators and increase the likelihood of long-term survival." Does this mean that every time good bone preservation is found bone discard took place a week or so prior to site abandonment? Certainly not, but to use this ethnoarchaeological observation we would need to be able to determine the relation between the time of bone deposition and site abandonment, a relation that surely is beyond the reach of modern archaeological analysis.

According to Gifford-Gonzalez (1991:218), the second reason the newly emergent phase of more complex analyses and attempts to derive inferences of hominid behaviors will be more difficult than its antecedant agent-identification phase is that the desired inferences involve "causality in biological systems ... because hominids and the contexts in which they exist are biological entities." Therefore, she turns to how paleobiologists and evolutionists deal with causality, and concludes that while there are no immutable laws of taphonomy that are not obvious or trivial (but see Gould 1986), most inferences these researchers derive are phrased in probabilistic terms; "explanation becomes a probabilistic account in which certain tactics may be employed to reduce uncertainty" (Gifford-Gonzalez 1991:241). The "certain tactics" involve "independent lines of evidence, derived from distinct systems of causation, mobilized to challenge and/or support one another, [and lead] to more strongly warranted inferences regarding the past life relations that produced certain configurations of [zooarchaeological and taphonomic trace] material in our sites" (Gifford-Gonzalez 1991:243). But, Gifford-Gonzalez (1991:243–245) cautions that we should not be misled into thinking that these tactics simply involve more agent-identification research. Rather, the "functional linkages" in the observed patterns of bone modification attributes need to be assessed ethnoarchaeologically and experimentally, and then "relational analogies" (see Chapter 3) between them and prehistoric cases are constructed (Gifford-Gonzalez 1991:245). As noted in earlier chapters (see especially the first two examples of taphonomic analysis in Chapter 3 and the "Butchering, breakage, and bone tools" section of Chapter 8), some of what I think is the best taphonomic research performed to date precisely follows Gifford-Gonzalez's (1991) recommended research program.

I liken modern taphonomic analysis to a huge analysis of variance, or ANOVA, in statistics. Do not be terrified by the metaphor, nor be misled into thinking taphonomic analysis necessarily involves statistics or even interval-scale measurements. The similarity of taphonomic analysis with ANOVA I note here is meant to underscore the fact that, like ANOVA, taphonomic analysis today involves the study of many samples of many variables, and comparisons between the means and variances of (taphonomic traces or) attributes displayed by faunal remains. ANOVA is meant to determine the statistical significance of variation displayed within and between groups (assemblages) of one or more variables measured in a set of multiple samples. And, simply put, that is what taphonomic analysis involves too, except the taphonomist is concerned with determining the taphonomic (rather than statistical) significance of the observed variation for the purpose of writing taphonomic histories, stripping away the taphonomic overprint, or ascertaining something about paleoecology from the taphonomic attributes (e.g., Wilson 1988).

In order to make the preceding argument clear, consider the basic analytic procedure followed during taphonomic analysis of a collection of vertebrate remains. First, we record various attributes displayed by the *individual specimens*. Is a bone gnawed, how weathered is it, is it burned, what was its orientation in the ground, is it an incomplete skeletal element (is it broken, and if so, how)? Generally, each variable that might be recorded has more than one possible value that it can display on a specimen: gnawed/not gnawed, weathering stage 0, 1, 2, 3, 4, or 5; etc. (For discussion, I ignore the fact that a specimen may display more than one possible value for a given variable, such as the skyward surface of a bone displaying weathering stage 3 whereas the groundward surface of that bone displays weathering stage 1.) And, each specimen may display one or more attributes of modification. A single specimen may be broken, butchery marked, and weathered, for example. The point here is, to paraphrase Gifford (1981:385–386), the condition of any individual fossil specimen is the end product of a series of taphonomic events whereas a fossil assemblage is an aggregate of individual end products. Regularities in taphonomic modifications displayed in specimen after specimen may be reflected in the assemblage, or regularities in proportions of specimens displaying a particular kind of modification may be found in assemblage after assemblage. These regularities are often found by inductive pattern recognition (Binford 1984b:9–15). Ascribing taphonomic meaning to the recognized patterns typically involves matching them with actualistically observed patterns.

The preceding paragraph is accurate but quite general and simplistic. If we add details, the magnitude of the complexity of the issue – the ANOVA-like character of multi-variate taphonomic analysis – becomes clear. Given that we record the taphonomic modification attributes displayed by each individual specimen, it should be obvious that any set of taphonomic data represents a palimpsest, even if the assemblage is one specimen displaying two or more attributes of modification (e.g., a butchering mark and an advanced weathering stage). For example, my typing the letter "F" in the first word of this sentence may be considered a rather fine-grained event at one scale, but at a finer scale I had to move one finger to the "shift" key, push that key down, hold it down, move a second finger to the "f" key on the keyboard, strike the "f" key, then release the "shift" key before typing the second letter of the sentence. The point here is that taphonomists must always be cognizant of the scale at which they conceive of fine-grained and coarse-grained or palimpsest assemblages of fossils and attributes of taphonomic modification.

Generally, archaeological taphonomists want (a) to identify the bone-accumulating agent, in order to assess differences between what is present in the assemblage and what might have been present (given known or suspected behaviors of the accumulator), and/or (b) to identify the taphonomic agents that modified the bone assemblage in order to assess differences between what was available for accumulation, what was deposited, what was buried, and what was collected, and/or (c) to tease out the human behavioral or paleoecological significance of modified faunal remains. The first two desires are interrelated because identified agents of bone modification are often taken also to be agents of bone accumulation. In fulfilling any or all three of these analytical desires, taphonomists deal with multiple attributes (e.g., Table 2.1). Because individual bone specimens may be modified by more than one taphonomic agent and process and thus may display more than one attribute of modification, it is important to note which specimens are, for example, both burned and butchery marked, or both weathered and broken. At its best, then, the potential modification attributes should be arranged paradigmatically in order that the analyst can tally how many humeri of a particular taxon are, for instance, butchery marked, burned, and broken, and how many femora of that taxon are not butchery marked, not burned, and complete; that is, to monitor covariation of the different taphonomic attributes.

A paradigmatic classification consists of *dimensions* which are sets of mutually exclusive variables. Each dimension consists of several attribute states or the values that a particular dimension or variable may display. For example, the dimension "burning" may have the attribute states "unburned, charred, calcined" and the dimension "weathering" could have the six weathering stages as its attribute states. Many of the dimensions of the paradigmatic classification along with some of the possible attribute states for each dimension are given in Table 13.1, and are variously supplemented by figures and tables cited therein. This classification can and should be modified for particular assemblages (e.g., if no burned specimens are found, then that dimension can be omitted). This classification allows the analyst to identify each specimen as belonging to a particular class, where each *class* is defined by intersection of the dimensions. The classification is in that regard, truly

Table 13.1 *Dimensions and attribute states for taphonomic analysis*

Dimension
Attribute states per dimension (partial)

Specimen attributes

a. taxon
    1. family, genus, species
    2. animal live weight size
    3. ethology and ecology of taxon
b. skeletal part (e.g., Table 4.1)
c. context/association
    1) horizontal and vertical position (e.g., depth from surface)
    2) depositional unit attributes (e.g., sediment texture, chemistry)
d. anatomical distribution of damage or attribute
    1) upward (skyward) or downward (groundward) surface
    2) anterior, posterior
    3) lateral, medial
    4) proximal, distal
e. weathering stage (Table 9.1)
f. gnawing damage: presence/absence
    1) carnivore
    2) rodent
    3) insect
    4) some combination of 1, 2, and/or 3
g. polish/abrasion/corrosion
h. fracture type (e.g., Figure 8.4)
i. completeness
    1) fractured or complete
    2) articulated with other bones *in situ*
    3) refits and conjoins
j. butchery marks (type, orientation, location, frequency or density per area)
k. azimuth/orientation (e.g., Figure 6.7)
l. plunge/dip (e.g., Figure 6.8)
m. burned/not burned

Assemblage attributes

n. frequencies (absolute and proportional) of b, d, e, f, etc., above
o. mortality patterns (e.g., Figure 5.1)
p. skeletal part frequencies
q. density per unit space

unwieldy. There are, for instance, 13 dimensions listed under "specimen attributes" in Table 13.1. There could be more dimensions, but let us assume for sake of discussion that this is a complete list. Let us further assume that each dimension has four attribute states; there could be fewer or there could be more for any given dimension. The 13 dimensions with four attribute states each means there are 4^{13}, or 67,108,864 possible classes of bones displaying various combinations of taphonomic traces! I know of no practicing taphonomist who records attributes of bone modification in this fashion, although an increasing number are beginning to approach this kind of system of data recording (i.e., more than two or three attributes per specimen are recorded). What Table 13.1 does, however, is underscore the multi-variate nature of modern taphonomic research, and the potential complexity of that research (see Stiner 1992; White 1992 for recent examples which approximate this multi-variate complexity).

Several questions are begged by Table 13.1. Which dimensions or attributes of bone modification are relevant or applicable to which questions? That is, how do we ensure concordance between the data we record, the analyses we perform, and the questions to which we seek answers? The answers reside in actualistic or neotaphonomic research: "[Taphonomic] agencies have been studied in modern contexts for the diagnostic patterning that each produces in the composition and character of bone accumulations. The diagnostic pattern-ing is in effect a signature of great interpretive value, because it is possible to identify the same patterning in ancient bone assemblages and to infer equiva-lent causative agencies" (Bunn and Kroll 1986:432). But as should be obvious from preceding chapters, there is a plethora of patterns that might be identified for any given variable that might be monitored, and there is also a plethora of variables. Given the facts that (a) a taphonomic history is cumulative in the sense that agents and processes that work late in the history may be mediated or influenced by agents and processes that work early in the history, and vice versa (e.g., obliteration of traces formed early by traces that form late), and (b) some of the variables are amenable to interval-scale measurement, others are amenable to only ordinal-scale measurement, and still others can only be measured in nominal-scale terms, what is the modern taphonomist to do?

Behrensmeyer (1991:317, 321) proposes a graphic solution to the question just posed (a similar solution for invertebrate taphonomy is illustrated by Parsons and Brett 1990). An example of her graphic technique is given in Figure 13.1, and the variables included there are defined and the plotted values are given in Table 13.2 for two fictional assemblages of vertebrate remains. This graphic technique "provides a visual overview [of taphonomic attributes displayed by an assemblage] that can be used as a basis for standardized comparisons among different accumulations," and "helps to summarize the complex array of taphonomic information available in each bone accumu-lation" (Behrensmeyer 1991:315). The graphs accomplish two goals: "(1) they show the behavior of different taphonomic variables in relation to one another within each assemblage, and (2) they encourage standardized representation of data, allowing comparative analysis of different assemblages" (Behrensmeyer 1991:315). Note, for example, that it is possible to determine by brief study of Figure 13.1 that Assemblage 1 is smaller, less taxonomically rich, more articulated, and has fewer skulls, fewer whole bones, fewer weathered bones, fewer burned bones, and more carnivore-gnawed bones than Assemblage 2.

Graphs like that in Figure 13.1 might be generated for any assemblage of

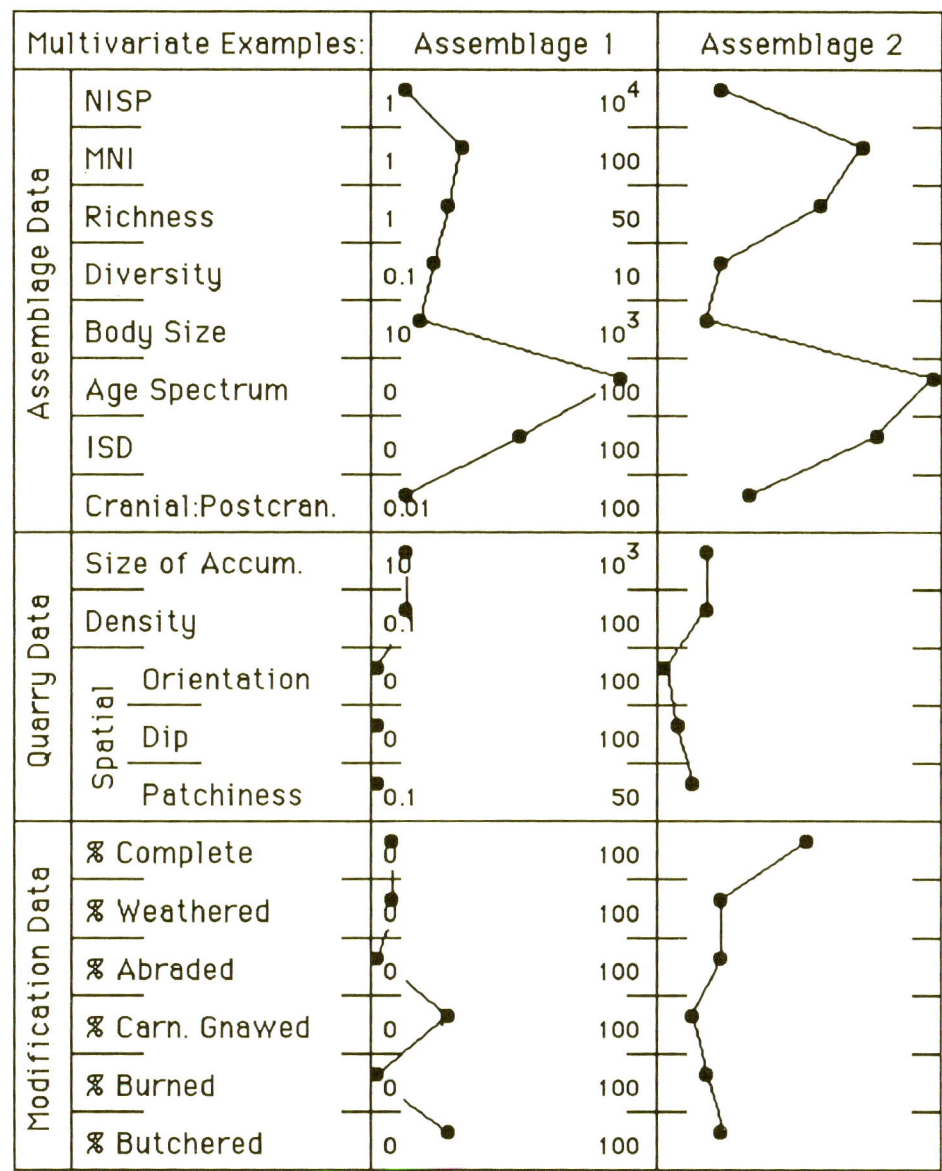

Figure 13.1. Example of graphic technique for summarizing and comparing taphonomic data for multiple assemblages (after Behrensmeyer 1991:317, Figure 2).

Table 13.2 *Definition of variables and listing of values per plotted variable in Figure 13.1*

Variable	Definition	Plotted values per assemblage 1	2
Assemblage data			
NISP	number of identified specimens	1000	2000
MNI	minimum number of individuals	30	70
Richness	number of taxa identified	15	27
Diversity	taxonomic diversity	1.5	1.7
Body size	body size (kg) of organisms with modal MNI	100	100
Age spectrum	% of MNI that are adults (skeletally mature)	85	100
ISD	index of skeletal disjunction (see Chapter 5)	50	75
Cranial:Postcran.	ratio of MNE of crania to MNE of post-cranial bones	0.1	25
Quarry data			
Size of accum.	area (m²) over which remains are recovered	100	150
Density	average NISP per m²	10	13.3
Orientation	% of long bones with similar orientation	0	0
Dip	% of long bones deviating $\geq 5°$ from horizontal	2	5
Patchiness	standard deviation of density	0.1	5
Modification data			
% complete	% of specimens that are whole skeletal elements	5	50
% weathered	% of specimens that display $\geq$ stage 2	5	20
% abraded	% of specimens that display abrasion	0	20
% carn. gnawed	% of specimens that display carnivore gnawing	25	10
% burned	% of specimens that have been burned	5	15
% butchered	% of specimens that have been butchered	25	20

vertebrate faunal remains. The key to comparative analysis of multiple assemblages is, of course, that the scales used for plotting the observed variables be identical for the compared assemblages. One of the advantages to such a graphic technique is that virtually any suite of variables might be plotted. For example, I have not discussed taxonomic richness or diversity in this volume, but these might be important variables for helping the analyst understand the taphonomic history of an assemblage. Nor have I discussed the "patchiness" of bone distributions (e.g., range and average of frequency of faunal remains per unit of space) but this variable also may be important. As well, the analyst might choose to plot the ratio of axial NISP to appendicular NISP rather than the ratio of cranial NISP to post-cranial NISP, and to omit the proportion of burned specimens and add the proportion of rodent-gnawed specimens. The set of African bovid size classes (e.g., Bunn and Kroll 1986; Klein 1989) could be used to develop a different scale for plotting the live weight size of taxa in the assemblage. One could include plots of the frequencies of specimens that are both burned and display butchering marks. The graphic technique is thus quite flexible in terms of the information that it might include or exclude. Comparison of a graph of the attributes of a neotaphonomic assemblage with a known taphonomic history with a graph of a prehistoric assemblage could help determine the taphonomic history of the latter, and may help the analyst ascertain which taphonomic traces are the result of diagenetic processes and which are the result of biostratinomic processes.

Graphs like that in Figure 13.1 necessarily simplify reality. They plot central tendencies (e.g., modes and averages) or proportions, and omit variation (e.g., ranges and standard deviations). My impression from the literature is that as more actualistic research is completed and reported, more and more variation is being documented (e.g., Lam 1992). That variation might eventually become equal in importance to the measures of central tendency plotted in Figure 13.1 and used by virtually all taphonomists who perform comparative analyses (e.g., Stiner 1992). Another bit of information that is not readily included in such graphs is the relation of some of the assemblage data and some of the modification data. It has been demonstrated time and again, for instance, that the taxonomic richness and diversity of an assemblage tend to correlate with the size of that assemblage measured as NISP (e.g., Grayson 1984) and measured as the size of the excavation (e.g., Lyman 1991a; Wolff 1975). Attempts to circumvent this problem have not yet been successful (see the discussion in Cruz-Uribe [1988] and Meltzer *et al.* [1992]). Care is called for then, when comparing assemblages of rather different size. There are several statistical techniques available for determining if the differences in proportions of modified specimens is a function of sample size, or if they validly measure differences in taphonomic histories (for example, see the discussion of the butchery mark data in Table 8.4).

Behrensmeyer's (1991; Figure 13.1) graphic solution is an intriguing one that can use nominal, ordinal, and interval scale data. Stiner (1992:436) uses a similar graphic technique for comparing eleven attributes displayed by 28 assemblages of bones. Her technique can employ only nominal and ordinal scale data, but accomplishes the same function as Behrensmeyer's. Stiner's technique, however, is noteworthy because she arranges the attributes she recorded into three categories: those only attributable to hominids, those only attributable to carnivores, and those potentially attributable to both hominids and carnivores. I have little doubt that this characterisitic of Stiner's graphic technique can be incorporated into Behrensmeyer's technique to provide summary illustrations of the taphonomic data recorded for archaeological assemblages.

The list of taphonomic variables in Table 13.1 is extensive, but it is incomplete. All of those variables could be structured so as to be included in graphs like that in Figure 13.1. Once these graphs are generated, they can be compared. To return to the analogy between multi-variate taphonomic analysis and ANOVA, each graph might be considered an ANOVA table (with the notable exception mentioned above that the graphs plot measures of central tendency rather than variation and it is the latter that is measured in ANOVA). Two questions must be answered at this point in the discussion. First, how do we *solve* the graph as certainly it is not an ANOVA table which can be solved and a summary statistic generated? And second, once the graph is solved, presuming it can be solved, what does the solution mean? Behrensmeyer (1991:318–319) argues that a solution is derived by comparison of graphs for prehistoric assemblages with graphs for neotaphonomic assemblages. Similar graphs suggest similar taphonomic histories. For example, most assemblages with bones displaying preferential orientations, abraded surfaces and rounded fracture edges, and a paucity of bones with low structural densities, have been fluvially transported (Behrensmeyer 1991:318). Thus, we can simultaneously *solve* the graphs and determine what the solution means in terms of taphonomic histories by comparison with actualistic assemblages and argument by analogy that similar *combinations* of traces or attributes of bone modification are indicative of similar taphonomic histories. This is, of course, the solution offered by Gifford-Gonzalez (1991) noted above, except that her solution explicitly considers the covariation of variables. The importance of this covariation cannot be overemphasized. For example, Yellen (1991b:186) suggests that "long bones broken during butchering lose their attraction [to carnivores] more rapidly than their complete counterparts and under certain circumstances are more likely to be preserved for potential archaeological discovery. Broken bones subjected to extended boiling are more completely rendered of marrow [and grease] than those which are not and consequently are of less interest to carnivores." There may be a simple rule here that broken bones will survive carnivore attrition better than unbroken bones (see also Blumenschine 1988; Marean *et al.* 1992), but what is important is the *covariation* of carnivore scavenging of bones and the presence of exploitable nutrients in bones: that covariation explains *why* the bones are or are not exploited by the scavenging carnivores (see below). To apply the rule in prehistoric settings the taphonomist must determine if the degreasing of the bones occurred prior to deposition (due, perhaps, to cooking and boiling) or after deposition (due to subaerial weathering or diagenetic processes such as leaching). The fact that the bones are broken (or not) and carnivore gnawed (or not) are only *two parts* of solving a multi-variate problem.

I suspect multi-variate analyses of taphonomic data will become more frequent and the number of variables included in these analyses will also increase as we learn more from actualistic research and find that two or three or four variables are simply not enough to unravel and write the complex taphonomic histories and infer the human behavioral or paleoecological significance of the coarse-grained zooarchaeological collections that many of us study. Solutions will be found using the comparative graphic technique only as the number and complexity of actualistic cases that are reported increase. If the majority of ethno-zooarchaeological research continues to focus on only two or three archaeologically visible variables, and to ignore the reality of the fossil record, we may find that we are recording much taphonomic data for prehistoric assemblages but that we do not know what those data signify in terms of the taphonomic history, human behavioral, and paleoecological significance of an assemblage. This admittedly personal observation prompts me to raise a final issue. Can we perhaps construct a general theory of taphonomy that might help us in our work with neotaphonomic and prehistoric assemblages alike? Such a theory would perhaps aid in the selection of variables for study, the variables that should covary in significant fashions, and the analytic techniques for manipulating the observations we make.

A general theory of taphonomy?

> Taphonomy is a science about the unlikely: the survival of organic materials and forms in spite of their general drive towards recycling by biological, physical and chemical degradation.
> (A. Seilacher 1992:109)

In an earlier paper I (Lyman 1987e) offered what I thought at the time was a reasonable outline of what might eventually, with some modification, constitute a general theory of taphonomy. Much of that outline is presented in the discussion of taphonomic histories ("On the structure of taphonomy: a personal view") in Chapter 2. With the benefit of hindsight I now perceive that discussion as constituting not a statement of theory, rather it is a description, in very general terms, of what taphonomic histories look like. Thus it is reasonable to wonder if a general theory of taphonomy, similar to the general theory of biological evolution, can be written.

Taphonomic research, as we have seen throughout this volume, focuses on the physical, chemical, and mechanical differences between fossil remains of organisms and what those organisms and their constituent parts looked like when the organism was alive. We can write laws (if this is too strong an epistemological term, substitute "rules") concerning cause–effect relations of various properties of an organism's tissues and how those tissues will respond to a particular taphonomic process. For example, I doubt that any vertebrate taphonomist would disagree that, given a set of skeletal elements representing a complete bovid skeleton, if those skeletal elements are subjected to gnawing by carnivores then the least structurally dense bones will be destroyed (consumed) first and the bones of greatest structural density will have the greatest chance to survive. This statement might be thought of as a taphonomic law or rule. There are no doubt other similar laws or rules concerning cause–effect relations; if a particular bone is subjected to a particular taphonomic process then a particular end-product will result. These are all laws concerning mechanical, chemical, and physical *properties* of vertebrate hard parts; they give modern taphonomy its decidedly "middle-range" flavor (e.g., Binford 1981b; Dunnell

1992; Simms 1992). The laws taphonomists tend to use (a) are *ahistorical* and produce little more than what might be termed a "physics of taphonomy" (after Dunnell 1992:212) and (b) do little to *explain* the fossil record (after Simms 1992:190–191) in the sense of helping us understand *why* taphonomic processes occur in the first place or *why* taphonomic processes operate the way and in the temporal order that they do. The former may be self-evident and the latter will probably be perceived as heresy in some quarters. Let me elaborate, then, on why I hold the view that taphonomy, as yet, does not seem to be explanatory in some ultimate sense.

We have come far in the past two decades; we can study a set of bones and posit a reasonable set of taphonomic agents and processes that acted on them based on the attributes they display (Gifford-Gonzalez 1991). We can say much about *what* happened to an assemblage of vertebrate remains and *how* it happened; I view these as *proximate causal explanations* of the fossil record. We can even, in some cases, write fairly detailed taphonomic histories although we may often be thwarted in our efforts to put the precise chronological sequence of taphonomic processes together, such as is found in the present debate over the superposition of carnivore gnawing marks and hominid-generated butchering marks on some Plio-Pleistocene aged bovid bones from Africa (e.g., Bunn and Kroll 1986, 1988; Potts 1988; Shipman 1986a, 1986b). Building the chronological sequence of taphonomic events a bone collection experienced is rather like the construction of a hominid phylogenetic lineage for the last 3–5 million years. Construction of the phylogenetic lineage is founded on principles of neo-Darwinian evolutionary theory, which unifies concepts such as natural selection, adaptation, inheritance, primitive characters, derived characters, and the like. A phylogenetic lineage arranges a series of fossils into a genetically related, time-transgressive series on the basis of evolutionary theory, the *ultimate causal explanation* for that arrangement. The taphonomist today has no such unifying theory and thus too often writes their particularistic taphonomic history with, it seems, little concern for the reasons why that history occurred in the first place and why that particular history has the particular form or chronology of events it does. I suspect this is so because we lack a coherent theory of taphonomy and thus we tend to focus on the what and how rather than the why of taphonomy.

While this is not the place to delve deeply into this issue, the reader certainly deserves an answer to the question: where do we find a starting point or frame of reference with which to build a general theory of taphonomy? I suggest basic ecological theory is a good place to start. As we have seen throughout this volume, many taphonomic processes concern the mechanical, chemical, and physical breakdown of animal carcasses and skeletal tissues. Animal carcasses are simply one stage in the biogeochemical cycle of matter and energy. A *biogeochemical cycle* can be defined as the "more or less circular paths chemical elements follow from the environment to organisms and back to the environment" (Odum 1971:86). After the death of a vertebrate, its constituent parts are "recycled" back into the environment, and it is that recycling that taphonomic research monitors. As Behrensmeyer and Hook (1992:90) note, "dead organisms are dynamic components of the ecological web that supports living plants and animals. Fossils are possible only when some of the organic remains are not recycled." Earlier in this volume, for instance, I suggested that burned bones may not be gnawed by carnivores because the organic chemical nutrients in bones that carnivores normally exploit – the marrow and grease – have already been cycled back into the environment by their combustion. It is, then, the particular biogeochemical path followed by those within-bone nutrients that results in their not following the different path of being consumed and metabolized into energy within another organism. Similarly, when bones are transported by geological processes, they are variously weathered and abraded as sedimentary particles the constituent parts of which become part of the potential geological reservoir of nutrients for feeding plants and animals. The recycling of chemical elements from the environment to organisms can be found in the feeding of organisms on animal carcasses, and the mechanical and chemical breakdown of those carcasses into more basic nutrients prior to their consumption.

The reader might protest that I have not truly suggested a way to shift our perspective from the middle range of chemical, physical, and mechanical laws underpinning a physics of taphonomy. The reader might well be correct in such a protest. I would nonetheless argue that what taphonomists are in fact studying are portions of such ecologically founded (and ecologically explainable) biogeochemical cycles. And that is, in fact, what underpins either studies that focus on the biasing aspects of taphonomic histories with regards to particular research questions, or studies that utilize taphonomic data to gain insights to the paleoecology of the past, whether or not that paleoecology concerns hominids. As we move into the twenty-first century, perhaps my suggestion regarding a general theory of taphonomy will be debated. Whether it is or not, given the present state of affairs in paleobiology and archaeology, I doubt that the necrology (and taphonomy) of taphonomy will ever be a subject pursued by historians of paleontology or zooarchaeology.

BIBLIOGRAPHY

Abler, W. L. 1984. A three-dimensional map of a paleontological quarry. *University of Wyoming Contributions to Geology* 23:9–14. Laramie.

1985. Skulls in fossil material: one mechanism contributing to their rarity. *Journal of Paleontology* 59:249–250.

Agenbroad, L. D. 1984. Hot Springs, South Dakota: entrapment and taphonomy of Columbian Mammoth. In (P. S. Martin and R. G. Klein, eds.) *Quaternary extinctions: a prehistoric revolution*, pp. 113–127. Tucson: University of Arizona Press.

1989. Spiral fractured mammoth bone from nonhuman taphonomic processes at Hot Springs Mammoth Site. In (R. Bonnichsen and M. H. Sorg, eds.) *Bone modification*, pp. 139–147. Orono: University of Maine Center for the Study of the First Americans.

1990. The mammoth population of the Hot Springs Site and associated megafauna. In (L. D. Agenbroad, J. I. Mead, and L. W. Nelson, eds.) *Megafauna and man: discovery of America's heartland*, pp. 32–39. Hot Springs: The Mammoth Site of Hot Springs, South Dakota, Inc., Scientific Papers Vol. 1.

Alexander, R. McN. 1985. Body support, scaling and allometry. In (M. Hildebrand, D. M. Bramble, K. F. Liem, and D. B. Wake, eds.) *Functional vertebrate anatomy*, pp. 26–37. Cambridge: Belknap Press.

Allison, P. A. 1990. Decay processes. In (D. E. G. Briggs and P. R. Crowther, eds.) *Palaeobiology: a synthesis*, pp. 213–216. Oxford: Blackwell Scientific Publications.

1991. Taphonomy has come of age! *Palaios* 6:345–346.

Allison, P. A. and Briggs, D. E. G. 1991a. Taphonomy of nonmineralized tissues. In (P. A. Allison and D. E. G. Briggs, eds.) *Taphonomy: releasing the data locked in the fossil record*, pp. 25–70. Topics in Geobiology Vol. 9. New York: Plenum Press.

1991b. (eds.) *Taphonomy: releasing the data locked in the fossil record.* Topics in Geobiology Vol. 9. New York: Plenum Press.

Anderson, K. M. 1969. Ethnographic analogy and archeological interpretation. *Science* 163: 133–138.

Andrews, P. 1990. *Owls, caves and fossils.* Chicago: University of Chicago Press.

1992. The basis for taphonomic research on vertebrate fossils. In (S. Fernández López, ed.) *Conferencias de la reunion de tafonomia y fosilizacion*, pp. 33–43. Madrid: Editorial Complutense.

Andrews, P. and Cook, J. 1985. Natural modifications to bones in a temperate setting. *Man* 20: 675–691.

Andrews, P. and Evans, E. N. 1983. Small mammal bone accumulations produced by mammalian carnivores. *Paleobiology* 9: 289–307.

Antia, D. D. J. 1979. Bone-beds: a review of their classification, occurrence, genesis, diagenesis, geochemistry, palaeoecology, weathering, and microbiotas. *Mercian Geologist* 7:93–174.

Ascenzi, A. 1969. Microscopy and prehistoric bone. In (D. Brothwell and E. Higgs, eds.) *Science in archaeology*, pp. 526–538. London: Thames and Hudson.

Ascher, R. 1961a. Analogy in archaeological interpretation. *Southwestern Journal of Anthropology* 17: 317–325.

466

1961b. Experimental archeology. *American Anthropologist* 63: 793–816.

1968. Time's arrow and the archaeology of a contemporary community. In (K. C. Chang, ed.) *Settlement archaeology*, pp. 43–52. Palo Alto: National Press Books.

Ascher, R. and Ascher, M. 1965. Recognizing the emergence of man. *Science* 147:243–250.

Avebury, Lord (John Lubbock). 1913. *Prehistoric times as illustrated by ancient remains and the manners and customs of modern savages*, 7th edition. New York: Henry Holt and Company.

Avery, G. 1984. Sacred cows or jackal kitchens, hyena middens and bird nests: some implications of multi-agent contributions to archaeological accumulations. In (M. Hall, G. Avery, D. M. Avery, M. L. Wilson, and A. J. B. Humphreys, eds.) *Frontiers: Southern African archaeology today*, pp. 344–348. British Archaeological Reports International Series 207.

Badgley, C. 1986a. Counting individuals in mammalian fossil assemblages from fluvial environments. *Palaios* 1: 328–338.

1986b. Taphonomy of mammalian fossil remains from Siwalik rocks of Pakistan. *Paleobiology* 12: 119–142.

Bailey, G. 1983. Hunter-gatherer behaviour in prehistory: problems and perspectives. In (G. Bailey, ed.) *Hunter-gatherer economy in prehistory*, pp. 1–6. Cambridge: Cambridge University Press.

Baker, J. and Brothwell, D. 1980. *Animal diseases in archaeology*. London: Academic Press.

Balcomb, R. 1986. Songbird carcasses disappear rapidly from agricultural fields. *Auk* 103: 817–820.

Balme, J. 1980. An analysis of charred bone from Devil's Lair, western Australia. *Archaeology and Physical Anthropology in Oceania* 15: 81–85.

Barlow, J. 1984. Mortality estimation: biased results from unbiased ages. *Canadian Journal of Fisheries and Aquatic Science* 41: 1843–1847.

Barnosky, A. D. 1985. Taphonomy and herd structure of the extinct Irish Elk (*Megaloceros giganteus*). *Science* 228: 340–344.

1986. 'Big game' extinction caused by late Pleistocene climatic change: Irish elk (*Megaloceros giganteus*) in Ireland. *Quaternary Research* 25: 128–135.

Bartram, L. E., Kroll, E. M., and Bunn, H. T. 1991. Variability in camp structure and bone food refuse patterning at Kua San hunter-gatherer camps. In (E. M. Kroll and T. D. Price, eds.) *The interpretation of archaeological spatial patterning*, pp. 77–148. New York: Plenum Press.

Bartsiokas, A. and Middleton, A. P. 1992. Characterization and dating of recent and fossil bone by X-ray diffraction. *Journal of Archaeological Science* 19:63–72.

Becking, J. H. 1975. The ultrastructure of avian eggshell. *Ibis* 117:143–151.

Behrensmeyer, A. K. 1975a. Taphonomy and paleoecology in the Hominid fossil record. *Yearbook of Physical Anthropology* 19: 36–50.

1975b. The taphonomy and paleoecology of Plio-Pleistocene vertebrate assemblages east of Lake Rudolf, Kenya. *Bulletin of the Museum of Comparative Zoology* 146: 473–578.

1978. Taphonomic and ecologic information from bone weathering. *Paleobiology* 4: 150–162.

1979. The habitat of Plio-Pleistocene hominids in East Africa: taphonomic and micro-stratigraphic evidence. In (C. Jolly, ed.) *Early hominids of Africa*, pp. 165–189. London: Duckworth.

1981. Vertebrate paleoecology in a recent East African ecosystem. In (J. Gray, A. J. Boucot, and W. B. N. Berry, eds.) *Communities of the past*, pp. 591–615. Stroudsburg: Hutchinson Ross Publishing Co.

1982. Time resolution in fluvial vertebrate assemblages. *Paleobiology* 8: 211–228.

1983. Patterns of natural bone distribution on recent land surfaces: implications for archaeological site formation. In (J. Clutton-Brock and C. Grigson, eds.) *Animals and archaeology: 1. hunters and their prey*, pp. 93–106. British Archaeological Reports International Series 163.

1984. Taphonomy and the fossil record. *American Scientist* 72: 558–566.

1987. Taphonomy and hunting. In (M. H. Nitecki and D. V. Nitecki, eds.) *The evolution of human hunting*, pp. 423–450. New York: Plenum Press.

1988a. The pull of the recent analogue. *Palaios* 3: 373.

1988b. Vertebrate preservation in fluvial channels. *Palaeogeography, Palaeoclimatology, Palaeoecology* 63: 183–199.

1990. Transport–hydrodynamics: bones. In (D. E. G. Briggs and P. R. Crowther, eds.) *Palaeobiology: a synthesis*, pp. 232–235. Oxford: Blackwell Scientific Publications.

1991. Terrestrial vertebrate accumulations. In (P. A. Allison and D. E. G. Briggs, eds.) *Taphonomy: releasing the data locked in the fossil record*, pp. 291–335. Topics in Geobiology Vol. 9. New York: Plenum Press.

Behrensmeyer, A. K. and Badgley, C. 1989. Foreword. In Weigelt 1989, pp. vii–x.

Behrensmeyer, A. K. and Boaz, D. E. D. 1980. The recent bones of Amboseli National Park, Kenya, in relation to East African paleoecology. In (A. K. Behrensmeyer and A. P. Hill, eds.) *Fossils in the making*, pp. 72–92. Chicago: University of Chicago Press.

Behrensmeyer, A. K., Gordon, K. D., and Yanagi, G. T. 1986. Trampling as a cause of bone surface damage and pseudo-cutmarks. *Nature* 319: 768–771.

1989. Nonhuman bone modification in Miocene fossils from Pakistan. In (R. Bonnichsen and M. H. Sorg, eds.) *Bone modification*, pp. 99–120. Orono: University of Maine Center for the Study of the First Americans.

Behrensmeyer, A. K. and Hill, A. P. 1980. (eds.) *Fossils in the making: vertebrate taphonomy and paleoecology*. Chicago: University of Chicago Press.

Behrensmeyer, A. K. and Hook, R. W. 1992. Paleoenvironmental contexts and taphonomic modes. In (A. K. Behrensmeyer, J. D. Damuth, W. A. DiMichele, R. Potts, H.-D. Sues, and S. L. Wing, eds.) *Terrestrial ecosystems through time: evolutionary paleoecology of terrestrial plants and animals*, pp. 15–136. Chicago: University of Chicago Press.

Behrensmeyer, A. K. and Kidwell, S. M. 1985. Taphonomy's contribution to paleobiology. *Paleobiology* 11: 105–119.

Behrensmeyer, A. K., Western, D., and Dechant Boaz, D. E. 1979. New perspectives in vertebrate paleoecology from a recent bone assemblage. *Paleobiology* 5: 12–21.

Bell, L. S. 1990. Paleopathology and diagenesis: an SEM evaluation of structural changes using backscattered electron imaging. *Journal of Archaeological Science* 17: 85–108.

Berger, J. 1983. Ecology and catastrophic mortality in wild horses: implications for interpreting fossil assemblages. *Science* 220: 1403–1404.

Berger, W. H. 1991. On the extinction of the mammoth: science and myth. In (D. W. Müller, J. A. McKenzie, and H. Weissert, eds.) *Controversies in modern geology*, pp. 115–132. London: Academic Press.

Bettinger, R. L. 1991. *Hunter-gatherers: archaeological and evolutionary theory*. New York: Plenum Press.

Bickart, K. J. 1984. A field experiment in avian taphonomy. *Journal of Vertebrate Paleontology* 4: 525–535.

Biddick, K. A. and Tomenchuck, J. 1975. Quantifying continuous lesions and fractures on long bones. *Journal of Field Archaeology* 2: 239–249.

Binford, L. R. 1964. A consideration of archaeological research design. *American Antiquity* 29: 425–441.

1967. Smudge pits and hide smoking: the use of analogy in archaeological reasoning. *American Antiquity* 32: 1–12.

1968. Archaeological perspectives. In (S. R. Binford and L. R. Binford, eds.) *New perspectives in archaeology*, pp. 5–32. Chicago: Aldine.

1972. *An archaeological perspective*. New York: Seminar Press.

1977. General introduction. In (L. R. Binford, ed.) *For theory building in archaeology*, pp. 1–10. New York: Academic Press.

1978. *Nunamiut ethnoarchaeology*. New York: Academic Press.

1980. Willow smoke and dogs' tails: hunter-gatherer settlement systems and archaeological site formation. *American Antiquity* 45: 4–20.

1981a. Behavioral archaeology and the "Pompeii Premise." *Journal of Anthropological Research* 37: 195–208.

1981b. *Bones: ancient men and modern myths*. New York: Academic Press.

1984a. Butchering, sharing, and the archaeological record. *Journal of Anthropological Archaeology* 3: 235–257.

1984b. *Faunal remains from Klasies River Mouth*. Orlando: Academic Press.

1986. Comment on Bunn and Kroll 1986. *Current Anthropology* 27: 444–446.

1987. Researching ambiguity: frames of reference and site structure. In (S. Kent, ed.) *Method and theory for activity area research: an ethnoarchaeological approach*, pp. 449–512. New York: Columbia University Press.

1988. Fact and fiction about the *Zinjanthropus* floor: data, arguments, and interpretations. *Current Anthropology* 29: 123–135.

Binford, L. R. and Bertram, J. B. 1977. Bone frequencies – and attritional processes. In (L. R. Binford, ed.) *For theory building in archaeology*, pp. 77–153. New York: Academic Press.

Blumenschine, R. J. 1986a. Carcass consumption sequences and the archaeological distinction of scavenging and hunting. *Journal of Human Evolution* 15: 639–659.

1986b. *Early hominid scavenging opportunities: implications of carcass availability in the Serengeti and Ngorongoro ecosystems*. British Archaeological Reports International Series 283.

1987. Characteristics of an early hominid scavenging niche. *Current Anthropology* 28: 383–407.

1988. An experimental model of the timing of hominid and carnivore influence on archaeological bone assemblages. *Journal of Archaeological Science* 15: 483–502.

Blumenschine, R. J. and Caro, T. M. 1986. Unit flesh weights of some East African bovids. *African Journal of Ecology* 24: 273–286.

Blumenschine, R. J. and Cavallo, J. A. 1992. Scavenging and human evolution. *Scientific American* 267(4): 90–96.

Blumenschine, R. J. and Selvaggio, M. 1988. Percussion marks on bone surfaces as a new diagnostic of hominid behavior. *Nature* 333: 763–765.

1991. On the marks of marrow bone processing by hammerstones and hyaenas: their anatomical patterning and archaeological implications. In (J. D. Clark, ed.) *Cultural beginnings: approaches to understanding early hominid life-ways in the African savanna*, pp. 17–32. Union Internationale des Sciences Préhistoriques et Protohistoriques Monographien Band 19.

Boaz, D. 1980. Fossils in the making: taphonomy and vertebrate paleoecology. *Current Anthropology* 21: 404–406.

1982. Modern riverine taphonomy: its relevance to the interpretation of Plio-Pleistocene hominid paleoecology in the Omo Basin, Ethiopia. Ph. D. dissertation, University of California, Berkeley. Ann Arbor: University Microfilms International.

Boaz, N. T. and Behrensmeyer, A. K. 1976. Hominid taphonomy: transport of human skeletal parts in an artifical fluviatile environment. *American Journal of Physical Anthropology* 45: 53–60.

Bohannan, P. and Glazer, M. 1988. (eds.) Introduction. In *High Points in Anthropology*, 2nd edition. New York: McGraw-Hill, Inc.

Bonnichsen, R. 1973. Some operational aspects of human and animal bone alteration. In (B. M.

Gilbert, ed.) *Mammalian osteo-archaeology: North America*, pp. 9–24. Columbia: Missouri Archaeological Society, University of Missouri.

1975. On faunal analysis and the australopithecines. *Current Anthropology* 16: 635–636.

1979. *Pleistocene bone technology in the Beringian Refugium.* Archaeological Survey of Canada Paper No. 89, Mercury Series. Ottawa: National Museum of Man.

1989. An introduction to taphonomy with an archaeological focus. In (R. Bonnichsen and M. H. Sorg, eds.) *Bone modification*, pp. 1–5. Orono: University of Maine Center for the Study of the First Americans.

Bonnichsen, R. and Sanger, D. 1977. Integrating faunal analysis. *Canadian Journal of Archaeology* 1: 109–133.

Bonnichsen, R. and Sorg, M. H. 1989. (eds.) *Bone modification.* Orono: University of Maine Center for the Study of the First Americans.

Bonnichsen, R. and Will, R. T. 1980. Cultural modification of bone: the experimental approach in faunal analysis. In (B. M. Gilbert, ed.) *Mammalian osteology*, pp. 7–30. Missouri Archaeological Society, University of Missouri-Columbia.

Borrero, L. 1990. Fuego-Patagonian bone assemblages and the problem of communal guanaco hunting. In (L. B. Davis and B. O. K. Reeves, eds.) *Hunters of the recent past*, pp. 373–399. London: Unwin Hyman.

Bower, J. 1986. *In search of the past.* Chicago: Dorsey Press.

Bower, J. R. F., Gifford, D. P., and Livingston, D. 1985. Excavations at the Loiyangalani Site, Serengeti National Park, Tanzania. *National Geographic Society Research Reports* 20: 41–56.

Bowyer, R. T. 1983. Osteophagia and antler breakage among Roosevelt elk. *California Fish and Game* 69: 84–88.

Boyd, D. W. and Newell, N. D. 1972. Taphonomy and diagenesis of a Permian fossil assemblage from Wyoming. *Journal of Paleontology* 46: 1–14.

Brain, C. K. 1967a. Bone weathering and the problem of bone pseudo-tools. *South African Journal of Science* 63: 97–99.

1967b. Hottentot food remains and their bearing on the interpretation of fossil bone assemblages. *Scientific Papers of the Namib Desert Research Station* 32: 1–7.

1969. The contribution of Namib Desert Hottentots to an understanding of australopithecine bone accumulations. *Scientific Papers of the Namib Desert Research Station* 39: 13–22.

1974. Some suggested procedures in the analysis of bone accumulations from southern African Quaternary caves. *Annals of the Transvaal Museum* 29: 1–8.

1976. Some principles in the interpretation of bone accumulations associated with man. In (G. L. Isaac and E. R. McCown, eds.) *Human origins: Louis Leakey and the East African evidence*, pp. 97–116. Menlo Park: W. A. Benjamin, Inc.

1980. Some criteria for the recognition of bone-collecting agencies in African caves. In (A. K. Behrensmeyer and A. P. Hill, eds.) *Fossils in the making*, pp. 107–130. Chicago: University of Chicago Press.

1981. *The hunters or the hunted? An introduction to African cave taphonomy.* Chicago: University of Chicago Press.

1989. The evidence for bone modification by early hominids in southern Africa. In (R. Bonnichsen and M. H. Sorg, eds.) *Bone modification*, pp. 291–297. Orono: University of Maine Center for the Study of the First Americans.

Brain, C. K. and Sillen, A. 1988. Evidence from the Swartkrans cave for the earliest use of fire. *Nature* 336: 464–466.

Bramwell, D., Yalden, D. W., and Yalden, P. E. 1987. Black grouse as the prey of the golden eagle at an archaeological site. *Journal of Archaeological Science* 14: 195–200.

Brett, C. E. 1990a. Destructive taphonomic processes and skeletal durability. In (D. E. G. Briggs and P. R. Crowther, eds.) *Palaeobiology: a synthesis*, pp. 223–226. Oxford: Blackwell Scientific Publications.

1990b. Obrution deposits. In (D. E. G. Briggs and P. R. Crowther, eds.) *Palaeobiology: a synthesis*, pp. 239–243. Oxford: Blackwell Scientific Publications.

Brett, C. E. and Speyer, S. E. 1990. Taphofacies. In (D. E. G. Briggs and P. R. Crowther, eds.) *Palaeobiology: a synthesis*, pp. 258–263. Oxford: Blackwell Scientific Publications.

Breuil, H. A. 1932. Le Feu et l'industrie de pierre et d'os dans le gisement du "Sinanthropus" a Chou Kou Tien. *Anthropologie* 42: 1–17.

1938. The use of bone implements in the Old Paleolithic period. *Antiquity* 12: 56–67.

1939. *Bone and antler industry of the Choukoutien* Sinanthropus *site*. Palaeontologica Sinica, Series D, New Series No. 6 (117). Peiking: Geological Survey of China.

Brewer, D. J. 1992. Zooarchaeology: method, theory, and goals. In (M. B. Schiffer, ed.) *Archaeological method and theory* vol. 4, pp. 195–244. Tucson: University of Arizona Press.

Briggs, D. E. G. 1990. Flattening. In (D. E. G. Briggs and P. R. Crowther, eds.) *Palaeobiology: a synthesis*, pp. 244–247. Oxford: Blackwell Scientific Publications.

Briggs, D. E. G. and Crowther, P. R. 1990. (eds.) *Palaeobiology: a synthesis*. Oxford: Blackwell Scientific Publications.

Brink, J. and Dawe, B. 1989. *Final report of the 1985 and 1986 field season at Head-Smashed-In buffalo jump, Alberta*. Archaeological Survey of Alberta Manuscript Series No. 16. Edmonton.

Briuer, F. L. 1977. Plant and animal remains from caves and rockshelters of Chevelon Canyon, Arizona: methods for isolating depositional processes. Ph.D. dissertation, University of California-Los Angeles. Ann Arbor: University Microfilms International.

Bromage, T. G. 1984. Interpretation of scanning electron microscopic images of abraded forming bone surfaces. *American Journal of Physical Anthropology* 64: 161–178.

Bromage, T. G. and Boyde, A. 1984. Microscopic criteria for the determination of directionality of cutmarks on bone. *American Journal of Physical Anthropology* 65: 359–366.

Brongersma-Sanders, M. 1957. Mass mortality in the sea. *Geological Society of America Memoir* 67(1): 941–1010.

Brothwell, D. 1976. Further evidence of bone chewing by ungulates: the sheep of North Ronaldsay, Orkney. *Journal of Archaeological Science* 3: 179–182.

Brown, R. D. 1983. (ed.) *Antler development in Cervidae*. Kingsville, TX: Caesar Kleberg Wildlife Research Institute.

Buckland, W. 1823. *Reliquiae diluvianae, or, observations on the organic remains contained in caves, fissures, and diluvial gravel, and on other geological phenomena, attesting to the action of an universal deluge*. London: Murray.

Buikstra, J. E. and Swegle, M. 1989. Bone modification due to burning: experimental evidence. In (R. Bonnichsen and M. H. Sorg, eds.) *Bone modification*, pp. 247–258. Orono: University of Maine Center for the Study of the First Americans.

Bunn, H. T. 1981. Archaeological evidence for meat-eating by Plio-Pleistocene hominids from Koobi Fora and Olduvai Gorge. *Nature* 291: 574–577.

1982. Meat-eating and human evolution: studies on the diet and subsistence patterns of Plio-Pleistocene hominids in East Africa. Ph. D. dissertation, University of California, Berkeley. Ann Arbor: University Microfilms International.

1983. Evidence on the diet and subsistence patterns of Plio-Pleistocene hominids at Koobi Fora, Kenya, and Olduvai Gorge, Tanzania. In (J. Clutton-Brock and C. Grigson, eds.) *Animals and archaeology: 1. hunters and their prey*, pp. 21–30. British Archaeological Reports International Series 163.

1986. Patterns of skeletal representation and hominid subsistence activities at Olduvai Gorge, Tanzania, and Koobi Fora, Kenya. *Journal of Human Evolution* 15: 673–690.

1989. Diagnosing Plio-Pleistocene hominid activity with bone fracture evidence. In (R. Bonnichsen and M. H. Sorg, eds.) *Bone modification*, pp. 299–315. Orono: University of Maine Center for the Study of the First Americans.

1991. A taphonomic perspective on the archaeology of human origins. *Annual Review of Anthropology* 20: 433–467.

Bunn, H. T., Bartram, L. E., and Kroll, E. M. 1988. Variability in bone assemblage formation from Hadza hunting, scavenging, and carcass processing. *Journal of Anthropological Archaeology* 7: 412–457.

Bunn, H. T. and Blumenschine, R. J. 1987. On "theoretical framework and tests" of early hominid meat and marrow acquisition: a reply to Shipman. *American Anthropologist* 89:444–448.

Bunn, H. T. and Kroll, E. M. 1986. Systematic butchery by Plio/Pleistocene hominids at Olduvai Gorge, Tanzania. *Current Anthropology* 27: 431–452.

1987. Reply to Potts. *Current Anthropology* 28: 96–98.

1988. Reply to Binford. *Current Anthropology* 29: 135–149.

Bunn, H. T., Kroll, E. M., and Bartram, L. E. 1991. Bone distribution on a modern East African landscape and its archaeological implications. In (J. D. Clark, ed.) *Cultural beginnings: approaches to understanding early hominid life-ways in the African savanna*, pp. 33–54. Union Internationale des Sciences Préhistoriques et Protohistoriques Monographien Band 19.

Burgett, G. R. 1990. The bones of the beast: resolving questions of faunal assemblage formation processes through actualistic research. Ph.D. dissertation, University of New Mexico, Albuquerque.

Burr, D. B. 1980. The relationships among physical, geometrical and mechanical properties of bone, with a note on the properties of nonhuman primate bone. *Yearbook of Physical Anthropology* 23: 109–146.

Butler, V. L. 1987. Distinguishing natural from cultural salmonid deposits in the Pacific Northwest of North America. In (D. T. Nash and M. D. Petraglia, eds.) *Natural formation processes and the archaeological record*, pp. 131–149. British Archaeological Reports International Series 352.

1990. Distinguishing natural from cultural salmonid deposits in Pacific Northwest North America. Ph.D. dissertation, University of Washington. Ann Arbor: University Microfilms.

1993. Natural versus cultural salmonid remains: origin of The Dalles Roadcut bones, Columbia River, Oregon, U.S.A. *Journal of Archaeological Science* 20: 1–24.

Butzer, K. W. 1982. *Archaeology as human ecology*. Cambridge: Cambridge University Press.

Cadée, G. C. 1990. The history of taphonomy. In (S. K. Donovan, ed.) *The processes of fossilization*, pp. 3–21. New York: Columbia University Press.

Cahen, D. and Moeyersons, J. 1977. Subsurface movements of stone artefacts and their implications for the prehistory of Central Africa. *Nature* 266: 812–815.

Carlson, S. J. 1990. Vertebrate dental structures. In (J. G. Carter, ed.) *Skeletal biomineralization: patterns, processes and evolutionary trends*, pp. 531–556. New York: Van Nostrand Reinhold.

Casteel, R. W. 1971. Differential bone destruction: some comments. *American Antiquity* 36: 466–469.

1976. *Fish remains in archaeology and environmental studies*. New York: Academic Press.

Caughley, G. 1966. Mortality patterns in mammals. *Ecology* 47: 906–918.

1977. *Analysis of vertebrate populations*. London: John Wiley & Sons.

Cavallo, J. A. and Blumenschine, R. J. 1989. Tree-stored leopard kills: expanding the hominid scavenging niche. *Journal of Human Evolution* 18: 393–399.

Chambers, A. L. 1992. Seal bone mineral density: its effect on specimen survival in archaeological sites. Unpublished Honors thesis, Department of Anthropology, University of Missouri, Columbia.

Chaplin, R. E. 1965. Animals in archaeology. *Antiquity* 39: 204–211.

1971. *The study of animal bones from archaeological sites.* London: Seminar Press.

Charlton, T. H. 1981. Archaeology, ethnohistory, and ethnology: interpretive interfaces. In (M. B. Schiffer, ed.) *Advances in archaeological method and theory* vol. 4, pp. 129–176. New York: Academic Press.

Chase, P. G. 1985. On the use of Binford's utility indices in the analysis of archaeological sites. *P.A.C.T.* 11: 287–302.

Chomko, S. A. and Gilbert, B. M. 1991. Bone refuse and insect remains: their potential for temporal resolution of the archaeological record. *American Antiquity* 56: 680–686.

Clark, J., Beerbower, J. R., and Kietzke, K. K. 1967. *Oligocene sedimentation, stratigraphy and paleoclimatology in the Big Badlands of South Dakota.* Fieldiana Geology Memoir 5.

Clark, J. and Guensberg, T. E. 1970. Population dynamics of *Leptomeryx*. *Fieldiana Geology* 16: 411–511.

Clark, J. and Kietzke, K. K. 1967. Paleoecology of the Lower Nodular Zone, Brule Formation, in the Big Badlands of South Dakota. In (J. Clark, J. R. Beerbower, and K. K. Kietzke) *Oligocene sedimentation, stratigraphy and paleoclimatology in the Big Badlands of South Dakota*, pp. 111–137. Fieldiana Geology Memoir 5.

Clark, J. D. 1972. Paleolithic butchery practices. In (P. J. Ucko, R. Tringham, and G. W. Dimbleby, eds.) *Man, settlement and urbanism*, pp. 149–156. Cambridge: Duckworth.

Clason, A. T. 1975. (ed.) *Archaeozoological studies.* Amsterdam: North Holland Publishing Co.

Coe, M. J. 1978. The decomposition of elephant carcasses in the Tsavo (East) National Park, Kenya. *Journal of Arid Environments* 1: 71–86.

1980. The role of modern ecological studies in the reconstruction of paleoenvironments in sub-Saharan Africa. In (A. K. Behrensmeyer and A. P. Hill, eds.) *Fossils in the making*, pp. 55–67. Chicago: University of Chicago Press.

Coe, R. J., Downing, R. L., and McGinnes, B. S. 1980. Sex and age bias in hunter-killed white-tailed deer. *Journal of Wildlife Management* 44: 245–249.

Colley, S. M. 1990a. Humans as taphonomic agents. In (S. Solomon, I. Davidson, and D. Watson, eds.) *Problem solving in taphonomy*, pp. 50–64. Tempus Vol. 2.

1990b. The analysis and interpretation of archaeological fish remains. In (M. B. Schiffer, ed.) *Archaeological method and theory* vol. 2, pp. 207–253. Tucson: University of Arizona Press.

Collier, S. and White, J. P. 1976. Get them young? Age and sex inferences on animal domestication in archaeology. *American Antiquity* 41: 96–102.

Conybeare, A. and Haynes, G. 1984. Observations on elephant mortality and bones in water holes. *Quaternary Research* 22: 189–200.

Cook, J. 1986. The application of scanning electron microscopy to taphonomic and archaeological problems. In (D. A. Roe, ed.) *Studies in the Upper Paleolithic of Britian and northwest Europe*, pp. 143–163. British Archaeological Reports International Series 296.

Cook, S. F., Brooks, S. T., and Ezra-Cohn, H. C. 1961. The process of fossilization. *Southwestern Journal of Anthropology* 17: 355–364.

Cornwall, I. W. 1956. *Bones for the archaeologist.* London: Phoenix House.

1968. *Prehistoric animals and their hunters.* London: Faber and Faber.

Crader, D. C. 1983. Recent single-carcass bone scatters and the problem of "butchery" sites in the archaeological record. In (J. Clutton-Brock and C. Grigson, eds.) *Animals and archaeology: 1. hunters and their prey*, pp. 107–141. British Archaeological Reports International Series 163.

Craig, G. Y. and Oertel, G. 1966. Deterministic models of living and fossil populations of animals. *Quarterly Journal of the Geological Society of London* 122: 315–355.

Crawford, H. 1982. Analogies, anomalies and research strategy. *Paléorient* 8: 5–9.

Cribb, R. 1985. The analysis of ancient herding systems: an application of computer simulation in faunal studies. In (G. Barker and C. Gamble, eds.) *Beyond domestication in prehistoric Europe*, pp. 75–106. London: Academic Press.

1987. The logic of the herd: a computer simulation of archaeological herd structure. *Journal of Anthropological Archaeology* 6: 376–415.

Cruz-Uribe, K. 1988. The use and meaning of species diversity and richness in archaeological faunas. *Journal of Archaeological Science* 15: 179–196.

1991. Distinguishing hyena from hominid bone accumulations. *Journal of Field Archaeology* 18: 467–486.

Currey, J. 1984. *The mechanical adaptations of bones*. Princeton: Princeton University Press.

1990. Biomechanics of mineralized skeletons. In (J. G. Carter, ed.) *Skeletal biomineralization: patterns, processes and evolutionary trends*, pp. 11–25. New York: Van Nostrand Reinhold.

Cussler, E. L. and Featherstone, J. D. B. 1981. Demineralization of porous solids. *Science* 213: 1018–1019.

Czaplewski, R. L., Crowe, D. M., and McDonald, L. L. 1983. Sample sizes and confidence intervals for wildlife population ratios. *Wildlife Society Bulletin* 11: 121–127.

Daly, P. 1969. Approaches to faunal analysis in archaeology. *American Antiquity* 34: 146–153.

Damuth, J. 1982. Analysis of the preservation of community structure in assemblages of fossil mammals. *Paleobiology* 8: 434–446.

D'Andrea, A. C. and Gotthardt, R. M. 1984. Predator and scavenger modification of recent equid skeletal assemblages. *Arctic* 37: 276–283.

Daniel, G. 1981. *A short history of archaeology*. London: Thames and Hudson.

Dart, R. A. 1949. The predatory implemental technique of the australopithecines. *American Journal of Physical Anthropology* 7: 1–16.

1956a. Cultural status of the South African man-apes. *Smithsonian Institution Annual Report for 1955*, pp. 317–338.

1956b. Myth of the bone-accumulating hyena. *American Anthropologist* 58: 40–62.

1957. *The osteodontokeratic culture of* Australopithecus prometheus. Transvaal Museum Memoir No. 10. Pretoria.

1958. Bone tools and porcupine gnawing. *American Anthropologist* 60: 715–724.

1960. The bone tool manufacturing ability of *Australopithecus prometheus*. *American Anthropologist* 62: 134–138.

David, B. 1990. How was this bone burnt? In (S. Solomon, I. Davidson, and D. Watson, eds.) *Problem solving in taphonomy*, pp. 65–79. Tempus vol. 2.

Davies, D. J., Powell, E. N., and Stanton, R. J., Jr. 1989. Relative rates of shell dissolution and net accumulation – a commentary: can shell beds form by the gradual accumulation of biogenic debris on the sea floor? *Lethaia* 22: 207–212.

Davis, K. L. 1985. A taphonomic approach to experimental bone fracturing and applications to several South African Pleistocene sites. Ph.D. dissertation, State University of New York at Binghamton.

Davis, S. J. M. 1987. *The archaeology of animals*. New Haven: Yale University Press.

Dawkins, W. B. 1869. On the distribution of British postglacial mammals. *Quarterly Journal of the Geological Society of London* 25: 192–217.

1874. *Cave hunting, researches on the evidence of caves respecting the early inhabitants of Europe*. London: Macmillan & Company.

Deevey, E. S., Jr. 1947. Life tables for natural populations of animals. *Quarterly Review of Biology* 22: 283–314.

de Rousseau, C. J. 1988. Bone biology. In (I. Tattersall, E. Delson, and J. van Couvering, eds.) *Encyclopedia of human evolution and prehistory*, pp. 95–96. New York: Garland Publishing.

d'Errico, F., Giacobini, B., and Puech, P.-F. 1984. An experimental study of the technology of bone-implement manufacture. *MASCA Journal* 3: 71–74.

Derry, D. E. 1911. Damage done to skulls and bones by termites. *Nature* 86: 245–246.

Dibble, D. S. and Lorrain, D. 1968. *Bonfire Shelter: a stratified bison kill site, Val Verde County, Texas.* Texas Memorial Museum Miscellaneous Papers No. 1.

Dixon, E. J. 1984. Context and environment in taphonomic analysis: examples from Alaska's Porcupine River caves. *Quaternary Research* 22: 201–215.

Dixon, E. J. and Thorson, R. M. 1984. Taphonomic analysis and interpretation in North American Pleistocene archaeology. *Quaternary Research* 22: 155–159.

Dodd, J. R. and Stanton, R. J., Jr. 1981. *Paleoecology: concepts and applications.* New York: Wiley-Interscience.

Dodson, P. 1971. Sedimentology and taphonomy of the Oldman Formation (Campanian), Dinosaur Provincial Park, Alberta (Canada). *Palaeogeography, Palaeoclimatology, Palaeoecology* 10: 21–74.

  1973. The significance of small bones in paleoecological interpretation. *University of Wyoming Contributions to Geology* 12: 15–19. Laramie.

  1980. Vertebrate burials. *Paleobiology* 6: 6–8.

Dodson, P and Wexlar, D. 1979. Taphonomic investigations of owl pellets. *Paleobiology* 5: 279–284.

Donovan, S. K. 1990. (ed.) *The processes of fossilization.* New York: Columbia University Press.

Doveton, J. H. 1979. Numerical methods for the reconstruction of fossil material in three dimensions. *Geological Magazine* 116: 215–226.

Duckworth, W. L. H. 1904. Note on the dispersive power of running water on skeletons: with particular reference to the skeletal remains of *Pithecanthropus erectus.* In (W. L. H. Duckworth, ed.) *Studies from the anthropological laboratory,* pp. 274–277. Cambridge: Cambridge University Press.

Dunnell, R. C. 1992. Archaeology and evolutionary science. In (L. Wandsnider, ed.) *Quandaries and quests: visions of archaeology's future,* pp. 209–224. Southern Illinois University Center for Archaeological Investigations Occasional Paper No. 20. Carbondale.

During, E. M. and Nilsson, L. 1991. Mechanical surface analysis of bone: a case study of cut marks and enamel hypoplasia on a Neolithic cranium from Sweden. *American Journal of Physical Anthropology* 84: 113–125.

Efremov, I. A. 1940. Taphonomy: a new branch of paleontology. *Pan-American Geologist* 74: 81–93.

  1958. Some considerations on biological bases of paleozoology. *Vertebrata Palasiatica* 2(2/3): 83–98.

Eickhoff, S. and Herrmann, B. 1985. Surface marks on bones from a Neolithic collective grave (Odagsen, Lower Saxony): a study on differential diagnosis. *Journal of Human Evolution* 14: 263–274.

Elder, R. L. and Smith, G. R. 1988. Fish taphonomy and environmental inference in paleolimnology. *Palaeogeography, Palaeoclimatology, Palaeoecology* 62: 577–592.

Elkin, D. C. and Zanchetta, J. R. 1991. Densitometria osea de camélidos – aplicaciones arqueológicas. *Actas del X Congreso Nacional de Arqueologia Argentina* 3: 195–204. Catamarca.

Emerson, A. M. 1990. Archaeological implications of variability in the economic anatomy of *Bison bison.* Ph.D. dissertation, Washington State University. Ann Arbor: University Microfilms.

Enlow, R. D. 1969. The bone of reptiles. In (C. Gans, ed.) *Biology of the Reptilia vol. 1, morphology A,* pp. 45–80. London: Academic Press.

Ericson, P. G. P. 1987. Interpretations of archaeological bird remains: a taphonomic approach.

Journal of Archaeological Science 14: 65–75.

Erzinçlioglu, Y. Z. 1983. The application of entomology to forensic medicine. *Medicine, Science and the Law* 23:57–63.

Evans, F. G. 1961. Relation of the physical properties of bone to fractures. *American Academy of Orthopedic Surgeons* 18: 110–121.

Everitt, B. S. 1977. *The analysis of contingency tables.* London: Chapman and Hall.

Fernandez-Jalvo, Y. and Andrews, P. 1992. Small mammal taphonomy of Gran Dolina, Atapuerca (Burgos), Spain. *Journal of Archaeological Science* 19: 407–428.

Finks, R. M. 1979. Fossils and fossilization. In (R. W. Fairbridge and D. Jablonski, eds.) *Encyclopedia of paleontology*, pp. 327–332. Stroudsburg: Dowden, Hutchinson & Ross, Inc.

Fiorillo, A. R. 1988. A proposal for graphic presentation of orientation data from fossils. *University of Wyoming Contributions to Geology* 26: 1–4. Laramie.

1989. An experimental study of trampling: implications for the fossil record. In (R. Bonnichsen and M. H. Sorg, eds.) *Bone modification*, pp. 61–71. Orono: University of Maine Center for the Study of the First Americans.

Fisher, D. C. 1981. Crocodilian scatology, microvertebrate concentrations, and enamel-less teeth. *Paleobiology* 7: 262–275.

1984a. Mastodon butchery by North American Paleo-Indians. *Nature* 308: 271–272.

1984b. Taphonomic analysis of late Pleistocene mastodon occurrences: evidence of butchery by North American Paleo-Indians. *Paleobiology* 10: 338–357.

1987. Mastodont procurement by Paleoindians of the Great Lakes region: hunting or scavenging? In (M. H. Nitecki and D. V. Nitecki, eds.) *The evolution of human hunting*, pp. 309–421. New York: Plenum Press.

Fisher, J. W., Jr. 1992. Observations on the late Pleistocene bone assemblage from the Lamb Spring Site, Colorado. In (D. J. Stanford and J. S. Day, eds.) *Ice Age hunters of the Rockies*, pp. 51–81. Niwot: Denver Museum of Natural History and University Press of Colorado.

Flinn, L., Turner, C. G., II, and Brew, A. 1976. Additional evidence for cannibalism in the Southwest: the case of LA4528. *American Antiquity* 41: 308–318.

Francillon-Vieillot, H., de Buffrénil, V., Castanet, J., Géraudi, J., Meunier, F. J., Sire, J. Y., Zylberberg, L., and de Ricqlès, A. 1990. Microstructure and mineralization of vertebrate skeletal tissues. In (J. G. Carter, ed.) *Skeletal biomineralization: patterns, processes and evolutionary trends*, pp. 471–530. New York: Van Nostrand Reinhold.

Freeman, L. G., Jr. 1968. A theoretical framework for interpreting archeological materials. in (R. B. Lee and I. DeVore, eds.) *Man the hunter*, pp. 262–267. Chicago: Aldine.

Fritz, J. M. 1972. Archaeological systems for indirect observation of the past. In (M. P. Leone, ed.) *Contemporary archaeology*, pp. 135–157. Carbondale: Southern Illinois University Press.

Frison, G. C. 1974. Archaeology of the Casper Site. In (G. C. Frison, ed.) *The Casper Site:a Hell Gap bison kill on the High Plains*, pp. 1–111. New York: Academic Press.

1982. Bone butchering tools in archaeological site. *Canadian Journal of Anthropology* 2: 159–167.

Frison, G. C. and Todd, L. C. 1986. *The Colby Mammoth Site: taphonomy and archaeology of a Clovis kill in northern Wyoming.* Albuquerque: University of New Mexico Press.

Frostick, L. and Reid, I. 1983. Taphonomic significance of sub-aerial transport of vertebrate fossils on steep semi-arid slopes. *Lethaia* 16: 157–164.

Fürsich, F. T. 1990. Fossil concentrations and life and death assemblages. In (D. E. G. Briggs and P. R. Crowther, eds.) *Palaeobiology: a synthesis*, pp. 235–239. Oxford: Blackwell Scientific Publications.

Galloway, A., Birkby, W. H., Jones, A. M., Henry, T. E., and Parks, B. O. 1989. Decay rates of human remains in an arid environment. *Journal of Forensic Sciences* 34: 607–616.

Gamble, C. 1978. Optimising information from studies of faunal remains. In (J. F. Cherry, C. Gamble, and S. Shennan, eds.) *Sampling in contemporary British archaeology*, pp. 321–353. British Archaeological Reports British Series 50.

Garland, A. N. 1987. A histological study of archaeological bone decomposition. In (A. Boddington, A. N. Garland, and R. C. Janaway, eds.) *Death, decay and reconstruction: approaches to archaeology and forensic science*, pp. 109–126. Manchester: Manchester University Press.

1988. Contributions to palaeohistology. In (E. A. Slater and J. O. Tate, eds.) *Science and archaeology: Glasgow 1987*, pp. 321–338. British Archaeological Reports British Series 196(ii).

Garland, A. N., Janaway, R. C., and Roberts, C. A. 1988. A study of the decay processes of human skeletal remains from the Parish Church of the Holy Trinity, Rothwell, Northamptonshire. *Oxford Journal of Archaeology* 7: 235–252.

Garvin, R. D. 1987. Research in Plains taphonomy: the manipulation of faunal assemblages by scavengers. Unpublished M.A. thesis, Department of Archaeology, University of Calgary, Calgary, Alberta.

Geniesse, S. C. 1982. A study of manufacture and wear on bone artifacts. Unpublished M.A. paper, Department of Anthropology, University of Washington, Seattle.

Gibbon, G. 1984. *Anthropological archaeology*. New York: Columbia University Press.

Gibert, J. and Jimenez, C. 1991. Investigations into cut-marks on fossil bones of Lower Pleistocene age from Venta Micena (Orce, Granada province, Spain). *Human Evolution* 6: 117–128.

Gifford, D. P. 1977. Observations of modern human settlements as an aid to archaeological interpretation. Ph.D. dissertation, University of California-Berkeley. Ann Arbor: University Microfilms International.

1981. Taphonomy and paleoecology: a critical review of archaeology's sister disciplines. In (M. B. Schiffer, ed.) *Advances in archaeological method and theory* vol. 4, pp. 365–438. New York: Academic Press.

1984. Taphonomic specimens, Lake Turkana. *National Geographic Research Reports* 17: 419–428.

Gifford, D. P. and Behrensmeyer, A. K. 1977. Observed formation and burial of a recent human occupation site in Kenya. *Quaternary Research* 8: 245–266.

Gifford, D. P., Isaac, G. L., and Nelson, C. M. 1980. Evidence for predation and pastoralism at Prolonged Drift: a pastoral Neolithic site in Kenya. *Azania* 15: 57–108.

Gifford-Gonzalez, D. P. 1989a. Ethnographic analogues for interpreting modified bones: some cases from East Africa. In (R. Bonnichsen and M. H. Sorg, eds.) *Bone modification*, pp. 179–246. Orono: University of Maine Center for the Study of the First Americans.

1989b. Modern analogues: developing an interpretive framework. In (R. Bonnichsen and M. H. Sorg, eds.) *Bone modification*, pp. 43–52. Orono: University of Maine Center for the Study of the First Americans.

1989c. Shipman's shaky foundations. *American Anthropologist* 91: 180–186.

1991. Bones are not enough: analogues, knowledge, and interpretive strategies in zooarchaeology. *Journal of Anthropological Archaeology* 10: 215–254.

Gifford-Gonzalez, D. P., Damrosch, D. B., Damrosch, D. R., Pryor, J., and Thunen, R. L. 1985. The third dimension in site structure: an experiment in trampling and vertical dispersal. *American Antiquity* 50: 803–818.

Gifford-Gonzalez, D. and Gargett, R. H. n.d. An exercise in utility: comments on bone density, *in situ* destruction, and utility indices. Unpublished manuscript.

Gilbert, A. S. 1979. Urban taphonomy of mammalian remains from the Bronze Age of Godin Tepe, western Iran. Ph.D. dissertation, Columbia University. Ann Arbor: University

Microfilms International.

Gilbert, A. S. and Singer, B. H. 1982. Reassessing zooarchaeological quantification. *World Archaeology* 14: 21–40.

Gilbert, B. M. 1969. Some aspects of diet and butchering techniques among prehistoric Indians of South Dakota. *Plains Anthropologist* 14: 277–294.

Gilbert, B. M. and Bass, W. M. 1967. Seasonal dating of burials from the presence of fly pupae. *American Antiquity* 32: 534–535.

Gilbow, D. W. 1981. Inference of human activity from faunal remains. Unpublished M.A. thesis, Department of Anthropology, Washington State University, Pullman.

Gilchrist, R. and Mytum, H. C. 1986. Experimental archaeology and burnt animal bone from archaeological sites. *Circaea* 4: 29–38.

Gilmore, R. M. 1949. The identification and value of mammal bones from archeologic excavations. *Journal of Mammalogy* 30: 163–169.

Goodman, N. 1967. Uniformity and simplicity. In (C. C. Albritton, Jr., ed.) *Uniformity and simplicity*, pp. 93–99. Geological Society of America Special Paper No. 89.

Gordon, B. C. 1976. Antler pseudo-tools made by caribou. In (J. S. Raymond, B. Loveseth, C. Arnold, and G. Reardon, eds.) *Primitive art and technology*, pp. 121–128. Calgary: University of Calgary Archaeological Association.

Gordon, C. C. and Buikstra, J. E. 1981. Soil pH, bone preservation, and sampling bias at mortuary sites. *American Antiquity* 46: 566–571.

Goss, R. J. 1983. *Deer antlers: regeneration, function, and evolution.* New York: Academic Press.

Gould, R. A. 1978. (ed.) *Explorations in ethnoarchaeology.* Albuquerque: University of New Mexico Press. 1980. *Living archaeology.* Cambridge: Cambridge University Press. 1990. *Recovering the past.* Albuquerque: University of New Mexico Press.

Gould, R. A. and Watson, P. J. 1982. A dialogue on the meaning and use of analogy in ethnoarchaeological reasoning. *Journal of Anthropological Archaeology* 1: 355–381.

Gould, S. J. 1965. Is uniformitarianism necessary? *American Journal of Science* 263: 223–228.

1967. Is uniformitarianism useful? *Journal of Geological Education* 15: 149–150.

1979. Agassiz's marginalia in Lyell's *Principles*, or the perils of uniformity and the ambiguity of heroes. *Studies in the History of Biology* 3: 119–138.

1982. Hutton's purposeful view. *Natural History* 91(5): 6–12.

1984. Toward the vindication of punctuational change. In (W. A. Berggren and J. A. van Couvering, eds.) *Catastrophes and earth history: the new uniformitarianism*, pp. 9–34. Princeton: Princeton University Press.

1986. Evolution and the triumph of homology, or why history matters. *American Scientist* 74: 60–69.

Graham, R. W., Holman, J. A., and Parmalee, P. W. 1983. *Taphonomy and paleoecology of the Christensen Bog mastodon bone bed, Hancock County, Indiana.* Illinois State Museum Reports of Investigations 38. Springfield.

Graham, R. W. and Kay, M. 1988. Taphonomic comparisons of cultural and noncultural faunal deposits at the Kimmswick and Barnhart sites, Jefferson County, Missouri. In (R. S. Laub, N. G. Miller, and D. W. Steadman, eds.) *Late Pleistocene and early Holocene paleoecology and archeology of the Eastern Great Lakes Region*, pp. 227–240. Bulletin of the Buffalo Society of Natural Sciences 33.

Graham, R. W. and Oliver, J. S. 1986. Taphonomy of ice-trapped coot population on Spring Lake, Tazewell County, Illinois. *American Quaternary Association 9th Biennial Meeting Program and Abstracts*, p. 84.

Grayson, D. K. 1978a. Minimum numbers and sample size in vertebrate faunal analysis. *American Antiquity* 43: 53–65.

1978b. Reconstructing mammalian communities: a discussion of Shotwell's method of paleoecological analysis. *Paleobiology* 4: 77–81.

1979. On the quantification of vertebrate archaeofaunas. In (M. B. Schiffer, ed.) *Advances in archaeological method and theory* vol. 2, pp. 199–237. New York: Academic Press.

1981. A critical view of the use of archaeological vertebrates in paleoenvironmental reconstruction. *Journal of Ethnobiology* 1: 28–38.

1983. The paleontology of Gatecliff Shelter: small mammals. In (D. H. Thomas, ed.) *The archaeology of Monitor Valley 2: Gatecliff Shelter*, pp. 99–135. American Museum of Natural History Anthropological Papers 59(1).

1984. *Quantitative zooarchaeology: topics in the analysis of archaeological faunas.* Orlando: Academic Press.

1986. Eoliths, archaeological ambiguity, and the generation of "middle-range" research. In (D. J. Meltzer, D. D. Fowler, and J. A. Sabloff, eds.) *American archaeology: past and future*, pp. 77–133. Washington: Smithsonian Institution Press.

1987. The biogeographic history of small mammals in the Great Basin: observations on the last 20,000 years. *Journal of Mammalogy* 68: 359–375.

1988. *Danger Cave, Last Supper Cave, and Hanging Rock Shelter: the faunas.* American Museum of Natural History Anthropological Papers 66(1): 1–130.

1989. Bone transport, bone destruction, and reverse utility curves. *Journal of Archaeological Science* 16: 643–652.

Greenfield, H. J. 1988. Bone consumption by pigs in a contemporary Serbian village: implications for the interpretation of prehistoric faunal assemblages. *Journal of Field Archaeology* 15: 473–479.

Guilday, J. E., Parmalee, P. W., and Tanner, D. P. 1962. Aboriginal butchering techniques at the Eschelman site (36LA12), Lancaster County, Pennsylvania. *Pennsylvania Archaeologist* 32: 59–83.

Guilday, J. E., Hamilton, H. W., Anderson, E., and Parmalee, P. W. 1978. *The Baker Bluff Cave deposit, Tennessee, and the late Pleistocene faunal gradient.* Bulletin of the Carnegie Museum of Natural History 11. Pittsburgh.

Gustafson, C. E., Gilbow, D., and Daugherty, R. D. 1979. The Manis Mastodon Site: early man on the Olympic Peninsula. *Canadian Journal of Anthropology* 3: 157–164.

Guthrie, R. D. 1967. Differential preservation and recovery of Pleistocene large mammal remains in Alaska. *Journal of Paleontology* 41: 243–246.

1990. *Frozen fauna of the Mammoth Steppe: the story of Blue Babe.* Chicago: University of Chicago Press.

Hackett, C. J. 1981. Microscopical focal destruction (tunnels) in exhumed human bones. *Medicine, Science and the Law* 21: 243–265.

Haglund, W. D. 1991. Applications of taphonomic models to forensic investigations. Ph.D. dissertation, University of Washington, Seattle.

Haneberg, W. C. 1983. A paradigmatic analysis of Darwin's use of uniformitarianism in *The origin of species. The Compass of Sigma Gamma Epsilon* 60: 89–94.

Hanson, C. B. 1980. Fluvial taphonomic processes: models and experiments. In (A. K. Behrensmeyer and A. P. Hill, eds.) *Fossils in the making*, pp. 156–181. Chicago: University of Chicago Press.

Hanson, D. B. and Buikstra, J. E. 1987. Histomorphological alteration in buried human bone from the Lower Illinois Valley: implications for palaeodietary research. *Journal of Archaeological Science* 14: 549–563.

Hare, P. E. 1980. Organic geochemistry of bone and its relation to the survival of bone in natural environments. In (A. K. Behrensmeyer and A. P. Hill, eds.) *Fossils in the making*, pp. 108–219. Chicago: University of Chicago Press.

Hargrave, L. L. 1970. *Mexican macaws: comparative osteology and survey of remains from the*

Southwest. Anthropological Papers of the University of Arizona No. 20. Tucson.

Haynes, G. 1980a. Evidence of carnivore gnawing on Pleistocene and Recent mammalian bones. *Paleobiology* 6: 341–351.

1980b. Prey bones and predators: potential ecologic information from analysis of bone sites. *Ossa* 7: 75–97.

1982. Utilization and skeletal disturbances of North American prey carcasses. *Arctic* 35: 266–281.

1983a. A guide for differentiating mammalian carnivore taxa responsible for gnaw damage to herbivore limb bones. *Paleobiology* 9: 164–172.

1983b. Frequencies of spiral and green-bone fractures on ungulate limb bones in modern surface assemblages. *American Antiquity* 48: 102–114.

1984. Age profiles in elephant and mammoth bone assemblages. *Quaternary Research* 24: 333–345.

1987. Proboscidean die-offs and die-outs: age profiles in fossil collections. *Journal of Archaeological Science* 14: 659–668.

1988a. Longitudinal studies of African elephant death and bone deposits. *Journal of Archaeological Science* 15: 131–157.

1988b. Mass deaths and serial predation: comparative taphonomic studies of modern large mammal death sites. *Journal of Archaeological Science* 15: 219–235.

1991. *Mammoths, mastodons, and elephants: biology, behavior, and the fossil record.* Cambridge: Cambridge University Press.

Haynes, G. and Stanford, D. 1984. On the possible utilization of *Camelops* by Early Man in North America. *Quaternary Research* 22: 216–230.

Hefti, E., Trechsel, U., Rüfenacht, H., and Fleisch, H. 1980. Use of dermestid beetles for cleaning bones. *Calcified Tissue International* 31: 45–47.

Henderson, J. 1987. Factors determining the state of preservation of human remains. In (A. Boddington, A. N. Garland, and R. C. Janaway, eds.) *Death, decay and reconstruction: approaches to archaeology and forensic science*, pp. 43–54. Manchester: Manchester University Press.

Hesse, B. and Wapnish, P. 1985. *Animal bone archeology: from objectives to analysis.* Manuals on Archeology No. 5. Washington: Taraxacum.

Hewson, R. and Kolb, H. H. 1976. Scavenging on sheep carcases by foxes (*Vulpes vulpes*) and badgers (*Meles meles*). *Journal of Zoology* 180: 496–498.

Hildebrand, M. 1968. *Anatomical preparations.* Berkeley: University of California Press.

1974. *Analysis of vertebrate structure.* New York: Wiley.

Hill, A. 1976. On carnivore and weathering damage to bone. *Current Anthropology* 17: 335–336.

1978. Taphonomical background to fossil man – problems in palaeoecology. In (W. W. Bishop, ed.) *Geological background to fossil man*, pp. 87–101. Edinburgh: Scottish Academic Press, Ltd.

1979a. Butchery and natural disarticulation: an investigatory technique. *American Antiquity* 44: 739–744.

1979b. Disarticulation and scattering of mammal skeletons. *Paleobiology* 5: 261–274.

1980. Early postmortem damage to the remains of some contemporary east African mammals. In (A. K. Behrensmeyer and A. P. Hill, eds.) *Fossils in the Making*, pp. 131–152. Chicago: University of Chicago Press.

1988. Taphonomy. In (I. Tattersall, E. Delson, and J. van Couvering, eds.) *Encyclopedia of human evolution and prehistory*, pp. 562–566. New York: Garland Publishing.

Hill, A. and Behrensmeyer, A. K. 1984. Disarticulation patterns of some modern East African mammals. *Paleobiology* 10: 366–376.

1985. Natural disarticulation and bison butchery. *American Antiquity* 50: 141–145.

Hill, A. and Walker, A. 1972. Procedures in vertebrate taphonomy: notes on a Uganda Miocene fossil locality. *Journal of the Geological Society of London* 128: 399–406.

Hiscock, P. 1990. A study in scarlet: taphonomy of inorganic artifacts. In (S. Solomon, I. Davidson, and D. Watson, eds.) *Problem solving in taphonomy*, pp. 34–49. Tempus vol. 2.

Hockett, B. S. 1989a. Archaeological significance of rabbit–raptor interactions in southern California. *North American Archaeologist* 10: 123–139.

   1989b. The concept of "carrying range:" a method for determining the role played by woodrats in contributing bones to archaeological sites. *Nevada Archaeologist* 7: 28–35.

   1991. Toward distinguishing human and raptor patterning on leporid bones. *American Antiquity* 56: 667–679.

Hodder, I. 1982. *The present past: an introduction to anthropology for archaeologists*. London: B. T. Batsford, Ltd.

Hoffman, R. 1988. The contribution of raptorial birds to patterning in small mammal assemblages. *Paleobiology* 14: 81–90.

Hoffman, R. and Hays, C. 1987. The eastern wood rat (*Neotoma floridana*) as a taphonomic factor in archaeological sites. *Journal of Archaeological Science* 14: 325–337.

Hofman, J. L. 1981. The refitting of chipped-stone artifacts as an analytical and interpretive tool. *Current Anthropology* 22: 691–693.

   1986. Vertical movement of artifacts in alluvial and stratified deposits. *Current Anthropology* 27: 163–171.

Holtzman, R. C. 1979. Maximum likelihood estimation of fossil assemblage composition. *Paleobiology* 5: 77–89.

Hooykaas, R. 1970. *Catastrophism in geology, its scientific character in relation to actualism and uniformitarianism*. Amsterdam: North Holland Publishing Company.

Horwitz, L. K. and Smith, P. 1988. The effects of striped hyaena activity on human remains. *Journal of Archaeological Science* 15: 471–481.

   1990. A radiographic study of the extent of variation in cortical bone thickness in Soay sheep. *Journal of Archaeological Science* 17: 655–664.

Howard, H. 1929. The avifauna of Emeryville Shellmound. *University of California Publications in Zoology* 32: 301–394. Berkeley.

Hubbert, M. K. 1967. Critique of the principle of uniformity. In (C. C. Albritton, ed.) *Uniformity and simplicity*, pp. 3–33. Geological Society of America Special Paper No. 89.

Hudson, J. L. 1990. Advancing methods in zooarchaeology: an ethnoarchaeological study among the Aka. Ph.D. dissertation, University of California, Santa Barbara.

   1991. Nonselective small game hunting strategies: an ethnoarchaeological study of Aka Pygmy sites. In (M. C. Stiner, ed.) *Human predators and prey mortality*, pp. 105–120. Boulder: Westview Press.

   1993. (ed.) *From bones to behavior: ethnoarchaeological and experimental contributions to the interpretation of faunal remains*. Center for Archaeological Investigations Occasional Paper No. 21. Carbondale: Southern Illinois University.

Huelsbeck, D. R. 1989. Zooarchaeological measures revisited. *Historical Archaeology* 23: 113–117.

   1991. Faunal remains and consumer behavior: what is being measured? *Historical Archaeology* 25: 62–76.

Hughes, A. R. 1954. Hyaenas versus australopithecines as agents of bone accumulation. *American Journal of Physical Anthropology* 12: 467–486.

Hughes, S. S. 1986. A modern analog to a bison jump. *The Wyoming Archaeologist* 29: 45–67.

Hulbert, R. C., Jr. 1982. Population dynamics of the three-toed horse *Neohipparion* from the Late Miocene of Florida. *Paleobiology* 8: 159–167.

Hutton, J. 1795. *Theory of the earth* (2 vols.). London: Cadell, Junior and Davies; Edinburgh: William Creech.

Irving, W. N. and Harington, C. R. 1973. Upper Pleistocene radiocarbon-dated artifacts from the northern Yukon. *Science* 179: 335–340.

Irving, W. N., Jopling, A. V., and Kritsch-Armstrong, J. 1989. Studies of bone technology and taphonomy, Old Crow Basin, Yukon Territory. In (R. Bonnichsen and M. H. Sorg, eds.) *Bone modification*, pp. 347–379. Orono: University of Maine Center for the Study of the First Americans.

Isaac, G. L. 1967. Toward the interpretation of occupational debris: some experiments and observations. *Kroeber Anthropological Society Papers* 37: 31–57.

James, S. R. 1989. Hominid use of fire in the Lower and Middle Pleistocene. *Current Anthropology* 30: 1–26.

Janaway, R. C. 1990. Experimental investigations of the burial environment of inhumation graves. In (D. E. Robinson, ed.) *Experimentation and reconstruction in environmental archaeology*, pp. 147–149. Oxford: Oxbow Books.

Jodry, M. A. and Stanford, D. J. 1992. Stewart's Cattle Guard site: an analysis of bison remains in a Folsom kill-butchery campsite. In (D. J. Stanford and J. S. Day, eds.) *Ice Age hunters of the Rockies*, pp. 101–168. Niwot: Denver Museum of Natural History and University Press of Colorado.

Johnson, D. 1975. Seasonal and microseral variation in the insect population on carrion. *American Midland Naturalist* 93: 79–90.

Johnson, D. L. and Haynes, C. V. 1985. Camels as taphonomic agents. *Quarternary Research* 24: 365–366.

Johnson, E. 1982. Paleoindian bone expediency tools: Lubbock Lake and Bonfire Shelter. *Canadian Journal of Anthropology* 2: 145–157.

   1983. A framework for interpretation in bone technology. In (G. M. LeMoine and A. S. MacEachern, eds.) *Carnivores, human scavengers & predators: a question of bone technology*, pp. 55–93. Calgary: University of Calgary Archaeological Association.

   1985. Current developments in bone technology. In (M. B. Schiffer, ed.) *Advances in archaeological method and theory* vol. 8, pp. 157–235. New York: Academic Press.

   1987. Cultural activities and interactions. In *Lubbock Lake: Late Quaternary studies on the Southern High Plains* (E. Johnson, ed.), pp. 120–158. College Station: Texas A&M University Press.

   1989. Human modified bones from early southern Plains Sites. In (R. Bonnichsen and M. H. Sorg, eds.) *Bone modification*, pp. 431–471. Orono: University of Maine Center for the Study of the First Americans.

Joly, N. 1887. *Man before metals*. London: Kegan Paul, Trench.

Jones, A. K. G. 1984. Some effects of the mammalian digestive system on fish bones. In (N. Desse-Berset, ed.) *2nd fish osteoarchaeology meeting, CNRS*, pp. 61–65. Center de recherches archéologiques. Notes et Monographies Techniques 16.

   1986. Fish bone survival in the digestive systems of the pig, dog and man: some experiments. In (D. C. Brinkhuizen and A. T. Clason, eds.) *Fish and archaeology*, pp. 53–61. British Archaeological Reports International Series 294.

   1990. Experiments with fish bones and otoliths: implications for the reconstruction of past diet and economy. In (D. E. Robinson, ed.) *Experimentation and reconstruction in environmental archaeology*, pp. 143–146. Oxford: Oxbow Books.

Jones, K. T. and Metcalfe, D. 1988. Bare bones archaeology: bone marrow indices and efficiency. *Journal of Archaeological Science* 15: 415–423.

Kay, R. F. 1988. Teeth. In (I. Tattersall, E. Delson, and J. van Couvering, eds.) *Encyclopedia of human evolution and prehistory*, pp. 571–578. New York: Garland Publishing.

Keepax, C. A. 1981. Avian egg-shell from archaeological sites. *Journal of Archaeological Science* 8: 315–335.

Kehoe, T. F. and Kehoe. A. B. 1960. Observations on the butchering technique at a prehistoric bison-kill in Montana. *American Antiquity* 25: 420–423.

Kelley, J. H. and Hanen, M. P. 1988. *Archaeology and the methodology of science*. Albuquerque: University of New Mexico Press.

Kent, S. 1981. The dog: an archaeologist's best friend or worst enemy – the spatial distribution of faunal remains. *Journal of Field Archaeology* 8: 367–372.

Kerbis-Peterhans, J. C. and Horwitz, L. K. 1992. A bone assemblage from a striped hyaena (*Hyaena hyaena*) den the in the Negev Desert, Israel. *Israel Journal of Zoology* 37: 225–245.

Kidwell, S. M. 1985. Paleobiological and sedimentological implications of fossil concentrations. *Nature* 318: 457–460.

  1986. Models for fossil concentrations: paleobiologic implications. *Paleobiology* 12: 6–24.

Kidwell, S. M. and Bosence, D. W. J. 1991. Taphonomy and time-averaging of marine shelly faunas. In (P. A. Allison and D. E. G. Briggs, eds.) *Taphonomy: releasing the data locked in the fossil record*, pp. 115–209. Topics in Geobiology vol. 9. New York: Plenum Press.

King, F. B. and Graham, R. W. 1981. Effects of ecological and paleoecological patterns on subsistence and paleoenvironmental reconstructions. *American Antiquity* 46: 128–142.

Kiszely, I. 1973. Derivatographic examination of subfossil and fossil bones. *Current Anthropology* 14: 280–286.

Kitching, J. M. 1980. On some fossil Arthropoda from the Limeworks, Makapansgat, Potgietersrus. *Palaeontologia Africana* 23: 63–68.

Kitts, D. B. 1977. *The structure of geology*. Dallas: Southern Methodist University Press.

Klein, R. G. 1975. Paleoanthropological implications of the nonarcheological bone assemblage from Swartklip I, south-western Cape Province, South Africa. *Quaternary Research* 5: 275–288.

  1976. The mammalian fauna of the Klasies River Mouth sites, southern Cape Province, South Africa. *South African Archaeological Bulletin* 31: 75–98.

  1982a. Age (mortality) profiles as a means of distinguishing hunted species from scavenged ones in Stone Age archaeological sites. *Paleobiology* 8: 151–158.

  1982b. Patterns of ungulate mortality and ungulate mortality profiles from Langebaanweg (Early Pliocene) and Elandsfontein (Middle Pleistocene), South-western Cape Province, South Africa. *Annals of the South African Museum* 90: 49–94.

  1989. Why does skeletal part representation differ between smaller and larger bovids at Klasies River Mouth and other archeological sites? *Journal of Archaeological Science* 16: 363–381.

Klein, R. G. and Cruz-Uribe, K. 1984. *The analysis of animal bones from archeological sites*. Chicago: University Chicago Press.

Klippel, W. E., Snyder, L. M., and Parmalee, P. W. 1987. Taphonomy and archaeologically recovered mammal bone from southeast Missouri. *Journal of Ethnobiology* 7: 155–169.

Knight, J. A. 1985. Differential preservation of calcined bone at the Hirundo Site, Alton, Maine. Unpublished M.S. thesis, Quaternary Science, University of Maine at Orono.

Koch, C. P. 1989. *Taphonomy: a bibliographic guide to the literature*. Orono: University of Maine Center for the Study of the First Americans.

Koike, H. and Ohtaishi, N. 1985. Prehistoric hunting pressure estimated by the age composition of excavated sika deer (*Cervus nippon*) using the annual layer of tooth cement. *Journal of Archaeological Science* 12: 443–456.

Kooyman, B. 1984. Moa utilisation at Owens Ferry, Otago, New Zealand. *New Zealand Journal of Archaeology* 6: 47–57.

  1990. Moa procurement: communal or individual hunting? In (L. B. Davis and B. O. K. Reeves, eds.) *Hunters of the recent past*, pp. 327–351. London: Unwin Hyman.

Korth, W. W. 1979. Taphonomy of microvertebrate fossil assemblages. *Annals of the Carnegie Museum* 48: 235–285.

Korth, W. W. and Evander, R. L. 1986. The use of age–frequency distributions of micromammals in the determination of attritional and catastrophic mortality of fossil assemblages. *Palaeogeography, Palaeoclimatology, Palaeoecology* 52: 227–236.

Kramer, C. 1979. (ed.) *Ethnoarchaeology: implications of ethnography for archaeology*. New York: Columbia University Press.

Kranz, P. M. 1974a. Computer simulation of fossil assemblage formation under conditions of anastrophic burial. *Journal of Paleontology* 48: 800–808.

1974b. The anastrophic burial of bivalves and its paleoecological significance. *Journal of Geology* 82: 237–266.

Krausman, P. R. and Bissonette, J. A. 1977. Bone-chewing behavior of desert mule deer. *Southwestern Naturalist* 22: 149–150.

Kreutzer, L. A. 1988. Megafaunal butchering at Lubbock Lake, Texas: a taphonomic reanalysis. *Quaternary Research* 30: 221–231.

1992. Bison and deer bone mineral densities: comparisons and implications for the interpretation of archaeological faunas. *Journal of Archaeological Science* 19: 271–294.

Krumbein, W. C. 1965. Sampling in paleontology. In (B. Kummel and D. Raup, eds.) *Handbook of paleontological techniques*, pp. 137–150. San Francisco: W. H. Freeman & Co.

Kuhn, T. S. 1970. *The structure of scientific revolutions*, 2nd edition. Chicago: University of Chicago Press.

Kummel, B. and Raup, D. 1965. (eds.) *Handbook of paleontological techniques*. San Francisco: W. H. Freeman & Co.

Kurtén, B. 1953. On the variation and population dynamics of fossil and recent mammal populations. *Acta Zoologica Fennica* 76: 1–122.

1958. Life and death of the Pleistocene cave bear. *Acta Zoologica Fennica* 95: 1–59.

1983. Variation and dynamics of a fossil antelope population. *Paleobiology* 9: 62–69.

Kusmer, K. D. 1986. Microvertebrate taphonomy in archaeological sites: an examination of owl deposition and the taphonomy of small mammals from Sentinel Cave, Oregon. Unpublished M.A. thesis, Department of Archaeology, Simon Fraser University, Burnaby, British Columbia.

1990. Taphonomy of owl pellet deposition. *Journal of Paleontology* 64: 629–637.

LaBarbera, M. 1989. Analyzing body size as a factor in ecology and evolution. *Annual Review of Ecology and Systematics* 20: 97–117.

Lam, Y. M. 1992. Variability in the behaviour of spotted hyaenas as taphonomic agents. *Journal of Archaeological Science* 19: 389–406.

Landals, A. 1990. The Maple Leaf site: implications of the analysis of small-scale bison kills. in (L. B. Davis and B. O. K. Reeves, eds.) *Hunters of the recent past*, pp. 122–151. London: Unwin Hyman.

Lange, F. W. 1980. Prehistory and hunter/gatherers: the role of analogs. *Mid-Continental Journal of Archaeology* 5: 133–147.

Lartet, E. 1860. On the coexistence of man with certain extinct quadrupeds, proved by fossil bones, from various Pleistocene deposits, bearing incisions made by sharp instruments. *M. G. S. Quarterly Journal of the Geological Society of London* 16: 471–479. (English translation published in 1969 [R. F. Heizer, ed.] *Man's discovery of his past*, pp. 122–131, Palo Alto: Peek Publications)

Lawrence, D. R. 1968. Taphonomy and information losses in fossil communities. *Geological Society of America Bulletin* 79: 1315–1330.

1971. The nature and structure of paleoecology. *Journal of Paleontology* 45: 593–607.

1979a. Biostratinomy. In (R. W. Fairbridge and D. Jablonski, eds.) *Encyclopedia of paleontology* , pp. 99–102. Stroudsburg: Dowden, Hutchinson & Ross, Inc.

1979b. Diagenesis of fossils – fossildiagenese. In (R. W. Fairbridge and D. Jablonski, eds.) *Encyclopedia of paleontology*, pp. 245–247. Stroudsburg: Dowden, Hutchinson & Ross, Inc.

1979c. Taphonomy. In (R. W. Fairbridge and D. Jablonski, eds.) *Encyclopedia of paleontology*, pp. 793–799. Stroudsburg: Dowden, Hutchinson & Ross, Inc.

LeMoine, G. M. and MacEachern, A. S. 1983. (eds.). *Carnivores, human scavengers & predators:*

a question of bone technology. Calgary: University of Calgary Archaeological Association.

Levine, M. A. 1983. Mortality models and the interpretation of horse population structure. In (G. Bailey, ed.) *Hunter-gatherer economy in prehistory: a European perspective*, pp. 23–46. London: Cambridge University Press.

Linz, G. M., Davis, J. E., Jr., Engeman, R. M., Otis, D. L., and Avery, M. L. 1991. Estimating survival of bird carcasses in cattail marshes. *Wildlife Society Bulletin* 19: 195–199.

Livingston, S. D. 1988. The avian and mammalian faunas from Lovelock Cave and the Humboldt Lakebed Site. Ph.D. dissertation, University of Washington. Ann Arbor: University Microfilms.

  1989. The taphonomic interpretation of avian skeletal part frequencies. *Journal of Archaeological Science* 16: 537–547.

Lowe, V. P. W. 1980. Variation in the digestion of prey by the tawny owl (*Strix aluco*). *Journal of Zoology* 192: 283–293.

Lucas, J. and Prévôt, L. E. 1991. Phosphates and fossil preservation. In (P. A. Allison and D. E. G. Briggs, eds.) *Taphonomy: releasing the data locked in the fossil record*, pp. 389–409. Topics in Geobiology vol. 9. New York: Plenum Press.

Lyman, R. L. 1978. Prehistoric butchering techniques in the Lower Granite Reservoir, southeastern Washington. *Tebiwa, Miscellaneous Papers of the Idaho State University Museum of Natural History* 13. Pocatello.

  1979a. *Archaeological faunal analysis: a bibliography*. Occasional Papers of the Idaho Museum of Natural History No. 31. Pocatello.

  1979b. Available meat from faunal remains: a consideration of techniques. *American Antiquity* 44: 536–546.

  1979c. Faunal analysis: an outline of method and theory with some suggestions. *Northwest Anthropological Research Notes* 13: 22–35.

  1980. Inferences from bone distributions in prehistoric sites in the Lower Granite Reservoir area, southeastern Washington. *Northwest Anthropological Research Notes* 14: 107–123.

  1982a. Archaeofaunas and subsistence studies. In (M. B. Schiffer, ed.) *Advances in archaeological method and theory* vol. 5, pp. 331–393. New York: Academic Press.

  1982b. The taphonomy of vertebrate archaeofaunas: bone density and differential survivorship of fossil classes. Ph.D. dissertation, University of Washington, Seattle. Ann Arbor: University Microfilms International.

  1984a. Bone density and differential survivorship of fossil classes. *Journal of Anthropological Archaeology* 3: 259–299.

  1984b. Broken bones, bone expediency tools, and bone pseudotools: lessons from the blast zone around Mount St. Helens, Washington. *American Antiquity* 49: 315–333.

  1985a. Bone frequencies: differential transport, *in situ* destruction, and the MGUI. *Journal of Archaeological Science* 12: 221–236.

  1985b. The paleozoology of the Avey's Orchard Site. In (J. R. Galm and R. A. Masten, eds.) *Avey's Orchard: Archaeological Investigations of a Late Prehistoric Columbia River Community*, pp. 243–319. Eastern Washington University Reports in Archaeology and History 100-42. Cheney.

  1987a. Archaeofaunas and butchery studies: a taphonomic perspective. In (M. B. Schiffer, ed.) *Advances in archaeological method and theory* vol. 10, pp. 249–337. San Diego: Academic Press.

  1987b. Hunting for evidence of Plio-Pleistocene hominid scavengers. *American Anthropologist* 89: 710–715.

  1987c. On the analysis of vertebrate mortality profiles: sample size, mortality type, and hunting pressure. *American Antiquity* 52: 125–142.

  1987d. On zooarchaeological measures of socioeconomic position and cost-efficient meat purchases. *Historical Archaeology* 21: 58–66.

1987e. Zooarchaeology and taphonomy: a general consideration. *Journal of Ethnobiology* 7: 93–117.

1988a. Was there a last supper at Last Supper Cave? In (D. K. Grayson, ed.) *Danger Cave, Last Supper Cave, and Hanging Rock Shelter: the faunas*, pp. 81–104. American Museum of Natural History Anthropological Papers 66(1).

1988b Zooarchaeology of 45DO189. In (J. R. Galm and R. L. Lyman, eds.) *Archaeological investigations at River Mile 590: the excavations at 45DO189*, pp. 97–141. Eastern Washington University Reports in Archaeology and History 100-61. Cheney.

1989a. Seal and sea lion hunting: a zooarchaeological study from the southern Northwest Coast of North America. *Journal of Anthropological Archaeology* 8: 68–99.

1989b. Taphonomy of cervids killed by the 18 May 1980 volcanic eruption of Mount St. Helens, Washington. In (R. Bonnichsen and M. Sorg, eds.) *Bone modification*, pp. 149–167. Orono: Center for the Study of the First Americans, University of Maine.

1991a. *Prehistory of the Oregon coast: the effects of excavation strategies and assemblage size on archaeological inquiry.* San Diego: Academic Press.

1991b. Subsistence change and pinniped hunting. In (M. C. Stiner, ed.) *Human predators and prey mortality*, pp. 187–199. Boulder: Westview Press.

1991c. Taphonomic problems with archaeological analysis of animal carcass utilization and transport. In (J. R. Purdue, W. E. Klippel, and B. W. Styles, eds.) *Beamers, bobwhites, and blue-points: tributes to the career of Paul W. Parmalee*, pp. 125–138. Illinois State Museum Scientific Papers vol. 23. Springfield.

1992a. Anatomical considerations of utility curves in zooarchaeology. *Journal of Archaeological Science* 19: 7–22.

1992b. Influences of mid-Holocene Altithermal climates on mammalian faunas and human subsistence in eastern Washington. *Journal of Ethnobiology* 12: 37–62.

1992c. Prehistoric seal and sea-lion butchering on the southern Northwest Coast. *American Antiquity* 57: 246–261.

1993a. A study of variation in the prehistoric butchery of large artiodactyls. In (E. Johnson, ed.) *Ancient peoples and landscapes* (in press). Lubbock: Texas Tech University Press.

1993b. Density–mediated attrition of bone assemblages: new insights. In (J. Hudson, ed.) *From bones to behavior*, pp. 324–341. Center for Archaeological Investigations Occasional Paper No. 21. Carbondale: Southern Illinois University Press.

n.d.a. Zooarchaeology of 45CH302. In (J. R. Galm, ed.) *Archaeological investigations at 45CH302*, Eastern Washington University Reports in Archaeology and History. Cheney (in press).

n.d.b. Zooarchaeology of 45GR445. In (J. R. Galm, ed.) *The archeology of Salishan Mesa*, Eastern Washington University Reports in Archaeology and History. Cheney (in press).

Lyman, R. L. and Fox, G. L. 1989. A critical evaluation of bone weathering as an indication of bone assemblage formation. *Journal of Archaeological Science* 16: 293–317.

Lyman, R. L. and O'Brien, M. J. 1987. Plow-zone zooarchaeology: fragmentation and identifiability. *Journal of Field Archaeology* 14: 493–498.

Lyman, R. L., Houghton, L. E., and Chambers, A. L. 1992a. The effect of structural density on marmot skeletal part representation in archaeological sites. *Journal of Archaeological Science* 19: 557–573.

Lyman, R. L., Savelle, J. M., and Whitridge, P. 1992b. Derivation and application of a food utility index for Phocid seals. *Journal of Archaeological Science* 19: 531–555.

Lyon, P. J. 1970. Differential bone destruction: an ethnographic example. *American Antiquity* 35: 213–215.

MacGregor, A. 1985. *Bone, antler, ivory and horn: the technology of skeletal materials since the Roman Period.* London: Croom Helm.

Mackin, J. H. 1963. Rational and empirical methods of investigation in geology. In (C. C. Albritton, Jr., ed.) *The fabric of geology*, pp. 135–163. Reading, MA.: Addison-Wesley Publishing Co.

Maguire, J. M., Pemberton, D., and Collett, M. H. 1980. The Makapansgat Limeworks Grey Breccia: hominids, hyaenas, hystricids or hillwash? *Paleontologia Africana* 23: 75–98.

Maltby, J. M. 1985a. Assessing variations in Iron Age and Roman butchery practices: the need for quantification. In (N. R. J. Fieller, D. D. Gilbertson, and N. G. A. Ralph, eds.) *Paleobiological investigations: research design, methods and data analysis*, pp. 19–30. British Archaeological Reports International Series 266.

   1985b. Patterns in faunal assemblage variability. In (G. Barker and C. Gamble, eds.) *Beyond domestication in prehistoric Europe*, pp. 33–74. London: Academic Press.

Marchiafava, V., Bonucci, E., and Ascenzi, A. 1974. Fungal osteoclasia: a model of dead bone resorption. *Calcified Tissue Research* 14: 195–210.

Marean, C. W. 1991. Measuring the post-depositional destruction of bone in archaeological assemblages. *Journal of Archaeological Science* 18: 677–694.

Marean, C. W. and Spencer, L. M. 1991. Impact of carnivore ravaging on zooarchaeological measures of element abundance. *American Antiquity* 56: 645–658.

Marean, C. W., Spencer, L. M., Blumenschine, R. J., and Capaldo, S. D. 1992. Captive hyaena bone choice and destruction, the schlepp effect and Olduvai archaeofaunas. *Journal of Archaeological Science* 19: 101–121.

Marshall, F. 1986. Implications of bone modification in a Neolithic faunal assemblage for the study of early hominid butchery and subsistence practices. *Journal of Human Evolution* 15: 661–672.

   1993. Food sharing and body part representation in Okiek faunal assemblages. *Journal of Archaeological Science* 20 (in press).

Marshall, F. and Pilgram, T. 1991. Meat versus within-bone nutrients: another look at the meaning of body part representation in archaeological sites. *Journal of Archaeological Science* 18: 149–163.

Marshall, L. G. 1989. Bone modification and "the laws of burial." In (R. Bonnichsen and M. H. Sorg) *Bone modification*, pp. 7–24. Orono: University of Maine Center for the Study of the First Americans.

Martill, D. M. 1990. Bones as stones: the contribution of vertebrate remains to the lithologic record. In (S. K. Donovan, ed.) *The processes of fossilization*, pp. 270–292. New York: Columbia University Press.

Martin, H. 1910. La percussion osseuse et les esquilles qui en dérivent expérimentation. *Société Préhistorique Française Bulletin* 7: 299–309.

Martin, P. S. and Klein, R. G. 1984. (eds.) *Quarternary extinctions*. Tucson: University of Arizona Press.

Matthews, W. H., III. 1962. *Fossils*. New York: Barnes & Noble.

McGrew, P. O. 1975. Taphonomy of Eocene fish from Fossil Basin, Wyoming. *Fieldiana Geology* 33: 257–270.

McKenna, M. C. 1962. Collecting small fossils by washing and screening. *Curator* 5: 221–235.

McKinney, M. L. 1990. Completeness of the fossil record: an overview. In (S. K. Donovan, ed.) *The processes of fossilization*, pp. 66–83. New York: Columbia University Press.

Meadow, R. H. 1978. Effects of context on the interpretation of faunal remains: a case study. In (R. H. Meadow and M. A. Zeder, eds.) *Approaches to faunal analysis in the Middle East*, pp. 15–21. Peabody Museum Bulletin 2. Cambridge.

   1981. Animal bones – problems for the archaeologist together with some possible solutions. *Paléorient* 6: 65-77.

Medlock, R. C. 1975. Faunal analysis. In (M. B. Schiffer and J. H. House, eds.) *The Cache River*

488   *Bibliography*

archeological project: an experiment in contract archeology, pp. 223–242. Fayetteville: Arkansas Archeological Survey Research Series No. 8.

Mehl, M. G. 1966. The Domebo mammoth: vertebrate paleomortology. In (F. C. Leonhardy, ed.) *Domebo: a Paleo-Indian mammoth kill in the Prairie-Plains*, pp. 27–30. Contributions of the Museum of the Great Plains 1. Lawton, Oklahoma.

Meinke, D. K. 1979. Bones and teeth. In (R. W. Fairbridge and D. Jablonski, eds.) *Encyclopedia of paleontology*, pp. 122–131. Stroudsburg: Dowden, Hutchinson & Ross, Inc.

Mellet, J. S. 1974. Scatological origins of microvertebrate fossils. *Science* 185: 349–350.

Meltzer, D. J., Leonard, R. D., and Stratton, S. K. 1992. The relationship between sample size and diversity in archaeological assemblages. *Journal of Archaeological Science* 19: 375–387.

Mengoni-Gonalons, G. L. 1991. La llama y sus productos primarios. *Arqueologia: Revista de la Sección Prehistoria* 1: 179–196.

Metcalfe, D. and Jones, K. T. 1988. A reconsideration of animal body part utility indices. *American Antiquity* 53: 486–504.

Meyer, C. A. 1991. Burial experiments with marine turtle carcasses and their paleoecological significance. *Palaios* 6: 89–96.

Micozzi, M. S. 1986. Experimental study of postmortem change under field conditions: effects of freezing, thawing, and mechanical injury. *Journal of Forensic Science* 31: 953–961.

  1991. *Postmortem change in human and animal remains: a systematic approach.* Springfield: Charles C. Thomas.

Miller, G. J. 1969. A study of cuts, grooves and other marks on recent and fossil bone: I. animal tooth marks. *Tebiwa, Journal of the Idaho State University Museum* 12: 20–26. Pocatello.

  1975. A study of cuts, grooves, and other marks on recent and fossil bones: II, weathering cracks, fractures, splinters, and other similar natural phenomena. In (E. Swanson, ed.) *Lithic technology*, pp. 212–226. The Hague: Mouton.

Miller, S. J. 1989. Characteristics of mammoth bone reduction at Owl Cave, the Wasden Site, Idaho. In (R. Bonnichsen and M. H. Sorg, eds.) *Bone modification*, pp. 381–393. Orono: University of Maine Center for the Study of the First Americans.

Milne, L. J. and Milne, M. 1976. The social behavior of burying beetles. *Scientific American* 230(2): 84–89.

Modell, W. 1969. Horns and antlers. *Scientific American* 220(4): 114–122.

Monks, G. G. 1981. Seasonality studies. In (M. B. Schiffer, ed.) *Advances in archaeological method and theory* vol. 4, pp. 177–240. New York: Academic Press.

Montelius, O. 1888. *The civilization of Sweden in heathen times.* London: Macmillan.

Moore, K. L. 1985. *Clinically oriented anatomy*, 2nd edition. Baltimore: Williams & Wilkins.

Morlan, R. E. 1980. *Taphonomy and archaeology in the upper Pleistocene of the northern Yukon Territory: a glimpse of the peopling of the New World.* Archaeological Survey of Canada Paper No. 94, Mercury Series. Ottawa: National Museum of Man.

  1983. Spiral fractures on limb bones: which ones are artificial? In (G. M. LeMoine and A. S. MacEachern, eds.) *Carnivores, human scavengers and predators: a question of bone technology*, pp. 241–269. Calgary: University of Calgary Archaeological Association.

  1984. Toward the definition of criteria for the recognition of artificial bone alterations. *Quarternary Research* 22: 160–171.

  1988. Pre-Clovis people: early discoveries of America? In (R. C. Carlisle, ed.) *Americans before Columbus: Ice-Age origins*, pp. 31–43. Ethnology Mongraphs 12. Pittsburgh: University of Pittsburgh Department of Anthropology.

Morlot, A. 1861. General views on archaeology. *Smithsonian Institution Annual Report for 1860*, pp. 284–343.

Morse, D. 1983. The skeletal pathology of trauma. In (D. Morse, J. Duncan, and J. Stoutmire, eds.) *Handbook of forensic archaeology and anthropology*, pp. 145–185. Tallahassee: Privately published.

Müller, A. H. 1951. *Grundlagen der biostratonomie*. Deutsche Akademie der Wissenschaften zu Berlin, Abhandlungen, 1950, No. 3.

  1963. *Lehrbuch de paläozoologie, band 1. allgemeine grundlagen*. Jena: Gustav Fischer Verlag.

Mundy, P. J. and Ledger, J. A. 1976. Griffon vultures, carnivores and bones. *South African Journal of Science* 72: 106–110.

Munson, P. J. 1969. Comments on Binford's "Smudge pits and hide smoking: the use of analogy in archaeological reasoning". *American Antiquity* 34: 83–85.

Munthe, L. and McLeod, S. A. 1975. *Collection of taphonomic information from fossil and recent vertebrate specimens with a selected bibliography*. PaleoBios 19. Contributions from the University of California-Berkeley Museum of Paleontology.

Münzel, S. 1986. Coding system for bone fragments. In (L. H. van Wijngaarden-Bakker, ed.) *Database management and zooarchaeology*, pp. 193–195. PACT 14.

Murray, T. and Walker, M. J. 1988. Like WHAT? a practical question of analogical inference and archaeological meaningfulness. *Journal of Anthropological Archaeology* 7: 248–287.

Myers, T., Voorhies, M. R., and Corner, R. G. 1980. Spiral fractures and bone pseudotools at paleontological sites. *American Antiquity* 45: 483–489.

Nairn, A. E. M. 1965. Uniformitarianism and environment. *Palaeogeography, Palaeoclimatology, Palaeoecology* 1: 5–11.

Nelson, D. E., Morlan, R. E., Vogel, J. S., Southon, J. R., and Harington, C. R. 1986. New dates on northern Yukon artifacts: Holocene not Upper Pleistocene. *Science* 232: 749–751.

Nimmo, B. W. 1971. Population dynamics of a Wyoming pronghorn cohort from the Eden-Farson Site, 48SW304. *Plains Anthropologist* 16: 285–288.

Nitecki, M. H. and Nitecki, D. V. 1987. (eds.) *The evolution of human hunting*. New York: Plenum Press.

Noe-Nygaard, N. 1974. Mesolithic hunting in Denmark illustrated by bone injuries caused by human weapons. *Journal of Archaeological Science* 1: 217–218.

  1975a. Bone injuries caused by human weapons in Mesolithic Denmark. In (A. T. Clason, ed.) *Archaeozoological studies*, pp. 151–159. Amsterdam: North Holland Publishing Co.

  1975b. Two shoulder blades with healed lesions from Star Carr. *Proceedings of the Prehistoric Society* 41: 10–16.

  1977. Butchering and marrow fracturing as a taphonomic factor in archaeological deposits. *Paleobiology* 3: 218–237.

  1987. Taphonomy in archaeology with special emphasis on man as a biasing factor. *Journal of Danish Archaeology* 6: 7–52.

  1989. Man-made trace fossils on bones. *Human Evolution* 4: 461–491.

Noe-Nygaard, N. and Richter, J. 1990. Seventeen wild boar mandibles from Sludegards Sømose – offal or sacrifice? In (D. E. Robinson, ed.) *Experimentation and reconstruction in environmental archaeology*, pp. 175–189. Oxford: Oxbow Books.

O'Connell, J. F., Hawkes, K., and Jones, N. B. 1988. Hadza hunting, butchering, and bone transport and their archaeological implications. *Journal of Anthropological Research* 44: 113–161.

  1990. Reanalysis of large mammal body part transport among the Hadza. *Journal of Archaeological Science* 17: 301–316.

  1992. Patterns in the distribution, site structure and assemblage composition of Hadza kill-butchering sites. *Journal of Archaeological Science* 19: 319–345.

O'Connell, J. F. and Marshall, B. 1989. Analysis of kangaroo body part transport among the Alyawara of Central Australia. *Journal of Archaeological Science* 16: 393–405.

Odum, E. P. 1971. *Fundamentals of ecology*, 3rd edition. Philadelphia: W. B. Saunders Company.

O'Gara, B. W. and Matson, G. 1975. Growth and casting of horns by pronghorns and exfoliation of horns by bovids. *Journal of Mammalogy* 56: 829–846.

Oliver, J. S. 1989. Analogues and site context: bone damages from Shield Trap Cave (24CB91),

Carbon County, Montana, U.S.A. In (R. Bonnichsen and M. H. Sorg, eds.) *Bone modification*, pp. 73–98. Orono: University of Maine Center for the Study of the First Americans.

Olsen, S. J. 1968. Fish, amphibian and reptile remains from archaeological sites, part I: southeastern and southwestern United States. *Papers of the Peabody Museum of Archaeology and Ethnology* 56(2). Cambridge: Harvard University.

Olsen, S. L. 1988. The identification of stone and metal tool marks on bone artifacts. In (S. L. Olsen, ed.) *Scanning electron microscopy in archaeology*, pp. 337–360. British Archaeological Reports International Series 452.

1989a. On distinguishing natural from cultural damage on archaeological antler. *Journal of Archaeological Science* 16: 125–135.

1989b. Solutré: a theoretical approach to the reconstruction of Upper Paleolithic hunting strategies. *Journal of Human Evolution* 18: 295–327.

Olsen, S. L. and Shipman, P. 1988. Surface modification on bone: trampling versus butchery. *Journal of Archaeological Science* 15: 535–553.

Olson, E. C. 1952. The evolution of a Permian vertebrate chronofauna. *Evolution* 6: 181–196.

1958. Fauna of the Vale and Choza: 14. summary, review, and integration of the geology and the faunas. *Fieldiana Geology* 10: 397–448.

1980. Taphonomy: its history and role in community evolution. In (A. K. Behrensmeyer and A. P. Hill, eds.) *Fossils in the making*, pp. 6–19. Chicago: University of Chicago Press.

Ortner, D. J. and Putschar, W. G. J. 1981. *Identification of pathological conditions in human skeletal remains*. Smithsonian Contributions to Anthropology No. 28.

Ortner, D. J., Von Endt, D. W., and Robinson, M. S. 1972. The effect of temperature on protein decay in bone: its significance in nitrogen dating of archaeological specimens. *American Antiquity* 37: 514–520.

Parker, A. J. 1988. The birds of Roman Britain. *Oxford Journal of Archaeology* 7: 197–226.

Parsons, M. K. and Brett, C. E. 1990. Taphonomic processes and biases in modern marine environments: an actualistic perspective on fossil assemblage preservation. In (S. K. Donovan, ed.) *The processes of fossilization*, pp. 22–65. New York: Columbia University Press.

Pate, F. D. and Hutton, J. T. 1988. The use of soil chemistry data to address post-mortem diagenesis in bone mineral. *Journal of Archaeological Science* 15: 729–739.

Payne, J. A. 1965. Summer carrion study of the baby pig *Sus scrofa. Ecology* 46: 592–602.

Payne, J. A., King, R. E., and Beinhart, G. 1968. Arthropod succession and decomposition of buried pigs. *Nature* 219: 1180–1181.

Payne, S. 1972a. On the interpretation of bone samples from archaeological sites. In (E. S. Higgs, ed.) *Papers in economic prehistory*, pp. 65–81. Cambridge: Cambridge University Press.

1972b. Partial recovery and sample bias: the results of some sieving experiments. In (E. S. Higgs, ed.) *Papers in economic prehistory*, pp. 49–64. Cambridge: Cambridge University Press.

Peale, T. R. 1871. On the uses of the brain and marrow of animals among the Indians of North America. *Smithsonian Institution Annual Report for 1870*, pp. 390–391.

Pei, W.-C. 1932. Preliminary note on some incised, cut and broken bones found in association with *Sinanthropus* remains and lithic artifacts from Choukoutien. *Geological Society of China Bulletin* 12: 105–108.

1938. *Le rôle des animaux et des causes naturelles dans la cassure des os*. Palaeontologica Sinica, Series D, New Series 7 (118). Nanking: Geological Survey of China.

Perkins, D. and Daly, P. 1968. A hunter's village in Neolithic Turkey. *Scientific American* 219(5): 96–106.

Perzigian, A. J. 1973. Osteoporotic bone loss in two prehistoric Indian populations. *American Journal of Physical Anthropology* 39: 87–96.

Piepenbrink, H. 1986. Two examples of biogenous dead bone decomposition and their consequences for taphonomic interpretation. *Journal of Archaeological Science* 13: 417–430.

Plotnick, R. E. and Speyer, S. E. 1989. (eds.) *Death, decay, disintegration: the newsletter for research on taphonomy* 1: 1–33. Chicago: University of Illinois.

    1990. (eds.) *Death, decay, disintegration: the newsletter for research on taphonomy* 2: 1–22.

Plotnick, R. E. and Walker, S. E. 1991. (eds.) *Death, decay, disintegration: the newsletter for research on taphonomy* 3: 1–18. Chicago: University of Illinois.

Plug, I. 1978. Collecting patterns of six species of vultures (Aves: Accipitridae). *Annals of the Transvaal Museum* 31(6): 51–63.

Polacheck, T. 1985. The sampling distribution of age-specific survival estimates from an age distribution. *Journal of Wildlife Management* 49: 180–184.

Potts, R. 1982. Lower Pleistocene site formation and hominid activities at Olduvai Gorge, Tanzania. Ph.D. dissertation, Harvard University. Ann Arbor: University Microfilms.

    1984. Hominid hunters? Problems of identifying the earliest hunter/gatherers. In (R. Foley, ed.) *Hominid evolution and community ecology*, pp. 129–166. London: Academic Press.

    1986. Temporal span of bone accumulations at Olduvai Gorge and implications for early hominid foraging behavior. *Paleobiology* 12: 25–31.

    1988. *Early hominid activities at Olduvai.* New York: Aldine de Gruyter.

Potts, R. and Shipman, P. 1981. Cutmarks made by stone tools on bones from Olduvai Gorge, Tanzania. *Nature* 291: 577–580.

Pozorski, S. G. 1979. Late prehistoric llama remains from the Moche Valley, Peru. *Annals of the Carnegie Museum* 48: 139–170.

Price, T. D., Blitz, J., Burton, J., and Ezzo, J. A. 1992. Diagenesis in prehistoric bone: problems and solutions. *Journal of Archaeological Science* 19: 513–529.

Rackham, J. 1983. Faunal sample to subsistence economy. In (M. Jones, ed.) *Integrating the subsistence economy*, pp. 251–277. British Archaeological Reports International Series 181.

Ragir, S. 1967. A review of techniques for archaeological sampling. In (R. F. Heizer and J. A. Graham, eds.) *A guide to field methods in archaeology: approaches to the anthropology of the dead*, pp. 181–197. Palo Alto: National Press Books.

Rapson, D. J. 1990. Pattern and process in intra-site spatial analysis: site structural and faunal research at the Bugas-Holding Site. Ph.D. dissertation, University of New Mexico, Albuquerque. Ann Arbor: University Microfilms International.

Read, C. E. 1971. Animal bones and human behavior: approaches to faunal analysis in archeology. Ph.D. dissertation, University of California-Los Angeles. Ann Arbor: University Microfilms International.

Read-Martin, C. E. and Read, D. W. 1975. Australopithecine scavenging and human evolution: an approach from faunal analysis. *Current Anthropology* 16: 359–368.

Reed, C. A. 1963. Osteo-archaeology. In (D. Brothwell and E. Higgs, eds.) *Science in Archaeology*, pp. 204–216. New York: Basic Books.

Reher, C. A. 1974. Population study of the Casper Site bison. In (G. C. Frison, ed.) *The Casper Site: A Hell Gap bison kill on the High Plains*, pp. 113–124. New York: Academic Press.

Reher, C. A. and Frison, G. C. 1980. *The Vore Site, 48CK302, a stratified buffalo jump in the Wyoming Black Hills.* Plains Anthropologist Memoir 16.

Reitz, E. J. 1986. Vertebrate fauna from Locus 39, Puerto Real, Haiti. *Journal of Field Archaeology* 13: 317–328.

Rensberger, J. M. and Krentz, H. B. 1988. Microscopic effects of predator digestion on the surfaces of bones and teeth. *Scanning Microscopy* 2: 1541–1551.

Rensch, B. 1960. *Evolution above the species level.* New York: Columbia University Press.

Retallack, G. 1984. Completeness of the rock and fossil record: some estimates using fossil soils. *Paleobiology* 10: 59–78.

1988. Down-to-earth approaches to vertebrate paleontology. *Palaios* 3: 335–344.

1990. *Soils of the past: an introduction to paleopedology.* Boston: Unwin Hyman.

Rich, P. V. 1980. Preliminary report on the fossil avian remains from late Tertiary sediments at Langebaanweg (Cape Province), South Africa. *South African Journal of Science* 76: 166–170.

Richardson, P. R. K. 1980. Carnivore damage to antelope bones and its archaeological implications. *Paleontologia Africana* 23:109–125.

Richardson, P. R. K., Mundy, P. J., and Plug, I. 1986. Bone crushing carnivores and their significance to osteodystrophy in griffon vulture chicks. *Journal of Zoology* 210: 23–43.

Richter, J. 1986. Experimental study of heat induced morphological changes in fish bone collagen. *Journal of Archaeological Science* 13: 477–481.

Richter, R. 1928. *Aktuopaläontologie und Paläobiologie, eine Abgrenzung.* Senkenbergiana 19.

Robison, N. D. 1978. Zooarchaeology: its history and development. *Tennessee Anthropological Association Miscellaneous Paper* 2: 1–22. Knoxville.

Rodriguez, W. C. and Bass, W. M. 1983. Insect activity and its relationship to decay rates of human cadavers in east Tennessee. *Journal of Forensic Sciences* 28: 423–430.

Rogers, R. R. 1992. Non-marine borings in dinosaur bones from the Upper Cretaceous Two Medicine Formation, northwestern Montana. *Journal of Vertebrate Paleontology* 12: 528–531.

Rolfe, W. D. I. and Brett, D. W. 1969. Fossilization processes. In (G. Eglinton and M. T. J. Murphy, eds.) *Organic geochemistry: methods and results*, pp. 213–244. Berlin: Springer-Verlag.

Romer, A. S. and Parsons, T. S. 1977. *The vertebrate body*, 5th edition. Philadelphia: W. B. Saunders Company.

Rosene, W., Jr. and Lay, D. W. 1963. Disappearance and visibility of quail remains. *Journal of Wildlife Management* 27: 139–142.

Rudwick, M. J. S. 1971. Uniformity and progression: reflections on the structure of geological theory in the age of Lyell. In (D. Roller, ed.) *Perspectives in the history of science and technology*, pp. 209–227. Norman: University of Oklahoma Press.

1976. *The meaning of fossils*, 2nd edition. New York: Neale Watson Academic Publications.

Russell, M. D. 1987. Mortuary practices at the Krapina Neandertal site. *American Journal of Physical Anthropology* 72: 381–397.

Russell, M. P. 1992. Review of "The processes of fossilization," edited by S. K. Donovan. *Palaios* 7: 331–332.

Sadek-Kooros, H. 1972. Primitive bone fracturing: a method of research. *American Antiquity* 37: 369–382.

1975. Intentional fracturing of bone: description of criteria. In (A. T. Clason, ed.) *Archaeozoological studies*, pp. 139–150. Amsterdam: North Holland Publishing Co.

Salmon, M. H. 1981. Ascribing functions to archaeological objects. *Philosophy of the Social Sciences* 11: 19–26.

Salmon, W. C. 1953. The uniformity of nature. *Philosophy and Phenomenological Research* 14: 39–48.

1963. *Logic.* Englewood Cliffs: Prentice-Hall, Inc.

1984. *Scientific explanation and the causal structure of the world.* Princeton: Princeton University Press.

Sander, P. M. 1992. The Norian *Plateosaurus* bonebeds of central Europe and their taphonomy. *Palaeogeography, Palaeoclimatology, Palaeoecology* 93: 255–299.

Sanders, H. L. 1968. Marine benthic diversity: a comparative study. *American Naturalist* 102: 243–282.

Saunders, J. J. 1977. Late Pleistocene vertebrates of the western Ozark Highland, Missouri. *Illinois State Museum Reports of Investigations* No. 33. Springfield.

Schäfer, W. 1962. *Aktuo-paläontologie nach Studien in der Nordsee*. Frankfurt am Main: Verlag Waldemar Kramer.

   1972. *Ecology and palaeoecology of marine environments*. Chicago: University of Chicago Press. (English translation of Schäfer 1962 by I. Oertel, edited by C. Y. Craig).

Scharnberger, C. K., Bushman, J. R., and Shea, J. H. 1983. Comments and reply on 'Twelve fallacies of uniformitarianism.' *Geology* 11: 312–313.

Schick, K. D., Toth, N., and Daeschler, E. 1989. An early paleontological assemblage as an archaeological test case. In (R. Bonnichsen and M. H. Sorg, eds.) *Bone modification*, pp. 121–137. Orono: University of Maine Center for the Study of the First Americans.

Schiffer, M. B. 1972. Archaeological context and systemic context. *American Antiquity* 37: 155–165.

   1976. *Behavioral archaeology*. New York: Academic Press.

   1978. Methodological issues in ethnoarchaeology. In (R. A. Gould, ed.) *Explorations in ethnoarchaeology*, pp. 229–247. Albuquerque: University of New Mexico Press.

   1983. Toward the identification of formation processes. *American Antiquity* 48: 675–706.

   1985. Is there a "Pompeii Premise" in archaeology? *Journal of Anthropological Research* 41: 18–41.

   1987. *Formation processes of the archaeological record*. Albuquerque: University of New Mexico Press.

Schild, R. 1976. The final Paleolithic settlements of the European Plain. *Scientific American* 234(2): 88–99.

Schneider, J. S. and Everson, G. D. 1989. The desert tortoise (*Xerobates agassizii*) in the prehistory of the southwestern Great Basin and adjacent areas. *Journal of California and Great Basin Anthropology* 11: 175–202.

Schopf, J. M. 1975. Modes of fossil preservation. *Review of Paleobotany and Palynology* 20: 27–53.

Schulz, P. D. and Gust, S. M. 1983. Faunal remains and social status in 19th century Sacramento. *Historical Archaeology* 17: 44–53.

Schumm, S. A. 1985. Explanation and extrapolation in geomorphology: seven reasons for geologic uncertainty. *Transactions of the Japanese Geomorphological Union* 6(1): 1–18.

Scott, G. H. 1963. Uniformitarianism, the uniformity of nature, and paleoecology. *New Zealand Journal of Geology and Geophysics* 6: 510–527.

Sdzuy, K. 1966. An improved method of analysing distortion in fossils. *Palaeontology* 9: 125–134.

Seilacher, A. 1990. Taphonomy of fossil-Lagerstätten: overview. In (D. E. G. Briggs and P. R. Crowther, eds.) *Palaeobiology: a synthesis*, pp. 266–270. Oxford: Blackwell Scientific Publications.

   1992. Dynamic taphonomy: the process-related view of fossil-lagerstätten. In (S. Fernández López, ed.) *Conferencias de la reunion de tafonomia y fosilizacion*, pp. 109–125. Madrid: Editorial Complutense.

Semenov, S. 1964. *Prehistoric technology*. Bath: Adams & Dart.

Shea, B. T. 1988a. Allometry. In (I. Tattersall, E. Delson, and J. van Couvering, eds.) *Encyclopedia of human evolution and prehistory*, pp. 20–22. New York: Garland Publishing.

   1988b. Ontogeny. In (I. Tattersall, E. Delson, and J. van Couvering, eds.) *Encyclopedia of human evolution and prehistory*, pp. 401. New York: Garland Publishing.

   1992. Developmental perspective on size change and allometry in evolution. *Evolutionary Anthropology* 1: 125–134.

Shea, J. H. 1982. Twelve fallacies of uniformitarianism. *Geology* 10: 455–460.

Shennan, S. 1988. *Quantifying archaeology*. San Diego: Academic Press.

Shipman, P. 1979. What are all these bones doing here? Confessions of a taphonomist. *Harvard Magazine* Nov.–Dec., pp. 42–46.

1981a. Applications of scanning electron microscopy to taphonomic problems. In (A. M. Cantwell, J. B. Griffin, and N. A. Rothschild, eds.) *The research potential of anthropological museum collections*, pp. 357–385. Annals of the New York Academy of Science 376.

1981b. *Life history of a fossil: an introduction to taphonomy and paleoecology.* Cambridge: Harvard University Press.

1983. Early hominid lifestyles: hunting and gathering or foraging and scavenging? In (J. Clutton-Brock and C. Grigson, eds.) *Animals and archaeology: 1. hunters and their prey*, pp. 31–49. British Archaological Reports International Series 163.

1986a. Scavenging or hunting in early hominids: theoretical framework and test. *American Anthropologist* 88: 27–43.

1986b. Studies of hominid–faunal interactions at Olduvai Gorge. *Journal of Human Evolution* 15: 691–706.

1987. Response to R. Lee Lyman. *American Anthropologist* 89: 715–717.

1988a. Actualistic studies of animal resources and hominid activities. In (S. L. Olsen, ed.) *Scanning electron microscopy in archaeology*, pp. 261–285. British Archaeological Reports International Series 452.

1988b. Diet and subsistence strategies at Olduvai Gorge. In (B. V. Kennedy and G. M. LeMoine, eds.) *Diet and subsistence: current archaeological perspectives*, pp. 3–12. Calgary: University of Calgary Archaeological Association.

1989. Altered bones from Olduvai Gorge, Tanzania: techniques, problems, and implications of their recognition. In (R. Bonnichsen and M. H. Sorg, eds.) *Bone modification*, pp. 317–334. Orono: University of Maine Center for the Study of the First Americans.

Shipman, P., Bosler, W., and Davis, K. L. 1981. Butchering of giant geladas at an Acheulian site. *Current Anthropology* 22: 257–268.

Shipman, P., Fisher, D. C., and Rose, J. 1984a. Mastodon butchery: microscopic evidence of carcass processing and bone tool use. *Paleobiology* 10: 358–365.

Shipman, P., Foster, G., and Schoeninger, M. 1984b. Burnt bones and teeth: an experimental study of color, morphology, crystal structure and shrinkage. *Journal of Archaeological Science* 11: 307–325.

Shipman, P. and Phillips, J. E. 1976. On scavenging by hominids and other carnivores. *Current Anthropology* 17: 170–172.

Shipman, P. and Phillips-Conroy, J. E. 1977. Hominid tool-making versus carnivore scavenging. *American Journal of Physical Anthropology* 46: 77–86.

Shipman, P. and Rose, J. 1983a. Early hominid hunting, butchering, and carcass-processing behaviors: approaches to the fossil record. *Journal of Anthropological Archaeology* 2: 57–98.

1983b. Evidence of butchery activities at Torralba and Ambrona: an evaluation using microscopic techniques. *Journal of Archaeological Science* 10: 465–474.

1984. Cutmark mimics on modern and fossil bovid bones. *Current Anthropology* 25: 116–117.

1988. Bone tools: an experimental approach. In (S. L. Olsen, ed.) *Scanning electron microscopy in archaeology*, pp. 303–335. British Archaeological Reports International Series 452.

Shipman, P. and Walker, A. 1980. Bone-collecting by harvesting ants. *Paleobiology* 6: 496–502.

Shotwell, J. A. 1955. An approach to the paleoecology of mammals. *Ecology* 36: 327–337.

Sillen, A. 1989. Diagenesis of the inorganic phase of cortical bone. In (T. D. Price, ed.) *The chemistry of prehistoric human bone*, pp. 211–229. Cambridge: Cambridge University Press.

Simms, S. R. 1992. Ethnoarchaeology: obnoxious spectator, trivial pursuit, or the keys to a time machine? In (L. Wandsnider, ed.) *Quandaries and quests: visions of archaeology's future*, pp. 186–198. Southern Illinois University Center for Archaeological Investigations Occasional Paper No. 20. Carbondale.

Simpson, G. G. 1961. Some problems of vertebrate paleontology. *Science* 133: 1679–1689.

1963. Historical science. In (C. C. Albritton, Jr., ed.) *The fabric of geology*, pp. 24–48. Stanford: Freeman, Cooper, & Co.

1970. Uniformitarianism: an inquiry into principle, theory, and method in geohistory and biohistory. In (M. K. Hecht and W. C. Steere, eds.) *Essays in evolution and genetics in honor of Theodosius Dobzhansky*, pp. 43–96. New York: Appleton.

Simpson, T. 1984. Population dynamics of mule deer. In (G. C. Frison and D. N. Walker, eds.) *The Dead Indian Creek Site: an Archaic occupation in the Absaroka Mountains of northeastern Wyoming*, pp. 83–96. The Wyoming Archaeologist 27.

Singer, D. A. 1985. The use of fish remains as a socio-economic measure: an example from 19th-century New England. *Historical Archaeology* 19: 132–136.

Skinner, J. D. and van Aarde, R. J. 1991. Bone collecting by brown hyaenas *Hyaena brunnea* in the Central Namib Desert, Namibia. *Journal of Archaeological Science* 18: 513–523.

Smith, B. D. 1974. Predator–prey relationships in the southeastern Ozarks – A.D. 1300. *Human Ecology* 2: 31–43.

  1976. "Twitching:" a minor ailment affecting human paleoecological research. In (C. E. Cleland, ed.) *Cultural change and continuity*, pp. 275–292. New York: Academic Press.

  1977. Archaeological inference and inductive confirmation. *American Anthropologist* 79: 598–617.

  1979. Measuring the selective utilization of animal species by prehistoric human populations. *American Antiquity* 44: 155–160.

Smith, G. E. 1908. The alleged discovery of syphilis in prehistoric Egyptians. *Lancet* 2: 521–524.

Smith, G. R., Stearley, R. F., and Badgley, C. E. 1988. Taphonomic bias in fish diversity from Cenozoic floodplain environments. *Palaeogeography, Palaeoclimatology, Palaeoecology* 63: 263–273.

Smith, R. J. 1984. Allometric scaling in comparative biology: problems of concept and method. *American Journal of Physiology* 246: R152–160.

Sokal, R. R. and Rohlf, F. J. 1969. *Biometry*. San Francisco: W. H. Freeman and Company.

Solomon, S. 1990. What is this thing called taphonomy? In (S. Solomon, I. Davidson, and D. Watson, eds.) *Problem solving in taphonomy*, pp. 25–33. Tempus vol. 2.

Solomon, S., Davidson, I., and Watson, D. 1990. (eds.) *Problem solving in taphonomy: archaeology and palaeontological studies from Europe, Africa and Oceania*. Tempus: Archaeology and Material Culture Studies in Anthropology vol. 2. St. Lucia: University of Queensland Anthropology Museum.

Solomon, S., Minnegal, M., and Dwyer, P. 1986. Bower birds, bones and archaeology. *Journal of Archaeological Science* 13: 307–318.

Solounias, N. 1988. Evidence from horn morphology on the phylogenetic relationships of the pronghorn (*Antilocapra americana*). *Journal of Mammalogy* 69: 140–143.

Speth, J. D. 1983. *Bison kills and bone counts*. Chicago: University of Chicago Press.

  1991. Taphonomy and early hominid behavior: problems in distinguishing cultural and non-cultural agents. In (M. C. Stiner, ed.) *Human predators and prey mortality*, pp. 31–40. Boulder: Westview Press.

Speth, J. D. and Spielmann, K. A. 1983. Energy source, protein metabolism, and hunter-gatherer subsistence strategies. *Journal of Anthropological Archaeology* 2: 1–31.

Spiess, A. E. 1979. *Reindeer and caribou hunters: an archaeological study*. New York: Academic Press.

Stahl, P. W. and Brinker, U. H. 1991. Removal of calcareous incrustation from bone and teeth with acid and ultrasound. *Journal of Field Archaeology* 18: 138–140.

Stahl, P. W. and Zeidler, J. A. 1990. Differential bone-refuse accumulation in food-preparation and traffic areas on an early Ecuadorian house floor. *Latin American Antiquity* 1: 150–169.

Stallibrass, S. 1984. The distinction between the effects of small carnivores and humans on post-glacial faunal assemblages. In (C. Grigson and J. Clutton-Brock, eds.) *Animals and archaeology: 4. husbandry in Europe*, pp. 259–269. British Archaeological Reports International Series 227.

1990. Canid damage to animal bones: two current lines of evidence. In (D. E. Robinson, ed.) *Experimentation and reconstruction in environmental archaeology*, pp. 151–165. Oxford: Oxbow Books.

Stein, B. R. 1989. Bone density and adaptation in semiaquatic mammals. *Journal of Mammalogy* 70: 467–476.

Stein, J. K. 1987. Deposits for archaeologists. In (M. B. Schiffer, ed.) *Advances in archaeological method and theory* vol. 11, pp. 337–395. San Diego: Academic Press.

Stevens, S. S. 1946. On the theory of scales of measurement. *Science* 103: 677–680.

Stewart, K. M. 1989. *Fishing sites of north and east Africa in the late Pleistocene and Holocene: environmental change and human adaptation.* British Archaeological Reports International Series 521.

1991. Modern fishbone assemblages at Lake Turkana, Kenya: a methodology to aid in recognition of hominid fish utilization. *Journal of Archaeological Science* 18: 579–603.

Stiles, D. 1977. Ethnoarchaeology: a discussion of methods and applications. *Man* 12: 87–103.

Stiner, M. C. 1990a. The ecology of choice: procurement and transport of animal resources by Upper Pleistocene hominids in west-central Italy. Ph.D. dissertation, University of New Mexico, Albuquerque. Ann Arbor: University Microfilms International.

1990b. The use of mortality patterns in archaeological studies of hominid predatory adaptations. *Journal of Anthropological Archaeology* 9: 305–351.

1991a. An interspecific perspective on the emergence of the modern human predatory niche. In (M. C. Stiner, ed.) *Human predators and prey mortality*, pp. 149–185. Boulder: Westview Press.

1991b. Food procurement and transport by human and non-human predators. *Journal of Archaeological Science* 18: 455–482.

1991c. (ed.) *Human predators and prey mortality.* Boulder, CO: Westview Press.

1991d. Introduction: actualistic and archaeological studies of prey mortality. In (M. C. Stiner, ed.) *Human predators and prey mortality*, pp. 1–13. Boulder: Westview Press.

1991e. The faunal remains from Grotta Guattari: a taphonomic perspective. *Current Anthropology* 32: 103–117.

1992. Overlapping species "choice" by Italian Upper Pleistocene predators. *Current Anthropology* 33: 433–451.

Stirton, R. A. 1959. *Time, life and man: the fossil record.* New York: John Wiley and Sons, Inc.

Stock, C. 1956. Rancho La Brea: a record of Pleistocene life in California. *Los Angeles County Museum of Natural History Science Series* No. 20, 6th edition.

Straus, L. G. 1990. Underground archaeology: perspectives on caves and rockshelters. In (M. B. Schiffer, ed.) *Archaeological method and theory* vol. 2, pp. 255–304. Tucson: University of Arizona Press.

Styles, B. W. 1981. *Faunal exploitation and resource selection: early Late Woodland subsistence in the Lower Illinois Valley.* Northwestern University Archaeological Program Scientific Paper No. 3, Evanston, Illinois.

Sutcliffe, A. J. 1970. Spotted-hyena: crusher, gnawer, digester and collector of bones. *Nature* 227: 1110–1113.

1973. Similarity of bones and antlers gnawed by deer to human artifacts. *Nature* 246: 428–430.

1977. Further notes on bones and antlers chewed by deer and other ungulates. *Deer* 4: 73–82.

Tappen, N. C. 1969. The relationship of weathering cracks to split-line orientation in bone. *American Journal of Physical Anthropology* 31: 191–198.

Tappen, N. C. and Peske, G. R. 1970. Microscopic examination of human femurs buried at Washington Island, Wisconsin. *American Antiquity* 35: 463–465.

Tedford, R. H. 1970. Principles and practices of mammalian geochronology in North America. In (E. L. Yochelson, ed.) *Proceedings of the North American paleontological convention*, pp. 666–703. Lawrence: Allen Press.

Thomas, D. H. 1971. On distinguishing natural from cultural bone in archaeological sites. *American Antiquity* 36: 366–371.

Thomas, D. H. and Mayer, D. 1983. Behavioral faunal analysis of selected horizons. In (D. H. Thomas) *The archaeology of Monitor Valley 2: Gatecliff Shelter*, pp. 353–391. American Museum of Natural History Anthropological Papers 59(1).

Thomas, R. D. K. 1986. Taphonomy: ecology's loss is sedimentology's gain. *Palaios* 1: 206.

Thurman, M. D. and Willmore, L. J. 1981. A replicative cremation experiment. *North American Archaeologist* 2: 275–283.

Tipper, J. C. 1979. Rarefaction and rarefiction – the use and abuse of a method in paleoecology. *Paleobiology* 5: 423–434.

Tobin, M. E. and Dolbeer, R. A. 1990. Disappearance and recoverability of songbird carcasses in fruit orchards. *Journal of Field Ornithology* 61: 237–242.

Todd, L. C. 1983a. Taphonomy: fleshing out the dry bones of Plains prehistory. *The Wyoming Archaeologist* 26(3–4): 36–46.

  1983b. The Horner Site: taphonomy of an early Holocene bison bonebed. Ph.D. dissertation, University of New Mexico, Albuquerque.

  1987a. Analysis of kill-butchery bonebeds and interpretation of Paleoindian hunting. In (M. H. Nitecki and D. V. Nitecki, eds.) *The evolution of human hunting*, pp. 225–266. New York: Plenum Press.

  1987b. Taphonomy of the Horner II bone bed. In (G. C. Frison and L. C. Todd, eds.) *The Horner Site: the type site of the Cody Cultural Complex*, pp. 107–198. Orlando: Academic Press.

Todd, L. C. and Rapson, D. J. 1988. Long bone fragmentation and interpretation of faunal assemblages: approaches to comparative analysis. *Journal of Archaeological Science* 15: 307–325.

Todd, L. C., Witter, R. V., and Frison, G. C. 1987. Excavation and documentation of the Princeton and Smithsonian Horner Site assemblages. In (G. C. Frison and L. C. Todd, eds.) *The Horner Site: the type site of the Cody Cultural Complex*, pp. 39–91. Orlando: Academic Press.

Toots, H. 1965a. Orientation and distribution of fossils as environmental indicators. *Nineteenth Field Conference of the Wyoming Geological Association*, pp. 219–229.

  1965b. Random orientation of fossils and its significance. *University of Wyoming Contributions to Geology* 4: 59–62. Laramie.

  1965c. Sequence of disarticulation in mammalian skeletons. *University of Wyoming Contributions to Geology* 4: 37–39. Laramie.

Torbenson, M., Aufderheide, A., and Johnson, E. 1992. Punctured human bones of the Laurel Culture from Smith Mound Four, Minnesota. *American Antiquity* 57: 506–514.

Tournal, M. 1833. General considerations on the phenomenon of bone caverns. *Annales de Chimie et de Physique* 25: 161–181. (English translation by A. B. Elsasser printed 1959 in *Kroeber Anthropological Society Papers* 21: 6–16; reprinted in 1969 in [R. F. Heizer, ed.] *Man's discovery of his past*, pp. 84–94, Palo Alto: Peek Publications).

Trigger, B. G. 1989. *A history of archaeological thought*. Cambridge: Cambridge University Press.

Tringham, R. 1978. Experimentation, ethnoarchaeology, and the leapfrogs in archaeological methodology. In (R. A. Gould, ed.) *Explorations in ethnoarchaeology*, pp. 169–199. Albuquerque: University of New Mexico Press.

Tucker, M. E. 1990. The diagenesis of fossils. In (S. K. Donovan, ed.) *The processes of fossilization*, pp. 84–104. New York: Columbia University Press.

Turner, A. 1983. The quantification of relative abundances in fossil and subfossil bone assemblages. *Annals of the Transvaal Museum* 33: 311–321.

  1984. Identifying bone-accumulating agents. In (M. Hall, G. Avery, D. M. Avery, M. L.

Wilson, and A. J. B. Humphreys, eds.) *Frontiers: Southern African archaeology today*, pp. 334–339. British Archaeological Reports International Series 207.

1989. Sample selection, schlepp effects and scavenging: the implications of partial recovery for interpretations of the terrestrial mammal assemblage from Klasies River Mouth. *Journal of Archaeological Science* 16: 1–11.

Turner, C. G., II. 1983. Taphonomic reconstructions of human violence and cannibalism based on mass burials in the American Southwest. In (G. M. LeMoine and A. S. MacEachern, eds.) *Carnivores, human scavengers & predators: a question of bone technology*, pp. 219–240. Calgary: University of Calgary Archaeological Association.

Tuross, N., Behrensmeyer, A. K., and Eanes, E. D. 1989. Strontium increases and crystallinity changes in taphonomic and archaeological bone. *Journal of Archaeological Science* 16: 661–672.

Tyler, C. 1969. Avian egg shells: their structure and characteristics. *International Review of Genetics and Experimental Zoology* 4: 81–130.

Uerpmann, H. P. 1973. Animal bone finds and economic archaeology: a critical study of "osteo-archaeological" method. *World Archaeology* 4: 307–332.

Van Couvering, J. A. H. 1980. Community evolution in east Africa during the late Cenozoic. In (A. K. Behrensmeyer and A. P. Hill, eds.) *Fossils in the making*, pp. 272–298. Chicago: University of Chicago Press.

Van Neer, W. and Muñiz, A. M. 1992. "Fish middens:" anthropogenic accumulations of fish remains and their bearing on archaeoichthyological analysis. *Journal of Archaeological Science* 19: 683–695.

Vehik, S. C. 1977. Bone fragments and bone grease manufacture: a review of their archaeological use and potential. *Plains Anthropologist* 22: 169–182.

Villa, P. 1991. Middle Pleistocene prehistory in southwestern Europe: the state of our knowledge and ignorance. *Journal of Anthropological Research* 47: 193–217.

Villa, P. and Mahieu, E. 1991. Breakage patterns of human long bones. *Journal of Human Evolution* 21: 27–48.

Villa, P., Bouville, C., Courtin, J., Helmer, D., Mahieu, E., Shipman, P., Belluomini, G., and Branca, M. 1986. Cannibalism in the Neolithic. *Science* 233: 431–437.

Von Endt, D. W. and Ortner, D. J. 1984. Experimental effects of bone size and temperature on bone diagenesis. *Journal of Archaeological Science* 11: 247–253.

Voorhies, M. 1969. *Taphonomy and population dynamics of an early Pliocene vertebrate fauna, Knox County, Nebraska.* University of Wyoming Contributions to Geology Special Paper No. 1. Laramie.

1970. Sampling difficulties in reconstructing late Tertiary mammalian communities. In (E. L. Yochelson, ed.) *Proceedings of the North American paleontological convention*, pp. 454–468. Lawrence: Allen Press.

1981. Ancient ashfall creates a Pompeii of prehistoric animals. *National Geographic* 159: 66–75.

Wainwright, S. A., Biggs, W. D., Currey, J. D., and Gosline, J. M. 1976. *Mechanical design in organisms.* New York: John Wiley & Sons.

Walker, P. L. and Long, J. C. 1977. An experimental study of the morphological characteristics of tool marks. *American Antiquity* 42: 605–616.

Wall, W. P. 1983. The correlation between high limb-bone density and aquatic habits in Recent mammals. *Journal of Paleontology* 57: 197–207.

Walters, I. 1984. Gone to the dogs: a study of bone attrition at a central Australian campsite. *Mankind* 14: 389–400.

1985. Bone loss: one explicit quantitative guess. *Current Anthropology* 26: 642–643.

Warme, J. E. and Häntzschel, W. 1979. Actualistic paleontology. In (R. W. Fairbridge and D.

Jablonski, eds.) *Encyclopedia of paleontology*, pp. 4–10. Stroudsburg: Dowden, Hutchinson & Ross, Inc.

Warrick, G. and Krausman, P. R. 1986. Bone-chewing by desert bighorn sheep. *Southwestern Naturalist* 31: 414.

Watson, J. A. L. and Abbey, H. M. 1986. The effects of termites (Isoptera) on bone: some archeological implications. *Sociobiology* 11: 245–254.

Watson, J. P. N. 1972. Fragmentation analysis of animal bone samples from archaeological sites. *Archaeometry* 14: 221–228.

Watson, P. J. 1979. The idea of ethnoarchaeology: notes and comments. In (C. Kramer, ed.) *Ethnoarchaeology: implications of ethnography for archaeology*, pp. 277–287. New York: Columbia University Press.

Watson, R. A. 1966. Discussion: is geology different: a critical discussion of "The fabric of geology." *Philosophy of Science* 33: 172–185.

1969. Explanation and prediction in geology. *Journal of Geology* 77: 488–494.

1976. Inference in archaeology. *American Antiquity* 41: 58–66.

Weigelt, J. 1927. *Rezente wirbeltierleichen und ihre paläobiologische bedeutung*. Leipzig: Max Weg Verlag.

1989. *Recent vertebrate carcasses and their paleobiological implications*. Chicago: University of Chicago Press. (English translation of Weigelt 1927, by J. Schaefer).

Wellman, H. W. 1962. A graphical method for analysing fossil distortion caused by tectonic deformation. *Geological Magazine* 99: 348–352.

Western, D. 1980. Linking the ecology of past and present mammal communities. In (A. K. Behrensmeyer and A. P. Hill, eds.) *Fossils in the making*, pp. 41–54. Chicago: University of Chicago Press.

Wheat, J. B. 1972. *The Olsen-Chubbuck site: a Paleo-Indian bison kill*. Society for American Archaeology Memoir 26.

1979. *The Jurgens Site*. Plains Anthropologist Memoir 15.

Wheeler, A. and Jones, A. K. G. 1989. *Fishes*. Cambridge: Cambridge University Press.

Wheeler, M. 1954. *Archaeology from the earth*. Middlesex: Penguin Books.

Whewell, W. 1832. Review of vol. 2 of Charles Lyell's *Principles of Geology*. *Quarterly Review* 47: 103–132.

White, E. M. and Hannus, L. A. 1983. Chemical weathering of bone in archaeological soils. *American Antiquity* 48: 316–322.

White, J. A., McDonald, H. G., Anderson, E., and Soiset, J. M. 1984. Lava blisters as carnivore traps. In (H. H. Genoways and M. R. Dawson, eds.) *Contributions in Quaternary vertebrate paleontology: a volume in memorial to John E. Guilday*, pp. 241–256. Carnegie Museum of Natural History Special Publication No. 8. Pittsburgh.

White, T. D. 1992. *Prehistoric cannibalism at Mancos 5MTUMR-2346*. Princeton: Princeton University Press.

White, T. E. 1952a. Animal bone and Plains archeology. *Plains Archaeological Conference Newsletter* 4: 46–48.

1952b. Observations on the butchering technique of some aboriginal peoples: no. 1. *American Antiquity* 17: 337–338.

1952c. Suggestions on the butchering technique of the inhabitants of the Dodd and Phillips Ranch sites in Oahe Reservoir Area. *Plains Archaeological Conference Newsletter* 5: 20–25.

1953a. A method of calculating the dietary percentage of various food animals utilized by aboriginal peoples. *American Antiquity* 19: 396–398.

1953b. Observations on the butchering technique of some aboriginal peoples: no. 2. *American Antiquity* 19: 160–164.

1953c. Studying osteological material. *Plains Archaeological Conference Newsletter* 6: 58–67.

1954. Observations on the butchering technique of some aboriginal peoples: nos. 3, 4, 5, and 6. *American Antiquity* 19: 254–264.

1955. Observations on the butchering technique of some aboriginal peoples: nos. 7, 8, and 9. *American Antiquity* 21: 170–178.

1956. The study of osteological materials in the Plains. *American Antiquity* 21: 401–404.

Whitmer, A. M., Ramenofsky, A. F., Thomas, J., Thibodeaux, L. J., Field, S. D., and Miller, B. J. 1989. Stability or instability: the role of diffusion in trace element studies. In (M. B. Schiffer, ed.) *Archaeological method and theory* vol. 1, pp. 205–273. Tucson: University of Arizona Press.

Wika, M. 1982. Antlers – a mineral source in *Rangifer*. *Acta Zoologica* 63: 7–10.

Wilkinson, P. F. 1976. "Random" hunting and the composition of faunal samples from archaeological excavations: a modern example from New Zealand. *Journal of Archaeological Science* 3: 321–328.

Will, R. T. 1985. Nineteenth Century Copper Inuit subsistence practices on Banks Island, N. W. T. Ph.D. dissertation, University of Alberta, Edmonton.

Willey, G. R. and Sabloff, J. A. 1980. *A history of American archaeology*, 2nd edition. San Francisco: W. H. Freeman and Company.

Willer, D. and Willer, J. 1973. *Systematic empiricism: critique of a pseudoscience*. Englewood Cliffs: Prentice-Hall, Inc.

Wilson, L. G. 1967. The origins of Charles Lyell's uniformitarianism. In (C. C. Albritton, Jr., ed.) *Uniformity and simplicity*, pp. 35–62. Geological Society of America Special Paper No. 89.

Wilson, M. V. H. 1988. Taphonomic processes: information loss and information gain. *Geoscience Canada* 15: 131–148.

Wilson, T. 1901. Arrow wounds. *American Anthropologist* 3: 513–531.

Wing, E. S. and Brown, A. B. 1979. *Paleonutrition: method and theory in prehistoric foodways*. New York: Academic Press.

Wintemberg, W. J. 1919. Archaeology as an aid to zoology. *The Canadian Field-Naturalist* 33: 63–72.

Wolberg, D. L. 1970. The hypothesized osteodontokeratic culture of the Australopithecinae: a look at the evidence and the opinions. *Current Anthropology* 11: 23–37.

Wolff, R. G. 1973. Hydrodynamic sorting and ecology of a Pleistocene mammalian assemblage from California (U.S.A.). *Paleogeography, Palaeoclimatology, Palaeoecology* 13: 91–101.

1975. Sampling and sample size in ecological analysis of fossil mammals. *Paleobiology* 1: 195–204.

Wood, W. R. 1962. Notes on the bison bone from the Paul Brave, Huff, and Demery Sites (Oahe Reservoir). *Plains Anthropologist* 7: 201–204.

1968. Mississippian hunting and butchering patterns: bone from the Vista Shelter, 23SR-20, Missouri. *American Antiquity* 33: 170–179.

Wood, W. R. and Johnson, D. L. 1978. A survey of disturbance processes in archaeological site formation. In (M. B. Schiffer, ed.) *Advances in archaeological method and theory* vol. 1, pp. 315–381. New York: Academic Press.

Wright, R. 1990. Taphonomy: its applications and implications for archaeology. In (S. Solomon, I. Davidson, and D. Watson, eds.) *Problem solving in taphonomy*, pp. 260–264. Tempus vol. 2.

Wylie, A. 1982a. An analogy by any other name is just as analogical: a commentary on the Gould–Watson dialogue. *Journal of Anthropological Archaeology* 1: 382–401.

1982b. Epistemological issues raised by a structuralist archaeology. In (I. Hodder, ed.) *Symbolic and structural archaeology*, pp. 39–46. Cambridge: Cambridge University Press.

1985. The reaction against analogy. In (M. B. Schiffer, ed.) *Advances in archaeological method and theory* vol. 8, pp. 63–111. New York: Academic Press.

1988. 'Simple' analogy and the role of relevance assumptions: implications of archaeological practice. *International Studies in the Philosophy of Science* 2: 134–150.

1989a. Archaeological cables and tacking: the implications of practice for Bernstein's 'options beyond objectivism and relativism.' *Philosophy of the Social Sciences* 19: 1–18.

1989b. The interpretive dilemma. In (V. Pinsky and A. Wylie, eds.) *Critical traditions in contemporary archaeology*, pp. 18–27. Cambridge: Cambridge University Press.

Wyman, J. 1868. An account of some kjoekkenmoeddings, or shell-heaps, in Maine and Massachusetts. *The American Naturalist* 1: 561–584.

Yellen, J. E. 1977. Cultural patterning in faunal remains: evidence from the !Kung Bushmen. in (D. Ingersoll, J. E. Yellen, and W. MacDonald, eds.) *Experimental archaeology*, pp. 271–331. New York: Columbia University Press.

1991a. Small mammals: !Kung San utilization and the production of faunal assemblages. *Journal of Anthropological Archaeology* 10: 1–26.

1991b. Small mammals: post-discard patterning of !Kung San faunal remains. *Journal of Anthropological Archaeology* 10: 152–192.

Zangerl, R. 1969. The turtle shell. In (C. Gans, ed.) *Biology of the Reptilia vol. 1, morphology A*, pp. 311–339. London: Academic Press.

Ziegler, A. C. 1965. The role of faunal remains in archeological investigations. *Sacremento Anthropological Society Papers* 3: 47–75.

1973. Inference from prehistoric faunal remains. *Addison-Wesley Module in Anthropology* No. 43.

Zierhut, N. W. 1967. Bone breaking activities of the Calling Lake Cree. *Alberta Anthropologist* 1(3): 33–36.

GLOSSARY

Palaeontologists apparently have never been very strict in the use of definitions and many words have evolved to have meanings quite different from the original.
(G. C. Cadée 1990:4)

These terms should be clearly defined by researchers who use them, taking into account both the data at hand and the questions being addressed.
(A. K. Behrensmeyer 1991:324)

One of the earliest lists of taphonomic terms was published by Clark *et al.* (1967:155). Since that time, the list has grown considerably. The following is no doubt incomplete, but I believe it contains most of the now regularly used terms. It is intended to serve as an introduction to the terms one will encounter when reading the literature on taphonomy, whether that literature concerns vertebrates or invertebrates, and also includes terms that are specifically archaeological but relevant to discussion of taphonomic problems in zooarchaeology. Different authors sometimes offer slightly different definitions, and thus where possible the source of a particular definition is given. Unattributed definitions are my own, gleaned from the literature. The order of multiple definitions for a term is not meant to denote primary and secondary, or preferred and nonpreferred definitions, rather the order of definitions is simply chronological. It is thus important to note that I have made no intensive effort to trace the etymology of each term and find its original definition. An analysis of the chronology of term introduction might illuminate various interesting aspects of the history of taphonomic research, but is beyond the scope of this volume.

abrasion: (1) "the result of any agent that erodes the bone surface through the application of physical force" (Bromage 1984:173); (2) "the removal of bone material caused by the impact of sedimentary particles" (Shipman and Rose 1988:323); (3) removal of edges and/or surfaces of animal remains by physical erosion (as opposed to chemical dissolution) (Behrensmeyer *et al.* 1989); (4) "physical grinding and polishing, resulting in rounding of skeletal elements and loss of surficial details" (Brett 1990a:224); (5) "the wearing-down of skeletons due to their differential movement with respect to sediments" (Parsons and Brett 1990:38); (6) "the loss of external surfaces and projections through physical or chemical erosion" (Behrensmeyer 1991:309); (7) see also "bioerosion," "corrasion," and "corrosion/dissolution"

actualism: "the methodology of inferring the nature of past events by analogy with processes observable and in action at the present" (Rudwick 1976:110)

actuopaleontology: (1) the study of modern organisms and environments for application to paleontological problems (Kranz 1974b; see also Schäfer 1962/1972; Richter 1928); (2) emphasizes the idea that understanding the post-mortem history of one fossil group often requires knowledge of the life histories of associated and interacting organisms (Lawrence 1968); (3) "the application of the uniformitarian principle to paleontological problems" (Warme and Häntzschel 1979); (4) see also "neotaphonomy" and "neontology"

agent or agency: (1) "the general group to which the phenomenon causing change in the original state of the bone or a bone assemblage can be classed (e.g., biological, geological, hominid)" (Johnson 1985:158); (2) the "immediate physical cause" of modification to bones (Gifford-Gonzalez 1991:228); (3) see also "effect" and "process"

allochthonous: (1) a fossil assemblage which has been transported from the area where the represented animals died and presumably lived; (2) "foreign, that part of the assemblage which did not inhabit the environment of deposition and arrived there after active or passive transport" (Saunders 1977:68); (3) "remains that have been moved from the site of death and out of the original [life] habitat" (Behrensmeyer and Hook 1992:19); (4) see also "autochthonous," "chthonic," and "parautochthonous"

anadiagenesis: (1) the second phase of diagenesis, involves "the deep burial phase of compaction and cementation ... inorganic chemical reactions predominate" (Rolfe and Brett 1969:233); (2) see also "diagenesis" and "syndiagenesis"

anastasionomy: "processes and events involved in exposure of fossils by natural or human means and subsequent material weathering or human processing;" from the Greek *anastasis* meaning "resurrection" (G. G. Simpson, cited in Saunders 1977:68)

anastrophe: a catastrophe of limited scope and area, generally producing mass mortality in the affected area (Kranz 1974b)

anataxic: (1) "the weathering and erosional factors which act to destroy a fossil during the degradation of the rocks in which it occurs" (Clark *et al.* 1967:155); (2) "the sum of the phenomena which operate to expose and destroy a fossil" (Clark and Kietzke 1967:118)

apolithosis: (1) "fossilization; a naturally occurring physico-chemical process that *preserves* the gross morphology [of bones] by ionic exchange and an increase in the size of the apatite crystals; also involves the loss of organic material of the bone and the filling of voids with minerals" (emphasis in original), from the Greek *apolelithomenon* meaning turned to stone, "apolithotic effects refer to the changes brought about by fossilization processes" (Bartsiokas and Middleton 1992:68); (2) see also "fossilization," "mineralization," "permineralization," and "petrifaction"

archaeofauna: (1) faunas recovered from archaeological sites (Grayson 1979); (2) see also "paleontological fauna" and "local fauna"

articulation: (1) "the occurrence of fossilized elements in the same anatomical sequence and proximity as in the living organism" (Badgley 1986a:332); (2) "articulated bones are those that retain their exact anatomical positions relative to one another" (Behrensmeyer 1991:302); (3) see also "disarticulation"

attrition: loss of fossil information by non-preservation (after Lawrence 1968)

attritional mortality: (1) diachronic death assemblage, the deaths of different-aged animals over a prolonged period, that indirectly reflects the age-specific survivorship of a population (Voorhies 1969; Gifford 1981); (2) "deaths (due to predation, disease, or other causes) that occur as a part of the normal, long-term turnover of natural animal populations over the areas inhabited by such populations" (Behrensmeyer 1983:94)

autochthonous: (1) a fossil assemblage which is found where the represented animals died and presumably lived; fossils that experienced life, death, and burial within the same place or locale; (2) "indigenous, that part of the assemblage which is fossilized *in situ* in the sediment" (Saunders 1977:68); (3) "skeletal remains are preserved close to the place of death, without significant transport by physical or biological processes" (Behrensmeyer 1983:94); (4) "remains are preserved at the death site of an organism or the site where parts are discarded; the death or discard site is usually within the original [life] habitat" (Behrensmeyer and Hook 1992:19); (5) see also "allochthonous," "chthonic," and "parautochthonous"

autolysis: enzymatic breakdown of soft tissue by the enzymes in the (once-living) organism, enzymes that assisted in metabolic functions (Haglund 1991)

background: (1) faunal materials contributed to geological deposits at carnivore kill sites and carnivore lairs and by "natural deaths" (Binford 1981b:20); (2) "the lowest density of bones [per unit of space, such as m² represented by attritional deaths and kills dispersed throughout an ecosystem" (Behrensmeyer 1987:429)

biocoenose: (1) the life assemblage of organisms; (2) consists of a living population; (3) "a biocoenose encompasses a biotope and a community of all organisms living in it" (Schäfer 1962/1972:453); (4) an ecological unit (or living community) consisting of an integrated living congregation of diverse organisms with both biotic and abiotic characteristics

bioerosion: (1) biological processes of erosion of skeletal material (Brett 1990a); (2) "corrasive processes by organisms [such as] boring and grazing; results from the search for food and/or shelter" (Parsons and Brett 1990:38, 45); (3) see also "abrasion," "corrasion," and "corrosion/dissolution"

biostratinomy: (originally biostratonomy) (1) the study of pre- and syn- burial interrelations between dead organisms and their external environment (Lawrence 1968); (2) the study of pre-burial taphonomic factors, e.g., those processes affecting an organism between death and final burial (Lawrence 1979a); (3) "the sedimentary history of the fossil until burial" (Cadée 1990:3); (4) see also "necrology" and "perthotaxic"

bioturbation: disturbance by organisms, e.g., churning, mixing, trampling, burrowing, root penetration (Behrensmeyer *et al.* 1989)

bone technology: broadly, "the hominid utilization of bone, the modification of which can be purposeful or accidental. The focus is on the techniques selected for modifying the inherent morphology of a bone" (Johnson 1985:162)

burial: (1) animal remains are interred below the ground surface and encased in sediment; remains may be subjected to alteration by sub-surface taphonomic processes, or exhumed/exposed and undergo further modification prior to reburial (Behrensmeyer *et al.* 1989); (2) see also "final burial"

catastrophic mortality: (1) synchronic death assemblage; (2) a representative sample of all living age classes killed more or less instantaneously, forming a "snapshot" of a living population structure (Voorhies 1969; Gifford 1981)

chrisocoenosis: the assemblage of bones created by the post-mortem use of bones by humans (A. S. Gilbert 1979)

chthonic: pertaining to things in the ground or in the earth (from the Greek "chthon" meaning earth)

coarse-grained: (1) "an assemblage that is the accumulated product of events spanning an entire year . . . resolution between archaeological remains and specific events is poor" (Binford 1980:17); (2) an assemblage that "accumulates over a considerable period of time and/or during period of rapid 'turnover' of events, resulting in the association of items, debris, features, land surfaces, and the like that were differential participants in different events during the course of the occupation" (Binford 1981b:20); (3) see also "fine-grained," "grain," and "palimpsest"

coprocoenosis: a fossil assemblage derived from scats and owl pellets (Mellet 1974)

corrasion: (1) any combination of mechanical abrasion, bioerosion, and corrosion/dissolution (Brett 1990a); (2) see also "abrasion," "bioerosion," and "corrosion/dissolution"

corrosion: (1) results from chemical instability of skeletal minerals [implied non-biological mechanism] (Brett 1990a); (2) see also "corrasion" and "dissolution"

cultural filter: human behaviors that contribute to or result in the formation of an archaeofaunal record (after Daly 1969)

death assemblage: (1) "the sum total of corpses which arrive upon a surface between the deposition of two successive increments of sediment, or between the inception of one period of incrementation and its termination" (Clark *et al.* 1967:155); (2) see "life assemblage"

decay: (1) utilization of dead organisms as a food source by microbes (e.g., fungi and bacteria) (Allison 1990); (2) decomposition of protein under aerobic conditions (Haglund 1991); (3) aerobic degradation of soft tissue (Micozzi 1991); (4) see also "scavenging"

deformation, plastic: post-depositional change in the shape of a bone without breakage, usually caused by compaction from the weight of overlying sediments (Shipman 1981b:200)

diagenesis: (1) "fossildiagenese" (Müller 1951); (2) study of the post-entombment histories of organic remains (Lawrence 1968); (3) the study of post-burial taphonomic factors, i.e., between burial and recovery (Lawrence 1979b); (4) chemical and physical changes that occur in animal remains as they alter from their original state and become fossilized (Behrensmeyer *et al.* 1989); (5) "the chemical and mechanical alterations within the sediment" (Cadée 1990:3); (6) "alteration after deposition; sometimes taken to mean alteration after burial" (Retallack 1990:129); (7) "all the processes affecting a sediment and its contained fossils from deposition until metamorphism takes over at elevated temperatures and pressures" (Tucker 1990:84); (8) see also "anadiagenesis," "anataxic," "syndiagenesis," and "taphic"

dip: "the angle of declination of an object or plane" (Shipman 1981b:200); *syn.* plunge

disarticulation: (1) the generic process and result of loss of anatomical integrity (Hill 1979b); (2) "the disintegration of multiple-element skeletons along pre-existing joints or articulations" (Brett 1990a:223); (3) "disarticulated but associated bones are separated from each other but in close proximity,

and they retain their identity as parts of individual animals" (Behrensmeyer 1991:302); (4) see also "dispersal" and "scattering"

dispersal: (1) the generic process and result of spatial movement of individual
skeletal elements from a single organism (note that two elements may
become more, or less, spatially contiguous) (Hill 1979b); (2) "dispersed
bones may be (a) associated, scattered over areas several times the size of
the articulated skeleton, but can be related to the same individual using
anatomical characteristics [e.g., anatomical refitting], or, (b) isolated,
widely separated from other bones belonging to the same skeleton"
(Behrensmeyer 1991:302); (3) see also "disarticulation" and "scattering"

dissolution: (1) "changes in chemical composition [of skeletal tissues] usually
appearing as pitting and corrosion of the skeletal surface" (Parsons and
Brett 1990:38, 43); (2) see also "corrosion" and "leaching"

distal community: (1) the species in an assemblage which lived far from the site
of deposition, these species have lower than average skeletal completeness
in the assemblage (after Shotwell 1955; but see Grayson 1978b); (2) see
also "autochthonous," "allochthonous," "parautochthonous," and
"proximal community"

durability: "the relative resistance of skeletons to breakdown and destruction
by physical, chemical, and biotic agents" (Brett 1990a:223)

effect: (1) "the modification or alteration one sees on a bone or in the bone
assemblage as a whole as the result of changes the bone or assemblage has
undergone" (Johnson 1985:159); (2) the static result or trace of a taphonomic process having acted on bones, including the physical and/or chemical
modification of a bone or bone assemblage (see Chapter 1); (3) see also
"agent" and "process"

element: an anatomically complete bone or tooth (a discrete anatomical organ)
in the skeleton of an animal

enrichment: (1) ["accumulation" is] "the deposition of any substance [either
organic or inorganic] at a level or concentration higher than that found
while the bones was in the living state" (Cook *et al.* 1961:360, 358); (2) the
addition of soluble matter; (3) see "mineralization" and
"permineralization"

equifinality: the property of allowing or having the same effect or result from
different events (*Webster's Third International Unabridged Dictionary*)

erosion: general term for the physical or chemical removal of original material
from edges and/or surfaces of animal remains (Behrensmeyer *et al.* 1989)

ethnoarchaeology: study of living peoples with the aim of elucidating archaeological problems; i.e., a discipline with the goal of establishing and

clarifying the relationships between material vestiges of human behavior and the living systems which generate them (Gifford 1977)

fauna: some specified set of animal taxa in close spatial and temporal association; usually qualified by some geographic, temporal and/or taxonomic criterion (after Odum 1971:366–367; see also Tedford 1970)

fidelity: the degree to which a fossil fauna accurately reflects a paleofauna, in terms of taxa represented, taxonomic richness, and/or taxonomic diversity (Kidwell and Bosence 1991)

final burial: (1) "operationally defined as burial to sufficient depth within a sedimentary substratum for a hardpart to both attain a refuge from further small-scale episodes of exhumation and exposure at the depositional interface and to escape destructive early diagenetic processes" (Kidwell 1986:7); (2) animal remains are interred below the ground surface, encased in sediment, and not exhumed/exposed to above-ground taphonomic processes (Behrensmeyer *et al.* 1989); (3) see also "burial"

fine-grained: (1) "an assemblage accumulated over a short period of time ... resolution between debris or by-products and events [is good]" (Binford 1980:17); (2) an assemblage in which "all the included items, features, and land surfaces relate to a very few events; that is, all associated archaeological characteristics of the deposit are the consequences of basically the same events" (Binford 1981b:20); (3) see also "coarse-grained," "grain," and "palimpsest"

fossil: (1) "fossils are the remains or traces of any recognizable organic structures preserved since prehistoric times" (Stirton 1959:18); (2) any contemporary trace or remain of an organism that died at some time in the past (Matthews 1962); (3) any specimen demonstrating physical evidence of the occurrence of ancient life; generally distinguished from Recent or non/sub-fossil remains on the basis of its (the fossil's) geologic mode of occurrence (Schopf 1975); (4) the identifiable remains of (once) living organisms or of their activities preserved in the sediments by natural processes (Finks 1979)

fossil assemblage: (1) an aggregate of individual elements (that interact with various modification agents in statistical fashion, with considerable potential for variation in traces they ultimately may bear) (Gifford 1981); (2) spatially associated fossils that became embedded together after biostratinomic processes affected them and that were influenced by diagenetic processes (Fürsich 1990:237)

fossilization: (1) "change is the essence of the process of fossilization" (Cook *et al.* 1961:356); (2) the maintenance or alteration of chemical properties of organic materials by natural processes (Finks 1979); (3) see also "apolithosis," "mineralization," "permineralization," "post-fossilization," and "pre-fossilization"

fossil-Lagerstätte (sing.; Lagerstätten – pl.): (1) "any rock [stratum] containing fossils which are sufficiently well preserved and/or abundant to warrant exploitation, if only for scientific purposes" (Seilacher 1990:266), (a) concentration Lagerstätten contain abundant fossils, (b) conservation Lagerstätten contain well-preserved fossils (Seilacher 1990); (2) see also "obrution deposits"

fossil record: (1) some set of remains of organisms, plants and/or animals, having a geological mode of occurrence in some defined geographic space *and* the geological context and spatial distributions of those fossils (modified from Lyman 1982a); (2) see also "local fauna"

fragmentation: mechanical disassociation of skeletal elements along non-articulation planes or non-joint-related planes

grain: (1) "the relative contextual complexity of an assemblage from the perspective of events occurring during the course of continuous occupation and derivative production of an archaeological assemblage" (Binford 1981b:20); (2) increasing coarseness of grain reduces interassemblage variability (Binford 1980:18); (3) see also "coarse-grained," "fine-grained," and "palimpsest"

hydrolysis: (1) often a leaching process involving the combination of water with other molecules; (2) see "leaching"

isotaphonomic: (1) "having the same taphonomic features and biases" (Behrensmeyer 1991:326); (2) a form of analysis involving the comparison of fossil assemblages from the same "taphonomic mode" (Behrensmeyer and Hook 1992:18)

lag deposit: (1) "an accumulation of bones that are too dense to be moved by wind or water current" (Shipman 1981b:201); (2) "the heavier, less [fluvially] transportable bones" (Behrensmeyer 1990:233); (3) see also "sorting," "Voorhies Group," and "winnowing"

leaching: (1) the removal of soluble matter; (2) see also "dissolution"

life assemblage: (1) "the total number of individuals within a specified area during a specified time, the term may be limited to any taxonomic group" (Clark *et al.* 1967:155); (2) see "death assemblage"

liptocoenosis: remnant (fossil) assemblage (Rolfe and Brett 1969)

local fauna: (1) the fauna represented by one or several geographically, geologically and taxonomically similar fossil samples; i.e., may be represented by fossil samples from a single site or several sites in close geographic and stratigraphic association (not necessarily representative of a biocoenose, and not necessarily implying any paleoecological reality) (Tedford 1970); (2) see also "fossil record"

MAU: the minimum number of animal units necessary to account for the specimens observed, calculated as

$$MNE_e \div \text{number of times } _e \text{ occurs in one skeleton} = MAU_e$$

where MNE is as defined in this glossary and $_e$ is a particular category of skeletal part such as the proximal half of a femur (portion of a skeletal element), a mandible (complete skeletal element), or the rib-cage (several anatomically related complete skeletal elements)

MGUI: the modified general utility index (one of a series of indices meant to measure the food value or utility of the soft tissues associated with or attached to the different skeletal parts in a carcass)

mineralization: (1) "the accumulation of inorganic solutes" (Cook *et al.* 1961:361); (2) the replacement of the original hard parts of an organism with other minerals; the skeletal tissue's chemical constituents are dissolved and removed while minerals dissolved in ground water are simultaneously deposited in their place (Matthews 1962); (3) also known as authigenic preservation or replacement (Schopf 1975); (4) see also "apolithosis," "fossilization," "permineralization," and "petrifaction"

MNE: (1) the minimum number of skeletal elements necessary to account for (to have contributed) the specimens observed; the category of skeletal element sometimes is a complete skeletal element or discrete anatomical organ and other times is a portion of a complete skeletal element such as the proximal half of the femur; (2) see also "element"

MNI: the minimum number of (complete) individual animals necessary to account for (to have contributed) the specimens observed

mummification: "all natural and artificial processes that bring about preservation of the body or its parts" (Micozzi 1991:17)

necrology: (1) alterations of animal carcasses prior to diagenesis (Warme and Häntzschel 1979:5–6); (2) study of "the death and decomposition of an organism" [necrosis, necrobiosis] (Cadée 1990:3); (3) see also "biostratinomy" and "perthotaxic"

neontology: (1) the "paleontology" of living animals including the "paleoecology" of modern environments (Warme and Häntzschel 1979); (2) see also "neotaphonomy" and "actuopaleontology"

neotaphonomy: (1) involves relevant experimentation or observations on the condition of modern vertebrate remains in closely defined environments, designed to test taphonomic conjectures and to suggest consequences for paleoecological interpretation not visible in the fossil record such as the absence of a taxon or the structure and composition of a paleocommunity from certain kinds of fossil remains (Hill 1978); (2) the study of contemporary processes of death, decay, and deposition of organisms (Shipman 1981b); (3) see also "actuopaleontology," "neontology," and "paleotaphonomy"

NISP: (1) the number of identified specimens in a collection, where "identified" usually means identified to taxon but may mean identified to skeletal element represented; (2) see also "specimens"

obrution deposits: (1) deposits of well-preserved and articulated fossils owing their preservation and articulation to rapid burial (Brett 1990b:239); (2) from the Latin *obruere*, meaning to smother (Seilacher 1992:116); (3) see also "fossil Lagerstätte"

orientation: "the azimuth, or compass direction, of the long axis of a bone, as measured on a horizontal plane" (Shipman 1981b:202)

oryctocoenose: remains that were found together in an outcrop (Lawrence 1979c)

paleoecology: (1) the study of environmental relations of fossil organisms between their birth and death (Lawrence 1968); (2) a discipline focusing on interrelations which occurred in the geologic past between living organisms and their surroundings (Lawrence 1971)

paleomortology: (1) the accumulation and interpretation of all data bearing on the death and postmortem events of skeletal finds that offer some chance of demonstrating "man–animal" association (Mehl 1966:27); (2) see also "archaeofauna"

paleontological fauna: (1) "the maximum geographic and temporal limits of a group of organisms sharing a suite of common species (as evidenced by the fossil record)" (Tedford 1970); (2) faunas recovered from paleontological sites; (3) see also "archaeofauna" and "local fauna"

paleotaphonomy: (1) observations on fossil assemblages (Hill 1978); (2) study of the fossil record (Shipman 1981); (3) see also "neotaphonomy" and "taphonomy"

palimpsest: (1) an accumulation of archaeological materials "characterized by an accretional formational history" (Binford 1980:9); (2) a deposit of archaeological materials "deriving from a variety of events or actions of both man and animals" (Binford 1981b:9); (3) see also "coarse-grained," "fine-grained," and "grain"

parautochthonous: (1) "remains that are transported from the death or discard site but stay within the original [life] habitat" (Behrensmeyer and Hook 1992:19); (2) "transport of remains within the boundaries of an organism's habitat" (Behrensmeyer and Hook 1992:78); (3) see also "allochthonous," "chthonic," and "autochthonous"

pedoturbation: various processes of homogenization (or haploidization), which impede soil horizon formation; soil mixing (Wood and Johnson 1978); may be mechanical or chemical (faunalturbation; floralturbation; cryoturbation; graviturbation; argilliturbation; aeroturbation; aquaturbation; crystalturbation; seismiturbation)

permineralization: (1) the infiltration of mineral-bearing solutions into pores in skeletal tissue and the deposition of minerals in those spaces (Cook *et al.* 1961:361; Matthews 1962); (2) also known as cellular permineralization or petrifaction, results in petrified fossils (Schopf 1975); (3) see also "apolithosis," "fossilization," "mineralization," and "petrifaction"

perthotaxic: (1) "the factors which act to destroy animal corpses within any particular environment" (Clark *et al.* 1967:155); (2) "climate and exposure are the chief variables controlling a perthotaxic system" (Clark and Kietzke 1967:117); (3) taphonomic factors which operate between the time of an organism's death and the time of its burial, including but not limited to scavenging and weathering; (4) see also "biostratinomy" and "necrology"

perthotaxis: "a death assemblage with the animal corpses in various stages of destruction by the set of processes normally operative under the environment concerned" (Clark *et al.* 1967:155)

perthotaxy: "the more or less orderly destruction of animals' corpses by the combined action of natural processes" (Clark *et al.* 1967:155)

petrification: (1) extreme permineralization (Cook *et al.* 1961:361); (2) "cellular permineralization" or permeation of cells and interstices (not replacement) by mineral matrix at or very soon after deposition (Schopf 1975); (3) see also "apolithosis," "fossilization," and "permineralization"

plunge: see "dip"

post-fossilization: (1) refers to the period in the taphonomic history after the original physical and chemical characteristics of animal remains have been altered by fossilization (Behrensmeyer *et al.* 1989); (2) see also "fossilization" and "pre-fossilization"

pre-fossilization: (1) refers to the period in the taphonomic history prior to alteration of the original characteristics by the process of fossilization

(Behrensmeyer *et al.* 1989); (2) see also "fossilization" and "post-fossilization"

preservation: (1) "duripatric (hard part) preservation"; original hard parts are preserved due to resistance to oxidation and physical damage (Schopf 1975); (2) "authigenic preservation"; fossil is encased by cementing minerals which preserve surface configuration of organic parts while internal organization is lost or degraded (Schopf 1975)

primary deposit: (1) deposits "produced by the primary concentration of disarticulated vertebrate material; (a) condensation deposits are formed over a long period of time by the withholding of sediment from an area [mortality is long term], (b) mass mortality deposits are formed over short periods of time by the mass mortality of vertebrate organisms" (Antia 1979: 108–109); (2) see "secondary deposit"

process: (1) "the 'how' a bone became modified (e.g., scored by a tooth scratching the cortical surface) and the 'what' that caused the modification (e.g., large carnivore)" (Johnson 1985:158–159); (2) the dynamic action of a taphonomic agent on bones (e.g., downslope movement, gnawing, or fracturing) (see Chapter 1); (3) see also "agent" and "effect"

proximal community: (1) the species of a community which lived in close spatial proximity to the site of the deposition of their remains, species of this community have greater than average skeletal completeness in the assemblage (Shotwell 1955; but see Grayson 1978b); (2) see also "autoch-thonous," "distal community," and "parautochthonous"

putrifaction: (1) bacterial breakdown of protein under anaerobic conditions (Haglund 1991); (2) anaerobic degradation of soft tissues (Micozzi 1991)

quarry site: localized concentrations of fossil bones; vary greatly in density of materials and total volume; vary in degree of representation of biocoenose (Shotwell 1955)

scattering: the increase in dispersion of skeletal parts (Hill 1979b); see also "dispersal" and "disarticulation"

scavenging: (1) the use of dead organisms as a source of food by "macro-organisms" (Allison 1990); (2) see also "decay"

scratch mark: general term for linear, sharply defined grooves in bone surfaces; usually less than 1 mm deep and 0.5 mm wide (Behrensmeyer *et al.* 1989)

secondary deposit: (1) deposit "produced by reworking and concentration from older sediments; (a) refers to a primary vertebrate deposit which has been buried, diagenetically altered, excavated, fragmented, and then further concentrated to form a bone-bed, (b) formed by the reworking of older vertebrate deposits" (Antia 1979:108, 111); (2) see also "primary deposit"

signature criterion/pattern: "a criterion that is constant and unique and that discriminates one modifying agent or set of agents from another" (Binford 1981b; Gould 1980)

skeletonization: (1) removal of soft tissue from a carcass to produce a skeleton; (2) see also "autolysis," "decay," "putrifaction," and "scavenging"

specimen: an archaeological/paleontological part of a skeleton that can consist of a complete bone or fragment thereof, a complete tooth or fragment thereof, or a bone (such as the mandible) with teeth in it

sorting: (1) the hydrodynamic process of separating bones of skeletons into those readily moved by water and those not readily moved by water (after Wolff 1973); (2) "separation of body parts according to [fluvial] transport rates" (Behrensmeyer 1990:233); (3) see also "Voorhies Groups"

sullegic: (1) "those factors influencing the collecting of fossils which determine whether or not any particular fossil at the surface will find its way into a collection" (Clark *et al.* 1967:155; Clark and Guensberg 1970:440); (2) factors influencing collections; i.e., whether or not a particular fossil is collected; includes area of site chosen and/or site chosen, sampling design (where you collect), collection procedures (e.g., hand-picking versus screening [and mesh size] versus flotation) (Meadow 1981)

syndiagenesis: (1) "the first phase of diagenesis, the early burial phase of bacterial activity in which organic matter provides the nutrient for bacterial metabolism ... occurs within the top few meters of a newly formed sediment" (Rolfe and Brett 1969:233); (2) not explicitly defined, but used by Andrews and Cook (1985; see also Andrews 1990; Cook 1986) apparently to signify taphonomic processes that occur simultaneous with burial (see Figure 2.8); (3) see also "diagenesis," "anadiagenesis," and "burial"

syntaphonomy: "the sum of taphonomic histories" of each specimen in a fossil assemblage (Seilacher 1992:110)

taphic: "factors determining whether or not a bone will be buried" (Clark *et al.* 1967:155); (2) the when, where, how and why of burial

taphocoenosis: (taphocoenose) (1) assemblage of organic materials which are buried together (Lawrence 1979c); (2) "death assemblage, grouping of organisms found together in fossil assemblages" (Brett and Speyer 1990:258); (3) "hard parts of organisms which became embedded together after having been subject to biostratinomic processes" (Fürsich 1990:237); (4) syn. = allochthonous thanatocoenosis taphofacies: (1) "suites of rock characterized by particular combinations of preservational features of the contained fossils; defined on the basis of consistent preservational features" (Brett and Speyer 1990: 258); (2) see also "taphonomic mode"

taphonomic mode: (1) "a recurring pattern of preservation of organic remains in a particular sedimentary context, accompanied by characteristic taphonomic features" (Behrensmeyer 1988:183); (2) "a set of fossil occurrences that result from similar physical, chemical, and biological processes; processes may be only broadly similar, or more specifically defined according to detailed sedimentological and taphonomic evidence" (Behrensmeyer and Hook 1992:15); (3) see also "taphofacies"

taphonomically active zone (TAZ): the sediment–water interface in aquatic settings, the sediment–air interface in terrestrial settings, where the majority of taphonomic processes are active (after Davies *et al.* 1989)

taphonomy: (1) the science of the laws of embedding or burial; the study of the transition, in all details, of organics from the biosphere into the lithosphere (Efremov 1940); (2) the study of differences between a fossil assemblage and the community(ies) from which it derived; the nebulous region of conjecture constituting hypothetical assertions about the causes of the observed bias in fossil assemblages (Hill 1978)

thanatic: (1) "the factors surrounding an animal's death which determine whether or not its body will arrive upon the surface as a member of a death assemblage" (Clark *et al.* 1967:155); (2) factors or variables pertaining to the death of an organism; (3) circumstances inducing death among individuals of a biocoenose

thanatococnose: (1) the death assemblage derived from a biocoenose (biocoenose ≠ thanatocoenose ≠ fossil assemblage); (2) may not be from one but several communities (Shotwell 1955); (3) organisms that died together (Lawrence 1979c); (4) "interpreted as variously biased derivatives of once living communities, or biocoenoses" (Brett and Speyer 1990:258); (5) "death assemblage, the preserved elements of a life assemblage after its death and decay" (Fürsich 1990:237)

time-averaged: (1) fossiliferous stratigraphic units that represent extended periods of time and mixing of organisms from different habitats (Behrensmeyer 1982); (2) "attritional mortality combined with continuous input of some skeletal remains into the sediment over long periods of time" (Behrensmeyer 1983:94); (3) time averaging "refers to the mixing of skeletal elements of non-contemporaneous populations or communities" (Fürsich 1990:237); (4) "taphonomic time averaging" concerns the fact that all species in a stratum, for example, must be "regarded as contemporaneous only within the inferred time interval" represented by the depositional time span of the stratum, whereas "analytical time averaging" results from lumping more or less contemporaneous collections or assemblages into one analytical unit (Behrensmeyer and Hook 1992:75–76)

transport: (1) loss of fossil information by physical movement of fossils away from the site of the original biocoenose (adapted from Lawrence 1968); (2) movement of animal remains by the physical action of water, wind, people, or carnivores (modified from Behrensmeyer *et al.* 1989)

trephic: (1) "factors incident to curating and identifying a specimen which determine whether or not a fossil in a collection becomes available for [analytical] use" (i.e., becomes or provides data) (Clark *et al.* 1967:155; Clark and Guensburg 1970:440); (2) includes determining which bones that were recovered have been identified and recorded (skill of analyst), analytic procedures (sampling), and the published form of data (Meadow 1981)

Voorhies Group: (1) a set of skeletal elements grouped on the basis of their similar potential for hydraulic transport (after Shipman 1981b:205); (2) "three distinct groups for medium to large mammals" of bones based on their potential for fluvial transport (Behrensmeyer 1990:233); (3) see also "lag deposit," "winnowing," and "sorting"

weathering: (1) refers to "the effects on bone of saturation, desiccation, and temperature changes" (Miller 1975:217); (2) "the process by which the original microscopic organic and inorganic components of bone are separated from each other and destroyed by physical and chemical agents operating on the bone *in situ*, either on the surface or with the soil zone" (Behrensmeyer 1978:153), "part of the normal process of nutrient recycling in and on soils" (Behrensmeyer 1978:150)

winnowing: (1) selective removal of animal remains from an assemblage due to current (water) action; can occur when all bones are in motion or when some are stationary (Behrensmeyer *et al.* 1989); (2) "removal [by fluvial processes] of lighter skeletal elements" (Behrensmeyer 1990:233); (3) see also "lag deposit," "sorting," and "Voorhies Group"

INDEX

517